FRONTIERS IN BIOENERGY AND BIOFUELS

Edited by **Eduardo Jacob-Lopes**
and **Leila Queiroz Zepka**

Frontiers in Bioenergy and Biofuels

http://dx.doi.org/10.5772/63185
Edited by Eduardo Jacob-Lopes and Leila Queiroz Zepka

Contributors

Ewa Pecka-Kiełb, Andrzej Zachwieja, Dorota Miśta, Wojciech Zawadzki, Anna Zielak-Steciwko, Patrick Mukumba, Masahide Yasuda, A. K. M. Aminul Islam, Md. Moniruzzaman, Muhammad Farooq, Gisela Montero, Conrado García, Margarita Stoytcheva, Marcos Coronado, Lydia Toscano-Palomar, Ricardo Torres, Ana MarÃa Vázquez, Daniela Montes, Vinicyus Rodolfo Wiggers, Laércio Ender, Edésio Simionatto, Henry Meier, Ramon Beims, Venko Beschkov, Elina Bastos Caramao, Javed Iqbal Qazi, Fouzia Tabssum, Valentin Antonovič, Jadvyga Kerienė, Rimvydas Stonys, Renata Boris, Jacek Szczerba, Eduardo Montero, Fernando Aguilar, Natalia Muñoz-Rujas, Fatima E. M. Alaoui, Henry Quesada Pineda, Jeremy Withers, Robert Smith, Valentin Kindomihou, Roland Ahouélété Yaovi Holou, Claudia Santibanez, Edson Luiz Silva, Paula Rúbia Ferreira Rosa, Romana Krištof, Jose Castanheiro, Salete Alves, Saima Mirza, Habib Ur Rehman, João Cajaiba Da Silva, Alex Queiroz, Alline Oliveira, Vinícius Kartnaller, Amornvadee Veawab, Hanifa Taher, Sulaiman Al-Zuhair, Elisabeth Jamet, Maria Juliana Calderan-Rodrigues, Juliana Fonseca, Carlos Alberto Labate, Eduardo Jacob-Lopes

CBS, Edition **2018**

Published by InTech

Janeza Trdine 9, 51000 Rijeka, Croatia

Publishing Process Manager Dajana Pemac

Technical Editor SPi Global

Cover Designer InTech Design Team

Additional hard copies can be obtained from orders@intechopen.com

Frontiers in Bioenergy and Biofuels , Edited by Eduardo Jacob-Lopes and Leila Queiroz Zepka
p. cm.
Print ISBN 978-953-51-2891-5
Online ISBN 978-953-51-2892-2

Contents

Preface

Human society will face enormous problems in the near future in order to cover the increasing demands of energy. The current ways these demands are covered by society are not sustainable and result in unacceptable changes in our environment.

To this end, this book aims to make a contribution to further exploring this area of bioenergy and biofuel research and development in the form of a compilation of topics covering the characterization, production, and uses of bioenergy, biofuels, and coproducts, summarizing a range of useful products and technologies applied to energy production.

We are convinced that this book will be an important resource for anyone who is interested in bioenergy and biofuels, and we express the hope that this book will stimulate and help researchers and industry professionals to move this field into new and improved applications.

Eduardo Jacob-Lopes
Federal University of Santa Maria,
Brasil

Leila Queiroz Zepka
Federal University of Santa Maria,
Brasil

Chapter 1

Introductory Chapter: Life Cycle Assessment as a Fundamental Tool to Define the Biofuel Performance

Mariany Costa Deprá, Leila Queiroz Zepka and
Eduardo Jacob-Lopes

Additional information is available at the end of the chapter

http://dx.doi.org/10.5772/64677

The widespread availability of inexpensive petroleum during the twentieth century, growing concerns of fossil fuel depletion, as well as stricter emission regulations and the search for alternative sources and economically viable substrates has been the main focus of researchers seeking to overcome the economic and environmental barriers to the renewable energy sector. The ideal source for production of biofuels mainly depends on its availability and cost. Thus, a need arises to address the current energy and environmental issues to produce biofuels [1, 2].

Biofuels have become an alternative source over the traditional energy sources. Therefore, the progress of knowledge through the establishment of more robust methods of analysis, such as the life cycle assessment (LCA), highlights the weaknesses of the systems, pressing the process engineering to develop sustainable solutions for application in production chains [3].

The life cycle assessment is a methodology to quantify the input and output streams of materials and energy throughout the production chain. Moreover, it is a useful tool to assess resource use and environmental burdens related to systems. According to **Figure 1**, four stages are used for conducting an LCA: (i) objective and scope definition; (ii) inventory analysis (LCI); (iii) impact assessment (LCIA); and (iv) interpretation [4].

The goal and scope definition stage includes the intended application, the reasons to carry out the study, the intended audience, and the use of the results. In addition, the system boundary and the functional unit should also be clearly defined. This stage is included in all the papers analyzed, although not always with the same level of detail. The system boundary defines the processes to be included in the analysis. The life cycle inventory (LCI) stage involves the compilation and quantification of inputs and outputs for each process included within the system boundary. The impact assessment categories are chosen to have an overview of the inventory data: energy balance, water footprint, global warming potential (GWP), potential of acidification, and eutrophication. The interpretation evaluates the inventory analysis results and impact analysis to select the favorite product or process, with a clear understanding of the uncertainties and assumptions used to generate results.

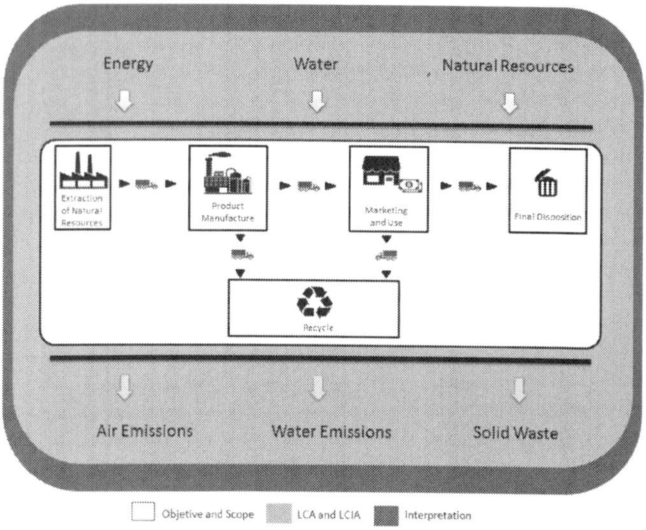

Figure 1. Stages for conducting an LCA.

The energy ratio (NER) is defined as the ratio of total energy produced (feedstock energy potential) over the energy content of construction and material, plus energy required for all plant operations. In case of the energy balance, the starting point for the economic and environmental viability of processes is the consolidation of a favorable energy balance (NER > 1) [5].

Moreover, water footprint (WF) of an enclosed area or process is determined by the sum of the water footprints of all processes. The blue WF refers to the amount of water incorporated in the product, which is determined by the evaporation rate plus incorporation and return flow. The green WF refers to the volume of water consumed in a production process, plus the water incorporated into the finish. The sum of all processes is determined per a volume of water per unit time [6].

Across the globe, there are two main public policy objectives driving the development of biofuel industries improving energy security and reducing global warming. Absorption capacity, concentration, and residence time of gases are used to evaluate the so-called global warming potential (GWP). The environmental impact generated by greenhouse gases, as well as the potential for acidification and eutrophication, in general can be quantified by the sum of the masses of the substances of gases (CO_2, CH_4, NO_x), multiplied by the characterization factors of these same substances. Once each of the factors will be different when related to the impactful gas to be measured [7].

Finally, including the life cycle assessment as a fundamental tool to define biofuel performance is a decision making that provides an understanding of the environmental impacts, and impacts on human health have traditionally not considered when selecting a product. This valuable tool should be used to expand the knowledge base of productive systems and their relationship with the environment, once can increase the efficiency of its processes, reduce the costs, and further promote marketing their products in such a sustainable way.

Author details

Mariany Costa Deprá, Leila Queiroz Zepka and Eduardo Jacob-Lopes*

*Address all correspondence to: jacoblopes@pq.cnpq.br

Food Science Technology Department, Federal University of Santa Maria, UFSM, Santa Maria, RS, Brazil

References

[1] Pragya, N. & Panved, K. Life cycle assessment of green diesel production from microalgae. Renewable Energy, v. 86, p. 623–632, 2016.

[2] Abu-Ghosh, S.; Fixler, D.; Dubinsky, N., Lluz, D.; Energy-input analysis of the life-cycle of microalgal cultivation systems and best scenario for oil-rich biomass production. Applied Energy, v. 154, p. 1082–1088, 2015.

[3] ISO 14040. Environmental management – life cycle assessment – principles and framework. International Organization for Standardization, Geneva, 2006.

[4] Corominas, L. I.; Foley, J.; Guest, J.S., Hospido, A.; Larsen, H.F.; Morera, S.; Shaw, A. Life cycle assessment applied to wastewater treatment: State of the art. Water research, v. 47, p. 5480–5492, 2013.

[5] Jorquera, O.; Kiperstok A.; Sales E. A.; Embiruçu, M.; Ghirardi, M. L. Comparative energy life-cycle analyses of microalgal biomass production in open ponds and photobioreactors. Bioresource Technology, v. 101, p. 1406–1413, 2010.

[6] Hoekstra, A.Y. A critique on the water-scarcity weighted water footprint in LCA. Ecological Indicators, v. 66, p. 564–573, 2016.

[7] Laratte B.; Guillaume B.; Kim, J.; Birregah, B. Modeling cumulative effects in life cycle assessment: The case of fertilizer in wheat production contributing to the global warming potential. Science of the Total Environment, v. 481, p. 588–595, 2014.

Chapter 2

Cell Wall Proteomics as a Means to Identify Target Genes to Improve Second-Generation Biofuel Production

Maria J. Calderan-Rodrigues, Juliana G. Fonseca, Carlos A. Labate and Elisabeth Jamet

Additional information is available at the end of the chapter

http://dx.doi.org/10.5772/66189

Abstract

Second-generation biofuels (B2G) generally uses residues composed of lignocellulosic materials to produce renewable energy (potentially up to 50%), without increasing the planted areas. However, the high cost of enzymes required for cell wall disassembly prior to the saccharification makes the B2G production more expensive yet, compared to the first-generation biofuels. Designing plants with less lignin, a barrier to B2G production, or facilitating cell wall disassembly by searching for the plant mechanisms can be the way to obtain B2G feasibility. Therewith, plant cell wall proteomics provides valuable information concerning the main cell wall proteins (CWPs) involved in its biosynthesis and rearrangements. Essentially, two plants of the grass family have been studied: sugarcane as a crop amenable to second-generation ethanol (E2G) production; and *Brachypodium distachyon* as a model plant amenable to genetic transformation. Cell wall proteomics has allowed the identification of numerous CWPs as well as their fine profiling in different organs and at various developmental stages. Proteins acting on carbohydrates, mostly glycosyl hydrolases, and oxidoreductases, including class III peroxidases and laccases, can be highlighted. Both kinds of CWPs are assumed to contribute to the remodelling of cell wall polysaccharides by enzymatic or non-enzymatic mechanisms. CWPs present in growing organs could also be attractive candidates since they greatly contribute to cell wall plasticity.

Keywords: *Brachypodium distachyon*, cell wall protein, grass, second generation ethanol, sugarcane

1. Introduction

Second-generation biofuels (B2G) are a promising renewable alternative to supply energy demand of fossil fuels worldwide, whose advantage is mostly due to the lower emission of greenhouse gases and the possibility to increase the production without widening the planted area. However, we are still far from producing B2G at an economically competitive way and reasonable amount to replace fossil fuels. B2G uses lignocellulosic material as substrates. Since sugarcane has been considered one of the best crops to produce bioethanol, its bagasse and straw have been studied as one of the main complementary sources of C_6 and C_5 sugars for B2G. One of the main constrains to its economic feasibility relies on the rate of success of the enzymatic saccharification enabling the conversion of the plant cell wall sugars into bioethanol [1]. Saccharification of the cell wall is the process of hydrolysis by which a complex carbohydrate, such as cellulose can be broken into monosaccharides. Thus, the production requires a pre-treatment of the biomass prior to expose the wall carbohydrates to substantial amounts of expensive enzymes in the industrial process.

Several strategies have been recently used to improve saccharification, mostly using microorganism enzymes. Different enzymes with cell wall polysaccharide degradation activity have been prospected from several organisms such as seaweed [2], termite stomach [3] and fungi [4]. However, even presenting some advances [5, 6], the cost of E2G is not competitive for first-generation ethanol production from sugarcane.

New approaches are emerging from the plant's perspective itself, which together may be the "eureka" to solve this puzzle. Presently applied research has been focusing on lowering or modifying the lignin content to allow its removal in the industrial production and thus increasing the access of carbohydrates to saccharification [7]. Indeed, lignin is frequently the major reason for biomass recalcitrance. However, several strategies that focused on diminishing the lignin content, and thus leading to improved saccharification, resulted in deleterious effect on plant development [8]. A different point-of-view based on lignin modification may be more effective, since even increased lignin content showed improved saccharification in *Brachypodium distachyon* [9]. Thereby, the expression of a bacterial enzyme into *Arabidopsis thaliana* altered lignin and improved saccharification, without lowering the lignin content [10].

Another strategy is to engineer the plant cell wall genes in order to enable the plant itself to produce easier breakable sugars. By producing cellulose with more adequate characteristics to allow a more efficient saccharification, such as cristallinity, the plant material showed to have improved saccharification efficiency in *A. thaliana* [11]. Genetic engineered rice and wheat also showed increased enzymatic saccharification when cell wall proteins (CWPs) acting on polysaccharides had their expression changed [1, 12].

The plant cell wall represents 50% of the organic carbon present on earth [13]. Cellulose is a major cell wall polysaccharide and the major second-generation ethanol (E2G) source. The biosynthesis of wall polymers and all the processes that occur in the plant cell wall are mediated by CWPs among which numerous enzymes. Prospective and directed studies to increase the

http://dx.doi.org/10.5772/66189

knowledge on CWPs both in model species and in plants of agricultural interest provide valuable information on target-proteins in order to direct the plant pathways and produce plant carbohydrates easily saccharified. Accordingly, the high potential of this research can be the key to B2G industrial production.

2. Plant cell wall proteomics

2.1. The plant cell wall

The plant cell wall was once considered as a static structure, but since the 1990s, it has been addressed as a dynamic part of the cell, more similar to an extracellular compartment [14]. It has to be strong and flexible at the same time to enable its several roles such as mechanical stability, osmotic control, signalling and defence against different types of stresses. Its composition varies according to the stage of development, cell types and environmental cues. As an example, epidermis cells have to be better prepared for water loss than inner cells [15].

Cell walls can be classified into two types: primary and secondary. The former is found in growing tissues, and thus extendable; and the latter type is formed after the end of cell growth. It can allow cells to resist to compression forces [16]. Cell wall composition includes cellulose, hemicelluloses, pectins, proteins [17] and lignin in some cell types [18].

Cellulose is a cell wall polysaccharide with a high molecular mass, formed by long linear chains of β-1,4-linked glucose residues forming microfibrils [19]. Primary walls contain around 20-30% cellulose, and secondary walls up to 50% [20]. Hemicelluloses are composed of β-1,4-linked monosaccharides with side chains [19]. The most present hemicelluloses in dicots and grasses are xyloglucan (XG) and β-(1,3-1,4)-mixed linked glucans, respectively. XG is probably involved in forming cross-links between cellulose microfibrils [21]. Pectic polysaccharides are formed by structures enriched by galacturonic acid with complex side chain structures [22]. Sugarcane and other grass family species cell walls present specific characteristics such as being poor in pectins and having no XG interlocking the cellulose microfibrils in dividing cells; this role is performed by glucuronoarabinoxylans (GAXs) [14]. Lignin is a phenolic polymer and confers rigidity to cellulose microfibrils, and thus, to the cell wall [23].

Cell wall biosynthesis seems to be specific for each cell type [21]. During this process, cellulose is synthesized at the level of the plasma membrane by specific protein complexes. Conversely, non-cellulosic polysaccharides, such as hemicelluloses and pectins, are synthesized in the secretion pathway and secreted to the apoplast, where they form the wall networks together with cellulose [24]. Cell expansion occurs with enzymatic or non-enzymatic cleavages of cell wall polymers and the osmotic pressure separating the microfibrils. Polymers are then deposited in the internal part of the cell wall, forming the new cross-linked network [14]. Several phytohormones are involved in cell expansion, acting specifically at the reorientation of the microtubules, which may reorient the cellulose deposition [21].

As widely known, sugarcane is the raw material for one of the largest bioethanol production. E2G production uses lignocellulosic material to convert into ethanol through the steps of pretreatment (to expose the cell wall polysaccharides to the enzymes), hydrolysis of the cellulosic and hemicellulosic polysaccharides into monomers and finally fermentation of these sugars into ethanol [25].

Over the years, the information regarding cell wall components from the chemical point-of-view has increased, enabling us to think about strategies to modulate the cell wall structure. There is knowledge available related to cellulose and hemicelluloses biochemical properties and to the pectic polysaccharides biochemistry [26]. However, less is known about the overall architecture of the cell wall. This knowledge should be enlarged to provide clues to engineer walls. Indeed, since the cell wall is constantly being modified either to respond to internal and external stimuli, this self-regulatory mechanism could be modulated to respond to commercial interests.

2.2. The plant cell wall proteome

The concept of CWPs includes not only the proteins present inside the cell wall structure but also those present in the apoplast. CWPs are essential to the wall functions such as modification of the cell wall components, its structure, signalling, interaction with the plasma membrane and response to stresses [27]. Several factors can modify the cell wall proteome content, such as development [28–31] and biotic or abiotic stresses [32, 33].

CWPs share three common characteristics: a signal peptide to be targeted to the secretory pathway, no intracellular retention motif and the absence of hydrophobic trans-membrane domains. The signal peptide presents a positive charge at its N-terminus, a hydrophobic central region and a polar C-terminus [34]. One of the best-described intracellular retention motif is the C-terminal H/KDEL, which maintains proteins inside the endoplasmic reticulum [35]. On the contrary, other sorting determinants are more complex. For example, vacuolar targeting routes are diverse and there seems to be different types of vacuole sorting determinants [36]. Bioinformatic programs can help predicting the subcellular location of proteins through protein amino acid sequences, but they rely on experimental evidence which can be incomplete [37].

Three types of CWPs can be considered according to their interaction with the cell wall matrix [27]. The labile proteins have little or no interaction with the cell wall polysaccharides and circulate in the extracellular matrix. They can be recovered by vacuum infiltration of tissues [38]. The weakly bound proteins can be linked to the wall components through Van der Waals interaction, hydrogen bonds, or ionic links and can be recovered with salt solutions. Strongly bound proteins such as structural proteins (SPs) are resistant to salt extractions and can be linked together or to polysaccharides by covalent bonds [39]. Regarding functions, CWPs can be divided into nine functional classes including a class of miscellaneous proteins (MPs) and a class of proteins yet unknown function (PUFs) [40]. As all classifications, this one has some drawbacks like the difficulty to classify proteins with dual functions such as protease possibly involved in protein turnover or in signalling, but it allows getting an overview of cell wall proteomes [41].

Proteins acting on carbohydrates (PACs) mostly comprise glycosyl hydrolases (GHs) and are involved in cell wall polysaccharides remodelling [42]. PACs belong to the most represented classes in cell wall proteomes. Cellulases and glucanases are examples of proteins that can be found in this family. These enzymes are used in enzymatic hydrolysis cocktails used in E2G production, so they could be targets for manipulation in the plant species. Oxidoreductases (ORs) are mostly class III peroxidases (Prxs). Prx activities are diverse, they can break cell wall polysaccharides in a non-enzymatic way and facilitate cell wall extension but they can also favour the cross-linking of cell wall components such as monolignols and SPs [43]. Proteins related to lipid metabolism (PLMs) are almost all lipid transfer proteins and lipases. Some of them could be involved in cell wall loosening through the bind of lipids to their hydrophobic cavity [44]. Proteases (Ps) can play roles in protein turnover, protein maturation, signalling or defence [45]. SPs, such as hydroxyproline-rich glycoproteins, proline-rich proteins and glycine-rich proteins can be cross-linked in cell walls and contribute to its architecture [46, 47]. Proteins with interaction domains with proteins or polysaccharides (PIDs) comprise lectins and enzyme inhibitors. There is a lack of knowledge regarding the role of lectins in plant cell walls [48]. Enzyme inhibitors play a critical role in the regulation of enzymatic activities. As an example, there is a subtle interplay between pectin methylesterase and pectin methylesterase inhibitors [49]. Proteins possibly involved in signalling (PSs) include arabinogalactan proteins which have been assumed to play diverse roles during plant development, and particularly in calcium signalling [50]. The miscellaneous proteins (MPs) contain many protein families which are not numerous enough to form a distinct class. The roles of proteins with domains of unknown function (PUFs) are mostly unknown, but this functional class offers potential for future research. Among PUFs, the DUF642 proteins have been shown to interact with cellulose *in vitro* [51]. They could also be involved in pectin methylesterification or in defence [52, 53].

Isolating and identifying CWPs is particularly challenging. Indeed, the difficulty begins with the extraction procedure. The cell wall is an open compartment and the polysaccharidic network can be a trap for intracellular contaminants. Either destructive (DP) or non-destructive (NDP) protocols have been used. DPs rely on grinding the tissues to isolate cell walls prior to the extraction of proteins with salt solutions [54]. The purification of cell walls relies on the fact that it is the denser cell compartment [55]. NDPs, using vacuum infiltration of tissues with mannitol or salt solutions, do not harm the cells and allow extraction of apoplastic proteins [56]. Usually, the salts used in the extraction protocols are $CaCl_2$ and LiCl. $CaCl_2$ extract CWPs through a competition mechanism [40] since pectins strongly chelate calcium ions [57]. An illustration of the effects of $CaCl_2$ has been provided by plasmolysis experiments performed on leaf tissues transiently expressing a CWP fused to the fluorescent TagRFP (red fluorescent protein) [38]. The fusion protein in displaced from the cell wall to the apoplastic space upon $CaCl_2$ application. On the other hand, LiCl is able to extract hydroxyproline-rich glycoproteins [58]. The use of both types of protocols to extract CWPs can be a good strategy to increase the coverage of the cell wall proteome [30]. However, some CWPs still escape because they are strongly bound to cell wall components [38]. At present, the cell wall proteomes are poor in SPs such as hydroxyproline-rich glycoproteins or proline-rich proteins. In addition, since some CWPs are heavily glycosylated, these post-translational modifications can be a problem for protein identification by mass spectrometry. Finally, proteomics studies of species that do not have a fully sequenced genome present an additional bottleneck because the precise identification of proteins cannot be achieved.

Even carefully performing all these protocols, the identification of proteins that are not secreted through the classical secretory pathway has been reported. These proteins can be predicted to belong to different cell compartments such as cytoplasm, nucleus, mitochondria, chloroplasts or vacuoles. The question of the existence of alternative routes of secretion is still a matter of debate [41].

3. A focus on *B. distachyon* and sugarcane cell wall proteomes

After designing several protocols to analyse the cell wall proteome of *A. thaliana* as a test case, around 700 CWPs have been identified in different organs such as leaves, stems, roots and etiolated hypocotyls as well as in cell suspension cultures, i.e. about one-third of the expected total number [59]. In order to widen the knowledge regarding CWPs targeted to find candidate routes to improve E2G production from the plant perspective, two additional species were studied: (i) *B. distachyon* as a model for grass species from temperate areas, amenable to genetic transformation and having a fully sequenced genome [60]; and (ii) sugarcane, only having a large EST collection, but being one of the major sources for E2G production.

3.1. Plant material

For *B. distachyon*, three types of organs were used: leaves, internodes and grains (**Figures 1A, B**). Two-month-old plants were used and the CWP extractions were performed in young or

Figure 1. *B. distachyon* and sugarcane plants used for proteomics studies: 2-month-old sugarcane plants (A), 4-month-old sugarcane plants (B, C), and 2-month-old *B. distachyon* plants (D, E). f (young leaves), g (mature leaves), h (apical internodes), and i (basal internodes).

mature leaves and apical or basal internodes [29]. These organs were studied in order to compare the differences between organs and to look for proteins possibly involved in cell wall extension and growth arrest. Grains were collected at different times after flowering (9, 13 or 19 days) [31, 61]. The aim of the study was to understand the modifications of cell wall polysaccharides during grain development and filling because they are key determinant of the size and mass of the grain.

In the case of sugarcane, three types of materials have been studied (**Figures 1C–E**): 11-day-old cell suspension cultures [62], 2-month-old stems [30], and 4-month-old young or mature leaves and apical or basal internodes [63]. The aim was to identify among CWPs possible targets for cell wall modification in order to facilitate E2G production.

3.2. Methods

3.2.1. Extraction procedures

In these experiments, different extraction techniques were used. For *B. distachyon*, a DP was used for all the materials [54]. It started with mixing the tissue in a 5 mM sodium acetate buffer, pH 4.6, 0.4 M sucrose and protease inhibitor cocktail. After that, the mixture had to be ground in a blender at full speed for about 15 min. PVPP was added to the homogenate, and it was stirred for 30 min at 4°C. To isolate cell walls, the mixture was submitted to several successive centrifugations (1000×*g*) in a solution of increasing sucrose concentration (0.6-1.0 M). The pellet was then extensively washed through a Nylon net (25 μm) to remove sucrose. The cell wall fraction was ground in liquid nitrogen. Then, proteins were extracted by different salt buffers prepared in 5 mM sodium acetate, pH 4.6: twice in 0.2 M $CaCl_2$, followed by twice in 2 M LiCl. Cell walls were resuspended in these buffers and centrifuged at high speed (40,000×*g*/15 min/4°C). The four supernatants were pooled.

The same DP with minor modifications was used for sugarcane cell suspension cultures and 2-month-old stems [30, 62]. Another extraction method was tested with young or mature leaves and basal or apical internodes. This method was based on vacuum infiltration [56], which is a NDP requiring working with fresh material only. The plant organs were cut to fit in a beaker and completely immersed in a solution of 3.0 M mannitol and 0.2 M $CaCl_2$ in a dessicator connected to a vacuum pump. The tissues were vacuum-infiltrated for 5 min. Plant organs were centrifuged in a swinging bucket rotor (200×*g*/15 min/20°C). The apoplastic fluids (released at the bottom of the tube) were collected and stored at low temperature. This procedure was repeated once with the same solution. Additional two rounds of vacuum infiltration were performed in a solution with 2 M LiCl instead of 0.2 M $CaCl_2$. All four extracts were pooled.

Samples resulted from DP and NDP were desalted, freeze-dried to concentrate proteins and then used in 1D-electrophoresis (1D-E) to check the quality of the protein extracts.

It should be mentioned that all the experiments have been repeated twice or thrice to take into account biological variation. Only CWPs identified in at least two biological replicates have been validated. A detailed description of these protocols can be found in Refs. [29–31, 61–63].

3.2.2. *Identification of proteins by mass spectrometry and bioinformatic analyses*

Then, proteins were identified by mass spectrometry (LC-MS/MS) and bioinformatics after tryptic digestion performed at 4°C, after separation by 1D-E or in solution. A detailed description of the parameters used for MS analysis can be found in [29–31, 61–63]. For *B. distachyon*, the genomic sequence data were used [64, 65]. For sugarcane, the SUCEST translated EST database was used [66]. The amino acid sequences of the identified proteins were systematically compared to those of *Sorghum bicolor*, the closest related species having a fully sequenced genome [64]. In case of partial EST sequence, this comparison allowed the bioinformatics prediction of sub-cellular localization and functional domains.

For both plant species, the bioinformatics analysis of the identified proteins was carried out *de novo* in the same way regarding the prediction of their subcellular localization and of functional domains using the ProtAnnDB annotation pipeline [67, 68]. All the experimental data were collected in the WallProtDB database [59, 69]. The Venn diagrams used in this chapter were made with the Venny online software [70].

3.2.3. *A comparative survey of B. distachyon and sugarcane cell wall proteomes*

As a key indicator of the quality of the protein extract, the percentage of proteins predicted to be secreted and not retained in an intracellular compartment can be calculated (**Figure 2**). The other proteins can be considered as intracellular contaminants. The highest proportion of proteins predicted to be intracellular has been found in sugarcane cell suspension cultures (82%). This could be explained by two facts: a DP was used thus increasing the chance for intracellular proteins to be trapped in the cell wall polysaccharidic matrix; and/or cell suspension cultures contain a certain proportion of dead cells whose content is released in the culture medium, so that intracellular proteins can interact with the cell walls of living cells. Such result has also been obtained with cell suspension cultures of *A. thaliana* [71]. Apart from this sample, the proportion of proteins predicted to be intracellular is above 40%. The highest proportion of CWPs was obtained with basal internodes of *B. distachyon*. In that case, we noticed that the

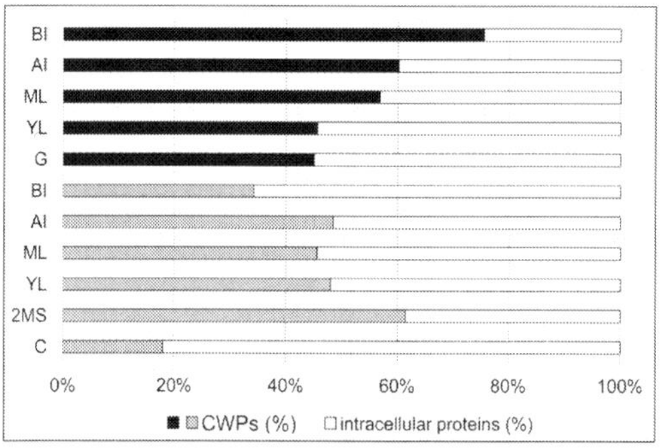

Figure 2. Percentage of CWPs and proteins predicted to be intracellular in each proteome. *B. distachyon* proteomes are in black and white, whereas sugarcane proteomes are in green and white. AI: apical internodes; BI: basal internodes; C: cell suspension cultures; G: grains; ML: mature leaves; YL: young leaves; 2MS: 2-month-old stems.

sedimentation of cell wall fragments were particularly easy for this sample, thus facilitating its purification [29].

Altogether, 567 and 273 different CWPs were identified in all mentioned experiments for *B. distachyon* and sugarcane, respectively. At present, these species, together with *Oryza sativa* (270 CWPs), have the largest cell wall proteomes among monocots [59].

The specific proteins found in each experiment, and the common ones are shown in **Figure 3** for both species. A first comparison can be made between the cell wall proteomes of the aerial parts of *B. distachyon* and sugarcane, the most amenable to E2G production. Sixty-three out of the 314 CWPs (20.1%) identified in *B. distachyon* leaves and internodes were common to both organs taken at two different stages of development (**Figure 3A**). The percentage of common proteins two by two was also homogenous, varying from 27% to 39%. This proportion was very different for sugarcane cell wall proteomes, with only 3.0% of the proteins common to all samples, i.e. 6 of 201 CWPs (**Figure 3C**). The comparison two by two reached a result similar to that obtained with *B. distachyon* only for CWPs present in apical and basal internodes (37.4%). The other duos have between 4.0% and 14.0% of common CWPs. This is probably related to the smaller size of the sugarcane cell wall proteomes of compared to those of *B. distachyon* and to the very different number of CWPs identified in leaves in comparison to stems for sugarcane. Using 2-month-old leaves, the difficulty in extracting proteins from cell walls was also observed (unpublished results). This might be inherent to the leather type of sugarcane leaves requiring a different extraction strategy. Another explanation could rely on the hexa- to octaploid genetic basis of sugarcane [72], which could lead to the expression of different sets of multigene family members at different developmental stages and in different organs.

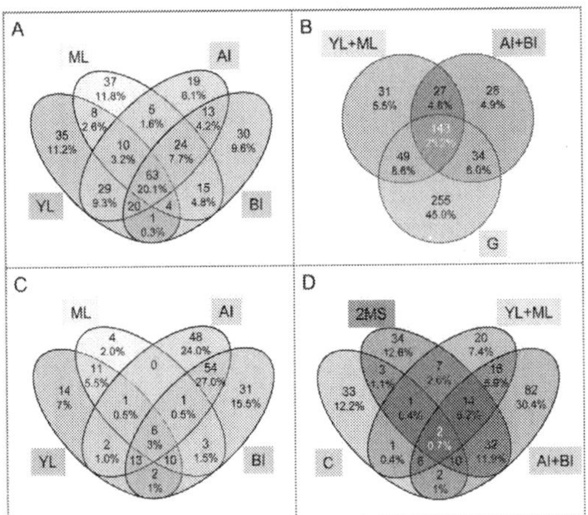

Figure 3. Venn diagrams showing common and specific CWPs for each experiment performed with *B. distachyon* (A and B) or sugarcane (C and D). AI: apical internodes; BI: basal internodes; C: cell suspension cultures; G: grains; ML: mature leaves; YL: young leaves; 2MS: 2-month-old stems.

Including the cell wall proteomes of *B. distachyon* grains, 25% of the CWPs were common to all organs (**Figure 3B**). It should be noted that the largest cell wall proteome was that of grains comprising 481 CWPs and that 45% of its CWPs were specific to this organ.

Now, looking at all the known cell wall proteomes of sugarcane, cell suspension cultures, leaves, 2- and 4-month-old stems only showed two common CWPs (**Figure 3D**). Eighty two of 273 CWPs (30.4%) were specific to 4-month-old basal and apical internodes.

These comparisons are of special interest because they allow identifying both CWPs specific to organ or developmental stages and CWPs common to all organs which may belong to a set of housekeeping CWPs essential for cell wall maintenance. For example, the set of proteins common to the 8 cell wall proteomes of *B. distachyon* comprises 42 CWPs among which 10 GHs, 4 Prxs, 8 proteases, 1 lipid transfer protein (LTP), 2 GDSL lipases and 1 DUF642 protein. In sugarcane, six CWPs were found to be common to 4-month-old leaves and internodes (**Figure 3C**): one GH, two Prxs, two proteinase inhibitors, and one subtilisin, whereas two CWPs were common to all six cell wall proteomes (**Figure 3D**): a protein of unknown function and a cys-protease. These CWPs would deserve functional studies to better understand their functions. The case of sugarcane seems more complex than that of *B. distachyon* with less putative housekeeping CWPs identified up to now.

Now, cell wall proteomes can be considered from the functional point of view. As explained above, it is possible to group proteins according to the prediction of functional domains [27, 56]. **Table 1** shows the distribution of *B. distachyon* and sugarcane CWPs into functional classes in the different cell wall proteomes. Some specific features can be noticed in *B. distachyon*: (i) PACs are less represented in basal internodes; (ii) ORs are more represented in internodes; (iii) PLMs are less represented in mature leaves; (iv) Ps are more represented in leaves; and (v) PIDs are less represented in mature leaves. Finally, SPs have been only found in grains with two leucine-rich extensins identified. In sugarcane, the situation is very different: (i) PACs are less represented in cell suspension cultures and in leaves; (ii) ORs are more represented in cell suspension cultures and in mature leaves; (iii) PLMs are less represented in cell suspension cultures and in internodes of 4-month-old plants; (iv) Ps are less represented in cell suspension cultures, but more in 4-month-old stems; (v) PIDs are poorly represented in 2-month-old stems, but more represented in cell suspension cultures and in mature leaves; and (vi) PSs are less represented in cell suspension cultures, 2-month-old stems, and in young leaves. In both plants, there are also variations in the contribution of MPs and PUFs to all cell wall proteomes.

This overview allows getting a profiling of the cell wall proteomes and to focus on specific functional classes of CWPs. Because of the variations observed in the contribution of each functional class to the whole proteomes, it also shows that each plant and each organ has to be studied in detail before choosing a strategy to modify its cell walls. For example, ORs includes mostly Prxs, but also blue copper-binding proteins, and multicopper oxidases. Prxs are involved in diverse physiological processes, such as signalling [43], lignification [73], and cross-linking of SPs [74]. Their roles in cell wall polysaccharide and protein network rearrangements could be the reason why they are more represented in *B. distachyon* stems. Curiously, the sugarcane cell wall proteomes exhibit the highest proportions of ORs compared to other plants. Such CWPs are interesting targets whose genes could be engineered for E2G production optimization.

Functional class	PACs	Ors	PLMs	Ps	PIDs	PSs	SPs	MPs	PUFs
B. distachyon									
All proteomes	**24.2**	**13.6**	**10.8**	**13.8**	**7.1**	**5.5**	**0.3**	**12.5**	**12.3**
YL	21.8	11.2	11.2	**18.2**	**3.5**	5.9		11.8	16.5
ML	26.5	15.1	**7.8**	**16.3**	4.8	**1.8**		13.9	13.9
AI	23.5	**18.0**	10.4	13.7	**2.7**	4.9		12.0	14.8
BI	**19.4**	**21.2**	10.0	12.9	5.3	4.7		10.0	16.5
G	24.1	11.6	10.6	14.5	7.3	5.8	**0.4**	12.5	12.3
Sugarcane									
All proteomes	**20.5**	**20.9**	**13.2**	**12.8**	**5.9**	**4.0**		**16.1**	**6.6**
C	**11.6**	**30.4**	**4.3**	**8.7**	**11.6**	**1.4**		20.3	11.6
2MS	20.2	21.4	16.7	13.1	**1.2**	**1.2**		11.9	14.3
YL	**8.5**	25.4	20.3	13.6	6.8	1.7		18.6	5.1
ML	**8.3**	**33.3**	19.4	11.1	**13.9**	0.0		11.1	2.8
AI	25.6	20.8	**4.8**	**16.8**	6.4	5.6		14.4	5.6
BI	24.2	20.8	**5.0**	**20**	5.8	5.0		12.5	6.7

Results are expressed as percentages of the number of CWPs identified in each proteome.

Values in bold are average values calculated with all proteomes data and values much different from these average values.

MPs: miscellaneous proteins; PLMs: proteins related to lipid metabolism; ORs: oxidoreductases; PACs: proteins acting on carbohydrates; PIDs: proteins with interaction domains; Ps: proteases; PSs: proteins involved in signalling; PUFs: proteins of unknown function; SPs: structural proteins; AI: apical internodes; BI: basal internodes; C: cell suspension cultures; G: grains; ML: mature leaves of 4-month-old plants; YL: young leaves of 4-month-old plants; 2MS: 2-month-old stems.

Table 1. Distribution of the CWPs found in each cell wall proteome of *B. distachyon* and sugarcane into functional classes.

PLMs are mostly represented by LTPs and GDSL lipases. LTPs exact biological roles are yet unknown, but they have been related to cell wall loosening and extension [44], pathogen response, and cutin assembly [75]. Since sugarcane at young developmental stages are similar to rolled leaves, this may explain the high proportion of LTPs, probably playing roles in cutin assembly of both sides leather-like leaves. Nevertheless, the better understanding of the mechanisms under this protein class may lead to the design of new strategies to increase biomass production.

The low percentage of PACs in sugarcane cell suspension cultures and leaves is also puzzling. PACs mostly include GHs, such as β-xylosidase, β-galactosidase and have been associated with cell wall loosening and expansion [76]. GH3, GH35, GH27, and GH51 can be of special interest since they show homology to enzymes of interest used for E2G production [3].

The two studied plant species, *B. distachyon* and sugarcane, appear to be complementary to identify CWPs and look for their functions. Although both plants are monocots and have similar cell wall composition, they seem to have different strategies to modulate cell wall structure during development. Combining genetics and biochemical approaches should allow getting insight in those mechanisms.

3.3. Perspectives and targets for E2G production

Changes in lignin composition have led to a subtle improved saccharification with no relevant deleterious effect [77]. However, for the cell wall polysaccharides, the challenge is still bigger since there is less knowledge regarding their synthesis. The main players able to modify cell wall polysaccharides are (i) the transcription factors that control the initial steps of gene expression and (ii) the enzymes and proteins involved in the biosynthesis of cell wall components and in their modifications *in muro* [78]. By altering transcription factors in *A. thaliana*, it was possible both to increase cellulose and decrease lignin content [79] and improve secondary cell wall synthesis in fibre cells [80]. In addition, the golden pot may be near; transgenic *A. thaliana* expressing microbial hydrolases showed no visible changes in phenotype and increased wall degradability [81]. An alternative to decrease the transgenic debate and perhaps optimize efficiency could be altering the expression of the own plant enzymes generating a genetically modified plant, but not a transgenic one. Besides hydrolases, another possibility is to consider the potential of the plant cell wall as a sensor to perceive changes and direct cell wall polysaccharides synthesis, such as in microorganisms [78]. Then, attention should be paid to the fasciclin arabinogalactan proteins, wall-associated kinases and other membrane proteins. Expressing carbohydrate-binding proteins such as expansins could facilitate cell loosening, and it may be a possibility to improve saccharification as well [82].

As can be seen, modulation of CWPs expression offers a wide range of possibilities to achieve a plant cell wall more cost-effective in terms of E2G production. Since some CWPs have been reported to act on cell wall remodelling or expansion, and we observed a different proportion of them in the several organs and developmental stages, we suggest focusing studies on some CWP families such as Prxs, GHs and LTPs, mostly those found in young and growing organs. By targeting the level of expression of these proteins or their spatial distribution, it may be possible to design plants with cell walls easily saccharified to E2G production. In order to achieve this goal, it is recommended to use tissue-specific and spatial regulation of gene expression using precise gene promotors, so that there will be no deleterious effect to the living plant. Notwithstanding, we highlight that more information on the modifications occurring on cell wall polysaccharides has to be collected in order to provide the basis for applied results.

Acknowledgements

This research was partially funded by USP-COFECUB (Proc. 10.1.1947.11.9), BIOEN/PRONEX FAPESP (Proc. 2008/56100-5), INCT do Bioetanol, CNPQ (Proc.: 142784/2007-9), and FAPESP (Proc. 2007/59327-8 and 2012/1212521-2). The authors are thankful to Dr. R Pont-Lezica who was at the origin of the projects. They also thank Dr MC Falco and CTC for providing the sugarcane plants and cells. They thank Dr GM Souza and Dr M Nishiyama for providing the SUCEST sequences. MJR, JGF, and CAL are thankful to the Max Feffer Laboratory of Plant

Genetics team, especially TR Cataldi and S Guidetti for handling the mass spectrometer and FE Morais and LM Franceschini for the bioinformatics assistance. EJ wishes to thank the CNRS, the Paul Sabatier-Toulouse III University and ANR (Grant Génoplante/PCS-08-KBBE-003/CELLWALL) for supporting her research.

Nomenclatures

B2G: Second-generation biofuel

CWP: Cell wall protein

DP: Destructive protocol

E2G: Second-generation ethanol

GAX: Glucuronoarabinoxylan

GH: Glycosyl hydrolase

MP: Miscellaneous protein

NDP: Non-destructive protocol

OR: Oxidoreductase

PAC: Protein acting on carbohydrates

PID: Protein with interaction domains with proteins or polysaccharides

PLM: Protein related to lipid metabolism

P: Protease

Prx: class III peroxidase

PUF: Protein of unknown function

SP: Structural protein

XG: Xyloglucan

Author details

Maria J. Calderan-Rodrigues[1], Juliana G. Fonseca[2], Carlos A. Labate[2] and Elisabeth Jamet[3*]

*Address all correspondence to: jamet@lrsv.ups-tlse.fr

1 Brazilian Bioethanol Science and Technology Laboratory (CTBE)/Brazilian Center of Research in Energy and Materials (CNPEM), Campinas, Brazil

2 Department of Genetics/Laboratory Max Feffer of Plant Genetics/Higher School of Agriculture "Luiz de Queiroz"/University of São Paulo, Piracicaba, Brazil

3 Plant Science Research Laboratory (LRSV)/University of Toulouse/CNRS/UPS, Auzeville, France

References

[1] Lionetti V, Francocci F, Ferrari S, Volpi C, Bellimcampi D, Galletti R, D'Ovidio R, De Lorenzo G, Cervone F. Engineering the cell wall by reducing. PNAS. 2010;107(2):616–621. doi:10.1073/pnas.0907549107

[2] Lee YC, Oh C, De Zoysa M, Kim H, Wickramaarachchi WDN, Whang I, Do-Hyung K, Lee J. Molecular cloning, overexpression, and enzymatic characterization. J Microbiol Biotechnol. 2013;23(7):901–910. doi:10.4014/jmb.1209.09009

[3] Franco-Cairo JPL, Leonardo FC, Alvarez TM, Ribeiro DA, Büchli F, Costa-Leonardo AM, Carazzolle MF, Costa FF, Paes Leme A, Gonçalo AGP, Squina FM. Functional characterization and target discovery of glycoside hydrolases from the digestome of the lower termite Coptotermes Gestroi Biotechnol Biofuels. 2011;4:50. doi:10.1186/1754-6834-4-50

[4] Kubicek CP, Starr TL, Glass NL. Plant cell wall-degrading enzymes and their secretion in plant-pathogenic fungi. Annu Rev Phytopathol. 2014;52:427–451. doi:10.1146/annurev-phyto-102313-045831

[5] Gao D, Uppugundla N, Chundawat SPS, Yu X, Hermanson S, Gowda K, Brumm P, Mead D, Balan V, Dale BE. Hemicellulases and auxiliary enzymes for improved conversion of lignocellulosic biomass to monosaccharides. Biotechnol Biofuels. 2011;4:5. doi:10.1186/1754-6834-4-5

[6] Zhang M, Su R, Qi W, He Z. Enhanced enzymatic hydrolysis of lignocellulose by optimizing enzyme complexes. Appl Biochem Biotechnol. 2010;160(5):1407–1414. doi: 10.1007/s12010-009-8602-3

[7] Boaretto LF, Mazzafera P. The proteomes of feedstocks used for the production of second-generation ethanol: a lacuna in the biofuel era. Ann Appl Biol. 2013;163(1):12–22. doi:10.1111/aab.12031

[8] Eudes A, Liang Y, Mitra P, Loque D. Lignin bioengineering. Curr Opin Biotechnol. 2015;26:189–198. doi:10.1016/j.copbio.2014.01.002

[9] Timpano H, Sibout R, Devaux M-F, Alvarado C, Looten R, Falourd X, Pontoire B, Martin M, Legée F, Cézard L, Lapierre C, Badel E, Citerne S, Vernhettes S, Höfte H, Guillon F, Gonneau M. Brachypodium cell wall mutant with enhanced saccharification potential despite increased lignin content. Bioenerg Res. 2015;8(1):53–67. doi:10.1007/s12155-014-9501-1

[10] Eudes A, George A, Mukerjee P, Kim JS, Pollet B, Benke PI, Yang F, Mitra P, Sun L, Cetinkol OP, Chabout S, Mouille G, Soubigou-Taconnat L, Balzergue S, Singh S, Holmes BM, Mukhopadhyay A, Keasling JD, Simmons BA, Lapierre C, Ralph J, Loqué D. Biosynthesis and incorporation of side-chain-truncated lignin monomers to reduce

lignin polymerization and enhance saccharification. Plant Biotechnol J. 2012;10(5):609–620. doi:10.1111/j.1467-7652.2012.00692.x

[11] Harris D, Stork J, DeBolt S. Genetic modification in cellulose-synthase reduces crystallinity and improves biochemical conversion to fermentable sugar. Glob Change Biol. 2009;1:51–61. doi:10.1111/j.1757-1707.2009.01000.x

[12] Nigorikawa M, Watanabe A, Furukawa K, Sonoki T, Ito Y. Enhanced saccharification of rice straw by overexpression of rice exo-glucanase. Rice. 2012;5:14. doi:10.1186/1939-8433-5-14

[13] Minic Z, Jouanin L. Plant glycoside hydrolases involved in cell wall polysaccharide degradation. Plant Physiol Biochem. 2006;44(7–9):435–449. doi:10.1016/j.plaphy.2006.08.001

[14] Carpita NC, Gibeaut DM. Structural models of primary cell walls in flowering plants: consistency of molecular structure with the physical properties of the walls during growth. Plant J. 1993;3(1):1–30. doi:10.1111/j.1365-313X.1993.tb00007.x

[15] Yu H, Liu R, Qiu L, Huang Y. Composition of the cell wall in the stem and leaf sheath of wheat straw. Appl Polymer. 2007;104(2):1236–1240. doi:10.1002/app.25755

[16] Lee JDK, Marcus SE, Knox JP. Cell wall biology: perspectives from cell wall imaging. Mol Plant. 2011;4(2):212–219. doi:10.1093/mp/ssq075

[17] Bashline L, Lei L, Li S, Gu Y. Cell wall, cytoskeleton, and cell expansion in higher plants. Mol Plant. 2014;7(4):586–600. doi:10.1093/mp/ssu018

[18] Zhao Q. Lignification: flexibility, biosynthesis and regulation. Trends Plant Sci. 2016;21(8):713–721. doi:10.1016/j.tplants.2016.04.006

[19] Endler A, Persson S. Cellulose synthases and synthesis in Arabidopsis. Mol Plant. 2011;4(2):199–211. doi:10.1093/mp/ssq079

[20] Albersheim P, Darvill A, Roberts K, Sederoff R, Staehelin A. Plant Cell Walls: From Chemistry to Biology. 1st ed. New York: Garland Science; 2010. 430 p. doi:10.1093/aob/mcr128

[21] Chandran AKN, Jeong HY, Jung K-H, Chanhui L. Development of functional modules based on co-expression patterns for cell-wall biosynthesis related genes in rice. J Plant Biol. 2016;59(1):1–15. doi:10.1007/s12374-016-0461-1

[22] Mohnen D. Pectin structure and biosynthesis. Curr Opin Plant Biol. 2008;11(3):266–277. doi:10.1016/j.pbi.2008.03.006

[23] Hu WJ, Harding SA, Lung J, Popko JL, Ralph J, Stokke DD, Tsai CJ, Chiang VL. Repression of lignin biosynthesis promotes cellulose accumulation and growth in transgenic trees. Nat Biotechnol. 1999;17(8):808–812. doi:10.1038/11758

[24] Driouich A, Faye L, Staehelin LA. The plant Golgi apparatus: a factory for complex polysaccharides and glycoproteins. Trends Biochem Sci. 1993;18(6):210–214.

[25] Kumar D, Murthy GS. Impact of pretreatment and downstream processing technologies on economics and energy in cellulosic ethanol production. Biotechnol Biofuels. 2011;4(27). doi:10.1186/1754-6834-4-27

[26] Lee TK, Tropini C, Hsin J, Desmarais SM, Ursell TS, Gong E, Gitai Z, Monds RD, Huang KC. A dynamically assembled cell wall synthesis machinery buffers cell growth. PNAS. 2014;111(12):4554–4559. doi:10.1073/pnas.1313826111

[27] Jamet E, Canut H, Boudart G, Pont-Lezica RF. Cell wall proteins: a new insight through proteomics. Trends Plant Sci. 2006;11(1):33–39. doi:10.1016/j.tplants.2005.11.006

[28] Kwon HK, Yokoyama R, Nishitani K. A proteomic approach to apoplastic proteins involved in cell wall regeneration in protoplasts of Arabidopsis suspension-cultured cells. Plant Cell Physiol. 2005;46(6):843–857. doi:10.1093/pcp/pci089

[29] Douché T, San Clemente H, Burlat V, Roujol D, Valot B, Zivy M, Pont-Lezica R, Jamet E. *Brachypodium distachyon* as a model plant toward improved biofuel crops: Search for secreted proteins involved in biogenesis and disassembly of cell wall polymers. Proteomics. 2013;13:2438–2454. doi:10.1002/pmic.201200507

[30] Calderan-Rodrigues MJ, Jamet E, Douché T, Bonassi MBR, Cataldi TR, Fonseca JG, San Clemente H, Pont-Lezica R, Labate CA. Cell wall proteome of sugarcane stems: comparison of a destructive and a non-destructive extraction method showed differences in glycoside hydrolases and peroxidases. BMC Plant Biol. 2016;16:14. doi:10.1186/s12870-015-0677-0

[31] Francin-Allami M, Lollier V, Pavlovic M, San Clemente H, Rogniaux H, Jamet E, Guillon F, Larré C. Understanding the remodelling of cell walls during *Brachypodium distachyon* grain development through a sub-cellular quantitative proteomic approach. Proteomes. 2016;4(3):21. doi:10.3390/proteomes4030021

[32] Guerra-Guimarães L, Vieira A, Chaves I, Pinheiro C, Queiroz V, Renaut J, Ricardo CP. Effect of greenhouse conditions on the leaf apoplastic proteome of *Coffea arabica* plants. J Proteomics. 2014;104:128–139. doi:10.1016/j.jprot.2014.03.024

[33] Bengtsson T, Weighill D, Proux-Wéra E, Levander F, Resjö S, Burra DD, Moushib LI, Hedley PE, Liljeroth E, Jacobson D, Alexandersson E, Andreasson E. Proteomics and transcriptomics of the BABA-induced resistance response in potato using a novel functional annotation approach. BMC Genomics. 2014;15:315. doi:10.1186/1471-2164-15-315

[34] von Heijne G. A new method for predicting signal sequence cleavage sites. Nucleic Acids Res. 1986;14(11):4683–4690.

[35] Denecke J, De Rycke R, Botterman J. Plant and mammalian sorting signals for protein retention in the endoplasmic reticulum contain a conserved epitope. EMBO J. 1992;11(6):2345–2355.

[36] Pereira C, Pereira S, Pissarra J. Delivering of proteins to the plant vacuole—an update. Int J Mol Sci. 2014;15(5):7611–7623. doi:10.3390/ijms15057611

[37] Emanuelsson O, Brunak S, von Heijne G, Nielsen H. Locating proteins in the cell using TargetP, SignalP and related tools. Nat Protoc. 2007;2(4):953–971. doi:10.1038/nprot.2007.131

[38] Albenne C, Canut H, Hoffmann L, Jamet E. Plant cell wall proteins: a large body of data, but what about runaways? Proteomes. 2014;2(2):224–242. doi:10.3390/proteomes2020224

[39] Tan L, Eberhard S, Pattathil S, Warder C, Glushka J, Yuan C, Hao Z, Zhu X, Avci U, Miller JS, Baldwin D, Pham C, Orlando R, Darvill A, Hahn MG, Kieliszewski MJ, Mohnen D. An Arabidopsis cell wall proteoglycan consists of pectin and arabinoxylan covalently linked to an arabinogalactan protein. Plant Cell. 2013;25(1):270–287. doi:10.1105/tpc.112.107334

[40] Jamet E, Canut H, Albenne C, Boudart G, Pont-Lezica R. Cell wall. In: Agrawal GK, Rakwal R, editors. Plant Proteomics: Technologies, Strategies and Applications. 1st ed. Hoboken: John Wiley & Sons, Inc; 2008. pp. 293–307. ISBN: 978-0-470-06976-9.

[41] Albenne C, Canut H, Jamet E. Plant cell wall proteomics: the leadership of Arabidopsis thaliana. Front Plant Sci. 2013;4:111. doi:10.3389/fpls.2013.00111

[42] Franková L, Fry SC. Biochemistry and physiological roles of enzymes that 'cut and paste' plant cell-wall polysaccharides. J Exp Bot. 2013;64(12):3519–3550. doi:10.1093/jxb/ert201

[43] Francoz E, Ranocha P, Nguyen-Kim H, Jamet E, Burlat V, Dunand C. Roles of cell wall peroxidases in plant development. Phytochemistry. 2015;112:15–21. doi:10.1016/j.phytochem.2014.07.020

[44] Nieuwland J, Feron R, Huisman BA, Fasolino A, Hilbers CW, Derksen J, Mariani C. Lipid transfer proteins enhance cell wall extension in tobacco. Plant Cell. 2005;17(7):2009–2019. doi:10.1105/tpc.105.032094

[45] van der Hoorn NA. Plant proteases: from phenotypes to molecular mechanisms. Annu Rev Plant Biol. 2008;59:191–223. doi:10.1146/annurev.arplant.59.032607.092835

[46] Ringli C, Keller B, Ryser U. Glycine-rich proteins as structural components. Cell Mol Life Sci. 2001;58:1430–1441. doi:10.1007/PL00000786

[47] Cassab GI. Plant cell wall proteins. Ann Rev Plant Physiol Plant Mol Biol. 1998;49:281–309. doi:10.1146/annurev.arplant.49.1.281

[48] Lannoo N, Van Damme EJM. Lectin domains at the frontiers of plant defense. Front Plant Sci. 2014;5:397. doi:10.3389/fpls.2014.00397

[49] Jolie RP, Duvetter T, Van Loey AM, Hendrickx ME. Pectin methylesterase and its proteinaceous inhibitor: a review. Carbohydr Res. 2010;345:2583–2595. doi:10.1016/j.carres.2010.10.002

[50] Lamport DTA, Varnai P, Charlotte ES. Back to the future with the AGP–Ca^{2+} flux capacitor. Ann Bot. 2014;114(6):1069–1085. doi:10.1093/aob/mcu161

[51] Vazquez-Lobo A, Roujol D, Zuñiga E, Albenne C, Piñero D, de Buen AG, Jamet E. The highly conserved spermatophyte cell wall DUF642 protein family: Phylogeny and first evidence of interaction with cell wall polysaccharides in vitro. Mol Phylogenet Evol. 2012;63(2):510–520. doi:10.1016/j.ympev.2012.02.001

[52] Zúñiga-Sánchez E, Soriano D, Martínez-Barajas Eleazar, Orozco-Segovia A, de Buen AG. BIIDXI, the At4g32460 DUF642 gene, is involved in pectin methyl esterase regulation during Arabidopsis thalianaseed germination and plant development. BMC Plant Biol. 2014;14:338. doi:10.1186/s12870-014-0338-8

[53] Xie X, Wang Y. VqDUF642, a gene isolated from the Chinese grape Vitis quinquangularis, is involved in berry development and pathogen resistance. Planta. Forthcoming. doi:10.1007/s00425-016-2569-4

[54] Feiz L, Irshad M, Pont-Lezica RF, Canut H, Jamet E. Evaluation of cell wall preparations for proteomics: a new procedure for purifying cell walls from Arabidopsis hypocotyls. Plant Methods. 2006;2:10. doi:10.1186/1746-4811-2-10

[55] Price CA. Plant cell fractionation. Methods Enzymol. 1974;31:501–519.

[56] Boudart G, Jamet E, Rossignol M, Lafitte C, Borderies G, Jauneau A, Esquerré-Tugayé MT, Pont-Lezica R. Cell wall proteins in apoplastic fluids of Arabidopsis thaliana rosettes: identification by mass spectrometry and bioinformatics. Proteomics. 2005;5(1): 212–221. doi:10.1002/pmic.200400882

[57] Angyal SJ, Craig DC. The composition and conformation of D-Threo-3,4-Hexodiulose in solution, and the X-ray crystal-structure of its beta-anomer. Carbohydr Res. 1989;194:21–29. doi:0008–6215

[58] Voigt J. Extraction by lithium chloride of hydroxyproline-rich glycoproteins from intact cells of Chlamydomonas reinhardii. Planta. 1985;164(3):379–389. doi:10.1007/BF00402950

[59] Cell wall Proteins and Development Team. WallProtDB [Internet]. 2008. Available from: http://www.polebio.lrsv.ups-tlse.fr/WallProtDB/[Accessed: 2016-08-12]

[60] Mur LA, Allainguillaume J, Catalán P, Hasterok R, Jenkins G, Lesniewska K, Thomas I, Vogel J. Exploiting the Brachypodium Tool Box in cereal and grass research. New Phytol. 2011;191(2):334–347. doi:10.1111/j.1469-8137.2011.03748.x

[61] Francin-Allami M, Merah K, Albenne C, Rogniaux H, Pavlovic M, Lollier V, Sibout R, Guillon F, Jamet E, Larré C. Cell wall proteomic of *Brachypodium distachyon* grains: a focus on cell wall remodeling proteins. Proteomics. 2015;15(13):2296–2306. doi:10.1002/pmic.201400485

[62] Calderan-Rodrigues MJ, Jamet E, Bonassi MB, Guidetti-Gonzalez S, Begossi AC, Setem LV, Franceschini LM, Fonseca JG, Labate CA. Cell wall proteomics of

sugarcane cell suspension cultures. Proteomics. 2014;14(6):738–749. doi:10.1002/pmic.201300132

[63] Fonseca JG. Characterization of the extracellular proteome of young and mature internodes of sugarcane [dissertation]. Piracicaba: Universidade de São Paulo, Escola Superior de Agricultura Luiz de Queiroz; 2015. 103 p. Available from: http://www.teses.usp.br/teses/disponiveis/11/11137/tde-23042015-162148/pt-br.php doi: 10.11606/D.11.2015.tde-23042015-162148

[64] Joint Genome Institute. Phytozome [Internet]. 1997. Available from: https://phytozome.jgi.doe.gov/pz/portal.html [Accessed: 2016-08-12]

[65] The International Brachypodium Initiative. Genome sequencing and analysis of the model grass *Brachypodium distachyon*. Nature. 2010;463(11):763–768. doi:10.1038/nature08747

[66] Vettore AL, da Silva FR, Kemper EL, Souza GM, da Silva AM, Ferro MI, Henrique-Silva F, Giglioti EA, Lemos MV, Coutinho LL, Nobrega MP, Carrer H, França SC, Bacci Júnior M, Goldman MH, Gomes SL, Nunes LR, Camargo LE, Siqueira WJ, Van Sluys MA, Thiemann OH, Kuramae EE, Santelli RV, Marino CL, Targon ML, Ferro JA, Silveira HC, Marini DC, Lemos EG, Monteiro-Vitorello CB, Tambor JH, Carraro DM, Roberto PG, Martins VG, Goldman GH, de Oliveira RC, Truffi D, Colombo CA, Rossi M, de Araujo PG, Sculaccio SA, Angella A, Lima MM, de Rosa Júnior VE, Siviero F, Coscrato VE, Machado MA, Grivet L, Di Mauro SM, Nobrega FG, Menck CF, Braga MD, Telles GP, Cara FA, Pedrosa G, Meidanis J, Arruda P. Analysis and functional annotation of an expressed sequence tag collection for tropical crop sugarcane. Genome Res. 2003;13(12): 2725–2735. doi:10.1101/gr.1532103

[67] Cell wall Proteins and Development Team. ProtAnnDB [Internet]. 2009. Available from: http://www.polebio.lrsv.ups-tlse.fr/ProtAnnDB/[Accessed: 2016-08-12]

[68] Clemente HS, Pont-Lezica R, Jamet E. Bioinformatics as a tool for assessing the quality of sub-cellular proteomic strategies and inferring functions of proteins: plant cell wall proteomics as a test case. Bioinform Biol Insights. 2009;3:15–28.

[69] San Clemente H, Jamet E. WallProtDB, a database resource for plant cell wall proteomics. Plant Methods. 2015;11(1):2. doi:10.1186/s13007-015-0045-y

[70] Oliveros JC. Venny [Internet]. 2007. Available from: http://bioinfogp.cnb.csic.es/tools/venny/index.html [Accessed: 2016-08-12]

[71] Bayer EM, Bottrill AR, Walshaw J, Vigouroux M, Naldrett MJ, Thomas CL, Maule AJ. Arabidopsis cell wall proteome defined using multidimensional protein identification technology. Proteomics. 2006;6(1):301–311. doi:10.1002/pmic.200500046

[72] Ming R, Liu SC, Lin YR, da Silva J, Wilson W, Braga D, van Deynze A, Wenslaff TF, Wu KK, Moore PH, Burnquist W, Sorrells ME, Irvine JE, Paterson AH. Detailed alignment

of saccharum and sorghum chromosomes: comparative organization of closely related diploid and polyploid genomes. Genetics. 1998;150(4):1663–1682. doi:PMC1460436

[73] Boerjan W, Ralph J, Baucher M. Lignin biosynthesis. Annu Rev Plant Biol. 2003;54:519–546. doi:10.1146/annurev.arplant.54.031902.134938

[74] Fry SC. Cross-linking of matrix polymers in the growing cell walls of angiosperms. Plant Biol. 1986;37:165–186. doi:10.1146/annurev.pp.37.060186.001121

[75] Yeats TH, Rose JK. The formation and function of plant cuticles. Plant Physiol. 2013;163(1):5–20. doi:10.1104/pp.113.222737

[76] Hrubá P, Honys D, Twell D, Capková V, Tupý J. Expression of beta-galactosidase and beta-xylosidase genes during microspore and pollen development. Planta. 2005;220(6): 931–940. doi:10.1007/s00425-004-1409-0

[77] Li X, Weng JK, Chapple C. Improvement of biomass through lignin modification. Plant J. 2008;54(4):569–581. doi:10.1111/j.1365-313X.2008.03457.x

[78] Doblin MS, Johnson KL, Humphries J, Newbigin EJ, Bacic A. Are designer plant cell walls a realistic aspiration or will the plasticity of the plant's metabolism win out? Curr Opin Biotechnol. 2014;26:108–114. doi:10.1016/j.copbio.2013.11.012

[79] Ambavaram MM, Krishnan A, Trijatmiko KR, Pereira A. Coordinated activation of cellulose and repression of lignin biosynthesis pathways in rice. Plant Physiol. 2011;155(2):916–931. doi:10.1104/pp.110.168641

[80] Yang F, Mitra P, Zhang L, Prak L, Verhertbruggen Y, Kim JS, Sun L, Zheng K, Tang K, Auer M, Scheller HV, Loqué D. Engineering secondary cell wall deposition in plants. Plant Biotechnol J. 2013;11(3):325–335. doi:10.1111/pbi.12016

[81] Pogorelko G, Fursova O, Lin M, Pyle E, Jass J, Zabotina OA. Post-synthetic modification of plant cell walls by expression of microbial hydrolases in the apoplast. Plant Mol Biol. 2011;77(4–5):433–445. doi:10.1007/s11103-011-9822-9

[82] Turumtay H. Cell wall engineering by heterologous expression of cell wall-degrading enzymes for better conversion of lignocellulosic biomass into biofuels. Bioenerg Res. 2015;8:1574–1588. doi:10.1007/s12155-015-9624-z

Chapter 3

Advances in the Application of Spectroscopic Techniques in the Biofuel Area over the Last Few Decades

João Cajaiba Da Silva, Alex Queiroz,
Alline Oliveira and Vinícius Kartnaller

Additional information is available at the end of the chapter

http://dx.doi.org/10.5772/65552

Abstract

Guided by the instability of the oil market, as well as limited availability of and, especially, the environmental impacts of fossil fuels, the needs of the market for environmental-friendly energy sources have increased. However, as with any other product that is intended to place on the market, it is essential to ensure the quality of the fuel for successful marketing and acceptance by consumers. Spectroscopic techniques have been widely used for different purposes in the literature for the past decades, from biological applications to the measurement of the elemental composition of planets. From studies focused on biodiesel, bioethanol, biomass and biofuel in general, different spectroscopic techniques have also been applied in the area. The focus of this chapter is to elucidate what has been published in the last few decades over the subject, detailing the basic concepts of the main spectroscopic techniques applied and showing the results and developments over biofuel. The aim of the chapter is to achieve a set of information that can be used as a bigger compile of information of the state of the art regarding the theme.

Keywords: biofuel, spectroscopy, infrared, Raman, NMR, UV/Vis, image analysis, ICP-OES

1. Introduction

Spectroscopy is a general term for the science that deals with the interaction between the various types of radiation and matter [1]. Therefore, these types of measures are only possible if the interaction between photons and matter causes some kind of change in one or more properties of the sample. The electromagnetic spectrum can be divided into regions according to the wavelength of the radiation, as shown in **Figure 1**.

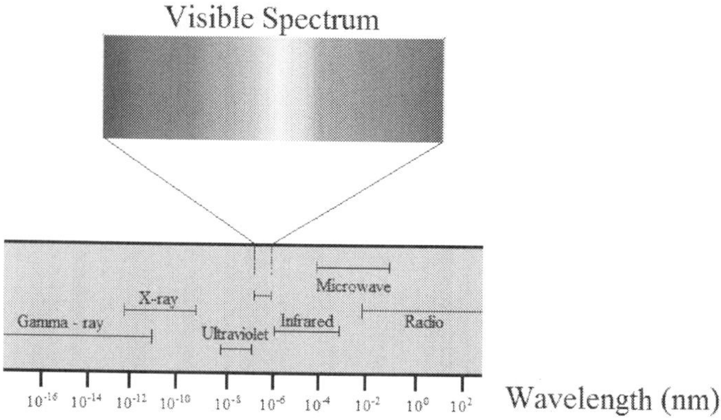

Figure 1. The electromagnetic spectrum showing the wavelength limits corresponding to each type of electromagnetic wave. The colored area shows the visible region.

According to the region of the electromagnetic spectrum, different types of spectroscopic techniques can be used, and the phenomena explored by them are of totally a different nature. This fact explains the applicability of each technique for specific cases. **Table 1** shows the spectroscopic methods according to the spectral region and the transition type.

Because of its versatility, the spectroscopic methods have been widely used for different purposes in literature for the past few decades, from biological applications [3, 4] to the measurement of elemental abundance in astronomical objects [5, 6]. Since the beginning of the century, the number of publications involving such techniques has grown exponentially.

At the same time, guided by the instability of the oil market, as well as limited availability and especially the environmental impacts of fossil fuels [7], the needs of the market for environmental-friendly energy sources have increased. However, as with any other product that is intended to place on the market, it is essential to ensure the quality of the fuel for successful marketing and acceptance by consumers [8]. For this reason, many countries have specific legislation that establishes desirable characteristics to minimize problems arising from the use of these compounds. Quality control, not only for biofuels but also for their increasingly used blends, is of paramount importance. The type (the chain length, degree of unsaturation, and presence of other chemical functions) and the concentration of the fatty acids used in the synthesis of biodiesel, for example, can influence the properties of the final product and its prerequisites in terms of storage and their tendency to oxidize [9]. The raw materials used and the process of obtaining biofuel can be decisive in terms of the introduction of some harmful contaminants in the final product [7]. Therefore, the monitoring of reactions is critical for quality control [10], and the monitoring of minor components, such as glycerol, water, residual alcohol, and heavy metals, is very relevant [11].

Type of energy transfer	Region of electromagnetic spectrum	Spectroscopic technique	Quantic transition type
Absorption	γ-ray	Mossbauer spectroscopy	Nuclear
	X-ray	X-ray absorption spectroscopy	Inner electrons
	Ultraviolet/Visible	UV/Vis spectroscopy	Bounding electrons
		Atomic absorption spectroscopy	Bounding electrons
	Infrared	Infrared spectroscopy	Rotation/vibration of molecules
		Raman spectroscopy	Rotation/vibration of molecules
	Microwave	Microwave spectroscopy	Rotation of molecules
	Radio wave	Electron spin resonance spectroscopy	Spin of electrons in a magnetic field
		Nuclear magnetic resonance spectroscopy	Spin of nuclei in a magnetic field
Emission (thermal excitation)	Ultraviolet/Visible	Atomic emission spectroscopy	Bounding electrons
Photoluminescence	X-ray	X-ray fluorescence	Inner electrons
	Ultraviolet/Visible	Fluorescence spectroscopy	Bounding electrons
		Phosphorescence spectroscopy	Bounding electrons
		Atomic fluorescence spectroscopy	Bounding electrons
Chemiluminescence	Ultraviolet/Visible	Chemiluminescence spectroscopy	Bounding electrons

[a] Adapted from Ref. [2].

Table 1. Spectroscopic methods based on electromagnetic radiation*.

Chromatographic and spectroscopic methods are the most popular ways for biofuel analysis [11]. Spectroscopic techniques, however, have attributes that make them ideal for use in these kinds of analytical measurements. These attributes include the following [12]:

- Fast and simple analysis, either qualitative or quantitative, is possible;
- Little or no sample preparation is usually required, and they do not use polluting solvents;
- It is easy to find appropriate calibration standards and methods of analysis that are validated by recognized institutes;
- Direct and non-invasive analyses are possible;
- Normally, non-destructive methods can be employed, with a few exceptions, such as atomic spectrometry methods;

- The observed response is usually directly proportional to the concentration of the species in the system, except for ^{13}C NMR spectroscopy.
- In-situ analysis can be conducted, for example, in the case of real-time monitoring of biofuel production reactions;
- They are relatively inexpensive with respect to analysis time and reagents.

For the above reasons, it is relevant to examine the recent advances provided by the introduction of spectroscopic methods for biofuel analysis. This chapter aims to provide the theoretical basis of the most widely used techniques for this purpose and to analyze the impact of their use on this area.

2. Vibrational spectroscopy: infrared and Raman spectroscopy for biofuel evaluations

The notion that systems comprised by molecules can undergo vibrational motions, in which their atoms are in constant movement around their equilibrium position, can be explained by vibrational spectroscopy [13]. Every molecule contains a ground state energy (including vibrational energy), which can be described as the sum of different components. This energy defines the minimum energy with which atoms in a molecule can move in a periodic motion, described mainly by the six movements shown in **Figure 2**.

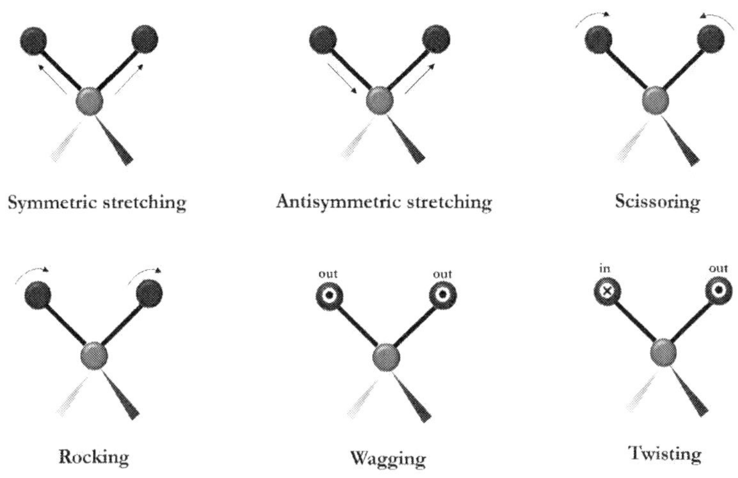

Figure 2. Main vibrational modes for molecules.

The principle of this spectroscopy type is that electromagnetic radiation can interact with molecules, leading them from their ground state energy to a vibrational excited state energy in which the atoms' movements have higher energy. The state transition follows quantum mechanical rules and obeys quantized energy levels; for a harmonic oscillator view of the vibration, the energy that takes a molecule from a vibrational state to an immediately higher state is $(\frac{1}{2}hv)$, where h is the Planck constant and v is the frequency of the vibration. Hence, the photon that can interact with a molecule must have an energy that equals this value in order to change its vibrational state; otherwise, the transition is not achieved. The portion of the electromagnetic spectrum that can lead to these transitions is the infrared region. This region can be divided into three different parts, with different applications for spectroscopic study: near infrared (NIR), 14,000–4000 cm^{-1}, allows the study of overtones and harmonic or combination vibrations; mid infrared (MIR), 4000–400 cm^{-1}, allows the study of the fundamental vibrations and the rotation-vibration structure of small molecules; and far infrared (FIR), 400–10 cm^{-1}, allows the study of low-heavy atom vibrations [14]. A polyatomic molecule can undergo many types of vibration, depending on their number of atoms and their degree of freedom. For nonlinear molecules, the number of possible vibrations can be defined by the $3N-6$ rule, where N is the total number of atoms. Hence, different molecules undergo different types of vibration, leading to different possible transitions by absorbing infrared radiation. This is the premise of this type of spectroscopy, since a spectrum achieved by such technique can be seen as the fingerprint of a molecule and can be used specially for functional group analysis [15].

The two major techniques to study these vibrations are infrared and Raman spectroscopy, and they have been studied to a greater degree than the other techniques for applications since the middle of the twentieth century. Although the theoretical knowledge was developed during the prior century, it was not until World War II that infrared spectroscopy had a breakthrough, with the number of instruments going from less than 20 prior to the war to 700 by 1947 [16, 17]. In fact, several UK/US programs during the war prompted research and development of the early commercially available infrared instruments, such as those used in petroleum analysis (e.g., to trace the origin of gasoline used by the German air force), in quality control for the production of synthetic rubber, and in resolution of the structure of penicillin [16–18]. Examples of early publications range from industrial applications to more academic studies For example, Downing et al. [19] used infrared spectroscopy to differentiate isomers of the molecule dichlorodiphenyltrichloroethane and to qualitatively detect the presence of impurities in its samples, and Pfann et al. [20] used infrared spectroscopy to evaluate the presence of molecular fragments of the initiators used to catalyze a polymerization reaction in the structure of the final polymer. Regarding the biofuels area, since it is a more recent subject, the first described applications of infrared spectroscopy started to appear only by the end of the 1990s. In 1996, Adjaye et al. [21] used it to characterize different mixtures of silica-alumina and HZSM-5 as catalysts for the catalytic conversion of a biofuel from the thermal processing of maple wood to liquid hydrocarbons; in the same year, Sanderson et al. [22] used the technique for the compositional analysis of biomass feedstocks. Raman spectroscopy, on the other hand, was applied for biofuel studies almost a decade later than infrared spectroscopy, as in the work of Oliveira et al. [23]. They used the technique to determine the adulteration of diesel/biodiesel blends by the addition of vegetable oil. In this study, both

Raman and near-infrared spectroscopy were applied for the determination, and both gave a final similar accuracy result for the quantification of the adulteration, dependent on the algorithm used for the calculation.

The two techniques, although related to the vibrational modes of the molecules, have different principles and are complementary. Infrared spectroscopy is based on radiation absorption, while Raman spectroscopy is based on the interaction via inelastic collisions. In infrared spectroscopy, an infrared photon of certain energy (with a frequency v), can be absorbed by a molecule if that energy is exactly the same as the energy difference between a vibrational ground state and a vibrational excited state; using a simplified harmonic model, the energy difference between the two states is ($\frac{1}{2}hv$), as mentioned before in the text. Hence, the photon energy (E_p) needed so that a transition is possible is defined by:

$$E_p = \Delta E = \frac{1}{2}hv. \quad (1)$$

However, even if a photon has such energy, this state transition may not exist due to the main selection rule of infrared spectroscopy: for absorption to occur (hence, transition), the vibration needs to lead to a change in the dipole moment. If no change occurs, then that transition is considered "infrared forbidden" [13]. The intensity in which the transitions occur also depends on the molecular bonds related to the vibration, since it is proportional to the square of this dipole moment [24].

For Raman spectroscopy, the dipole moment change by the vibration is not the cutting rule, and vibration transitions that are not allowed in infrared spectroscopy may be allowed in Raman spectroscopy. This is due to the different phenomena taking place in this spectroscopy; as said, the technique takes advantage of an inelastic collision between the photon and the molecule, leading to the scattering of the electromagnetic radiation with different wavelength than the irradiated. The electric field in the electromagnetic radiation can interact with the molecule, creating an additional induced dipole moment as a response of the electrons and nuclei moving in opposite directions of the field, in accordance with Coulomb's law [25]. The dependency in this case relates to the ability of the molecule to be polarized, measured by its polarizability, that is, the deformability of the electron cloud around the molecule by an external electric field. In fact, the selection rule for a vibration transition in Raman spectroscopy is that it causes a change in the polarizability of the molecule. The incident photon is then momentarily absorbed by the molecule, leading to a transition from the ground state into a virtual state; a new photon is created and scattered by a transition from this virtual state to a lower energy vibrational state, as shown in **Figure 3**. Most of the scattered light is the same frequency as the initial light, as described by Rayleigh scattering, which does not contain any information regarding the molecule. The information comes from the Stokes and anti-Stokes scattering, in which the molecule goes from a ground state to a virtual state or from an excited state to a virtual state; in the first case, the scattered light has a frequency smaller than the initial light, while for the second, the scattered light has a frequency higher than the initial light. Hence, the Raman spectra can be divided by both types of scattering spectra, and the frequency difference is equivalent to the one in infrared spectroscopy.

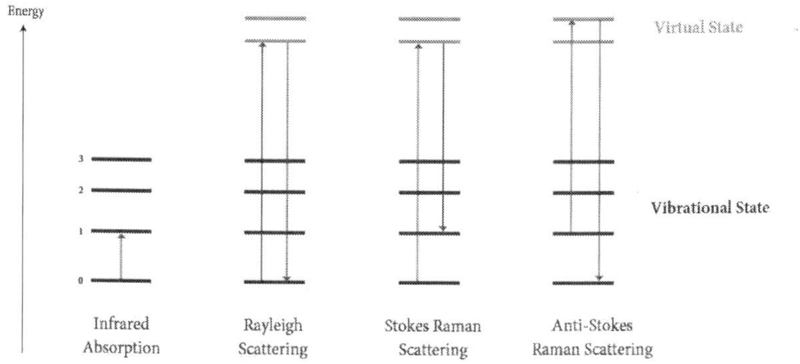

Figure 3. Energy-level diagram for the possible transitions in infrared and Raman spectroscopy.

For both techniques, characterization is one of the main possibilities, since the frequencies of the molecular vibrations depend on the masses of the atoms, their geometric arrangement, and the strength of their chemical bonds. The spectra provide information on the molecular structure, dynamics, and environment [24].

In terms of application, both techniques have different apparatus schemes, which will not be dealt by the authors here, but can be easily found elsewhere. Regarding the difference between the techniques for comparison purposes, Peter Larkin has described it in his book *"Infrared and Raman Spectroscopy: Principles and Spectral Interpretation,"* as shown in **Table 2**.

	Raman	Infrared	Near-IR
Ease of Sample Preparation	Very simple	Variable	Simple
Liquids	Very simple	Very simple	Very simple
Powders	Very simple	Simple	Simple
Gases	Simple	Very simple	Simple
Fingerprinting	Excellent	Excellent	Very good
Best vibrations	Symmetric	Assymetric	Comb/Overtone
Group frequencies	Excellent	Excellent	Fair
Aqueous solutions	Very good	Very difficult	Fair
Quantitative analysis	Good	Good	Excellent
Low-frequency modes	Excellent	Difficult	No

Table 2. Comparison of Raman, mid-IR and near-IR spectroscopy, taken from [24].

The difference between the spectra of samples achieved by infrared spectroscopy and Raman spectroscopy is illustrated by Corsetti et al. [26], in which both techniques were used to quantitative measurements in systematically varied blends of ethanol and a gasoline, a subject of interest due to the production of bioethanol. **Figure 4** shows a comparison between the two techniques for the mixture and for pure ethanol and gasoline.

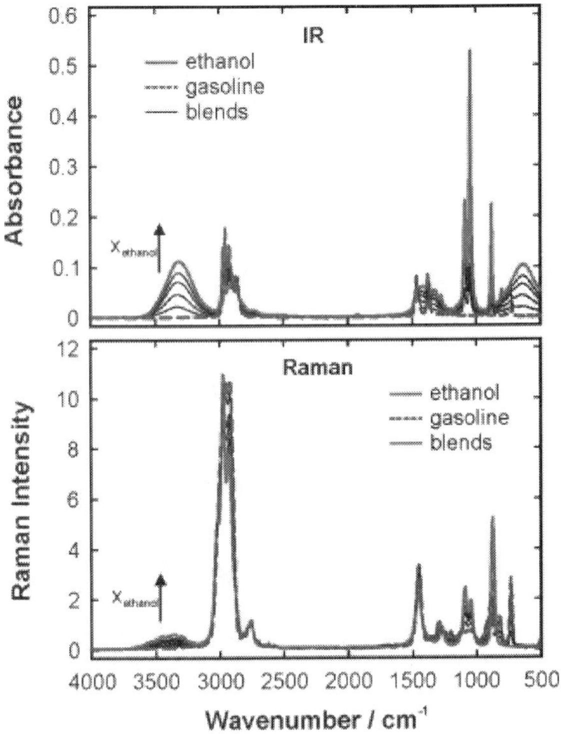

Figure 4. IR and Raman spectra of pure ethanol (red), pure gasoline (dashed blue), and blends (black), presented in [26].

A comparison between the spectra can be very direct, in which bands that are very intense for IR spectroscopy have low intensity in Raman spectroscopy, and vice versa. The vibrational bands related to alcohol molecules, which are the stretches of the O—H (3600–3000 cm^{-1}) and the C—O (1000–1100 cm^{-1}) bands, are intense in the IR spectra due to the electronegativity that causes great change in the dipole moment; however, these vibrations are not very polarizable, so they are shown with low intensity in the Raman spectra. On the other hand, the bands related to the stretches and bending of the C—H bands (3000–2800 cm^{-1} and 1600–1200 cm^{-1}) do not cause much change in the dipole moment and are not very intense in the IR spectra, while they tend to be more polarizable and have greater intensity in the Raman spectra.

As mentioned previously, the first applications of the two techniques in the biofuel area were for the characterization and quantification of components/properties of biofuels. These types of applications are very common, and a great amount of work from the past two decades can be found in the literature regarding it. Characterization is the principal feature of these techniques, which have been applied in different situations. One of them is the evaluation of catalysts, as in the study of heterogeneous biocatalysts from sucrose, saw dust, and chicken egg shells for biodiesel production by Wembabazi et al. [27]. Another study investigated the production of hydrocarbon biofuels through hydroprocessing of carinata oil [28], while other looked at the production of bioadditive fuels from bioglycerol under eco-friendly conditions [29]. Many other works can be found that deal with this theme [21, 30–32] Characterization has been applied not only to the catalysts used for biofuel processes but also to the biofuel itself, such as the structural analysis of bio-oils from the subcritical and supercritical hydrothermal liquefaction of a plant from Turkey in a study by Durak and Aysu [33], or the characterization of insoluble material isolated from Colombian palm oil biodiesel in Plata et al. [34]. Also, Nanda et al. evaluated horse manure through catalytic supercritical water gasification as a possible next generation feedstock for biofuel production [35]. Many others have used this approach [36–42].

Another great feature of these techniques is their concentration dependency. The intensity of the signals is related to the characteristic of the band vibration and to the number of molecules needed to make the state transition (i.e., molecular concentration). This dependence can be defined by the Lambert-Beer law, which states that there is a linear relationship between signal and concentration. Hence, this can be used for the estimation of the concentration of components present in mixtures or for other properties in different systems. As commented earlier, one of the first articles published using infrared spectroscopy aimed to achieve a compositional analysis of biomass feedstock, where they determined the chemical composition of several woody and herbaceous feedstocks (121 samples in total) using NIR spectroscopy. Samples were analyzed for ethanol extractives, ash, lignin, uronic acids, arabinose, xylose, mannose, galactose, glucose, and C, H, N, and O, and all those responses were quantified using spectroscopy. The results showed great correlations for most responses, as depicted in **Figure 5**, suggesting that infrared spectroscopy was able to quantify components related to biomass feedstocks with high accuracy responses.

For quantification or classification purposes, many chemometric algorithms can be applied to vibrational data to transform the data to more relevant quantitative information. In the case of the Sanderson's work presented above, partial least squares regression (PLSR) was used to convert the spectra data to the responses. Several other works in the biofuel area have used different algorithms for the quantitative evaluation of different responses or for quality control and classification, such as multivariate linear regression (MLR), principal component regression (PCR), support vector machines (SVM), artificial neural network (ANN), partial least squares-discriminant analysis (PLS-DA), principal component analysis (PCA), K-nearest neighbors (KNN), linear discriminant analysis (LDA), quadratic discriminant analysis (QDA), regularized discriminant analysis (RDA), amongst others [39, 43–60] (see **Figure 6**).

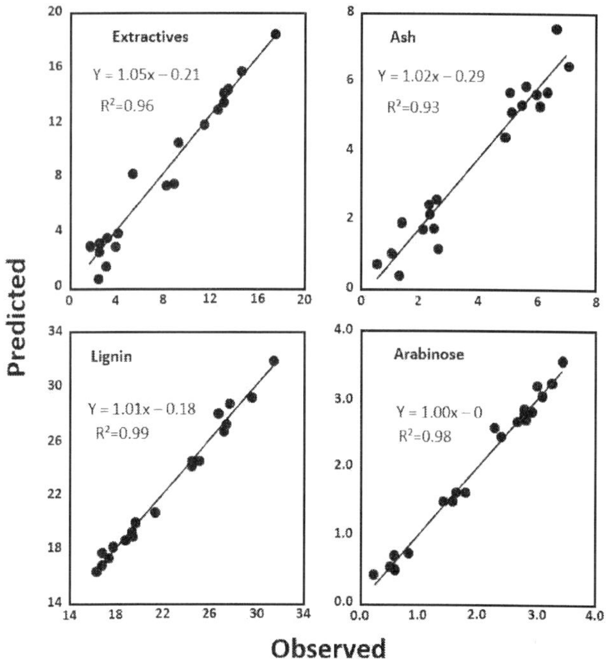

Figure 5. Graph of predicted vs. observed values of the responses studied in [22] for the quantification of components present in biomass feedstock.

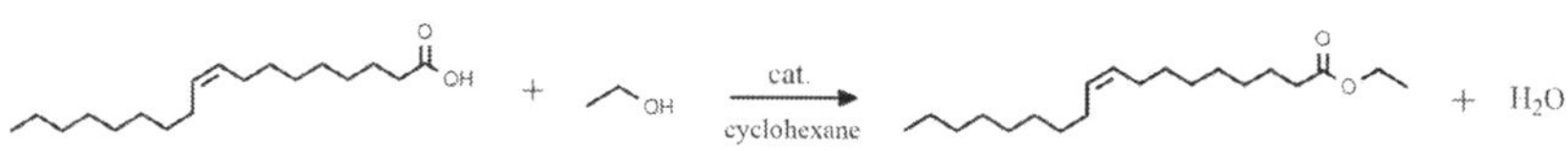

Figure 6. Scheme of the biodiesel production reaction made by [79].

Other applications in the area using the vibrational techniques are the evaluation of biofuel degradation [61–64] and process evaluation [65–70]. Regarding the latter subject, a new strategy for process optimization and control emerged in the last decade, namely, the process analytical technology (PAT) initiative [71]. PAT can be defined as the optimal application of technologies in feedback process-control strategies, information management tools, and/or product-process optimization strategies, as described by Chadwick et al. [72]. To meet its objectives, real-time sensing techniques are employed to control and monitor parameters throughout the processing chain, achieving a great amount of data. In the literature, PAT has been applied to follow reaction paths, monitor conversion over time, and even quantitatively evaluate catalyst efficiencies for biodiesel production [73–79]. As an example, from our research group, Kartnaller et al. have used infrared spectroscopy and chemometrics to follow the esterification reaction of a free fatty acid to biodiesel, quantifying each component in the mixture and evaluating different enzymatic catalysts [79]. **Figure 7a** shows the change in the spectra of the reaction mixture in which the bands related to different components change as the reaction progress. As discussed, the spectroscopic signal is concentration dependent per the Lambert-Beer law; hence, if a product in a mixture changes in its concentration, so will the intensity of its bands. With the advance of technology, the infrared equipment has changed in the past few decades and new types have been shown in the market. This has allowed the equipment to acquire more information in less time and in an automated manner, which may generate a great deal of information and allow easy monitoring of reaction paths, as shown in **Figure 7b**.

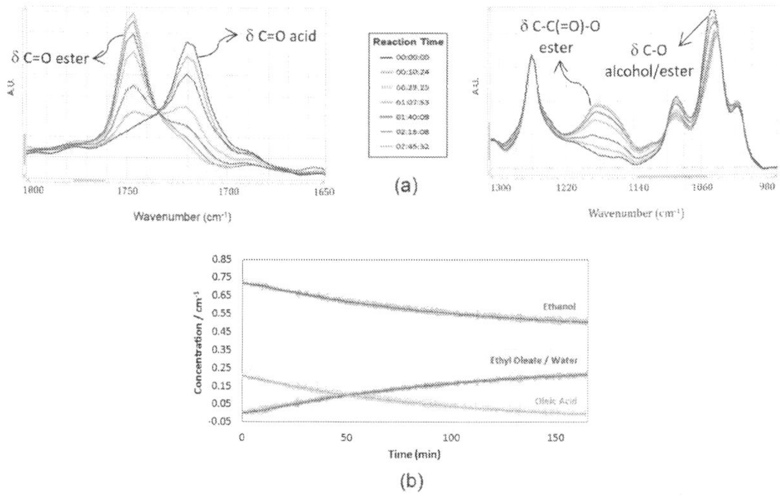

Figure 7. (a) Change in the intensity of infrared bands from different components in a mixture due to reaction and (b) quantitative monitoring of the different components of the reaction over time, calculated by the Lambert-Beer law, from [79].

Hence, it is easy to conclude that vibrational spectroscopy has a large range of applications in general, and the biofuel area has used it for many types of goals in the past years. With the increase of technological advances, both from the technical perspective and the biofuel research, the use of the technique is bound to increase even more.

3. Nuclear magnetic resonance and biofuel applications

The theory of nuclear magnetic resonance (NMR) spectroscopy was proposed by Pauli in 1924. He suggested that when certain atomic nuclei have spin and a nuclear magnetic moment, in the presence of an applied magnetic field (Bo), they could be induced to absorb energy and change their spin orientation with respect to this applied field [1, 15].

The independent works of Felix Block and Eduard Purcell in 1946 experimentally showed that by absorbing electromagnetic radiation in the presence of an intense magnetic field, the nucleus experiences an unfolding of its spin state energy levels, as shown in **Figure 8**, which is dependent on the spin itself. In addition, the energy absorbed must equal the energy difference between the two states involved; in this case, the energy absorption is a quantized process [15]. The stronger the applied magnetic field, the greater the energy difference between the two inverse spin states, as shown in **Figure 9**.

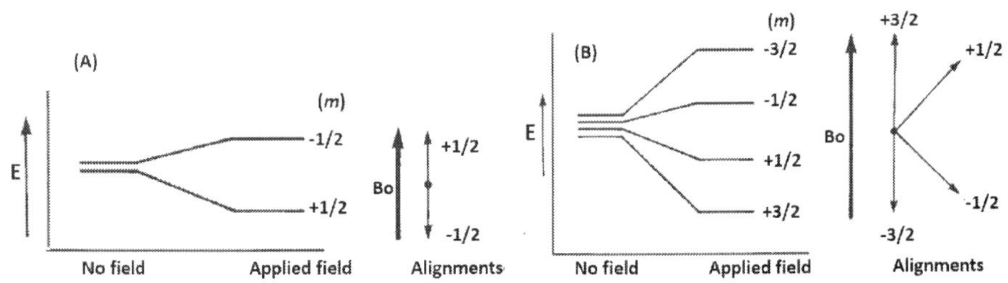

Figure 8. Energy levels for a nucleus with spin quantum number; (a) $I = 1/2$ and (b) $I = 3/2$ in the absence and in the presence of an applied magnetic field (Bo).

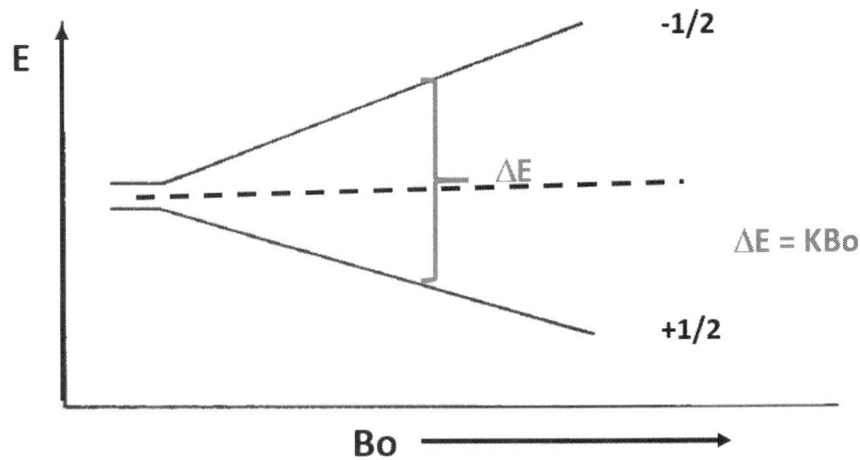

Figure 9. Spin state energy separation as a function of the increase in the magnetic field.

NMR spectroscopy is based on the measurement of the absorption of electromagnetic radiation in the radiofrequency region (4–900 MHz). The applied magnetic field differentiates the energy

http://dx.doi.org/10.5772/65552

of the two inverse spins of the nuclei that was previously degenerated, allowing the state transition to be achieved by the absorption of radiation and generating the signal. Hence, in this technique, the nuclei of the atoms are the main parts involved in the absorption process, while there is a great dependence on the applied magnetic field.

After the studies performed by Block and Purcell, other scientists discovered that the frequency of absorbed radiation by certain nuclei is strongly affected by their chemical environment, which is influenced by their electrons and nuclei neighbors. Hence, this phenomenon can associate the absorption spectra with the molecular structure. This discovery showed the great utility of NMR, since it is based on spectrum information, it is possible to obtain the structural elucidation of a molecule.

Figure 10 shows the predicted ^{1}H NMR spectra for ethanol, in which the nucleus evaluated is of hydrogen-1. Since the hydrogen nucleus is composed solely on a proton, the spectroscopy is commonly referred to as proton NMR. As was said before, not all nuclei in a molecule have a transition resonance at the same frequency, since the nuclei in a molecule are surrounded by electrons and exist in slightly different electronic environments. For the molecule of ethanol, there are three different chemical environments, which generate small differences in the absorption frequency of the nuclei. This is defined by the chemical shift (δ). The chemical shift can be used to identify functional groups and can help in the elucidation of the structure arrangements of groups. These applications are based on empiric correlations between structure and shift. Several tables have been published with values of the chemical shifts related to functional groups and molecules [15, 80]. In addition, the exact values of chemical shift depend on the solvent type and solution concentration. For the proton NMR spectra, the chemical shift values are defined in relation to an internal standard molecule, tetramethylsilane (TMS), and they vary between 0 at 20 ppm, while TMS is defined as zero.

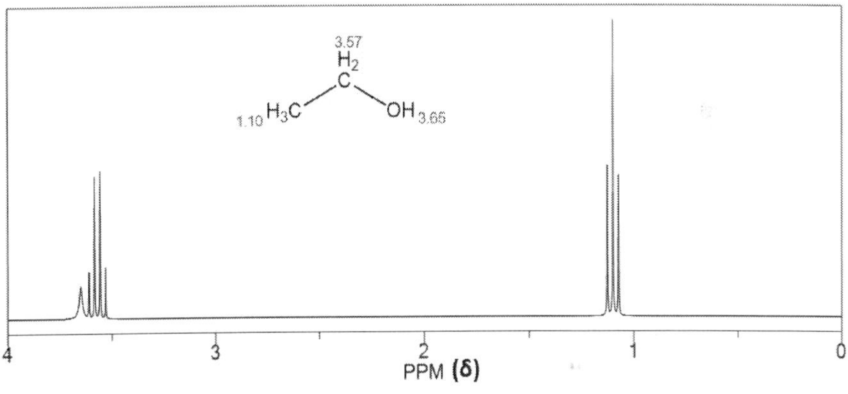

Figure 10. Predicted ^{1}H NMR spectra of the ethanol molecule.

Nowadays, two types of spectrometers are commercially available: the continuous-wave (CW) and the pulsed or Fourier transform spectrometer (FT-NMR). Historically, CW instruments were mainly used for spectroscopic studies. Today, FT-NMR instruments dominate the market. For the CW instruments, the absorption sign is monitored while the frequency source is varied slowly. In some instruments, the frequency source stays constant, and the intensity of the magnetic field is varied. For the pulsed instruments, the sample is irradiated with a periodic energy pulse of radiofrequency, which leads to a signal in the time domain that decays in the interval between the pulses. This decay is known as the free induction decay (FID), and the signal is converted to a sign frequency domain using the Fourier transform (a mathematical operation), resulting in a spectrum, as shown in **Figure 11**.

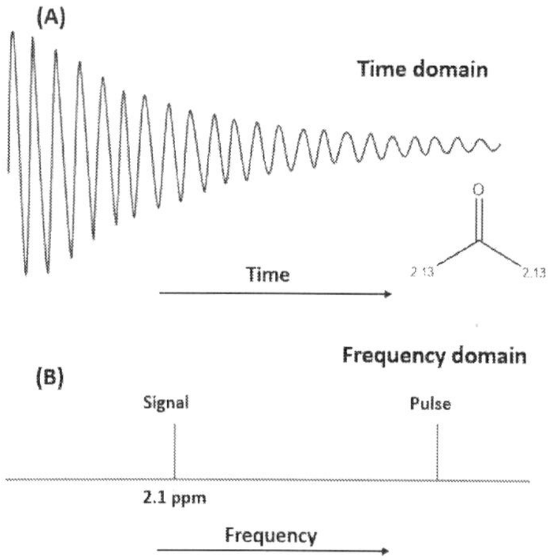

Figure 11. (a) Time domain. FID Curve for ^{1}H NMR for acetone; (b) spectra created using the Fourier transformation to convert the time domain to a frequency domain.

The integrated peak areas of ^{1}H NMR are proportional to the number of nuclei present in the molecule; that means, for the molecule of ethanol shown in **Figure 10**, the three different types of hydrogen have signals in which the areas are proportional to 3:2:1. However, for ^{13}C NMR spectra (another very common NMR technique in which the nuclei analyzed are from the carbon-13), the signals are not proportional to the number of nuclei present, and they cannot be used for this type of proportional characterization. Other applications use proton NMR for the determination of functional groups within the molecules, such as hydroxyl in alcohols and phenols, aldehydes, carboxylic acids, olefins, amines, and amides [1]. NMR can also be correlated to the concentration of the molecule itself, not only to the amount of nuclei within it. The technique can be used for the quantification of compounds in mixtures, especially when there are different types of functional groups [81, 82].

^{13}C NMR, as with ^{1}H NMR, can be used for the determination of organic and biochemical structures. These determinations are based on the chemical shifts and, as for the ^{1}H spectra, these values are relative to tetramethylsilane, with values between 0 and 200 ppm. In general, the effects of the environment are analogous to the ones observed for the ^{1}H NMR. The applications of Fourier transform methodology and broadband decoupling of protons have accelerated the development of, and enhanced the interest in, ^{13}C NMR spectroscopy. Techniques such as distortionless enhancement by polarization transfer (DEPT) and ^{1}H-^{13}C COSY (HETCOR) have further endeared ^{13}C NMR spectroscopy to organic chemists. In general, ^{13}C NMR spectroscopy allows chemists to directly observe the carbon framework of organic moieties and to make (collaborative) inferences about active nuclei attached to the carbon backbone. The technique is particularly valuable in research on natural products, pharmaceutical drugs, and biochemical processes, where complex cyclic species are common and often difficult to identify with only ^{1}H NMR [82].

This section illustrates the utility and application of NMR spectroscopy as a probative tool in the field of biomass, bio-oils and biofuels. Examples of its application include describing the characterization of the pyrolysis of biomass to achieve bio-oils, analyzing oil quality using quantification methodologies, and real-time monitoring of reaction paths for biofuel production [83–87].

Many reports in the literature describe the techniques used to obtain liquid fuels from biomass, including fractionation, liquefaction, pyrolysis, hydrolysis, fermentation, and gasification. Biomass pyrolysis is the cheapest way to obtain these products [88], and therefore it is the most studied methodology for such conversion [89], either in terms of physical properties or characterization of the resulting components [84, 90–92]. Compounds such as acids, esters, alcohols, ketones, aldehydes, anhydrosugars, furans, and phenols are usually found in bio-oil [93, 94], and its exact composition is strongly dependent on a number of factors, including the biomass origin and composition, the reaction temperature and heating rates to which the biomass is submitted, and the residence time on reactors [95]. During the pyrolysis process, the oil fractions are condensed, collected in a sample tube, and posteriorly analyzed. NMR spectroscopy is a suitable method for biomass pyrolysis studies. Different from other spectroscopic techniques, which can be limited for only the qualitative analysis of bio-oil samples, ^{1}H and ^{13}C NMR can be used as a way to obtain approximate ratios of the chemical environments of protons and carbon atoms. Besides, it can be useful to determine the approximate aromatic/aliphatic ratios. Another advantage is that NMR is a nondestructive technique, and the sample fractions can normally be recovered after the procedure.

Mullen et al., for example, described a method using ^{1}H and ^{13}C NMR to characterize bio-oils fast pyrolysis from numerous types of biomass [95]. Based on the proton NMR spectra of six different bio-oils and the integral values of selected regions, the ^{1}H and ^{13}C NMR spectra were used to determine the functional groups present in each oil. Using the DEPT spectra that were enabled, it was possible to evaluate the extent of branching in the molecules. These procedures provided important structural data about the analyzed samples. The work of Ben and Ragauskas [96] described an in-situ investigation of the relationship between the structures of pyrolysis oil at various time points during the accelerated aging process at 80°C, using ^{1}H and ^{13}C NMR. It was possible to detect a reduction in the percentage of aldehyde, carboxylic acid, and ether compounds, while the level of aromatic, alkene, and alkane compounds increased. In other words, it was possible to use NMR spectroscopy in the investigation of structural changes of pyrolysis oil compounds before and after an accelerated aging process.

The NMR technique has an advantage with respect to other common methods, such as infrared spectroscopy or chromatographic techniques, as it has high resolution compared to the others, and it also offers the possibility of obtaining valuable structural information. Besides that, the accelerated aging process of pyrolysis oil is carried out inside the NMR equipment, so the analysis is made in situ and in real time [96].

NMR spectroscopy can also be used for quality control either of the vegetable oils used as raw material for biofuel production or the final obtained product [38]. The quality of the vegetable oil influences many properties in the produced biodiesel, such as viscosity, lubricity oxidative stability, cold flow, ignition quality, and the heat of combustion. The type of fatty acid also influences the final product. Monounsaturated and saturated fatty acids are more stable than highly unsaturated ones, although they show higher viscosity and a bigger tendency to solidify in cold weather.

Analytical techniques for fatty acid determination, such as gas chromatography and near-infrared spectroscopy, need calibration models using some standards with similar chemical composition related to the analyzed sample [38]. High-resolution NMR has an advantage above the others because calibration models are not necessary. It is possible to quantify compounds in the sample through observing the integrated area of peaks with the chemical shift corresponding to a specific type of fatty acid [97]. Usually, the acquisition time for this type of analysis is quite long. However, Prestes et al. developed a method that is useful in determining the fatty acid profile in intact seeds in a fast and automated way using low-resolution NMR. It was possible to analyze more than a thousand samples per hour, thus working as a powerful tool to speed up the selection of oilseeds that are suitable for biodiesel applications [98]. Garro Linck et al. successfully analyzed biodiesel obtained via the transesterification of different raw materials using low-field spectrometers [99].

Another NMR application is as the PAT tool (as described for infrared spectroscopy), which can be used for the *in-situ* monitoring of the biodiesel production process, where sampling is not necessary [85]. As was discussed, ^{1}H NMR spectroscopy can be a robust, rapid, and quantitative method that can be applied for determining the presence of multiple components due to specific chemical shifts in the spectrum, and reaction monitoring can be applied over time, based on the integration of individual proton signals. An example of this application is the work of Anderson and Franz, where they used high-resolution NMR equipment for the monitoring of the biodiesel production reaction, evaluating the transesterification of triacylglycerol (TAG) and the resulting products, including diacylglycerol (DAG), monoacylglycerol (MAG), glycerol, and fatty acid methyl esters (FAME) [100]. Due to different molecular structures and different environments and different forms of hydrogen, it is possible to differentiate the signals from each molecule and evaluate them over time, as shown in **Figure 12**.

Anderson and Franz were able to see, then, the decrease of the TAG bands due to the biodiesel production and also the isomers of intermediates, such as DAG, throughout the reaction. This is a statement of the sensibility and sensitivity of the NMR spectroscopy technique in achieving information from the molecules, which could not be achieved by other techniques.

http://dx.doi.org/10.5772/65552

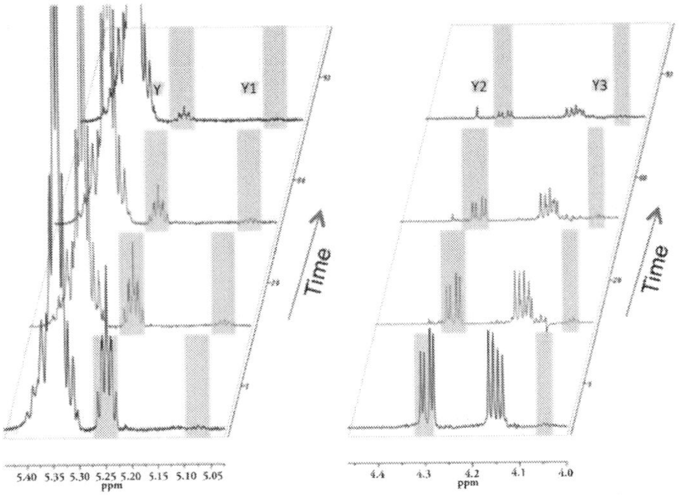

Figure 12. ^{1}H NMR stacked spectra over time of transesterification for biodiesel production, where Y indicates the TAG glyceryl methine at 5.27 ppm, Y1 indicates 1,2-DAG glyceryl methine at 5.08 ppm, Y2 indicates TAG glyceryl methylene at 4.29 ppm and Y3 indicates 1,3-DAG glyceryl methine at 4.09 ppm [100].

4. UV/Vis spectroscopy and image analysis for the quantification and classification of biofuels

The absorption of radiation in the ultraviolet/visible (UV/Vis; 200–700 nm) region results from the excitation of bounding electrons. The UV/Vis radiation has enough energy to promote electronic transitions, and this is the main principle investigated by UV/Vis absorption spectroscopy. This technique is very useful in identifying functional groups in a molecule because a correlation between the absorption bands and the functional group can be done. While inorganic compounds normally absorb light in the visible part of the spectrum, organic molecules usually present some functional groups capable of absorbing radiation from UV light sources. These functional groups must be unsaturated or have a heteroatom with non-bonding electrons, such as oxygen, sulfur, or halogens. **Table 3** shows some functional groups and the radiation wavelength absorbed [1].

According to the Lambert-Beer law, the intensity of the absorption of radiation by the species present in the sample is directly proportional to its concentration in the system. Thus, quantitative determination of compounds containing absorbing groups can be easily made. UV/Vis spectroscopy is widely used for many applications, including in the biofuel area, since it is low cost and allows the analyst to perform qualitative and quantitative analysis in a fast and reliable way.

Functional groups	Example	λ_{max} (nm)
Alkene	$C_6H_{13}CH{=}CH_2$	177
Alkyne	$C_5H_{11}{-}C{\equiv}C{-}CH_3$	178
		196
		225
Carbonyl	$CH_3\overset{\overset{O}{\|}}{C}CH_3$	186
		280
Carboxyl	CH_3COOH	204
Aromatic	C_6H_6	255
Alcohol	CH_3OH	167
Esther	$C_6H_{10}O_3$	250
Halogen	CH_3I	258
Sulfur	$(CH_3)_2S$	229

Table 3. Ultraviolet absorption of some organic molecules.

UV/Vis spectroscopy is very useful, for example, for the evaluation and quantification of the blend level in biodiesel/diesel mixtures. The utilization of blends between biodiesel and conventional diesel is very common, and the blend level can vary. Usually, pure biodiesel is referred to as B100, while some blends are classified according to its percentage of biodiesel, such as B20, B5, and B2 (respectively presenting 20, 5, and 2% of biodiesel in conventional diesel). The blend level directly influences some important characteristics of the fuel, such as its lubricity and tail pipe emissions [101]. Thus, it is very important to develop adequate methods for such applications. For example, Zawadzki et al. studied different types of biodiesel blends, employing an UV-Vis spectrometer as a low-cost detector to identify the range of UV frequencies suitable for detecting the biodiesel blend level [102]. The proposed method is robust, even with changes in the biodiesel feedstock and the fuel diesel origin.

UV/Vis detectors are also widely employed in high-performance liquid chromatography (HPLC) [103–105]. Many papers describing the use of HPLC-UV/Vis for biofuel analysis can be found in the literature. An interesting example is a statistical study performed by Foglia et al. [106], in which the precision and accuracy of HPLC-UV/Vis were evaluated with good results. This system was employed to determinate the level of simple alkyl esters of fats and oils (biodiesel) blended in petroleum diesel and to validate this method for this kind of analysis using different biodiesel feedstocks.

Another interesting application involving biodiesel/diesel blends is the detection of adulterations using vegetable oil. Commonly, near/mid infrared spectroscopy and spectrofluorimetry are used for this kind of analysis together with chemometric tools for multivariate classification/calibration [104]. These are expensive methods, so it would be interesting to evaluate the possibility of detecting adulterated blends using UV/Vis spectroscopy, which is simpler and cheaper. A study performed by Fernandes et al. employed UV/Vis and chemometric tools to detect soybean oil in biodiesel/diesel blends, focusing on biodiesel/diesel blend (B5) adulterations with soybean oil percentages from 0.5 to 2.5% (v/v) [105].

Another application that is indirectly correlated to the concept of visible light spectroscopy and that has been applied throughout the past few years is image analysis. The simplest definition for the word *image* can be understood as an optical replica of a luminous or illuminated object formed by a mirror of lens. Therefore, the object that gives rise to an image does so through an interaction with electromagnetic radiation, either by emission or absorption processes. Over the years, a wide variety of analytical methods involving image analysis as the main tool have been introduced, due to the ease with which they are carried out in a fast, noninvasive, and inexpensive way compared to more advanced spectroscopic techniques. New fields were opened with the introduction of chemometric methods to analyze image analysis results, such as exploratory image analysis, multivariate statistical process monitoring, multivariate regression, and image resolution [107]. The versatility of this technique has been explored in several areas, from flame determination of elements [108] to biochemistry analysis [109], providing fast and cheap results in a simple way.

In the field of renewable fuels, image analysis has been also explored for applications in various stages of production. For example, for the application of microalgae for biofuel production, an area that has attracted increasing attention recently, digital image processing has been successfully used as a way for monitoring and quantifying the amount of biomass present on photobioreactors, using the RGB color system [110], and also to build the light distribution profile in microalgae cultivation [111], a key factor for mobile productivity. Another interesting application related to microorganisms with potential use in biodiesel production is the determination of the intracellular accumulation level of lipids in yeast cells, which can be done in a dynamic and nondestructive manner via high-content images [112].

Because biodiesel quality control is becoming increasingly important, given the high market demand for renewable energy sources, methods such as image analysis have emerged as fast and reliable alternatives for the evaluation of some parameters. RGB/gray scale histograms obtained from digital images have been used together with statistical methods in biodiesel classification as an efficient way to determine the feedstock used in its production, since this influences the properties of the final product [113] without, however, employing time-consuming techniques, such as chromatographic methods. Another example is the application of image analysis together with thin layer chromatography, a very simple method, for glycerol detection in biodiesel [114]. This organic compound is a byproduct of the manufacturing process and is considered a contaminant in the final product. Similarly to fossil fuels, the burning of biodiesel, for example, causes the emission of pollutants, such as NOx species. Image analysis has proved useful in the prediction of the level of these emissions by the use of flame radical imaging to monitor the biomass combustion process [115].

In the coming years, researchers expect that image analysis will be explored in other stages of the production process because of its potential for easily obtaining relevant results for the biofuel area.

5. Atomic absorption and emission: trace element determination in biofuel samples

The determination of trace elements in biofuels is a key parameter for their use, and the importance of conducting such analyses reliably and quickly grows as the global demand for renewable energy sources becomes greater. The main spectroscopic techniques currently used for trace elements determination in biofuels are based on the phenomena of emission and absorption of electromagnetic radiation by atoms or elementary ions. In the case of the atomic absorption principle, electronic excitation occurs in atoms from the ground state to more energetic electronic levels because of energy absorption at specific wavelengths that are characteristic of each element. The excited electrons tend to quickly return to the ground state, releasing energy at wavelengths also characteristic of each element. This is the concept explored by atomic emission methods. Both of the concepts are shown in **Figure 13**.

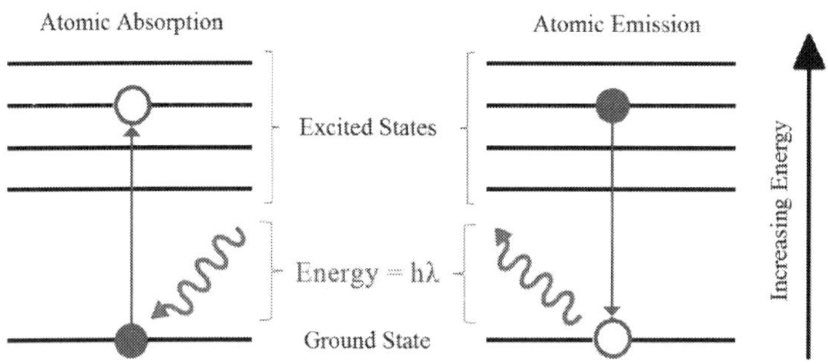

Figure 13. Atomic energy absorption/emission. The blue circles represent electrons, and the blue vacancies represent the final electron state after the absorption/emission occurs; h is the Planck constant and λ represents the wavelength in which the transition takes place.

The most common techniques are inductively coupled plasma-optical emission spectrometry (ICP-OES) and atomic absorption spectrometry (AAS) [116]. ICP-OES employs a highly energetic plasma, formed by an electrically neutral gas, usually argon, converted to positive ions and electrons. Such plasma has enough energy to atomize, ionize, and virtually excite all the elements of the periodic table to more energetic electronic states [117]. These species tend to return rapidly to the ground state, releasing energy at characteristic wavelengths depending on the elements present in the analyzed sample, and the radiation intensity is directly proportional to the concentration of the element in the sample [1]. AAS also employs atomization methods as well as emission techniques. Commonly, this process is carried out by means of a flame, in which desolvation, evaporation, and dissociation of the molecules into atoms take place [1]. Another common atomizing method is electrothermal atomization, in which the energy for volatilization and atomization is provided by means of an electric current applied to a graphite furnace [118]. Two more specific atomization techniques are hydride atomization by heating a quartz tube, a very common method for the analysis of some metals, such as

http://dx.doi.org/10.5772/65552

arsenic, and the cold vapor technique, which is widely used for mercury determination. After atomization, the analytes are submitted to radiation from a source, whether a single emission line source (e.g., hollow cathode lamps, multielement lamps or electrodeless discharge lamps) or a continuous source (e.g., halogen or deuterium lamps) that needs an auxiliary monochromator system to select the desired lines [119]. The analytes in the ground state absorb energy at specific wavelengths corresponding to their more favorable electronic transitions, thus generating an absorption spectrum whose intensity depends on the population of the atoms in the ground state, directly proportional to their concentration in the sample.

Trace elements may cause significant problems with serious implications for the use of biofuels, such as biodiesel and bioethanol. Such elements can be present in original vegetable oils subjected to transesterification processes due to absorption from the soil by the plant used as a feedstock, or they may even be incorporated in the vegetable matrix by means of the catalysts used in the biofuel synthesis or extraction/refining procedures [120, 121]. Sodium and potassium, for example, are used in the form of hydroxides in the synthesis of biodiesel, and together with Al, Ca, Fe, Mg, and Ti, they form a group of elements that tend to form a large amount of ash (metal oxides) after the fuel is burned [122, 123]. It causes difficulties for the operation of the gears, reducing the longevity of the engines. Furthermore, these elements are also involved in corrosion processes [121]. Some transition metals, such as Cu, Pb, Cd, and Zn, can cause biodiesel oxidation [124], resulting in residues that may be deposited in the engines. Moreover, they can contribute to air pollution and cause environmental damage due to their toxicity [116]. The sulfur level in the produced biofuels also must be carefully monitored because of emission legislation. Sulfur has been related to the depletion of the Earth's ozone layer, acid rain incidence, and chronic respiratory diseases. Low sulfur levels are also needed for good performance in modern engines [125, 126]. Iron and vanadium can act as catalyst poisoners and may reduce the efficiency of advanced catalysts that are commonly used in gears [127]. Some additives that are added to biofuels to improve physical or burning characteristics also contain metals, and their levels must be monitored. Because of these problems, strict laws have been adopted for the maximum level of metal contaminants in commercial biodiesel. **Table 4** presents the tolerated level of some metals established by ASTM standards [128].

Biofuel	Element	Maximum level	ASTM standard (year)
Biodiesel	Na+K	5 ppm	D6571 (2012)
	Ca+Mg	5 ppm	D6571 (2012)
	S	Biodiesel S15: 0.0015% Biodiesel S500: 0.05%	D6571 (2012)
	P	0.001%	D6571 (2012)
Fuel ethanol	S	30 ppm	D4806 (2014)
Bioethanol	Cu	0.1 ppm	D1688 (2012)
	P	0.5 ppm	D3231 (2013)
	S	10 ppm	D3231 (2013)

Table 4. Maximum concentrations allowed for some trace elements in biofuels (ASTM standard).

It has become essential to develop precise, accurate, and sensitive methods that are applicable for monitoring metal contaminants to qualify biofuels according to the established standards for their use. The quantification of metals in bioethanol and biodiesel, for example, is associated with a number of difficulties, such as their low concentration in the organic matrix (in the range of μg/L), the limited number of certified standards, the sample complexity, and the dependence of the final product on the raw material used [128].

Despite ICP-OES and AAS being the most widely used techniques, the details involved in the determination of metals in biofuel samples must be observed. In general, organic matter decomposition procedures (often microwave assisted) can be applied to convert the matter to a simple aqueous matrix, therefore reducing the chances of spectral interference with the measurements and avoiding difficulties in the injection of these samples [129–134]. This process is also necessary because of the difficulties in introducing the samples in atomizers and in selecting appropriating calibration standards for the analyzed system [135].

In some cases, however, direct injection of samples of biodiesel, bioethanol or fuel ethanol may be advantageous in order to reduce the number of steps of the analysis procedure. This operation brings some drastic consequences. The injection, for example, can be greatly hindered due to the physical-chemical characteristics of the sample, such as viscosity and surface tension, thus modifying the ease of suction of the components. Both in the case of the ICP-OES technique (in which a nebulizer is used for generating an aerosol) [128] and in the case of the sample introduction systems used in flame atomic absorption equipment, problems in the injection of organic samples affect the efficiency and reproducibility of the analytical methods. Burning organic compounds normally generates instability in the plasma with ICP-OES [136], and the generated pyrolysis products cause some spectral interference. Another problem may be the deposition of material originating in the pyrolysis processes on the torch or other spectrometer facilities [128, 137]. When using a graphite furnace as an atomization method for AAS, organic samples usually cause excessive material spreading during volatilization [116].

Due to the importance of determining trace elements in biofuels, and taking into account the difficulties mentioned above, several research groups have been working on the development of new methods for this purpose. Some interesting approaches have been described in the past few years. Dilution with an organic solvent or water can be an interesting way to minimize the errors observed in the analysis of biodiesel samples [127, 138–142]. Methods involving pre-emulsification of the sample using surfactants have also achieved good reproducibility [121, 123, 143–148]. Furthermore, the extraction of analytes from a sample for further analysis provides good results in some cases [149–151].

Surely, the spectroscopic methods, whether based on atomic emission or absorption principles, are the most popular for elemental determination in biofuel samples due to their high sensibility and reliability, despite the specific characteristics of the analytical methods involving organic samples.

6. Summary

To conclude, biofuels have been presented in the last few decades as alternatives or substitutes for fossil fuels to decrease the amount of fossil fuels used. With the increase in awareness of the need to develop more sustainable ways to support the earth's energy system in the future, much study is still needed, mainly in the areas of production, characterization, and quality control. This chapter described several methods that involve spectroscopic techniques, including infrared, Raman, NMR, UV/Vis absorption, image analysis, ICP-OES, and AAS. With these techniques, applications can be performed to characterize organic components in biofuels, to study the development of methods for controlling quality (e.g., the validation of biodiesel blends or detection of adulterations), to develop methods for discerning trace elements in biofuel, and even to monitor the production of biofuels in real time.

The referenced works in this chapter represent only a brief summary of the uses of spectroscopic techniques, depicting their importance in terms of fostering new developments in the biofuels area.

Author details

João Cajaiba Da Silva, Alex Queiroz, Alline Oliveira and Vinícius Kartnaller*

*Address all correspondence to: vkartnaller@yahoo.com.br

Federal University of Rio de Janeiro, Chemistry Institute, Cidade Universitária, Rio de Janeiro, RJ, Brazil

References

[1] Skoog DA, Holler FJ, Crouch SR. Principles of Instrumental Analysis. Thomson Brooks/Cole; Belmonte, CA, USA; 2007.

[2] Harvey D. Modern Analytical Chemistry. McGraw-Hill Higher Education; USA; 2000.

[3] Beauvoit B, Chance B. Time-resolved spectroscopy of mitochondria, cells and tissues under normal and pathological conditions. Molecular and Cellular Biochemistry 1998;184:445–55.

[4] Diamandis EP. Immunoassays with time-resolved fluorescence spectroscopy: Principles and applications. Clinical Biochemistry 1988;21:139–50.

[5] Jura M. The elemental compositions of extrasolar planetesimals. Proceedings of the International Astronomical Union 2012;8:219–28.

[6] Orton GS, Moses JI, Fletcher LN, Mainzer AK, Hines D, Hammel HB, et al. Mid-infrared spectroscopy of Uranus from the Spitzer infrared spectrometer: 2. Determination of the mean composition of the upper troposphere and stratosphere. Icarus 2014;243:471–93.

[7] Stauffer E, Byron D. Alternative fuels in fire debris analysis: biodiesel basics. Journal of Forensic Sciences 2007;52:371–9.

[8] Knothe G. Analyzing biodiesel: standards and other methods. Journal of the American Oil Chemists' Society 2006;83:823–33.

[9] Pinto AC, Guarieiro LLN, Rezende MJC, Ribeiro NM, Torres EA, Lopes WA, et al. Biodiesel: an overview. Journal of the Brazilian Chemical Society 2005;16:1313–30.

[10] Fernando S, Karra P, Hernandez R, Jha S. Effect of incompletely converted soybean oil on biodiesel quality. Energy 2007;32:844–51.

[11] Knothe G. Analytical methods used in the production and fuel quality assessment of biodiesel. Transactions of the ASAE 2001;44:193–200.

[12] Baeten V, Dardenne P. Spectroscopy: Developments in Instrumentation and Analysis. Grasas y Aceites 2002;53:45–63.

[13] Diem M. Modern Vibrational Spectroscopy and Micro-Spectroscopy: Theory, Instrumentation and Biomedical Applications. John Wiley & Sons; Chichester, UK; 2015.

[14] Theophanides T. Infrared Spectroscopy – Materials Science, Engineering and Technology. Croatia: Intech; 2012.

[15] Pavia DL, Lampman GM, Kriz GS, Vyvyan JR. Introduction to Spectroscopy: Cengage Learning; USA; 2013.

[16] Rabkin YM. Technological innovation in science: the adoption of infrared spectroscopy by chemists. Isis 1987;78:31–54.

[17] Barth A, Haris PI. Biological and Biomedical Infrared Spectroscopy. Amsterdam: IOS Press BV; 2009.

[18] Miller FA. Reminiscences of Pioneers and Early Commercial IR Instruments. Analytical Chemistry 1992;64:824A–31A.

[19] Downing JR, Walker IF, Patterson GD, Freed WV. Applications of infrared spectroscopy to DDT. Industrial & Engineering Chemistry Analytical Edition 1946;18:461–7.

[20] Pfann HF, Williams VZ, Mark H. Mechanism of peroxide-initiated styrene polymerization. Infrared absorption of some polymer samples. Journal of Polymer Science 1946;1:14–21.

[21] Adjaye JD, Katikaneni SPR, Bakhshi NN. Catalytic conversion of a biofuel to hydrocarbons: effect of mixtures of HZSM-5 and silica-alumina catalysts on product distribution. Fuel Processing Technology 1996;48:115–43.

[22] Sanderson MA, Agblevor F, Collins M, Johnson DK. Compositional analysis of biomass feedstocks by near infrared reflectance spectroscopy. Biomass and Bioenergy 1996;11:365–70.

[23] Oliveira FCC, Brandão CRR, Ramalho HF, da Costa LAF, Suarez PAZ, Rubim JC. Adulteration of diesel/biodiesel blends by vegetable oil as determined by Fourier transform (FT) near infrared spectrometry and FT-Raman spectroscopy. Analytica Chimica Acta 2007;587:194–9.

[24] Larkin P. Infrared and Raman Spectroscopy: Principles and Spectral Interpretation. USA: Elsevier; 2011.

[25] Bernath PF. Spectra of Atoms and Molecules: Oxford University Press; New York, USA; 2005.

[26] Corsetti S, McGloin D, Kiefer J. Comparison of Raman and IR spectroscopy for quantitative analysis of gasoline/ethanol blends. Fuel 2016;166:488–94.

[27] Wembabazi E, Mugisha PJ, Ratibu A, Wendiro D, Kyambadde J, Vuzi PC. Spectroscopic analysis of heterogeneous biocatalysts for biodiesel production from expired sunflower cooking oil. Journal of Spectroscopy 2015;2015:1-8.

[28] Zhao X, Wei L, Cheng S, Kadis E, Cao Y, Boakye E, et al. Hydroprocessing of carinata oil for hydrocarbon biofuel over Mo-Zn/Al2O3. Applied Catalysis B: Environmental 2016;196:41–9.

[29] Mallesham B, Sudarsanam P, Reddy BM. Production of biofuel additives from esterification and acetalization of bioglycerol over SnO2-based solid acids. Industrial & Engineering Chemistry Research 2014;53:18775–85.

[30] Kumar P, Srivastava VC, Mishra IM. Dimethyl carbonate synthesis from propylene carbonate with methanol using Cu–Zn–Al catalyst. Energ Fuel 2015;29:2664–75.

[31] Cheng S, Wei L, Zhao X, Kadis E, Julson J. Conversion of Prairie Cordgrass to hydrocarbon biofuel over Co-Mo/HZSM-5 using a two-stage reactor system. Energy Technology 2016;4:706–13.

[32] Mandey R, Wijaya K, Triyono, Syoufian A. Chrome impregnated on Bentonite (Cr-Bentonite) as hydrocracking catalyst of used cooking oil. Asian Journal of Chemistry 2016;28:282–6.

[33] Durak H, Aysu T. Structural analysis of bio-oils from subcritical and supercritical hydrothermal liquefaction of Datura stramonium L. The Journal of Supercritical Fluids 2016;108:123–35.

[34] Plata V, Gauthier-Maradei P, Romero-Bohórquez AR, Kafarov V, Castillo E. Characterization of insoluble material isolated from Colombian palm oil biodiesel. Biomass and Bioenergy 2015;74:6–14.

[35] Nanda S, Dalai AK, Gokalp I, Kozinski JA. Valorization of horse manure through catalytic supercritical water gasification. Waste management 2016;52:147–58.

[36] Zou S, Wu Y, Yang M, Li C, Tong J. Thermochemical catalytic liquefaction of the marine Microalgae Dunaliella tertiolecta and characterization of bio-oils. Energ Fuel 2009;23:3753–8.

[37] Shuping Z, Yulong W, Mingde Y, Kaleem I, Chun L, Tong J. Production and characterization of bio-oil from hydrothermal liquefaction of microalgae Dunaliella tertiolecta cake. Energy 2010;35:5406–11.

[38] Ahmad M, Ullah K, Khan MA, Ali S, Zafar M, Sultana S. Quantitative and qualitative analysis of sesame oil biodiesel. Energy Sources, Part A: Recovery, Utilization, and Environmental Effects 2011;33:1239–49.

[39] Corsetti S, Zehentbauer FM, McGloin D, Kiefer J. Characterization of gasoline/ethanol blends by infrared and excess infrared spectroscopy. Fuel 2015;141:136–42.

[40] Rambo MKD, Alexandre GP, Rambo MCD, Alves AR, Garcia WT, Baruque E. Characterization of biomasses from the north and northeast regions of Brazil for processes in biorefineries. Food Science and Technology (Campinas) 2015;35:605–11.

[41] Grilc M, Likozar B, Levec J. Simultaneous liquefaction and hydrodeoxygenation of lignocellulosic biomass over $NiMo/Al_2O_3$, Pd/Al_2O_3, and zeolite Y catalysts in hydrogen donor solvents. ChemCatChem 2016;8:180–91.

[42] Xu Y, Hu X, Yu H, Wang K, Cui Z. Hydrothermal catalytic liquefaction mechanisms of agal biomass to bio-oil. Energy Sources, Part A: Recovery, Utilization, and Environmental Effects 2016;38:1478–84.

[43] Bertschi IT, Yokelson RJ, Ward DE, Christian TJ, Hao WM. Trace gas emissions from the production and use of domestic biofuels in Zambia measured by open-path Fourier transform infrared spectroscopy. Journal of Geophysical Research: Atmospheres 2003;108.

[44] Lestander TA, Rhen C. Multivariate NIR spectroscopy models for moisture, ash and calorific content in biofuels using bi-orthogonal partial least squares regression. The Analyst 2005;130:1182–9.

[45] Oliveira JS, Montalvão R, Daher L, Suarez PAZ, Rubim JC. Determination of methyl ester contents in biodiesel blends by FTIR-ATR and FTNIR spectroscopies. Talanta 2006;69:1278–84.

[46] Baptista P, Felizardo P, Menezes JC, Correia MJ. Multivariate near infrared spectroscopy models for predicting the methyl esters content in biodiesel. Anal Chim Acta 2008;607:153–9.

[47] Balabin RM, Safieva RZ. Biodiesel classification by base stock type (vegetable oil) using near infrared spectroscopy data. Anal Chim Acta 2011;689:190–7.

[48] Balabin RM, Safieva RZ, Lomakina EI. Gasoline classification using near infrared (NIR) spectroscopy data: comparison of multivariate techniques. Anal Chim Acta 2010;671:27–35.

[49] Satyarthi JK, Srinivas D. Fourier Transform infrared spectroscopic method for monitoring hydroprocessing of vegetable oils to produce hydrocarbon-based biofuel. Energy Fuel 2011;25:3318–22.

[50] Ault AP, Pomeroy R. Quantitative investigations of biodiesel fuel using infrared spectroscopy: an instrumental analysis experiment for undergraduate chemistry students. Journal of Chemical Education 2012;89:243–7.

[51] Pinzi S, Alonso F, García Olmo J, Dorado MP. Near infrared reflectance spectroscopy and multivariate analysis to monitor reaction products during biodiesel production. Fuel 2012;92:354–9.

[52] Baum A, Hansen PW, Meyer AS, Mikkelsen JD. Simultaneous measurement of two enzyme activities using infrared spectroscopy: a comparative evaluation of PARAFAC, TUCKER and N-PLS modeling. Anal Chim Acta 2013;790:14–23.

[53] Corgozinho CNC, Barbeira PJS. Quantification of biodiesel in biodiesel-diesel blends using spectrofluorimetry and multivariate calibration. Journal of Analytical Chemistry 2015;70:936–42.

[54] Mata Sánchez J, Pérez Jiménez JA, Díaz Villanueva MJ, Serrano A, Núñez N, López Giménez J. Assessment of near infrared spectroscopy for energetic characterization of olive byproducts. Renewable Energy 2015;74:599–605.

[55] Mazivila SJ, de Santana FB, Mitsutake H, Gontijo LC, Santos DQ, Neto WB. Discrimination of the type of biodiesel/diesel blend (B5) using mid-infrared spectroscopy and PLS-DA. Fuel 2015;142:222–6.

[56] Mitsutake H, Guimarães E, Freitas HC, Gontijo LC, Santos DQ, Borges Neto W. Multivariate control charts based on NAS and mid-infrared spectroscopy for quality control of B5 blends of methyl soybean biodiesel in diesel. Journal of Chemometrics 2015;29:411–9.

[57] Paiva EM, Rohwedder JJR, Pasquini C, Pimentel MF, Pereira CF. Quantification of biodiesel and adulteration with vegetable oils in diesel/biodiesel blends using portable near-infrared spectrometer. Fuel 2015;160:57–63.

[58] Alves JCL, Poppi RJ. Quantification of conventional and advanced biofuels contents in diesel fuel blends using near-infrared spectroscopy and multivariate calibration. Fuel 2016;165:379–88.

[59] Duca D, Pizzi A, Mancini M, Rossini G, Mengarelli C, Ilari A, et al. Fast measurement by infrared spectroscopy as support to woody biofuels quality determination. Journal of Agricultural Engineering 2016;47:17.

[60] Sitoe BV, Mitsutake H, Guimarães E, Gontijo LC, Santos DQ, Borges Neto W. Quality control of biodiesel content of B7 blends of Methyl Jatropha and methyl crambe biodiesels using mid-infrared spectroscopy and multivariate control charts based on net analyte signal. Energy & Fuel 2016;30:1062–1070.

[61] Neyestanaki AK, Klingstedt F, Lindfors L-E. Catalytic removal of NOx from gas mixtures simulating emissions from the combustion of biofuels. Reaction Kinetics and Catalysis Letters 2000;70:17–25.

[62] Yang H, Yan R, Chin T, Liang DT, Chen H, Zheng C. Thermogravimetric analysis: Fourier transform infrared analysis of palm oil waste pyrolysis. Energy & Fuel 2004;18:1814–21.

[63] Das B, Mandal TK, Patra S. A comprehensive study on Chlorella pyrenoidosa for phenol degradation and its potential applicability as biodiesel feedstock and animal feed. Applied Biochemistry and Biotechnology 2015;176:1382-401.

[64] Li H, Niu S, Lu C, Wang Y. Comprehensive investigation of the thermal degradation characteristics of biodiesel and its feedstock oil through TGA–FTIR. Energy & Fuel 2015;29:5145-53.

[65] Zhang T, Zhou Y, Liu D, Petrus L. Qualitative analysis of products formed during the acid catalyzed liquefaction of bagasse in ethylene glycol. Bioresource Technology 2007;98:1454-9.

[66] Gillespie GD, Everard CD, McDonnell KP. Prediction of biomass pellet quality indices using near infrared spectroscopy. Energy 2015;80:582-8.

[67] Kiran B, Pathak K, Kumar R, Deshmukh D, Rani N. Influence of varying nitrogen levels on lipid accumulation in Chlorella sp. International Journal of Environmental Science and Technology 2016;13:1823-32.

[68] Mishra S, Gupta S, Raghuvanshi S, Pal P. Energetic assessment of fixation of CO_2 and subsequent biofuel production using B. cereus SM1 isolated from sewage treatment plant. Bioprocess and biosystems engineering 2016;39:1247-58.

[69] Ortiz EG, Quintero I, Arévalo K. Biodiesel production from three mixes of oils with high free fatty content: quality evaluation and variable analysis. International Journal of Environmental Science and Technology 2016;13:1367-76.

[70] Ruan T, Zeng R, Yin X-Y, Zhang S-X, Yang Z-H. Water Hyacinth (Eichhornia crassipes) Biomass as a Biofuel Feedstock by Enzymatic Hydrolysis. BioResources 2016.

[71] Chen Z, Lovett D, Morris J. Process analytical technologies and real time process control a review of some spectroscopic issues and challenges. Journal of Process Control 2011;21:1467-82.

[72] Chadwick DT, McDonnell KP, Brennan LP, Fagan CC, Everard CD. Evaluation of infrared techniques for the assessment of biomass and biofuel quality parameters and conversion technology processes: a review. Renewable and Sustainable Energy Reviews 2014;30:672-81.

[73] Killner MHM, Rohwedder JJR, Pasquini C. A PLS regression model using NIR spectroscopy for on-line monitoring of the biodiesel production reaction. Fuel 2011;90: 3268–73.

[74] de Souza AVA, Cajaiba da Silva JF. Biodiesel Synthesis Evaluated by Using Real-Time ATR-FTIR. Organic Process Research & Development 2012;17:127–32.

[75] Natalello A, Sasso F, Secundo F. Enzymatic transesterification monitored by an easy-to-use Fourier transform infrared spectroscopy method. Biotechnology Journal 2013;8:133–138.

[76] Richard R, Dubreuil B, Thiebaud-Roux S, Prat L. On-line monitoring of the transesterification reaction carried out in microreactors using near infrared spectroscopy. Fuel 2013;104:318–25.

[77] de Lima SM, Silva BFA, Pontes DV, Pereira CF, Stragevitch L, Pimentel MF. In-line monitoring of the transesterification reactions for biodiesel production using NIR spectroscopy. Fuel 2014;115:46–53.

[78] Reyman D, Saiz Bermejo A, Ramirez Uceda I, Rodriguez Gamero M. A new FTIR method to monitor transesterification in biodiesel production by ultrasonication. Environmental Chemistry Letters 2014;12:235–40.

[79] Kartnaller V, Junior II, de Souza AVA, Costa ICR, Rezende MJC, da Silva JFC, et al. Evaluating the kinetics of the esterification of oleic acid with homo and heterogeneous catalysts using in-line real-time infrared spectroscopy and partial least squares calibration. Journal of Molecular Catalysis B: Enzymatic 2016;123:41–6.

[80] Silverstein RM, Webster FX, Kiemle DJ. Spectrometric Identification of Organic compounds Wiley; USA; 2005.

[81] Evilia R. Quantitative NMR spectroscopy. Analytical Letters 2001;34:2227.

[82] Darbeau RW. Nuclear magnetic resonance (NMR) spectroscopy: a review and a look at its use as a probative tool in deamination chemistry. Applied Spectroscopy Reviews 2006;41:401–25.

[83] Monteiro MR, Ambrozin ARP, Lião LM, Ferreira AG. Determination of biodiesel blend levels in different diesel samples by 1H NMR. Fuel 2009;88:691–6.

[84] Vispute TP, Zhang HY, Sanna A, Xiao R, Huber GW. Renewable chemical commodity feedstocks from integrated catalytic processing of pyrolysis oils. Science 2010;330: 1222–7.

[85] Andrade DF, Mazzei JL, Kaiser CR, d'Avila LA. Assessment of different measurement methods using 1H-NMR data for the analysis of the transesterification of vegetable oils. Journal of the American Oil Chemists' Society 2012;89:619–30.

[86] Flores IS, Godinho MS, de Oliveira AE, Alcantara GB, Monteiro MR, Menezes SMC, et al. Discrimination of biodiesel blends with 1H NMR spectroscopy and principal component analyses. Fuel 2012;99:40–4.

[87] Killner MHM, Garro Linck Y, Danieli E, Rohwedder JJR, Blümich B. Compact NMR spectroscopy for real-time monitoring of a biodiesel production. Fuel 2015;139:240–7.

[88] Sharifzadeh M, Wang L, Shah N. Decarbonisation of olefin processes using biomass pyrolysis oil. Applied Energy 2015;149:404–14.

[89] Sharifzadeh M, Richard CJ, Liu K, Hellgardt K, Chadwick D, Shah N. An integrated process for biomass pyrolysis oil upgrading: a synergistic approach. Biomass and Bioenergy 2015;76:108–17.

[90] Venderbosch RH, Prins W. Fast pyrolysis technology development. Biofuels, Bioproducts and Biorefining 2010;4:178–208.

[91] Akhtar J, Saidina Amin N. A review on operating parameters for optimum liquid oil yield in biomass pyrolysis. Renewable and Sustainable Energy Reviews 2012;16: 5101–9.

[92] Pradhan D, Singh RK, Bendu H, Mund R. Pyrolysis of Mahua seed (Madhuca indica): Production of biofuel and its characterization. Energy Conversion and Management 2016;108:529–38.

[93] Junming X, Jianchun J, Yanju L, Jie C. Liquid hydrocarbon fuels obtained by the pyrolysis of soybean oils. Bioresource Technology 2009;100:4867–70.

[94] Fagernäs L, Kuoppala E, Tiilikkala K, Oasmaa A. Chemical composition of Birch Wood slow pyrolysis products. Energy & Fuel 2012;26:1275–83.

[95] Mullen CA, Strahan GD, Boateng AA. Characterization of various fast-pyrolysis bio-oils by NMR spectroscopy. Energy & Fuel 2009;23:2707–18.

[96] Ben HX, Ragauskas AJ. In situ NMR characterization of pyrolysis oil during accelerated aging. Chemsuschem 2012;5:1687–93.

[97] Joseph J, Baker C, Mukkamala S, Beis SH, Wheeler MC, DeSisto WJ, et al. Chemical shifts and lifetimes for nuclear magnetic resonance (NMR) analysis of biofuels. Energy & Fuel 2010;24:5153–62.

[98] Prestes RA, Colnago LA, Forato LA, Vizzotto L, Novotny EH, Carrilho E. A rapid and automated low resolution NMR method to analyze oil quality in intact oilseeds. Anal Chim Acta 2007;596:325–9.

[99] Garro Linck Y, Killner MHM, Danieli E, Blümich B. Mobile low-field 1H NMR spectroscopy desktop analysis of biodiesel production. Applied Magnetic Resonance 2013;44:41–53.

[100] Anderson LA, Franz AK. Real-time monitoring of transesterification by 1H NMR spectroscopy: catalyst comparison and improved calculation for biodiesel conversion. Energy & Fuel 2012;26:6404–10.

[101] Shahid EM, Jamal Y. A review of biodiesel as vehicular fuel. Renewable and Sustainable Energy Reviews 2008;12:2484–94.

[102] Zawadzki A, Shrestha DS, He B. Biodiesel blend level detection using ultraviolet absorption spectra. Transactions of the Asabe 2007;50:1349–53.

[103] Brandão LFP, Braga JWB, Suarez PAZ. Determination of vegetable oils and fats adulterants in diesel oil by high performance liquid chromatography and multivariate methods. Journal of Chromatography A 2012;1225:150–7.

[104] de Oliveira RR, de Lima KMG, Tauler R, de Juan A. Application of correlation constrained multivariate curve resolution alternating least-squares methods for determination of compounds of interest in biodiesel blends using NIR and UV–visible spectroscopic data. Talanta 2014;125:233–41.

[105] Fernandes DDS, Gomes AA, de Fontes MM, da Costa GB, de Almeida VE, de Araujo MCU, et al. UV-Vis spectrometric detection of biodiesel/diesel blend adulterations with soybean oil. Journal of the Brazilian Chemical Society 2014;25:169–75.

[106] Foglia TA, Jones KC, Phillips JG. Determination of biodiesel and triacylglycerols in diesel fuel by LC. Chromatographia 2005;62:115–9.

[107] Prats-Montalban JM, de Juan A, Ferrer A. Multivariate image analysis: a review with applications. Chemometr Intell Lab 2011;107:1–23.

[108] Silva Lyra W, dos Santos VB, Dionízio AGG, Martins VL, Almeida LF, Nóbrega Gaião E, et al. Digital image-based flame emission spectrometry. Talanta 2009;77:1584–9.

[109] Alexandre I, Hamels S, Dufour S, Collet J, Zammatteo N, De Longueville F, et al. Colorimetric silver detection of DNA microarrays. Analytical Biochemistry 2001;295:1–8.

[110] Sarrafzadeh MH, La H-J, Lee J-Y, Cho D-H, Shin S-Y, Kim W-J, et al. Microalgae biomass quantification by digital image processing and RGB color analysis. Journal of Applied Phycology 2014;27:205–9.

[111] Chen X, Goh QY, Tan W, Hossain I, Chen WN, Lau R. Lumostatic strategy for microalgae cultivation utilizing image analysis and chlorophyll a content as design parameters. Bioresource Technology 2011;102:6005–12.

[112] Capus A, Monnerat M, Ribeiro LC, de Souza W, Martins JL, Sant'Anna C. Application of high-content image analysis for quantitatively estimating lipid accumulation in oleaginous yeasts with potential for use in biodiesel production. Bioresource Technology 2016;203:309-17.

[113] Costa GB, Fernandes DD, Almeida VE, Araujo TS, Melo JP, Diniz PH, et al. Digital image-based classification of biodiesel. Talanta 2015;139:50–5.

[114] Bansal K, McCrady J, Hansen A, Bhalerao K. Thin layer chromatography and image analysis to detect glycerol in biodiesel. Fuel 2008;87:3369–72.

[115] Li N, Lu G, Li X, Yan Y. Prediction of NOx emissions from a biomass fired combustion process based on flame radical imaging and deep learning techniques. Combustion Science and Technology 2015;188:233–46.

[116] Lepri FG, Chaves ES, Vieira MA, Ribeiro AS, Curtius AJ, DeOliveira LCC, et al. Determination of trace elements in vegetable oils and biodiesel by atomic spectrometric techniques: a review. Applied Spectroscopy Reviews 2011;46:175–206.

[117] Taylor HE. Inductively Coupled Plasma-Mass Spectrometry: Practices and Techniques. Academic Press; USA; 2001.

[118] Schlemmer G, Radziuk B. Analytical Graphite Furnace Atomic Absorption Spectrometry: A Laboratory Guide. Birkhäuser Verlag; Basel, Switzerland; 1999.

[119] Welz B, Sperling M. Atomic Absorption Spectrometry. Wiley-VCH; Weinheim, Germany; 2005.

[120] Balat M. Production of biodiesel from vegetable oils: a survey. Energy Sources, Part A: Recovery, Utilization, and Environmental Effects 2007;29:895–913.

[121] Lisboa MT, Clasen CD, Vellar DCdS, Oreste EQ, Saint'Pierre TD, Ribeiro AS, et al. An easy and fast procedure for the determination of Ca, Cu, Fe, Mn, Mg, Na, K and Si in biodiesel by ICP OES using emulsification as sample preparation strategy. Journal of the Brazilian Chemical Society 2013;25:143–151.

[122] Baernthaler G, Zischka M, Haraldsson C, Obernberger I. Determination of major and minor ash-forming elements in solid biofuels. Biomass and Bioenergy 2006;30:983–97.

[123] de Jesus A, Silva MM, Vale MG. The use of microemulsion for determination of sodium and potassium in biodiesel by flame atomic absorption spectrometry. Talanta 2008;74:1378–84.

[124] Sarin A, Arora R, Singh NP, Sharma M, Malhotra RK. Influence of metal contaminants on oxidation stability of Jatropha biodiesel. Energy 2009;34:1271–5.

[125] Balcaen L, Woods G, Resano M, Vanhaecke F. Accurate determination of S in organic matrices using isotope dilution ICP-MS/MS. Journal of Analytical Atomic Spectrometry 2013;28:33–9.

[126] Amais RS, Long SE, Nobrega JA, Christopher SJ. Determination of trace sulfur in biodiesel and diesel standard reference materials by isotope dilution sector field inductively coupled plasma mass spectrometry. Analytica Chimica Acta 2014;806:91–6.

[127] Woods GD, Fryer FI. Direct elemental analysis of biodiesel by inductively coupled plasma-mass spectrometry. Analytical and Bioanalytical Chemistry 2007;389:753–61.

[128] Sánchez R, Sánchez C, Lienemann C-P, Todolí J-L. Metal and metalloid determination in biodiesel and bioethanol. Journal of Analytical Atomic Spectrometry 2015;30:64–101.

[129] Saleh MI, Murray RS, Chin CN. Ashing techniques in the determination of iron and copper in palm oil. Journal of the American Oil Chemists' Society 1988;65:1767–70.

[130] Carbonell V, Mauri AR, Salvador A, de la Guardia M. Direct determination of copper and iron in edible oils using flow injection flame atomic absorption spectrometry. Journal of Analytical Atomic Spectrometry 1991;6:581–4.

[131] Matos Reyes MN, Campos RC. Determination of copper and nickel in vegetable oils by direct sampling graphite furnace atomic absorption spectrometry. Talanta 2006;70:929–32.

[132] Cindric IJ, Zeiner M, Steffan I. Trace elemental characterization of edible oils by ICP–AES and GFAAS. Microchemical Journal 2007;85:136–9.

[133] de Oliveira AP, Villa RD, Antunes KCP, de Magalhães A, Silva ECe. Determination of sodium in biodiesel by flame atomic emission spectrometry using dry decomposition for the sample preparation. Fuel 2009;88:764–6.

[134] Gonzalvez A, Ghanjaoui ME, El Rhazi M, de la Guardia M. Inductively coupled plasma optical emission spectroscopy determination of trace element composition of argan oil. Food Science and Technology International 2010;16:65–71.

[135] Aucélio RQ, de Souza RM, de Campos RC, Miekeley N, da Silveira CLP. The determination of trace metals in lubricating oils by atomic spectrometry. Spectrochimica Acta Part B: Atomic Spectroscopy 2007;62:952–61.

[136] Blades MW, Caughlin BL. Excitation temperature and electron density in the inductively coupled plasma: aqueous vs organic solvent introduction. Spectrochimica Acta Part B: Atomic Spectroscopy 1985;40:579–91.

[137] Jai Kumar S, Gangadharan S. Determination of trace elements in naphtha by inductively coupled plasma mass spectrometry using water-in-oil emulsions. Journal of Analytical Atomic Spectrometry 1999;14:967–71.

[138] Karadjova I, Zachariadis G, Boskou G, Stratis J. Electrothermal atomic absorption spectrometric determination of aluminium, cadmium, chromium, copper, iron, manganese, nickel and lead in olive oil. Journal of Analytical Atomic Spectrometry 1998;13:201–4.

[139] Edlund M, Visser H, Heitland P. Analysis of biodiesel by argon-oxygen mixed-gas inductively coupled plasma optical emission spectrometry. Journal of Analytical Atomic Spectrometry 2002;17:232–5.

[140] Canário CM, Katskov DA. Direct determination of Cd and Pb in edible oils by atomic absorption spectrometry with transverse heated filter atomizer. Journal of Analytical Atomic Spectrometry 2005;20:1386.

[141] Saint'Pierre TD, Tormen L, Frescura VLA, Curtius AJ. The direct analysis of fuel ethanol by ICP-MS using a flow injection system coupled to an ultrasonic nebulizer for sample introduction. Journal of Analytical Atomic Spectrometry 2006;21:1340–4.

[142] dos Santos EJ, Herrmann AB, Chaves ES, Vechiatto WWD, Schoemberger AC, Frescura VLA, et al. Simultaneous determination of Ca, P, Mg, K and Na in biodiesel by axial view inductively coupled plasma optical emission spectrometry with internal standardization after multivariate optimization. Journal of Analytical Atomic Spectrometry 2007;22:1300.

[143] Ibrahim H. Determination of lead in frying oils by direct current plasma atomic emission spectrometry. Journal of the American Oil Chemists Society 1991;68:678–9.

[144] Murillo M, Benzo Z, Marcano E, Gomez C, Garaboto A, Marin C. Determination of copper, iron and nickel in edible oils using emulsified solutions by ICP-AES. Journal of Analytical Atomic Spectrometry 1999;14:815-20.

[145] de Jesus A, Zmozinski AV, Barbará JnA, Vale MGR, Silva MrM. Determination of calcium and magnesium in biodiesel by flame atomic absorption spectrometry using microemulsions as sample preparation. Energy & Fuel 2010;24:2109–12.

[146] de Souza RM, Leocádio LG, da Silveira CLP. ICP OES Simultaneous Determination of Ca, Cu, Fe, Mg, Mn, Na, and P in biodiesel by axial and radial inductively coupled plasma-optical emission spectrometry. Analytical Letters 2008;41:1615–22.

[147] Silva JSA, Chaves ES, Santos ÉJd, Saint'Pierre TD, Frescura VLA, Curtius AJ. Calibration techniques and modifiers for the determination of Cd, Pb and Tl in biodiesel as microemulsion by graphite furnace atomic absorption spectrometry. Journal of the Brazilian Chemical Society 2010;21:620–6.

[148] Lima AS, Silva DG, Teixeira LS. Determination of cadmium in biodiesel using microemulsion and electrothermal atomization atomic absorption spectrometry. Environmental monitoring and assessment 2015;187:4122.

[149] De Leonardis A, Macciola V, De Felice M. Copper and iron determination in edible vegetable oils by graphite furnace atomic absorption spectrometry after extraction with diluted nitric acid. International Journal of Food Science & Technology 2000;35:371–5.

[150] Kowalewska Z, Izgi B, Saracoglu S, Gucer S. Application of liquid-liquid extraction and adsorption on activated carbon of the determination of different forms of metals present in edible oils. Chem Anal (Warsaw) 2005;50.

[151] Tokay F, Bağdat S. Determination of iron and copper in edible oils by flame atomic absorption spectrometry after liquid–liquid extraction. Journal of the American Oil Chemists' Society 2015;92:317–22.

Chapter 4

Liquid Scintillation Spectrometry as a Tool of Biofuel Quantification

Romana Krištof and Jasmina Kožar Logar

Additional information is available at the end of the chapter

http://dx.doi.org/10.5772/65549

Abstract

Biofuels are common addition or substitute for fossil fuels, applied as an attempt to decrease the impact of transport on the environment. Because of a large variety of already known biofuels and intensive research in the field, there is a high demand for analytical techniques for their quantification in fuels. Liquid scintillation counting (LSC) is one of the ideal candidates for this kind of measurements because the measured substance is radiocarbon found in all biofuels. This chapter describes the fundamental feature of LSC measurements and possible sample preparation steps. One of the methods (direct LSC method) is highlighted. One of the method's advantages is simple sample preparation, thus suitability for every LSC laboratory. Calibration and validation results of three types of biocomponents, i.e., bioethanol, synthetic biodiesel [hydrogenated vegetable oil (HVO)], and conventional biodiesel [fatty acid methyl esters (FAME)], are presented. All results show that the described method is suitable for routine analysis of various biocomponents.

Keywords: LSC, liquid scintillation spectrometry, ^{14}C, diesel, bioethanol, synthetic diesel, fossil fuels

1. Introduction

Recently observed environmental changes and increased exploitation of fuels lead to a concern and emphasis of substituting fossil fuels by biofuels. With numerous types of renewables with very different characteristics, from bioethanol, ETBE (ethyl tetra butyl ether), biodiesel or FAME (fatty acid methyl esters), synthetic diesel or HVO (hydrogenated vegetable oil), Fischer-Tropsch products, etc., the need for several methods for biofuel quantification has aroused [1]. Several analytical techniques are in use, some of them have a status of standardized methods. For instance, in EN 14078 [2], near-infrared spectrometry with trans-reflectance fiber optics for determination of methyl esters is mentioned while oxygenates, i.e., ethanol and ETBE are determined using an oxygenate flame ionization detector (EN 1601) or gas chromatography using column switching (EN 13132) [3, 4].

For differentiation between ethanol and bioethanol, common analytical techniques cannot be used since both components are chemically identical. Measurement of ^{14}C activity of such fuel mixture presents a solution because radiocarbon is only present in recently grown components (biocomponent) while in fossil component all ^{14}C has already decayed with radionuclide's half-life (5760 years). Radiocarbon or ^{14}C is a cosmogenic radionuclide produced in the atmosphere by neutron capture from ^{14}N. Since production and decay of ^{14}C in the atmosphere is in equilibrium, living organisms who uptake ^{14}C via photosynthesis, ingestion or inhalation have closely related activity as its environment [5]. The principle of radiocarbon dating is therefore usable also for quantification of ^{14}C in fuels. Genuine standard ASTM D6866 describes the accelerator mass spectrometry (AMS) and liquid scintillation counting (LSC) methods. Sample treatment is needed for these two methods. Additionally, a so-called direct LSC method that does not require special sample preparation before measurement is included in standard DIN 51637 [6]. It is only mentioned for limited types of biocomponents.

The following section explains the detection principle of the LSC system. Possible sample preparation techniques and their differences will be explained. Finally, the chapter shows some biofuel measurements using the direct LSC method. Results of ethanol, synthetic biofuel, and FAME measurements as well as validation parameters are presented and discussed.

2. Liquid scintillation counting (LSC)

Liquid scintillation spectrometry is one of the techniques for radioactivity measurements; especially suitable for the detection of β-emitting isotopes such as ^{14}C. As previously stated, isotope ^{14}C is naturally produced in the atmosphere by neutron capture. When radiocarbon decays it releases electron and antinevtrino (see Eq. (1)), produced energy is distributed among created particles what makes distinguishing continuous spectra of the β particle.

$$^{14}\mathrm{C} \xrightarrow{\text{decay}} {}^{14}\mathrm{N} + \beta^{-} + \overline{\nu} \tag{1}$$

where: $\overline{\nu}$ is Antinevtrino.

In order to detect β decay by means of LSC, a scintillation cocktail consisted of three types of chemicals: solvent, emulsifier, and scintillator (fluorescent material) has to be mixed with sample. Decay energy is absorbed in the emulsifier and via solvent transferred to the scintillator. The scintillator then emits energy in the form of light and so produced scintillations are detected by the photomultiplier tubes (PMT) in which conversion to an electrical pulse occurs (see **Figure 1**). If measurement is conducted without interferences, the measured count rate is directly proportional to the activity of the sample. In the case of biofuels, counting efficiency should be taken into account for proper transformation of counts to activity because of substantial effects of quenching.

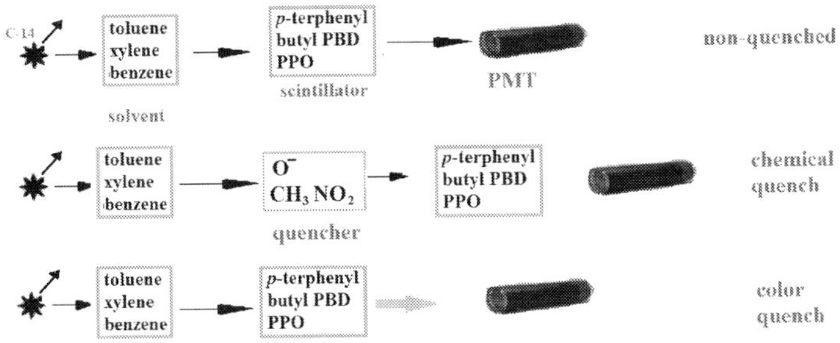

Figure 1. Quenching processes [7].

The quenching occurs when energy transfer between the radioactive isotope and the scintillation cocktail is disturbed. Reduced photon production results in a reduced number of counts as well as shifted spectrum toward lower energies. There are several types of quenching but two the most important ones have also an effect on biofuel measurements, i.e., chemical and color quench. As shown in **Figure 1**, chemical quench disturbs energy transmission between solvent and the scintillator while color quench affects transmission of energy between scintillator and PMT by attenuation of produced light in the sample.

A level of quenching can be evaluated through quench indication parameters describing the path by which they were obtained, i.e., the spectral quench parameter of the external standard (SQP(E)), spectral index of the sample (SIS), and spectral quench parameter of the internal standard (SQP(I)). A calibration curve describes the relation between the quenching parameter and counting efficiency. The most common approach to obtain the calibration curve goes through the preparation of a spiked set of samples in which various quantities of quenchers such as nitromethane is added. Efficiency is calculated by easy division of a count rate with known activity of the sample. There is also a possibility of quench set made from samples which by their nature have a variety of quenching; biodiesel blends have various quench levels (depending on feedstock and quantity of the biofuel) what can be useful in quench curve sample preparation.

2.1. Sample preparation

There are several sample preparation methods in combination with the LSC technique. In the so-called LSC-A method, also known as the CO_2 or carbamate method, organic carbon in the sample has to be converted to CO_2. This process is conducted in a special apparatus where sample is combusted to the form of CO_2 under a controlled environment. The gas is trapped or absorbed on an absorbent or on one of the components of specially designed scintillation cocktails. In the LSC community, various mixtures are known [8, 9]. In the LSC-B method, the combusted sample in the form of CO_2 is further carbonized to benzene [10]. Benzene synthesis consists of three steps: forming of carbide (usually lithium carbide), hydrolyzation to acetylene and trimerization into benzene [8]. The LSC-A and LSC-B methods are used for environmental sample measurements for the purposes of age determination, monitoring of nuclear site, etc. The third option for the sample preparation is the direct LSC-C method that does not need any sample pretreatment. The liquid scintillation (LS) sample is prepared by simple mixing of the sample with a suitable scintillation cocktail. Since there is a lack of sample pre-treatment, various matrixes and thus rather more complicated calibration with careful quench correction are needed. In the case of biofuels, extensive quench is induced by samples' yellow color. Namely, yellowish samples are actually very unfavorable for LSC due to complementarity to blue, which is the typical scintillation color. Some authors have reported attempts to degrade and limit the color of the LS sample [11, 12]. Biodegradation and low oxidative stability of biocomponent presents another difficulty in calibration process. The most important features of all three mentioned LSC methods are summarized in **Table 1**.

	Advantages	Disadvantages
LSC-A (CO_2 method)	- Uniform matrix of LS sample - Colorless LS sample - Suitable for liquid, gaseous, and solid samples	- Complicated sample preparation with several possibilities for error - Repeatability of CO_2 absorption - Higher limits of detection
LSC-B (Benzene method)	- Uniform matrix of LS sample - Colorless LS sample - Suitable for liquid, gaseous, and solid samples - Low detection limits	- Long and expensive sample preparation with several possibilities for errors - Demands highly trained personnel - Preparation time
LSC-C (direct method)	- Fast sample preparation - Repeatability, accuracy - General laboratory practice trained personnel	- Complicated calibration - Various matrices

Table 1. Comparisons of different LSC methods for quantification of biocomponents in fuels.

3. Biofuel measurements with direct LSC

3.1. Bioethanol measurements

Bioethanol is used as an additive to gasoline or substitution of ethanol. Several authors have already reported measurements of bioethanol using various LSC methods [8, 13–17]. Since ethanol/bioethanol matrices are colorless, the calibration can easily be made also directly from a count rate. This is a so-called one-step calibration curve, conducted as a correlation between the count rate and the biofuel content. In blends of ethanol/bioethanol/gasoline, a two-step calibration is advised. The two-step calibration consists of efficiency correction due to matrix variation and expected chemical quench.

Calibration samples can easily be made by blending certified fossil ethanol (in our case Fisher Scientific) or gasoline (provided by local petroleum industry) with certified bioethanol (like Carbo Elba ethanol). In each matrix, at least 10 LS samples were made and analysis was conducted. The obtained spectra of the sample can be seen in **Figure 2**. We found that it is useful if blends were made in accordance with market demands; that is several blends in the ranges from 0–to 10% and 80 to 100% and the rest of LS samples in blends in between. Measurements show that the difference in counting efficiency among ethanol/bioethanol Blends does not exceed 1% what is within uncertainty of a typical quench curve. However, in ethanol/bioethanol/gasoline blends, the difference among counting efficiencies can be up to 10% (between 82 and 72%); therefore, quench affects the counting efficiency and it has to be taken into account.

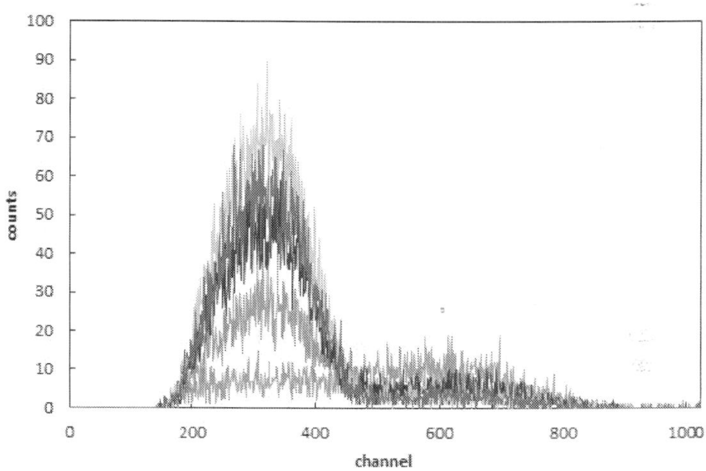

Figure 2. The set of bioethanol/ethanol blend spectra.

3.2. Synthetic biodiesel (HVO) measurements

Synthetic biodiesel or hydrotreated vegetable oil has recently been introduced into fuel market as a substitution of classic biodiesel (FAME) due to its oxidative stability and similarity to fossil diesel. The fuel itself can be mixed with diesel in various quantities or can be even substitution of fossil fuel. The latter is a problem with FAME since changes in automotive engine has to be made, while HVO can be used without any consequences to engine. Several authors have reported measurements of HVO blends in the range currently reasonable for fuel market, which is up to 20% [14, 18, 19]. In the same range, also a standardized method with direct LSC and FT-IR methods is available [6].

The HVO is colorless so one can expect only chemical quench when blends with diesel are made. Furthermore, our measurements have shown that HVO can work as a slight reducer of quench (see **Figure 3**) and counting efficiencies were higher (from 74 to 83%). As in the case of bioethanol blends, blends varying from fossil diesel up to 100% HVO were made and analyzed. Since there is significant difference among counting efficiencies of various blends, a two-step calibration is advised. Counting efficiency can be evaluated with the same quench curve as bioethanol blends in the case of the same range. HVO can be produced from various feedstocks, but according to our results and experiences, the activity of various blends is similar regardless of the initial feedstock or preparation procedure of HVO.

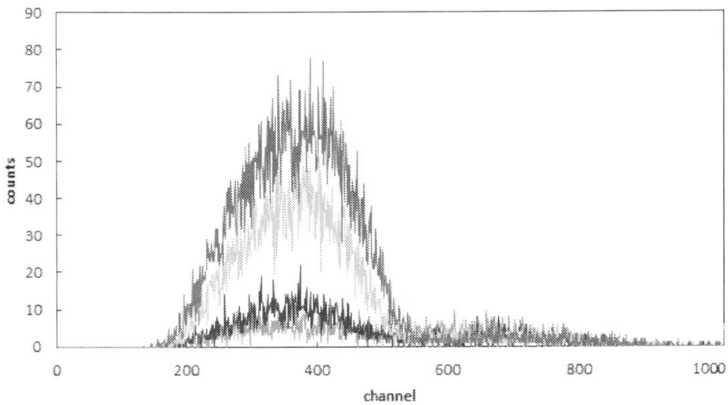

Figure 3. The set of HVO blend spectra.

3.3. Fatty acid methyl ester (FAME) measurements

As one of the most important parts of diesel's biofuels, FAME is referred to as biodiesel. Characteristics such as chemical form, color and oxidative stability depends on feedstock oils, what can affect analysis by the direct LSC method. As was explained in the sample preparation (Section 2.1.), the color of biodiesel has an effect on counting and thus besides chemical also color quench can be observed. Biodiesel aging or oxidative instability is one of the drawbacks that affect the use of biocomponent in fuel market, but as research had shown, it has positive effects on the LSC measurement [12]. That occurs due to decomposition of fatty acid esters while radiocarbon is still present in the sample (see **Figure 4**). However, forced oxidation was not shown as a promising step in sample treatment because of big differences among biodiesels from various feedstocks.

Several authors have reported measurements of biodiesel in various blends but mainly in level up to 20% of biodiesel what is limit of current fuel market [19–21]. However, our research shows that reliable analyzes of biodiesel in quantities up to 100% of biocomponent can be conducted as well. Some changes of preprepared setups of LS counter Quantulus™ (PerkinElmer) and counting protocol gave promising results, together with careful and precise calibration [14]. Hence, changes in coincidence circuit enabled analysis of various feedstock biodiesels in the whole range (from 0 to 100% biodiesel). **Figure 5** presents spectra that were obtained measuring various feedstock biodiesel ranging in quench levels from 360 up to 850, where the FAME from waste oil presents the sample with the lowest quenching number and thus the highest observed quenching while sample with the lowest observed quenching was made from sunflower oil.

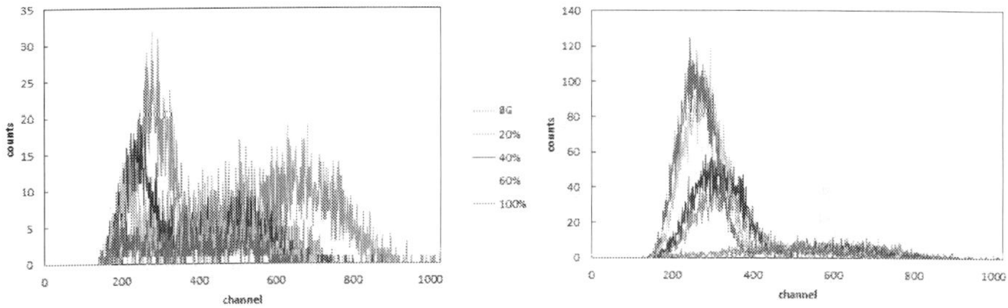

Figure 4. Biodiesel blend spectra. The spectra set of the first measurement (left) belongs to the fresh biocomponent, while the set of second measurement (right) represents the spectra of already oxidized biocomponent.

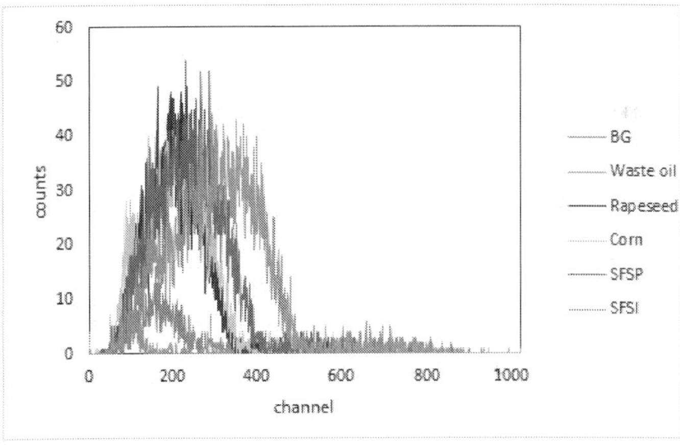

Figure 5. Biodiesel spectra after protocol changes. Legend: BG: background; SFE: sunflower (Spain), SFSI: sunflower (Slovenia).

3.4. Validation parameters

All analytical methods need to go through validation in order to use them in routine analysis. Uncertainty, detection limit, linearity, repeatability, and sensitivity were evaluated. Law of uncertainty propagation was followed while contributions of variable parameters were evaluated using GUM guidelines [22].

It was found that the uncertainty results are directly affected by the uncertainty of the balance, sample and background count rate, counting efficiency, and uncertainty of calibration. The indirect contributions are represented with uncertainty of pipette, temperature variation, and luminescence of the LS samples. It was found that uncertainties of sample and background count rate represent the largest contribution in the measurements near detection limits. In analysis of blends with biocomponents quantity higher than 10%, the largest contribution of uncertainty causes counting efficiency determination. In both cases, the uncertainty of the balance presents negligible part of the uncertainty budget.

In recent years, a new approach in determination of limits of detection is taken; the standard ISO 11929 applies a null measurements uncertainty [23]. Besides background count rate, the background and sample counting time present variables of limit of detection calculation. Although a long-term average of background is taken in our routine analysis, the detection limits were evaluated conservatively, and thus 1000 min of background and 500 min of sample counting time were taken. Obtained limits of detection (see **Table 2**) are comparable to those obtained by other laboratories and methods [2–4, 6, 8, 13, 15–21].

Parameter\component	bioEtOH	HVO	FAME	
Range	DL−100%	DL−100%	DL−10%	DL−100%
Limit of detection	0.63%	1.13%	0.66%	1.67%
Linearity	0.997	0.998	0.999	0.998
Repeatability	0.54%	0.40%	0.5%	0.44%
Sensitivity	17.5	9.156	12.546	10.882

Table 2. Validation parameters summary.

Linearity of calibration curves is demonstrated using the least-square method and the correlation coefficient R^2 while uncertainties of individual calibration curve parameters were used for calculation of calibration uncertainty. As can be seen in **Table 2**, excellent linearity of calibration curves was obtained in all biocomponents measurements. It has to be noted that the individual calibration curve is a result of at least 20 blends analysis.

Repeatability was tested by several measurements (at least 10) of the same sample; standard deviation of analysis results was calculated. A test was conducted on various blends in the full range of the calibration. As can be seen in **Table 2**, results of the analysis with the direct LSC method are repeatable within 1 standard deviation of results; furthermore, it never exceeds 0.54% that was achieved with bioethanol calibration.

Steepness of the calibration curve slope was taken as a parameter for determination of sensitivity; factor *k* from linear regression line was compared. The most sensitive calibration was shown to be for bioethanol analysis (17.5) while the least sensitive was the HVO calibration curve with 9.156, respectively.

4. Conclusions

Biocomponents in world fuel market differ in their chemical characteristics. As a consequence, several analytical techniques have to be applied for their quantification in fuel blends. Their maintenance needs a lot of human and financial sources, especially in the form of equipment and knowledge. Liquid scintillation spectrometry is a good alternative. Namely, the measured quantity is always the same. Radiocarbon, ^{14}C is found in all biocomponents regardless the type of biofuel.

Three approaches can be applied as the sample preparation step before LSC analysis. Two of them, LSC-A and LSC-B consist of several sample preparation phases, from raw fuel to CO_2 production and benzene synthesis. The final result of sample preparation is the same matrix regardless the initial biocomponent. On the other hand, the LS sample in the LSC-C method can appear in many different forms. The matrix is not constant and predictable. Calibration and validation processes of the method are therefore extended and need expert knowledge. But, it should be done only once.

The maintenance of the already calibrated and validated LSC-C direct method is simple. It demands only periodical and simple check-ups of calibration parameters. The method does not need any special sample preparation steps. Routine analysis with this method is therefore very fast, cheap and does not need highly trained experts.

Author details

Romana Krištof[1,2*] and Jasmina Kožar Logar[1,2]

*Address all correspondence to: romana.kristof@ijs.si

1 Jožef Stefan Institute, Department for Low and Medium Energy Physics, Ljubljana, Slovenia

2 University of Ljubljana, Faculty of Health Sciences, Ljubljana, Slovenia

References

[1] Soetaert, Wim and Vandamme, Erick J. Biofuels. Wiltshire, UK : John Wiley & Sons Ltd., 2009. ISBN 978-0-470-02674-8.

[2] EN 14078:2014 Liquid petroleum products – Determination of fatty acid methyl ester (FAME) content in middle distillates - Infrared spectrometry method. 2014. European Committee for Standardization. ISBN 978 0 580 81183 8.

[3] EN 1601:2014 Liquid petroleum products. Unleaded petrol. Determination of organic oxygenate compounds and total organically bound oxygen content by gas chromatography (O-FID). 2014. European Committee for Standardization. ISBN 978 0 580 88552 5.

[4] EN 13132:2000 Liquid petroleum products – Unleaded petrol – Determination of organic oxygenate compounds and total organically bound oxygen content by gas chromatography using column switching. 2000. European Committee for Standardization. ISBN 0 580 36216 7.

[5] L'annunziata, Michael F. Radioactivity: Introduction and history. First edit. Amsterdam, The Netherlands : Elsevier, 2007. ISBN 978-0-444-52715-8.

[6] DIN 51637:2014-02 Liquid petroleum products - determination of the bio-based hydrocarbon content in diesel fuels and middle distillates using liquid scintillation method. 2014. Deutsches Institut für Normung.

[7] Krištof, Romana. Quantification of biocomponents in fuels by 14C. Nova Gorica, Slovenia : University of Nova Gorica, 2015.

[8] Culp, Randy, Cherkinsky, Alex and Prasad, Ravi G. V. Comparison of radiocarbon techniques for the assessment of biobase content in fuels. Applied Radiation and Isotopes: Including Data, Instrumentation and Methods for Use in Agriculture, Industry and Medicine. 2014. Vol. 93, pp. 106-109. DOI 10.1016/j.apradiso.2014.01.007.

[9] Varlam, Carmen, Stefanescu, Ioan, Varlam, Mihai, Popescu, Irina and Faurescu, Ionut. Applying the direct absorption method and LSC for C-14 concentration measurements in aqueous samples. Radiocarbon. 2007. Vol. 49, no. 2, pp. 281–289.

[10] Krajcar Bronić, Ines, Horvatincić, Nada, Baresić, Jadranka and Obelić, Bogomil. Measurement of 14C activity by liquid scintillation counting. Applied Radiation and Isotopes: Including Data, Instrumentation and Methods for Use in Agriculture, Industry and Medicine. 2009. Vol. 67, pp. 800–804.

[11] Zhang, HaiXia, Nie, KaiLi, Wang, Fang, Tan, TianWei and Deng, Li. Bleaching of deep color biodiesel made from waste cooking oils by oxidation. Journal of Beijing University of Chemical Technology (Natural Science Edition). 2009. Vol. 36, pp. 96–99.

[12] Krištof, Romana and Kožar Logar, Jasmina. Quenching parameter in the measurement of biodiesel by liquid scintillation counting. In : Cassette, Philippe (ed.), LSC 2010, Advances in Liquid Scintillation Spectrometry. Paris, France, 2010. pp. 35–39.

[13] Noakes, John E., Cherkinsky, Alexander and Culp, Randy. Comparative radiocarbon analysis for artificial mixes of petroleum and biobased products. In : Cassette, Philippe (ed.), LSC 2010, Advances in Liquid Scintillation Spectrometry. Paris, France, 2010. pp. 15–21.

[14] Krištof, Romana and Kožar Logar, Jasmina. Direct LSC method for measurements of biofuels in fuel. Talanta. 2013. Vol. 111, pp. 183–188.

[15] Yunoki, Shunji and Saito, Masaaki. A simple method to determine bioethanol content in gasoline using two-step extraction and liquid scintillation counting. Bioresource Technology. 2009. Vol. 100, pp. 6125–6128.

[16] Edler, Ronald and Kaihola, Lauri. Differentiation between fossil and biofuels by liquid scintillation beta spectrometry – direct method. Nukleonika. 2010. Vol. 55, no. September 2009, pp. 127–131.

[17] Dijs, Ivo J, Van der Windt, Eric, Kaihola, Lauri and Van der borg, Klaas. Quantitative determination by component in fuels 14 c analysis of the biological component in fuels. Radiocarbon. 2006. Vol. 48, no. 3, pp. 315–323.

[18] Norton, Glenn A. and Woodruff, Mary X. Simplified radiocarbon analysis procedure for measuring the renewable diesel concentration in diesel fuel blends. Journal of the American Oil Chemists' Society. 2011. Vol. 89, no. 5, pp. 797–803.

[19] Idoeta, Raquel, Pérez, Elena, Herranz, Margarita and Legarda, Fernando. Characteristic parameters in the measurement of (14)C of biobased diesel fuels by liquid scintillation. Applied Radiation and Isotopes: Including Data, Instrumentation and Methods for Use in Agriculture, Industry and Medicine. 2014. Vol. 93, pp. 110–113. DOI 10.1016/j.apradiso.2014.01.019.

[20] Takahashi, Yui, Sakurai, Hirohisa, Inui, Emiko, Namai, Saori and Sato, Taiichi. Radiocarbon measurement of biodiesel fuel using the liquid scintillation counter quantulus. In: Cassette, Philippe (ed.), LSC 2010, Advances in Liquid Scintillation Spectrometry. Paris, France, 2010. pp. 41–46.

[21] Norton, Glenn A., Cline, Anna M. and Thompson, Greg C. Use of radiocarbon analyses for determining levels of biodiesel in fuel blends – Comparison with ASTM method D7371 for FAME. Fuel. 2012. Vol. 96, pp. 284–290.

[22] Taylor, Barry N. and Kuyatt, Cris E. Guidelines for evaluating and expressing the uncertainty of NIST measurement results. 1994. Gaithersburg : NIST.

[23] ISO 11929: Determination of the characteristic limits (decision threshold, detection limit and limits of the confidence interval) for measurements of ionizing radiation – Fundamentals and application. 2010. International Organization for Standardization. ISBN 978 0 580 59044 3.s

Chapter 5

Chromatographic Methods Applied to the Characterization of Bio-Oil from the Pyrolysis of Agro-Industrial Biomasses

Maria Silvana A. Moraes, Débora Tomasini,
Juliana M. da Silva, Maria Elisabete Machado,
Laíza C. Krause, Claudia A. Zini,
Rosângela A. Jacques and Elina B. Caramão

Additional information is available at the end of the chapter

http://dx.doi.org/10.5772/66326

Abstract

Biomass conversion into solid, liquid and gaseous products by pyrolytic technology is one of the most promising alternative to convert the biomass into useful products and energy. The total characterization of the products from the pyrolysis of biomass is one of the great challenges in this field, mainly due to their molecular complexity. Pyrolysis is a process that causes degradation of biomass in a non-oxidative atmosphere, at relatively high temperatures, producing a solid residue rich in carbon and mineral matter, gases and bio-oil. The yield and properties of the products depend on the nature of the biomass and the type of the pyrolysis process (type of reactor, temperature, gas flow, catalyst). Due to the high molecular complexity of bio-oil, many different technical had been developed to their complete characterization. This chapter describes the principles of the techniques and main application of chromatographic methods (GC, LC, GC × GC, LC × LC, Nano-LC) in the analysis of bio-oils derived from thermo-degradation of biomasses. Especial attention is carried out to two-dimensional techniques that represent the state of the art in terms of separation, sensibility, selectivity and velocity of data acquisition for characterization of complex organic mixtures. For proper use of bio-oil in the chemical industry, it is essential the identification and unambiguous determination of its major constituents. Only then, it is possible to propose a recovery route of some of these components for the development of an industry dedicated to a bio-refinery. For this, chromatographic methods, especially GC × GC/MS, are fundamental because they allow analysis with high sensitivity and accuracy in identifying each constituent of the bio-oil.

Keywords: Chromatography, bio-oil, biomass, pyrolysis

1. Introduction

The current world energy scenario shows a tendency to decrease the use of mineral resources source, considering its environmental impact, the socioeconomical and market problems if compared to renewable sources [1–3]. In addition, the diversification of energy sources is necessary in order to meet the growing energy demand. In this context, it arises the biomass coming from different sources of natural resources, which is a renewable alternative, abundant and suitable for competition with conventional fossil fuels [3, 4]. The biomass conversion into solid, liquid and gaseous products by pyrolytic technology is a promising alternative to convert the biomass into useful products and energy [5–7]. The total characterization of the products from the pyrolysis of biomass is one of the great challenges in this field, mainly due to their molecular complexity [8].

Pyrolysis is a process that causes degradation of biomass in a non-oxidative atmosphere, at relatively high temperatures, producing a solid residue rich in carbon and mineral matter, gases and bio-oil [6, 7, 9, 10]. The yield and properties of the products depend on the nature of the biomass and the type of the pyrolysis process (type of reactor, temperature, gas flow, catalyst) [7, 11–13].

One of the main products of biomass pyrolysis is the bio-oil and, due to its high molecular complexity, it has been subject of several characterization studies in order to identify its compounds and indicate their best uses [8, 10, 14, 15].

This chapter describes the principles of the techniques and main application of chromatographic methods (GC, LC, GC × GC, LC × LC and Nano-LC) in the analysis of bio-oils derived from thermo-degradation of biomasses. Especially, attention is done to two-dimensional techniques that represent the state of the art in terms of separation, sensibility, selectivity and velocity of data acquisition for characterization of complex organic mixtures.

2. Pyrolysis of biomass

Energy production in the twentieth century was dominated by fossil fuels (coal, oil and gas) that represented, still in the beginning of the century, about 80% of all energy produced in the world. Nuclear power, hydropower and renewable energy sources (solar, wind, geothermal and small hydropower plants), which are the most attractive from an environmental point of view, represented only 1.5% of world production in those years [1]. Currently, the depletion of natural resources of coal and oil, together with the greenhouse effect, has attracted great interest for the production of sustainable energy [2]. The search for alternative fuels has led some countries to opt for biofuels, which in turn increased interest in energy from biomass [3].

The conversion of biomass into fuels and chemicals can economize fossil reserves and thus boost research, social and economic activities, especially in countries where the biomass is an abundant raw material in the agro-industrial sector [4]. The use of renewable sources for obtaining chemicals such as bio-plastics, bio-fertilizers and bio-polyesters can help reducing the demand of insumes from petrochemical origin.

The biomass can be converted into energy using thermal, biological, mechanical or physical processes. The biological processing (biological catalysis) is very selective and produces a discrete number of products in high yield, but it requires a raw material containing sugar or carbohydrates and a water content exceeding 40% [5]. The thermochemical methods are more suited to dry biomass (moisture content <10%) and rich in lignin (which is less suitable for biochemical conversion, since it is hardly broken through enzymatic activity), such as wood and agricultural waste. Furthermore, the thermal conversion process provides often multiple products in a short reaction time, typically using inorganic catalyst to improve the quality of the product. The main processes used for the thermal conversion of biomass into a more useful form of energy are combustion, gasification and pyrolysis [5, 6].

Pyrolysis is defined as a thermal decomposition of the biomass that occurs in the absence of oxygen at temperatures between 350 and 700°C, producing gases, liquid and solid products [5, 7]. Processes using lower temperatures and long residence time favor the production of charcoal, while processes using higher temperatures and long residence time increase the conversion of biomass into gas. A great production of liquids is favored using processes at moderate temperature and short residence time [9, 10]. It should be noted that the three products are always generated, but the proportions may be varied by adjusting the process parameters such as heating rate, gas flow, pressure, particle size of biomass and residence time of the biomass in the reactor [6, 11]. Also in accordance with these parameters, the pyrolysis can be classified as slow, fast, flash, catalytic or vacuum [12, 13].

The slow pyrolysis, or carbonization, employs low temperatures and long residence times, favoring the production of charcoal. In the case of fast pyrolysis, the heating occurs at higher rates (above 20°C min^{-1}), while in the flash pyrolysis, heating rates are between 500 and 1000°C s^{-1} [14].

Fast pyrolysis produces 60–75% by weight of bio-oil (liquid) 15–25% by weight of biochar (solid) and 10–20% by weight of noncondensable gases, depending on conditions and the biomass used. The solid residue from this process can be charcoal or ash only, depending on the final temperature and on the content of mineral matter originally present in the biomass used. This residue can be used as fuel or as a soil additive or still for the production of ceramic material. The gas produced can be recycled back to the process and facilitating cleavage reactions of the original biomass [14]. The main objective of fast pyrolysis is to prevent the breakdown of primary products in small noncondensable gas molecules, or even prevent them from being recombined and polymerized. In any of these cases, it obtains a smaller yield of bio-oil [10].

The order of the reactions that occur during the pyrolysis process and the yield of obtained compounds will depend on parameters such as the heating rate, the pyrolysis temperature, pretreatment of biomass and catalyst effects. The study of these reactions and the effect of these parameters are important for obtaining high yields of the desired products [5]. To perform the pyrolysis, it can be used different types of reactors, according to the main purpose of the process.

2.1. Pyrolysis reactors

2.1.1. Fixed bed reactors (FxBR)

In this type of pyrolysis reactor, the fluid phase (gas) passes through the particulate solid phase (biomass). This FxBR aims to promote intimate contact between the phases involved in the process—the gaseous fluid phase with the stationary phase (particles of biomass) [15–22]. The FxBR is constituted by tubular structures of stainless steel or quartz, being used as an oven or grill support during the controlled heating of the system. The inert gas passes through the compartment with the biomass (in steady state) carrying the products out of the reaction bed during the pyrolysis [15]. Typically, units for feeding biomass, ash removal and outlet or collecting gases are added to these reactors. Such reactors operate with high residence time of the biomass in the oven and low inert gas velocities [16–22]. In this way, they are considered suitable for research in laboratory scale or bench. Some units operating in China can convert up to 600 kg/h of biomass to bio-oil [23].

Pyrolysis in FxBR is very used to the slow pyrolysis, with the main purpose of producing coal or ashes.

2.1.2. Fluidized bed reactors (FzBR)

In this type of reactor, a fluid (gas) is passed through a granular solid material at high speed, sufficient to suspend the solid material and cause it to behave like a fluid. This process known as fluidization provides extensive use in studies on the pyrolysis of biomass and is ideal for the technique of fast pyrolysis because it can achieve the necessary requirements for its realization and is virtually the only ones used in the world on a commercial scale for the production of biofuels and chemicals [24–27].

The FzBR operates with suspended solids by the action of rising gases, which are introduced from the bottom of the reactor. To promote more effective heat transfer, there is used a bed of solid particles, generally consisting of sand finely dispersed in the biomass itself [23, 28].

The rapid exchange of heat between the heat source and the biomass is one of the most important points in the pyrolytic process. In this context, in FzBR, the dried and comminuted biomass is introduced into the reactor, wherein, in the heating zone, the material remains in a continuous movement, promoted by the carrier gas flow (inert gas), which maintaining the reactor with a low oxygen content as it is heated to high temperatures (500–900°C) using a heating rate between 100 and 500°C min^{-1} [23, 29, 30].

Among the many advantages of this model, we can mention its ease of construction and operation, good temperature control, the operation at atmospheric pressure, the easy scale-up and operation possibility with fine particle size biomass, which is common in agriculture, in forestry and industry. Beside this, the reactor has an excellent heat exchange between the heat source and the biomass, and an efficient gas-solid contact due to the movement of the particle bed. This effective contact simulates an isotherm condition, which implies an operation more secure and with optimum performance, generating yields of liquid products of approximately 70–75% by weight (dry basis), minimizing side reactions [24, 26].

But, the heterogeneity of the residence times in the reactor due to the agitation of the solids in the bed can compromise the uniformity of the products. Moreover, the wear caused by the moving particles and agglomeration of ash produced with the inert bed material may lead to loss of fluidization and therefore the pyrolytic process [23, 25, 28]. Therefore, the successful application of this technology depends on understanding and overcoming their disadvantages and thus the development of a reactor that meets the needs of the pyrolysis process in a whole.

Several research groups in the development of fast pyrolysis use the fluidized bed reactor. Among the many examples, one can cite: the Union Fenosa located in Meirama, Spain, which has a pilot plant with biomass feed capacity of approximately 250 kg h^{-1}. Still in Europe, the Wellman Company in the UK, which has a pilot plant with a capacity of 200 kg h^{-1}, in Canada, the DynaMotive, which employs a pilot plant with capacity of 8000 kg h^{-1} [23, 24].

2.1.3. *Continuously feeding reactors (CFRs)*

Continuously feeding reactors (CFRs) are those which operate at all times with an input a specific substrate(s) and output of product(s). These types of reactors are widely used in industrial scale [31].

In general, CFRs offer reduced fixed and operating costs and improve heat exchange capability [31, 32]. Furthermore, they provide an increase in quality of the final product, since the variations that exist between batches are eliminated [33]. Continuous processes are still able to reduce losses caused by operational problems during the process, and it is not necessary to interrupt the production line [33] (start-up and shutdown). The applications of these reactors are aimed to minimize the difficulties encountered in the process on an industrial level using reactors in pilot scale [31].

2.2. Catalytic pyrolysis

Studies of the composition of bio-oils obtained by pyrolysis of various biomasses (rice husk, coconut husk fiber, core tropical fruits, straw sugarcane, wood residues, etc.) found that the volatile fractions of bio-oils consist of a complex mixture of different classes of compounds such as, ketones, phenols, aldehydes, hydrocarbons [8, 15, 34–40]. These bio-oils have physicochemical characteristics that avoid their direct use as fuel, without any treatment. One of the reasons is its high oxygen content, thus leading to a high chemical instability during storage hindering their production on an industrial scale [41].

For industrial use of bio-oil, some enhancement process is needed (upgrading). The catalytic pyrolysis has emerged as an alternative for improving the quality of liquid products of pyrolysis acting as an upgrading process of bio-oil. The main variable of this process is the type of catalyst used, in particular those able to considerably reduce its oxygen content [41].

A method used worldwide is the hydrodeoxygenation (HDO), which converts and fragments heavy molecules in the biomass into hydrocarbons with lower oxygen content by the catalytic addition of hydrogen [42]. This process has been studied with the aim of producing a liquid mixture of hydrocarbons that do not have the undesirable properties of bio-oil and thus can be used as fuel. This method is considered the most efficient for the upgrading of bio-oils [42].

Several catalysts can be applied for HDO, being necessary the use of high temperatures and pressures, under an atmosphere of H_2. The most used are those consisting of cobalt and nickel and may be supported on alumina (Al_2O_3), silica (SiO_2), carbon, zeolites (ZSM-5), among others [43, 44].

The H-ZSM-5 catalysts, for example, have strongly acidic active sites, which can supply hydrogen for the pyrolytic reactions, favoring the deoxygenation of bio-oil. Thus, the great advantage of the use of zeolites for production of bio-oils is the possibility of working on hydrogen-free atmosphere and the use of ambient pressure during the catalytic pyrolysis process [45].

Zeolites have been shown to be highly effective for converting lignocellulosic biomass into aromatics through catalytic pyrolysis. Whereas this type of biomass has low amounts of hydrogen in their composition (low H/C ratio), the maximum yield of hydrocarbons that can be obtained in the absence of H_2 as external reagent is limited. Thus, an upgrading using zeolites generally results in yields of aromatic hydrocarbons and olefinic near 30% [46, 47]. In addition, in the catalytic pyrolysis occurs greater coke formation when compared with conventional pyrolysis [48].

Catalysts conventional hydrotreating as Co-Mo and Ni-Mo and supported noble metal catalysts have been studied to produce stable products with high calorific value from pyrolysis [49], however, require high pressures of H_2 that lead to hydrogenation of the aromatic ring, resulting in reduction in calorific value and increase in H_2 consumption [50].

2.2.1. *Catalytical pyrolysis in-situ versus ex-situ*

Depending on the method of contact of catalyst and vapors of the pyrolysis, catalytic pyrolysis can be classified as in situ or ex situ.

In the in situ catalytic pyrolysis, the catalyst is mixed with the biomass to be pyrolyzed. Thus, the most suitable type of reactor for this is the fluidized bed reactor since this biomass is directly mixed with the catalyst. As the catalyst is exposed to a concentrated stream of vapors generated by depolymerization of the biomass components, the catalysis reactions are facilitated [51–54].

For the ex situ catalytic pyrolysis, biomass is pyrolyzed in a separate compartment of the catalyst. The pyrolysis vapors generated in the first compartment are diluted in an inert gas which is inserted between the two compartments being transported into a second compartment which is filled with the catalyst. As the second gas flow between the compartments, the contact time with the catalyst decreases [51]. An advantage of this technique is the use of different temperatures for both the pyrolysis reactor and for the catalysis reactor, thus allowing the use of catalysts which are sensitive to high temperatures [51–55].

2.3. Microwave assisted pyrolysis (MWAP)

The different pyrolysis processes have various conventional and unconventional heating methods. The development of an efficient heating method, with a precise control of heating parameters and with a reduction of adverse effects on the quality of the product is one of the challenges to be overcome in the development of efficient pyrolytic processes [56]. The use of microwave can be an efficient way of heating the biomass in a thermochemical conversion processes. Tech-En Ltd (UK) did the first study of the use of microwave pyrolysis in early 1990 [57, 58]. This is a technology which can be a very effective alternative for pyrolysis of biomass, presenting several advantages such as reduction in waste volume, fast heating, better chemical reactivity, ease of control, energy saving, overall cost-effectiveness, portability of equipment and processes, a cleaner source of energy compared to conventional systems [59].

Microwave heating allows a more careful control of the parameters of the pyrolysis process, enabling the maximization of the production of liquids or gases, once these parameters may induce or modify specific chemical reactions resulting in different product profiles. The process can be modulated, aiming the product optimization in accordance with conditions of temperature, power and residence time used [60].

The biggest advantage in the use of microwave heating as compared with the conventional process is the significant reduction in temperature and consequent energy gain in the pyrolysis process [61].

Another important aspect of the heating by microwave is capability to obtain basically the volatile organic compounds at lower temperatures when compared to conventional heating. Moreover, obtaining bio-oil and gases is nearly synchronized. This heat and mass transfer characteristics of the process are related to the selective heating of the components to absorb the microwave with more intensity [60].

Obtaining a greater heating uniformity in the process may be possible if the temperature is homogenized at some point during the process, so that, without this, different product compositions are obtained due to the temperature profile formed [60].

Furthermore, the temperature selection depends on the desired product. Processes at low temperatures provide greater bio-oil yields and lower energy cost. The determination and use of the appropriate power to microwave are also important. Lower powers, with lower heating rates favor formation of biochar, whereas higher power with higher heating rates favors the gasification reaction. In both cases, there is a reduction in the yield of bio-oil [60, 61].

The use of microwaves also has some challenges for the applying in the pyrolysis such as the need for different systems for measuring the temperature [62], the limitation of usable materials for propagation of microwave [63] and obtaining heating equipment compatible with the process and with an efficient scale [64].

2.4. Pilot plants of pyrolysis

The pilot plant consists of many components (steps/processes) that together form a unique process, which enables testing technologies to evaluate the quantity and quality of products desired [65, 66]. According to RESEM®, a Sino-American Company specialized in equipment for pyrolysis [61], there are different sizes of pilot plants, reactors differentiated according to the type and number of samples to be processed or products to be produced [67, 68].

The development of new technologies for the production from clean energy sources [69, 70], are associated with the emergence of pilot plants for the production of bio-oil through pyrolysis of biomass in both, laboratory and industrial scale, mainly because it is a simple, reproductive and fast process [71].

Pilot plants are equipments that consist of a closed or open system with physical, chemical and thermodynamic operation, in order to perform a technological process on a small scale. So, a prototype, designed for industrial processes, can be installed either in research laboratories or in industries [66]. In these prototypes, new and different technologies, shapes, sizes and physical structures for generating and processing information for use in pyrolytic systems can be tested before the scale-up [67].

According to **Figure 1** that shows one graphic with the research scenario related to global patent on the theme in pyrolysis plant, from 1940 to this year, they were deposited 673 patents worldwide [72, 73].

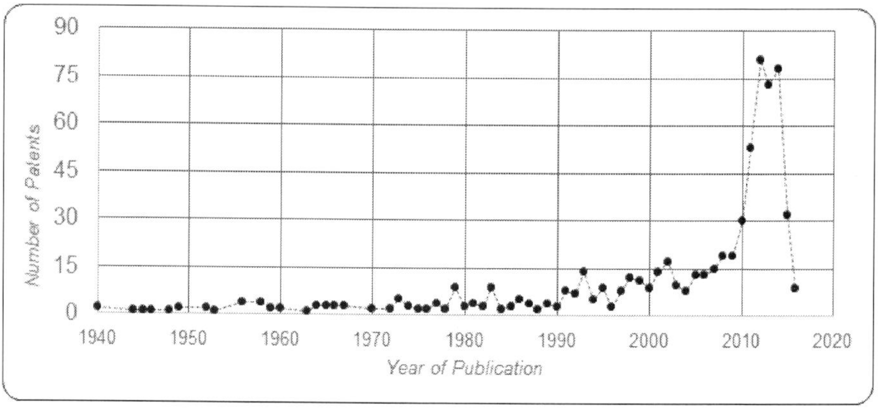

Figure 1. Distribution of patents related to pyrolysis pilot plants in the last 80 years (according https://patentes.google.com and https://patentscope.wipo.int [72, 73]).

Few of these patents actually generated commercial pyrolysis plant, indicating the need a lot of investment in this regard, especially in countries with great potential for use of biomass and lower oil reserves. As already described, in recent decades, there has been a growing concern about the processing of biomass through pyrolysis due to the interest in their products, both biochar [74], as bio-oil. In this sense, grow-related searches to biomass conversion technologies studies from laboratory scale to the development of bio-refinery [75] involving processes for the production of fuels and chemicals [76, 77].

The experimental setup on a pilot scale is based on the adjustment process through the system variables, and especially in the reactor used technology.

The reactors in pilot plants can be classified in two systems [78]:

- Batch system: The biomass is loaded at the beginning and from there it collects the products and the flow is not continuous.
- Continuous system: The biomass remains in continuous flow and the products generated are also continuously collected.

With the technological advancement of pyrolysis technique, some models of reactors have been exploited to optimize the process, cost and quality of the products generated, and the main ones are the fluidized bed (bubbling and circulating), in addition to these, some others can be found like fixed bed, ablative, vacuum, rotary cone, plasma, microwave and solar, among others [79–99].

2.4.1. Fixed bed pilot reactor

The pilot fixed bed system, similar to the bench reactors (Section 2.1.1), consists in fixed reactor with an inert carrier gas, similar to the bench scale. The technology of the fixed bed reactor is considered simple, reliable [100], and efficient for the production of bio-oil in the pilot plant. In these reactors, the solids move down and collide in counter current with a heated inert gas. They are used in small-scale productions [101]. The cooling and cleaning gas system consist in a filtration and in the use of separators cyclones. The biggest problem is associated with the removal of bio-oil and losses involved in this process.

The fixed-bed pilot plants developed from the late 1980s, with a final capacity of 5 tons per day are considered small, from the point of view of the transfer for industry, considering that these technologies are still in development stages for commercial applications.

2.4.2. Fluidized bed pilot reactor

As mentioned in Section 2.1.2, in the pilot fluidized bed reactor, a mixture of fluid and solid is obtained by introducing a pressurized fluid through the solid particles with smaller diameter [102, 103]. The general scheme of this type of reactor is very similar to that used in fluidized bed in bench scale.

Two main types of fluidized bed reactors are used in pilot plant:

Bubbling fluidized bed reactor: According to the literature [104], the bubbling bed reactors are simple to construct and operate. They provide a better ability to control the temperature, fluid-solid contact, heat transfer and capacity of storage of solids. Heated sand is used as a bed solid phase, rapidly heating the biomass in an oxygen-absent atmosphere. The biomass is decomposed into coal, steam, gas and aerosols. The fluidized gas stream entrains the compounds produced out of the reactor [105].

After the pyrolytic reaction, biochar is removed by a cyclone separator and stored. An important factor for the full operation of these reactors is that the biomass needs to be in small particles (less than 2–3 mm) to achieve high heating rates in the oven.

Circulating fluidized bed reactor: It has similar characteristics to the reactor in bubbling fluidized bed except for the shorter residence time. This design results in higher speed and higher yield of bio-oil compared to fluidized bed reactors [106]. The reactor moves around its main axis. One advantage is its high performance even with a more complex hydrodynamics.

2.5. Bio-oil from pyrolysis of biomass

The bio-oil, or pyrolysis oil, is a dark brown color liquid with a characteristic smell and comprising a complex mixture of hydrocarbons and oxygenated compounds with an appreciable amount of water, originated from the natural moisture of the biomass as well as a product of reactions that occurred during the pyrolysis process [26, 41, 107].

The literature registers that the bio-oil can contain more than 400 different chemical compounds from different chemical classes, varying among organic acids, aldehydes, ketones, alcohols, esters, furans, sugar derivatives, phenols, among others [108, 109]. In addition to the oxygenated compounds, many aliphatic and aromatic hydrocarbons can be found [8].

The anhydrous-sugar levoglucosan (1, 6-anhydro-β-D-glucopyranose) is the main component of bio-oil, being derived from the thermal depolymerization of the cellulose and from it, many other sugar derivatives can be formed. The yield of these anhydrous-sugars is affected by the source of biomass and by the experimental pyrolysis conditions. The increase in pyrolysis temperature reduces the concentration of levoglucosan, in contrast to other products because it stimulates breakdown of this molecule [110].

Mixtures of compounds as phenols, cresols and catechols (monomers and oligomers) are formed from the lignin [111, 112]. Phenolic compounds found in bio-oils are composed mainly of simple phenols with a hydroxyl and which may contain other substituent group on the benzene ring, forming mixed functions (carbonyl, carboxyl, alkyl or aryl radical) [108].

Due to the presence of acids, especially acetic and formic, bio-oil may show pH values in the range 2–4 [41, 113] that constitutes a problem, since it will affect storage conditions (equipment, transport) and its processing [113, 114].

The oxygen content in bio-oil is approximately 35–40% by weight. The specific composition depends mainly on the type of biomass used, the pyrolysis conditions (temperature, residence time and heating rate) and the storage conditions of bio-oil [10, 41, 113]. The high oxygen concentration results in a low energy density that is less than 50% of the value for conventional oils. It also affects the bio-oil miscibility with other petroleum fuels and the stability of bio-oil [41, 114].

The water constitutes 10–30% bio-oil, and their quantity depends on the original biomass, and the pyrolysis conditions, since the moisture is coming from biomass and also from dehydration reactions taking place during pyrolysis [41, 108, 113]. Depending on the feedstock and process conditions, the ratio of aqueous and oil phase can vary from 50:50 to 30:70, and the presence of two phases can hinder the application of bio-oil. This high water content may cause problems in the ignition engines by reducing the rate of vaporization of the oil, which prevents its application directly as fuel [115]. In many instances, drying the biomass prior to the pyrolysis is sufficient to reduce this problem.

The instability of bio-oil is mainly due to the presence of highly reactive organic compounds (ketones, aldehydes, organic acids), which can undergo reactions to form ethers, acetals or hemiacetals [41, 116]. This kind of reactions may increase the average molecular weight oil, water content and its viscosity, resulting in a low quality oil and that, when stored, results in phase separation. However, the addition of polar solvents such as methanol or acetone can significantly reduce the viscosity of the bio-oil [41].

The ash content of the bio-oil can also cause problems in some applications. The composition of the ash is dominated by alkali metals (potassium and sodium) that are responsible for severe corrosion and deposition turbines on heating surfaces during combustion [115, 117].

However, the crude bio-oil, before use as fuel, must be chemically modified through complex processes such as hydrodeoxygenation, hydrocracking, decarbonylation or decarboxylation to reduce oxygen content, which is the main unwanted constituent in the bio-oil for energy purposes [113]. Another alternative improving (upgrading) of the bio-oil is catalytic pyrolysis using zeolites alumina or metals as catalysts [114].

For the use and recovery of chemicals from bio-oil can be used conventional separation techniques, such as solvent extraction, column chromatography and distillation (single, fractionated, or under vacuum). Solvents commonly used for extraction of compounds of interest in bio-oil include water, alcohols, ethyl acetate, hydrocarbons such as toluene and mixtures thereof [108].

Fractionation in open or pressurized column has also been used as a pretreatment for the separation of compounds from bio-oil [118–120].

Among its uses, bio-oils are potential fuels to diesel engines, gas turbines and boilers. They can be used also as raw materials for obtaining hydrocarbons by catalytic conversion or hydrotreating [121]. Considering its phenolic fraction, bio-oil appears as a substituent for petrochemical phenol in the production of phenolic resins (phenol-formaldehyde) or can be used in pharmaceutical, food or fine chemicals industries [108, 122]. Furthermore, bio-oil may be fractionated to obtain many other products of commercial interest, such as abrasives, filter elements, battery separators, electric components, refractory materials, adhesives for wood, paints, varnishes, enamels, etc. [8, 123].

The reaction of bio-oil with ammonia, urea or other amino compounds produces amides and amines stable, nontoxic to plants and can be used as organic fertilizer. The bio-oils derived from wood residues can be commercially applied in smoking food [10]. In case of bio-oils with high concentrations of hydrolyzable sugars, it may be favorable to the production of ethanol by fermentation, whereas bio-oils with high phenol content are mentioned as attractive starting material for the production of adhesive [123].

As application examples of some of the most important bio-oil compounds can be mentioned: the levoglucosan (food additive, pharmaceutical industry); the levoglucosenone (synthesis of antibiotics and rare sugars); furfural (pharmaceuticals, pesticides); acetic acid (chemical industry); formic acid (wood preservatives, antibacterial agents); and hydroxyacetaldehyde (pharmaceuticals, fragrances) [124].

2.5.1. Aqueous phase of bio-oil

As mentioned above, the amount of aqueous phase in the bio-oil will depend on the original biomass composition, its original moisture and the pyrolysis conditions [108, 113]. However, it cannot be removed by conventional methods such as distillation [10]. The phase separation will occur when the amount of water exceeds the maximum level in bio-oil (usually above 30–45%) or by extraction methods [125].

The addition of water allows to readily separating the bio-oil into organic and aqueous phase. The aqueous phase contains mainly components of higher polarity, such as levoglucosan and other anhydrous-sugars, furan, furfural, organic acids of low molecular weight, hydroxyacetone, hydroxyacetaldehyde and guaiacol [126–128]. The separation of the aqueous phase of the bio-oil is also commonly performed using dichloromethane and sodium bicarbonate solution, obtaining an acid extract [129, 130]. Although solvent extraction is widely used in phase separation of bio-oil, it can affect the qualitative and quantitative composition of the extracted sample due to the different affinity of the solvent for each class of chemical compounds present in the sample [131].

The aqueous phase of the bio-oil cannot be directly discarded as wastewater, since some compounds may be above discharge limits. Different processes may be applied for the treatment of wastewater together with the pyrolysis process or subsequently in a wastewater treatment plant [132, 133].

Currently, several studies have been conducted for the treatment of aqueous phase and the application of its compounds as industrial raw material. Upgrading processes such as the hydrodeoxygenation and moderate catalytic cracking allow the production of hydrogen, hydrocarbons, alcohols and olefins from the aqueous phase [127, 130, 134].

Due to high amount of oxygenated compounds from C2 to C6 such as aldehydes, ketones, acids, and carbohydrates, several gasoline additives, alcohols and diols can be produced from the aqueous phase by hydrogenation processes [135, 136]. Light alkanes C1 to C6 and liquid alkanes C7 to C15 may also be produced from the aqueous phase carbohydrates through upgrading processes as dehydration and hydrogenation [137].

The sugars present in the aqueous phase (levoglucosan, pentoses and hexoses) are recognized as key compounds for the production of furan derivatives, such as furfural and 5-hydroxymethylfurfural [126]. The conversion of sugars into furan derivatives can improve the economic perspective for using the aqueous phase as source of raw material for a wide variety of chemicals [124, 126].

The acidity of the aqueous phase, as already mentioned [127], can cause corrosion in equipments (based on carbon steel); however, these organic acids may also be valuable byproducts that can be used in industry as solvents and wood preservatives [124, 127].

2.6. Analytical techniques applied to bio-oil and aqueous phase

The full chemical characterization of pyrolysis oils is very complex, mainly because they are formed by the degradation of carbohydrates and lignin, that have an abundant content of water and a great variety of organic functions associated a small amount of inorganic material [138].

The composition of bio-oils can be divided into four distinct fractions: moderately polar monomers, detectables by GC (40%); polar monomers directly detectables by HPLC or by GC after derivatization (12%); water (28%); lignin and pyrolytic materials (20%) [139] not detectables by GC.

Gas chromatography is the most widely used technique in chemical analysis of bio-oils and other complex mixtures. Despite its wide application, the GC [7, 107, 110, 111] has some limitations when applied to mixtures with a great variety of compounds and different concentration ranges. Co-elution is a major impediment for the separation and unambiguous identification of a compound. In this sense, there was developed the comprehensive two-dimensional gas chromatography (GC × GC) which is being applied to this kind of sample [8, 15, 119, 140], which greatly reduces the number of co-elutions in a chromatogram, through two dimensions separations.

Known since the 90s, the GC × GC is an analytical tool that differs from other techniques due to the sequential use of two chromatographic columns, which allows a significant increase in selectivity, favoring the structuring of the peaks in the chromatographic space. Regarding the dimensional gas chromatography, GC × GC shows most significant increase in the sensitivity and resolution, allowing a higher peak capacity, that is, a higher number of peaks separated and identified [141, 142].

Studies employing liquid chromatography for characterization of bio-oils have also been recently undertaken [143–149]. In view of the complexity of the samples, the comprehensive two-dimensional liquid chromatography (LC × LC) [150] in the same way as it happens for GC × GC becomes an important tool for characterizing bio-oils, because there is a large increase in the resolving power when compared to one-dimensional methods. Another great alternative is the use of micro HPLC columns that allow very small mobile phase flows, facilitating the coupling to more efficient detectors than conventional ones [151–153].

3. Gas chromatographic methods applied to the analysis of bio-oils

3.1. One-dimensional gas chromatography (1D-GC)

One-dimensional gas chromatography (1D-GC) coupled to mass spectrometry (GC/MS) or flame ionization detector (GC-FID) is an important analytical tool that provides chemical profile information of bio-oil, aiming its correct destination as fuel or in the chemical industry. There are several studies in the literature about the use of 1D-GC to chemical characterization of bio-oils from pyrolysis of various biomasses [12, 154, 155]. The large Brazilian biodiversity contributes to many options of biomasses, significantly increasing the total number of identified chemical compounds and the potential use of these materials. There are several studies using 1D-GC to analysis of bio-oil from Brazilian biomasses such as straw and sugarcane bagasse (*Saccharum officinarum*) [38, 70, 156], eucalyptus sawdust (*Eucalyptus globulus*) [70, 156], Amazon tucumã (*Astrocaryum aculeatum*) [157], mangaba seed (*Hancornia speciosa*) [39], coconut fiber (*Cocos nucifera*) [34], fruit of palm (*Arecaceae*) and pine wood *(Pinus)* [158], among others. A large part of this work has employed the GC/MS to identify families and major compounds, usually containing oxygen in its constitution, as phenols, furans, alcohols, ketones, aldehydes, esters and carboxylic acids, regardless of the biomass used. As an illustration of this type of analysis, **Figure 2** shows the chromatogram of a sample of sugarcane straw bio-oil using the GC/MS system with DB-5 column (60 m × 0.25 mm × 0.25 μm), developed in our laboratory. In this work, 208 and 336 compounds were detected in the analyzed bio-oil sample directly and by SPME with PDMS fiber, respectively. Among these, 33 and 35 compounds were identified in two cases.

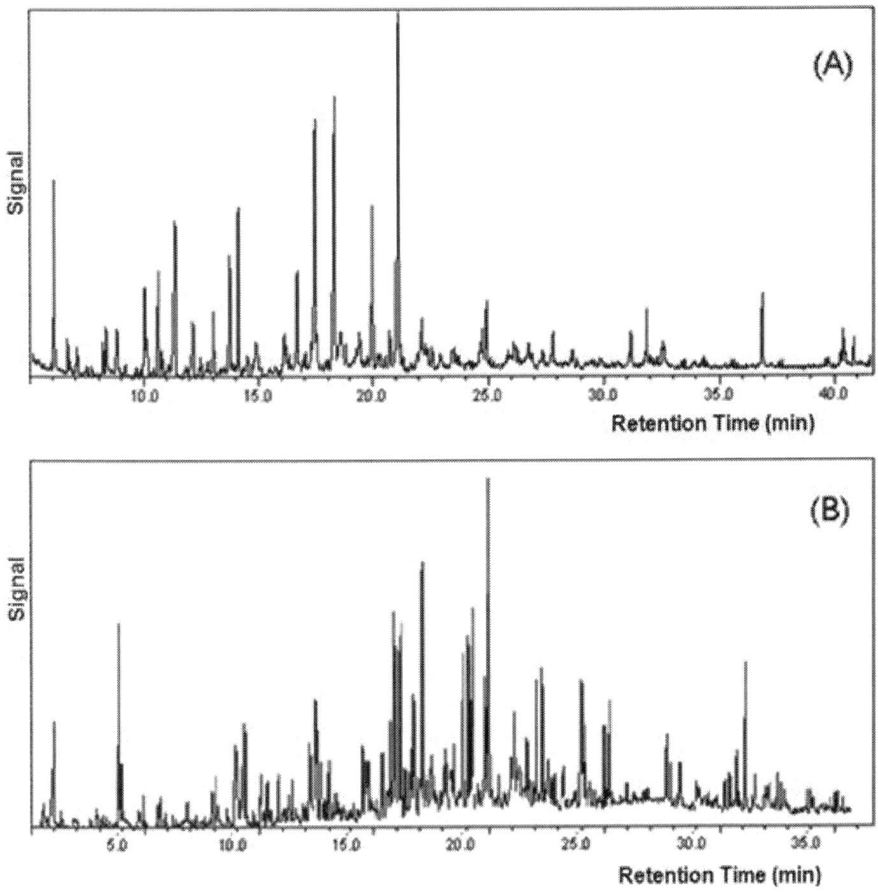

Figure 2. Total ion chromatogram (GC/MS) for bio-oil (A) and SPME (B) of the pyrolysis of straw sugarcane. Chromatographic Conditions: DB-5 column (60 m × 0.25 mm × 0.25 μm); oven temperature: initial oven temperature was 40°C, hold for 2 min, heating to 280°C at a 5°C/min, where it stayed for 2 min.

In general, phenolics compounds are the most family in the majority of bio-oils. Phenols are widely used in fine chemical industry, food processing, pharmaceutical and production of phenolics resins [34, 108, 158].

Hydrocarbons (saturated, unsaturated and aromatic) are also identified by GC/MS in most bio-oils; however, their percentage is very small compared to other components, with few exceptions. In the study of Santos et al. [39] were identified various hydrocarbons (saturated, unsaturated and aromatic) through GC/MS in the bio-oil from pyrolysis of mangaba seed (*Hancornia speciosa*). The total relative area of peaks which is one of the ways used to estimate the quantitative composition of bio-oil ranged from 8.1 to 13.8% of total compounds tentatively identified in this bio-oil. In this same work, carboxylic acids were the family of major compounds found (from 72.5 to 84.4%). Carboxylic acids, although lower quality becomes the bio-oil, are possible hydrocarbon precursors, enabling further study of the bio-oil upgrading.

Patel et al. [54] characterized by GC/MS the crude bio-oil from sugarcane bagasse produced by fast pyrolysis in a fluidized bed and evaluated the efficiency of Mo_2C/Al_2O_3 catalyst in the deoxygenation and quality of bio-oil. The catalyst resulted in an increase in the content of phenolics and furans, which arouse great industrial interest.

Spilã et al. [159] developed a method for separation and analysis of the aqueous fraction of bio-oil by adding water to raw bio-oil followed by extraction with ion exchange resin and ethyl ether. The ether soluble compounds were analyzed by GC/MS, since the insoluble fraction was evaporated and solubilized in methanol for analysis (GC/MS, CHN and pyrolysis-GC/MS). The characterization method was applied to bio-oils derived from wood, scots pine and wheat straw. Further studies have applied the same separation method to different biomass such as forest waste [160] and wood [161] with subsequent chromatographic analysis (GC/MS and GC-FID).

Wiggers et al. [162] performed a study on pyrolysis of soybean oil at pilot scale, in continuous system aiming higher yield of bio-oil. The authors conducted a prior distillation to separate light bio-oil fractions (LBO) and heavy bio-oil fractions (HBO). Various hydrocarbons were found in bio-oil and benzene, toluene, ethylbenzene, p-xylene, o-xylene and linear hydrocarbons (C_7 to C_{12}) being the majority for LBO; while for HBO the majority were fatty acids, toluene, ethylbenzene and linear hydrocarbons (C_8 to C_{17}).

Owing to the industrial importance of phenols and furfural, the author highlights the refining of the aqueous phase can extract these aromatic compounds of high value to the industry.

GC/MS can be also coupled directly to one pyrolyzer system since it is required to just check the potential of the chosen biomass to generate bio-oil, since the amount of sample used in this case is very short not allowing further analysis with the bio-oil produced. This technique is known as analytic pyrolysis and represented by Py-GC/MS.

3.2. Pyrolysis-gas chromatography/mass spectrometry (Py-GC/MS)

Analytical pyrolysis (Py-GC/MS) is an important characterization tool of the products generated by the pyrolysis of lignocellulosic material, in addition to allowing a better understanding of the function and effects of catalysts for the production of hydrocarbons or other desirable compounds prior to the pyrolysis in analytical laboratory scale. Direct analysis of condensable gases is held in a pyrolyzer coupled to a gas chromatograph with mass spectrometry detection (Py-GC/MS). In the pyrolyzer, a small amount of biomass is subjected to a heat treatment and condensable gases are simultaneously separated and identified by their mass spectra and retention times [163–165].

The Py-GC/MS have advantages, such as the use of small quantities of biomass, which allows the realization of several pyrolysis under various operating conditions (ratio catalyst/biomass, temperature, heating rate, etc.), in a simple and fast way [166, 167]. Through Py-GC/MS can also predict what kinds of industrial interest compounds are produced during the process, which the effect of catalysts, and if inhibition occurs, etc.

Liaw et al. [165] studied the optimal pyrolysis temperature for two standards cellulose and two different types of wood by Py-GC/MS, and yields variations were evaluated by PCA (principal component analysis). The PCA showed that pyrolysis products can be divided into groups and are strongly influenced by the nature of the raw material studied. However, the levoglucosan represents those few compounds that are influenced by temperature because as the temperature increases, its concentration decreases.

Barbosa et al. [168] developed method of determining the ratio between syringyl and guaiacyl content (S/G) in lignins from eucalyptus wood by Py-GC/MS using different markers of syringyl and guaiacyl units. Traditional methods, which involve chemical degradation, generally are laborious, time consuming and require large sample amounts. However, the method developed by the authors is fast, uses very small amount of sample and highly sensitive.

The analysis of macauba fruit (*Acrocomia sclerocarpa M.*) [169] by Py-GC/MS revealed the formation of alkanes, alkenes, dienes and cycloalkanes, being the main products the 2-propenal and acrolein (triglycerides derived). In the bio-oil from the oil pulp, most of the compounds found (71–79%) were aldehydes, cycloalkanes, alkenes and dienes. Through this method, it was possible to observe the differences in composition of biomass and its influence on the quality of bio-oil formed.

Py-GC/MS enables understanding of formation of fatty acids obtained by pyrolysis of seed and babassu oil, studying their composition and fragmentation mechanisms of the compounds produced by thermal degradation of the samples [170].

The evaluation of the use of different catalysts (ZnO, CaO, $ZnCl_2$ and $MgCl_2$) to obtain bio-oil from several biomasses [171–173] was also assessed by Py-GC/MS.

Despite numerous studies reported in the literature, 1D-GC has limitations when applied to the characterization of complex sample, such as low resolution and co-elution, which lead to incorrect identification of some analytes [112, 174, 175]. A wide range of chemical classes present in the matrix of bio-oil also complicates the choice of chromatographic columns proper polarity at all [165]. Therefore, currently, most jobs that show a characterization of bio-oil by one-dimensional chromatography also make use of the characterization by chromatography comprehensive two-dimensional gas (GC × GC).

3.3. Comprehensive two-dimensional gas chromatography (GC × GC)

The comprehensive two-dimensional gas chromatography (GC × GC) is a multi-dimensional chromatographic technique that emerged in the early 90s [176]. Due to its high separation power [177], GC × GC has been widely used use in the analyzes of pyrolytic bio-oils, once these are complex mixtures and their composition can present more than 400 organic compounds belonging to a great number of distinct chemical classes. The bio-oil composition may vary according to the pyrolysis conditions and the kind of biomass chosen [178, 179].

This technique provides many advantages in relation to 1D-GC in the elucidation of the composition of complex samples [177]. Among these advantages one can be highlighted the increase in peak capacity, which leads to a better separation, not only between analytes, but also between them and the matrix. An increase in detectability, due to the narrow chromatographic bands resulting from the modulation, may also be considered another advantage.

Furthermore, the technique GC × GC compared to conventional gas chromatography also provides an increase in sensitivity and a generation of structured diagrams that facilitate the identification of unknown compounds [180–183].

The two-dimensional system consists of two chromatographic columns connected in series, one with a standard size (normally 30 or 60 m length and 0.25 mm internal diameter) and other shorter and with smaller diameter (~2 m length and 0.15 mm i.d.). A column set commonly referred as conventional consists of a nonpolar column in the first dimension (^{1}D) and a polar or with intermediate polarity in the second dimension (^{2}D). The two columns have different separation mechanisms (orthogonal separation), that is, the first column performs separation of compounds according to molecular weight or boiling point and second column by polarity, providing a major breakthrough in the separation of complex samples [141, 177, 184–188].

Figure 3 shows a GC × GC system where is possible to observe the modulator between both columns and some details of the peak processing. The modulators have the main function of to continuously collect fractions from the first column, re-concentered them and to inject in the second one [190–192]. This procedure is responsible for the increase in the signal-to-noise and the decreasing in detection limits, if compared to 1D-GC [188].

In the graphical representation of GC × GC, the register of the detector signal versus retention time is a continuous sequence of short chromatograms of each eluted fraction from the second column. After this, tridimensional graphics can be constructed considering the detector signal and the retention times in the first and second dimensions (^1t_R and ^2t_R) [191, 192].

Data acquisition from 2D-GC can be better explained viewing **Figure 3b**. In this figure, one co-eluted peak corresponding to three analytes) eluted from the first dimension pass through the modulator being fractionated and eluted from the second column. The crude chromatogram originated corresponds to the sum of all the chromatograms obtained on ^{2}D. The second step (**Figure 2c** is the transformation (by and adequate software) of these data in a two-dimensional diagram ($^1t_R \times {}^2t_R$). The last step is the tri-dimensional visualization of the results (**Figure 2c** that was also performed by adequate software [182].

There are two main kinds of commercial modulators: thermal and with valves [193–196]. The thermal modulators (as by heating or by freezing) present more utilization in GC × GC [193, 195]. The cryogenic modulators can use liquid nitrogen or liquid CO_2 and act as a cold trap for the analytes [196].

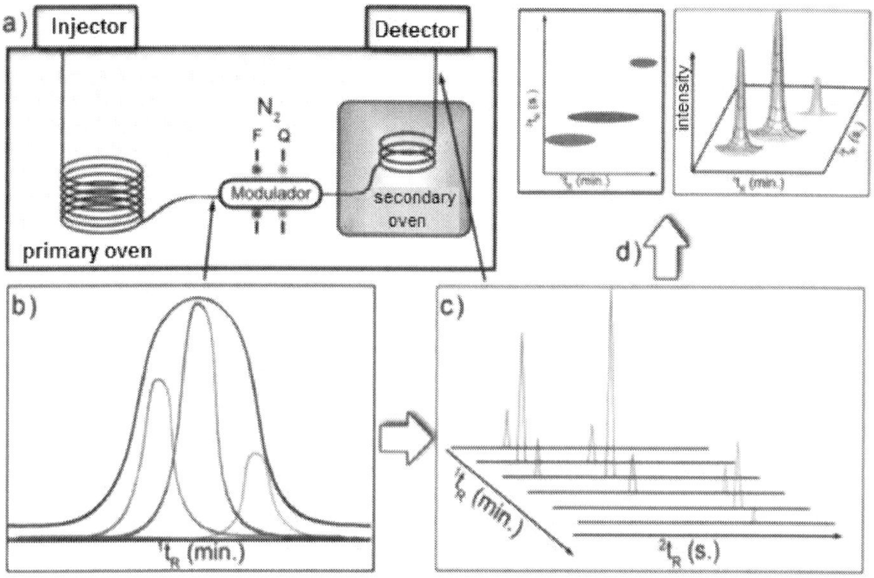

Figure 3. (a) GC × GC system; (b) three peaks co-eluted in the first dimension (^{1}D); (c) transformation of raw data into a two-dimensional chromatogram ($^{1}t_R \times {}^{2}t_R$); (d) reconstruction process of chromatographic peaks forming the color diagrams (two-dimensional and three-dimensional) [189].

GC × GC allows the utilization of several detectors, with few modifications to adapt them to the low volume and high velocity of data processing. One can be highlight the flame ionization detector (FID), time-of-flight mass spectrometry (TOF-MS) and quadrupole mass spectrometry (qMS) as the more used [197–203].

FID has been associated to GC × GC since the first works found in the literature, including the initial researches in the bio-oil characterization [204–207].

GC × GC promotes high speed separations, providing very narrow peaks, requiring detectors with equally rapid acquisition rates to get a sufficient number of points per peak [208, 209] to permit quantitation. The detector scan should be short and its internal volume has to be small [194]. The high acquisition rate of the FID (up to 200 Hz) [210] allied to a good response for almost all organic compounds and its good performance in quantitative analysis justifies the wide use of this detector in two-dimensional chromatographic analysis. The main difficulty in the FID employment is that it does not provide structural information about the separate peaks. Thus, detectors with mass analyzers gain space for identification and confirmation of the separate compounds.

The TOFMS is particularly effective for GC × GC, since it presents acquisition rates between 50 and 500 Hz [180, 182, 197–203], making it the preferred choice among researchers for these studies. The coupling of separation GC × GC with efficient detection TOFMS has an additional advantage, which is its higher sensitivity than the full scan mode over conventional mass spectrometry detectors with quadrupole analyzer (qMS). Consequently, TOFMS outperforms the other detectors in qualitative and quantitative analysis. The disadvantages of systems of this type analyzer are its relatively high cost, the need for proper training and specific operating conditions for daily operation, especially due to its high sensitivity [191].

The GC × GC-qMS system is also showing its potential to analyze complex samples [211–213]. Initially, quadrupole mass spectrometers showed very low data acquisition rates (up to 20 Hz) which made them too slow to use in GC × GC systems [214]. However, decreasing the mass range investigated or monitoring only a few ions during the run, it is possible to obtain a higher scan rate [191]. This has been the subject of research of several research groups [180, 208, 214, 215], that is, to develop quadrupole mass analyzers faster and comparable to TOFMS. From this, it was introduced on the market a fast quadrupole system, which allows achieving data acquisition rates above 50 Hz allowing their use coupled to GC × GC system [209, 216].

In recent years, it is possible to find a lot of research in the literature applying GC × GC in the analysis of the chemical composition of bio-oils.

Among the initial studies of bio-oils through GC × GC, it can be highlighted the Works of Marsman [205–207] and Sfetsas [174]. Marsman et al. [206, 207] evaluated the compounds presented in the bio-oil from beech using GC × GC with FID and TOFMS detectors. Authors used GC × GC-FID and GC × GC/TOFMS for the identification of approximately 248 and 368 compounds, respectively, with concentration higher than 0.3% in beech (*Fagussylvatica*) hydrodesoxygenated (HDO). In these studies, it was also made a classification for these compounds, according to their chemical class into nine groups (acids, aldehydes, ketones, furans, guaiacols, syringols, sugars, alkyl phenols, alkyl-benzenes). The major compounds found in beech bio-oil were levoglucosan, hydroxymethyl furfural, furanone, furfural, mequinol and butanediol. Similarly, Sfetsas et al. [174] used the GC × GC technique for analyzing three oils pyrolysis, in which were tentatively identified, approximately 96 compounds with concentration higher than 0.3%, classified in acids, esters, aldehydes and ketones, hydrocarbons, aromatic hydrocarbons, phenols, sugars and other compounds not classified. Acetic acid, levoglucosan and hydroxy-propanone were the majority compounds.

Bio-oils derived from the fast pyrolysis of several Brazilian residual biomasses as orange bagasse [217], peach core [15, 189], rice husk [15] and sugarcane straw [8] were recently characterized by GC × GC. The analysis of bio-oil from orange bagasse [15], without any pretreatment, by GC × GC/FID and GC × GC/TOFMS was compared. The last one showed better results and 167 compounds were identified, belonging to acids, aldehydes, ketones, phenols, esters, ethers and some nitrogen-compounds. From these, 26 compounds appeared in concentration above 1%.

Bio-oils from peach core and rice husk analyzed by GC × GC/TOFMS in a conventional set of columns showed the presence of ketones, phenols, alcohols, ethers, acids, aldehydes sugar derivatives and hydrocarbons, with around 500 peaks in each sample [15].

Studying the bio-oil from peach core, by comparison with 1D-GC/qMS and GC × GC/TOFMS, Migliorini et al. [189] observed the superiority of the multidimensional technique. Another observation was the spatial structuration of the GC × GC color diagram, which allowed the identification of all the isomers of C1 to C4 alkyl phenols.

This group of researchers also applies another tool for the identification: the dispersion graphics (DG). These graphics, constructed using Excel™ software clarifying the distribution of compounds and allow to preview the presence of others homologues in a series of compounds, like alkyl substitutes in phenols or in aromatic hydrocarbons. **Figure 4** shows examples of DG for the separation of alkyl phenols. The results from GC/MS and GC × GC/TOFMS showed, for the same sample, 51 and 220 components, respectively. The chemical classes found were alcohols, aldehydes, anhydrides, ketones, esters, ethers and phenols and were observed by both techniques employed. However, using GC × GC were also found carboxylic acids, hydrocarbons and sugar derivatives such as levoglucosan [189].

The analysis of the aqueous extract of peach core pyrolysis is illustrated in **Figure 5**. This figure has been an example of the spatial distribution of the constituents in a GC × GC/TOFMS (**Figure 5A**) and an illustration of separation capacity offered by the second dimension (**Figure 5B**). In this last one, it is observed a separation of four peaks that co-eluted in the first dimension and is adequately separated in the second one, giving mass spectra of high purity.

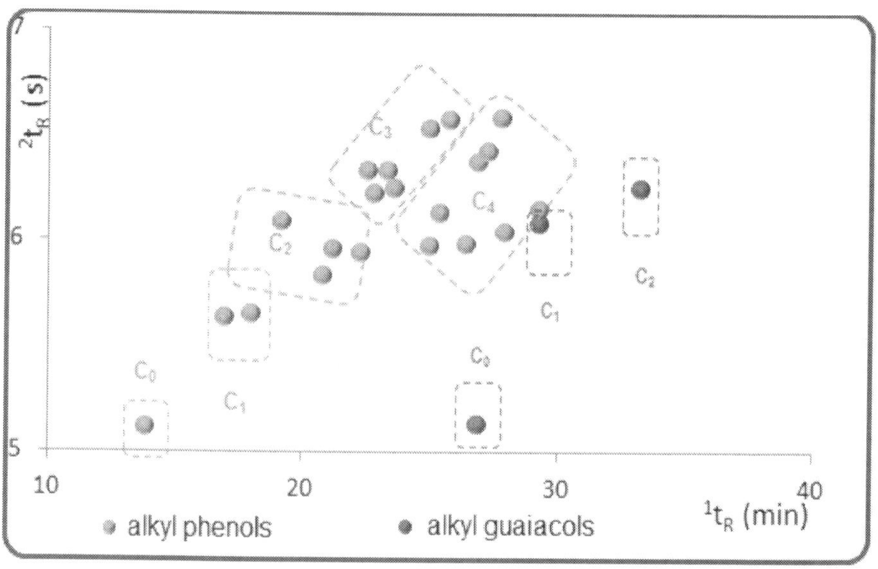

Figure 4. Dispersion graphic of the separation of phenols in the bio-oil from pyrolysis of peach core. Legend: C_x represent the side alkyl chain on the aromatic main chain of the phenols were x is equal to the number of carbon in the side chain. Based on Ref. [91].

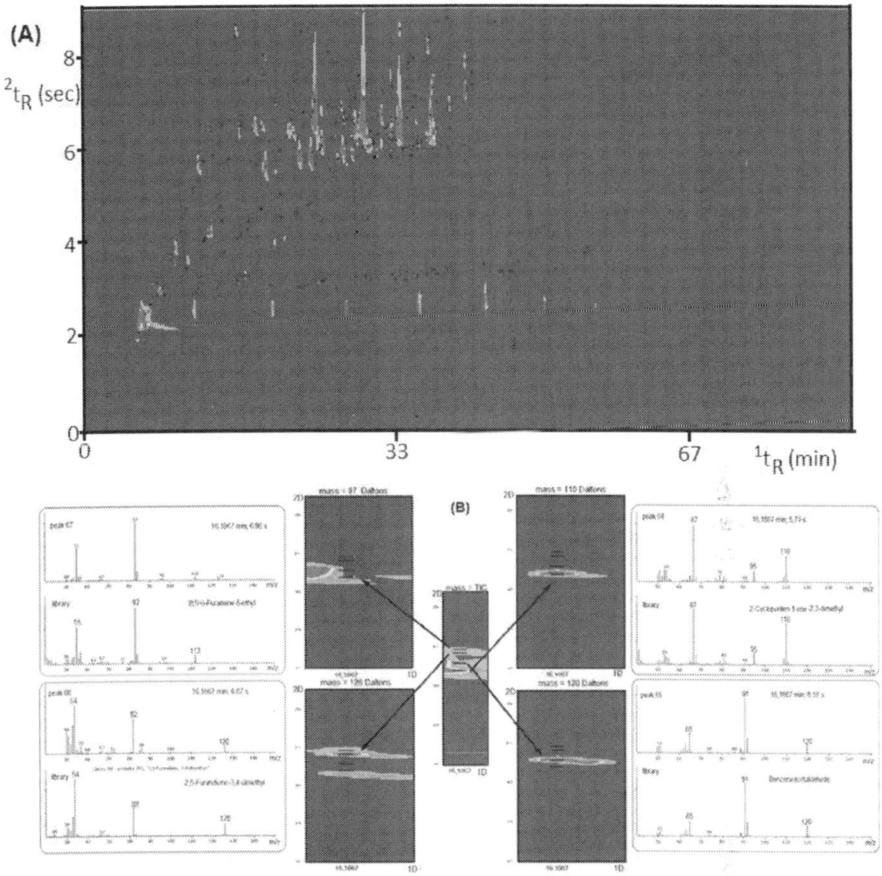

Figure 5. (A) Color two-dimensional diagram for analysis by GC × GC/TOFMS of the organic extract of aqueous fraction of the pyrolysis of peach core. (B) Example of chromatographic separation by GC × GC using the total ion current chromatogram (TIC) and the extracted ion chromatogram (EIC) for the following compounds (a) C2-cyclopentenone, (b) benzene acetaldehyde, (c) ethyl furandione and (d) ethyl furanone and their respective mass spectra. Chromatographic conditions: column set: DB5 (60 m × 0.25 mm × 0.25 μm) and DB17 (1.94 m × 0.18 mm × 0.18 μm); heating ramp: 40.0°C (5.0 min)–3°C/min–270°C.

In the study of fast pyrolysis of sugarcane straw, Moraes and coworkers [8] used GC × GC/TOFMS, infrared spectroscopy with Fourier transform (FTIR) and scanning electron microscopy (SEM) to fully characterizing the products from the pyrolysis of sugarcane straw. The results of the analysis of bio-oils have demonstrated efficiency in the combination of techniques, especially, GC × GC/TOFMS, showed the presence of 123 compounds belonging, mostly, to the aldehydes and carboxylic acids. Maciel et al. [218] also studied the fast pyrolysis of sugarcane straw by GC × GC/TOFMS, but researching the aqueous phase of this process. They found that this phase is very similar to the bio-oil but enriched more soluble phenols, such as *ortho, meta* and *para* cresols.

The technique using GC × GC with detector quadrupole mass detector (qMS) is growing and showing the efficiency of this detector coupled to two-dimensional gas chromatography. In work carried out by Da Cunha et al. [119] and Schneider et al. [219], using a set of conventional speakers have shown the potential of the technique to evaluate the straw pyrolysis product of sugar cane and forest wood sawdust (lignocel). The bio-oil from pyrolysis of sugarcane straw was fractionated on a silica column with pressurized liquid, being separated hydrocarbons of other polar fractions. In this sample, 166 compounds was to find including carboxylic acids, aldehydes, ketones, esters, phenols, ethers, alcohols and sugar derivatives in the polar fraction, and, the nonpolar fraction, were formed from aromatic, aliphatic, cyclic and olefinic hydrocarbons [119]. The polar compounds of the bio-oil from lignocel sample (forestry wood sawdust) [219] were extracted with alkaline solution before chromatographic analysis, and 130 compounds were identified by GC × GC/qMS among phenols, ethers, ketones, aldehydes, carboxylic acids, alcohols and aromatic hydrocarbons. This analysis is illustrated in **Figure 6** demonstrating the quality of data generated using a mass spectrometry detector with quadrupole analyzer [119].

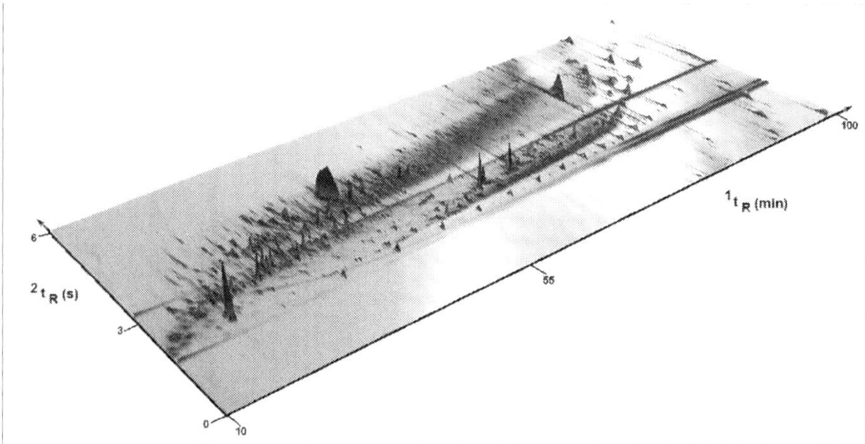

Figure 6. Color diagram (GC × GC/qMS) of the bio-oil from sugarcane straw. Chromatographic conditions: set of columns: OV–5 (60 m × 0.25 mm × 0.25 μm) and DB-17ms (2.15 m × 0.18 mm × 0.18 μm). Heating ramp: 40°C–3°C/min to 120°C, then 2°C/min to 200°C [119].

In recent years, the literature has been publishing a wide range of papers in the pyrolysis of many biomasses with lignocellulosic or residual origin. Eucalyptus [220], mango seed [35], coconut fiber [221], residual cake of crambe seed [37] and castor seed [222], waste of forest industry [36], fruits of palm and pine wood chips [223], sugarcane bagasse [61] sugarcane straw [119] and forest wood sawdust (lignocel) [219] were some of these. These biomasses were submitted to different kinds of pyrolysis (slow [220, 222], fast [35, 37, 177, 221, 223], intermediate [36] and catalyst [34, 218]).

Many other analytical methods can be used to improve the quality of bio-oil and to facilitate its characterization. In this context, extraction methods have been used to isolated fractions and better analyze their components. Some examples of application of extraction techniques are solid phase micro extraction (SPME) [220], liquid-liquid extraction (LLE) [35, 219, 221], mechanical press extraction [37], soxhlet extraction [37], pressurized fluid extraction (PFE) [37] and pressurized liquid fractionation (PLF) [119] among others.

The use of another analytical method beside GC or GC × GC is also a good choice for a more complete characterization of these products. According to reports in the literature GC × GC, especially with TOFMS detector, has been widely used in association with other techniques like GC/qMS [36, 221], FTIR [35], FT-ICR MS [223] and 1H NMR [46] for determining the composition of the biomass pyrolysis products.

The results of the characterization of bio-oil only by GC × GC and the association thereof with other techniques show that different classes of chemical compounds form these samples. The differences in the pyrolysis conditions [14, 224] or in the chemical composition of the biomass can influence the constitution of the composition of bio-oils [178]. The main compounds belonging to the chemical class of phenols, esters, ethers, acids, aldehydes, ketones, alcohols, hydrocarbons (saturates, olefinics and aromatics), sugar derivatives, N-compounds (nitriles, anilines, quinolines, pyridines, indoles, pyrazines, pyrroles, carbazoles and acridines) and sulfur compounds (disulfides and thiophenes) [35–37, 46, 119, 218–223].

As with any other chemical analysis, no single technique is sufficient for complete identification of bio-oil samples. However, the GC × GC has demonstrated its high potential for use combined with other techniques such as infrared spectroscopy, elemental analysis and liquid chromatography with the use of mass detectors among others. The development of rapid chromatographic processes (columns in micro scale) and multidimensional systems (especially comprehensive) allow full characterization of the samples for both constituents in greater proportion as for those at trace levels.

4. Liquid chromatographic methods applied to the analysis of bio-oils

Liquid chromatography (LC) techniques are important tools for the separation and identification of compounds present in bio-oil fractions that are not analyzable by gas chromatography (GC). The high performance liquid chromatography (HPLC) analysis is widely used in different types of samples mainly polar and thermally labile compounds [155]. The main advantage of the LC techniques is the possibility of direct injection of aqueous phases obtained without extraction and sample preparation step. Studies employing such techniques for the characterization of bio-oils and aqueous phases have been recently reported in the literature and are briefly summarized in this chapter.

Similarly as for the GC, the development of LC has been, especially, toward miniaturization and improved chromatographic resolution. Then, one can classify the liquid chromatographic methods in one-dimensional LC and two-dimensional LC.

4.1. One-dimension high performance liquid chromatography

The system can use many detectors, but, the main used for pyrolysis products are UV, RID and MS.

HPLC-UV uses a simple UV detector or diode array detectors (DAD detector or more specifically HPLC PDA detector) especially in the determination of polar compounds containing carbonyl, carboxyl and hydroxyl groups in aqueous samples derived from pyrolysis. The identification of aldehydes in biomass derivatives is the main application of this technique because these compounds are very soluble in water and the recovery of their extraction using conventional techniques such as liquid-liquid extraction or solid phase extraction is very low.

Successful applications of HPLC-UV in the identification of furfural and hydroxymethylfurfural (HMF) in the aqueous phase of the bio-oil obtained by pyrolysis of agroindustrial biomasses have been described recently [126, 143, 225]. In some cases [225], aldehydes were confirmed and quantified in bio-oil using a GC/MS system.

The HPLC-RID is considered standard for analysis of sugars in aqueous samples. The RID detector is used for detection of compounds that do not absorb in the UV or visible because it is based on measurement of the difference in refractive index between the pure mobile phase and the eluent coming out of the column containing the sample components. As some biomasses (like sugarcane bagasse and sugarcane straw) contain a great amount of sugars, this technique can be important in the characterization of compounds formed during the pyrolysis of these biomasses.

The HPLC-RID has been used, normally, as a complementary technique The bio-oil obtained from red oak fast pyrolysis in a fluidized bed reactor was characterized and sugar derivatives were identified in the water-soluble fraction [148]. In this work, levoglucosan, maltosan celobiosan, xylose and cellobiose were quantified. Johnston and Brown [147] also analyzed glucose and xylose in switchgrass bio-oil samples using HPLC-RID.

HPLC-MS is a technique for the analysis of polar or thermolabile fractions not analyzable by GC on samples of bio-oil and water fraction derived from pyrolysis of biomass. The main difference between the MS in a GC and the MS in a LC system is that LC-MS performs the ionization of analytes in atmospheric pressure (API) with a low-energy fragmentation ("soft"), allowing identification of the molecular ion. The fragmentation can only be done for selected ions and there is no library for identification of compounds. The analyzer must use standard compounds and has to study the entire fragmentation pattern for each peak. HPLC-MS technique is compatible for volatiles and nonvolatiles in a wide range of polarity [226, 227].

On Py-HPLC-MS (online system of pyrolysis and liquid chromatography-mass spectrometry) was recently developed [228] and applied to the analysis of lignin isolated from forest waste sample. In MS, compounds were detected with molecular masses in the range of 250–500 Daltons. The major compounds identified were syringol and resorcinol derivatives.

A method based on the pyrolysis online with HPLC-UV was developed for analyzing the bio-oil derived from polymers [229]. The bio-oil was fractionated and the resulting fractions were analyzed by mass spectrometry ionization and laser desorption matrix assisted (MALDI-MS) system and finally HPLC-MS using electrospray ionization (ESI). This system is mainly appropriate for the measurements of oligomeric products, being tested in polymer samples forming less volatile pyrolyzates such as poly(butylene terephthalate) and poly(2, 6-dimethyl-1, 4-phenylene ether). Another utility of this method was in the characterization of the cross-linking sequences in some polymeric resins.

HPLC-MS technique in combination with GC-FID and GC/MS was used for characterization of bio-oils from different forest residues [149], showing a wide range of compounds with masses of between 100 and 400 Da. The major compounds identified in all bio-oils were cyclohexane carboxylic acid, 1, 2, 4-trimethoxy benzene and 2, 6-dimethyl phenol.

4.2. Two-dimensional LC (heart-cutting (LC-LC)) and comprehensive two-dimensional liquid chromatography (LC × LC)

Two-dimensional liquid chromatography techniques involve two distinct separations, which can be classified as heart-cutting (LC-LC) or comprehensive (LC × LC). In heart-cutting 2D-LC, only relevant parts of the effluent, containing the target compounds, are directed to the second dimension. The main applications of LC-LC are the analytes purification, improvement of the separation efficiency and the sensitivity of analysis [150]. Bio-oil obtained from pyrolysis of pine sawdust was analyzed by LC-LC after the gel permeation chromatography (which made a lean up of the high molecular weight lignin derivatives) allowing the separation of phenolic fraction [144]. The results were compared with earlier analyzes by GC/MS. Among the phenols quantified in this work, the major compounds were guaiacol, vanillin, o-cresol e catechol. LC-LC technique proved to be a faster analysis, with a minimal sample preparation and with less loss of analytes than GC/MS.

Similarly to GC, LC techniques with a higher resolution power are also required due to the complexity of the bio oil and aqueous phase samples, and therefore the comprehensive two-dimensional liquid chromatography (LC × LC) becomes also important for the characterization of this kind of sample. This technique involves the coupling of two independent mechanisms of separation, through a high-pressure switching valve, and it is able to provide a complete separation of the whole sample, since all fractions eluting from the first dimension are subjected to a second separation [150, 230–239]. The use of LC × LC solves co-eluting problems due to increased peak capacity, which results in a larger number of identified compounds. This technique represents a great improvement in the analysis of organic samples mainly due to enhanced of the separation power and the resolution. Carr and Stoll [239] wrote an excellent chapter, edited by Agilent, with theory, instrumental and applications of 2D-LC.

Le Masle et al. [151] used LC × LC with detection by photodiode array (PDA) for the separation of compounds from the aqueous phase formed during the pyrolysis of oak. Using a standard solution with 38 compounds (phenols, acids, ketones, aldehydes, alcohols and furans), the authors developed a separation method, evaluating the peak capacity and the orthogonality of different sets of columns. However, the compounds in the aqueous phase samples have not been identified, since a more informative detector as MS would be required for this. This work was after compared with the LC × SFC technique (on-line comprehensive liquid chromatography × supercritical fluid chromatography) for the analysis of the aqueous phase samples by the same researchers, with the aim of evaluating the two-dimensional system. The new method showed a larger peak capacity in comparison with the previous method [152].

Tomasini et al. [153] described a method for the characterization of aqueous phases from bio-oils of coconut fiber, sugarcane straw and sugarcane bagasse using comprehensive two-dimensional liquid chromatography with detection by diode array followed by mass spectrometry with atmospheric pressure chemical ionization. Using this system, it was identified 26 compounds belonging to the classes of phenols, ketones, furans and alcohols. Phenol and 2-hydroxy-3-methyl-2-cyclopenten-1-one were found in greater abundance for all samples. Belonging to furans, the furfural was detected in higher concentrations in the aqueous phases of coconut fiber and sugarcane bagasse and the 1-(2-furanyl)-ethanone was detected in higher concentrations in the aqueous phase of the sugarcane straw. Belonging to alcohols, the phenyl propanol was detected in higher concentrations in the aqueous phase of coconut fiber, while for samples of sugarcane (straw and bagasse), the coniferyl alcohol had the highest concentration.

4.3. NanoLC

Among the innovations that LC has been showing in last years, it stands out techniques that substantially reduce the volume of solvent employed [240, 241]. The NanoLC is one of these innovations. It consists in the use of a column with micro-dimensions and low solvent flow, producing separations in short time but with high performance [242]. As the number of columns manufactured for LC techniques using micro to nano flows is very small (compared to conventional HPLC), there is a limited number of studies described in the literature on this subject [242, 243]. However, due to the positive results obtained, the NanoLC has been successfully applied in many fields such as biomedical, pharmaceutical, agrochemical and food [244].

The low flow used due to the nano-dimensions allows better coupling to mass detectors, including those with electron impact ionization, normally used for GC [153, 244]. In the case of coupling a NanoLC with a mass spectrometer by electron impact (EI-MS), there is the advantage of compounds identification be performed by direct comparison with mass spectral libraries [153, 244, 245].

As a complementary part to the analysis by two-dimensional liquid chromatography of aqueous phases obtained from pyrolysis from Brazilian biomasses, Tomasini et al. [153] applied the NanoLC-EI-MS to confirm the identification of compounds through the mass available libraries in the same samples before cited. The analysis showed a similar composition, with compounds belonging to the classes of ketones, phenols, and furans. The aqueous phase (AP) from coconut fiber presented the phenol as major compound, while the AP from sugarcane bagasse presented a higher amount of furfural and AP from sugarcane straw presented a lower concentration of almost all the compounds.

The use of a mass spectrometry detector with ionization by electron impact (EI-MS) shows an enrichment of information for characterization of the samples, since the obtained mass spectra can be identified by comparison to spectra available in libraries of software used. In addition to the advantages presented by the detection of NanoLC system, it is important to emphasize the "green chemistry" approach due to the reduced volume of solvent used in the analysis of this technique.

In general, the LC is necessary for the characterization of polar compounds remain in the aqueous phase, which normally would be discarded. This discard would cause environmental and economic damage, since the identified compounds can have some kind of industrial application. Furthermore, the injection of aqueous samples is not possible without a prior step of extraction, which may have different yields due to the presence of compounds belonging to different chemical classes or can result in contamination of the sample, besides the higher use of materials and longer time.

Although there is a still reduced number of works applying LC for the analysis of bio-oil compared to GC, it should be considered the importance of using new analytical techniques for a complete characterization of bio-oils and aqueous phases from pyrolysis of biomass.

5. Conclusions

The initially suggested use for bio-oil was as an alternative biofuel to diesel and petroleum. However, this route has proved to be nonviable, due to the high oxygen content of bio-oils and operating cost of deoxygenation. Currently, the major studies indicate the use of bio-oil in the chemical industry, particularly the chemistry of phenols, furfural and levoglucosan. The bio-oil can be considered an alternative for crude oil for fine chemical industry.

For proper use of bio-oil in the chemical industry, it is essential the identification and unambiguous determination of its major constituents. Only then, it is possible to propose a recovery route of some of these components for the development of an industry dedicated to a bio-refinery. For this, chromatographic methods, especially GC × GC/MS, are fundamental because they allow analysis with high sensitivity and accuracy in identifying each constituent of the bio-oil.

Acknowledgements

Authors would like to thank to CNPq, CAPES, FINEP and Petrobras for the financial support to this work through scholarships, grants and direct funding.

Author details

Maria Silvana A. Moraes[1,2], Débora Tomasini[1], Juliana M. da Silva[1,3], Maria Elisabete Machado[1,4], Laíza C. Krause[4,5], Claudia A. Zini[1,5], Rosângela A. Jacques[1,5] and Elina B. Caramão[1,4,5*]

*Address all correspondence to: elina@ufrgs.br

1 University of Rio Grande do Sul (UFRGS), Porto Alegre, Rio Grande do Sul, Brazil

2 Pampa University, Rio Grande do Sul, Brazil

3 Liberato Salzano Vieira da Cunha, Federal Technical School, Novo Hamburgo, Rio Grande do Sul, Brazil

4 Tiradentes University, Aracaju, Sergipe, Brazil

5 INCT-E&A, Brazil

References

[1] Mythili R, Venkatachalam P, Subramanian P, Uma D. Characterization of bioresidues for bio-oil production through pyrolysis. Bioresour Technol. 2013, **138**: 71–78.

[2] Thangalazhy-Gopakumar S, Adhikari S, Gupta RB, Fernando SD. Influence of pyrolysis operating conditions on bio-oil components: a microscale study in a pyroprobe. Energy Fuel. 2011, **25**: 1191–1199.

[3] Nzihou A, Flamant G, Stanmore B. Synthetic fuels from biomass using concentrated solar energy—a review. Energy. 2012, **42**: 121–131.

[4] Akhtar J, Amin NAS. A review on process conditions for optimum bio-oil yield in hydrothermal liquefaction of biomass. Renew Sust Energy Rev. 2011, **15**: 1615–1624.

[5] Van de Velden M, Baeyens J, Brems A, Janssens B, Dewil R. Fundamentals, kinetics and endothermicity of the biomass pyrolysis reaction. Renew Energy. 2010, **35**: 232–242.

[6] Bridgwater AV, Peacocke GVC. Fast pyrolysis processes for biomass. Renew Sust Energy Rev. 2000, **4**: 1–73.

[7] Cao JP, Xiao XB, Zhang SY, Zhao XY, Sato K, Ogawa Y, Wei XY, Takarada T. Preparation and characterization of bio-oils from internally circulating fluidized-bed pyrolysis of municipal, livestock, and wood waste. Bioresour Technol. 2011, **102**: 2009–2015.

[8] Moraes MSA, Georges F, Almeida SR, Damasceno FC, Maciel GPS, Zini CA, Jacques RA, Caramão EB. Analysis of products from pyrolysis of Brazilian sugar cane straw. Fuel Process Technol. 2012, **101**: 35–43.

[9] Akhtar J, Amin NS. A review on operating parameters for optimum liquid oil yield in biomass pyrolysis. Renew Sust Energ Rev. 2012, **16**: 5101–5109.

[10] Venderbosch RH, Prins W. Biofuels. Fast pyrolysis technology development. Bioprod Bioref. 2010, **4**: 178–208.

[11] Tsai WT, Lee MK, Chang YM. Fast pyrolysis of rice husk: product yields and compositions. Bioresour Technol. 2007, **98**: 22–28.

[12] Ferdinand FW, Steene LV, Blaise KK, Siaka T. Prediction of pyrolysis oils higher heating value with gas chromatography–mass spectrometry. Fuel. 2012, **96**: 141–154.

[13] Zhang K, Chang J, Guan Y, Chen H, Yang Y, Jiang J. Renew Energy. 2013, **49**: 175.

[14] Mohan D, Pittman CU, Steele PH. Pyrolysis of wood/biomass for bio-oil:a critical review. Energy Fuel. 2006, **20**: 848–889.

[15] Moraes MSA, Migliorini MV, Damasceno FC, Georges F, Almeida S, Zini CA, Jacques RA, Caramão EB. Qualitative analysis of bio oils of agricultural residues obtained through pyrolysis using comprehensive two dimensional gas chromatography with time-of-flight mass spectrometric detector. J Anal Appl Pyrolysis. 2012, **98**: 51–64.

[16] Jahirul MI, Rasul MG, Chowdhury AA, Ashwath N. Biofuels production through biomass pyrolysis—a technological review. Energies. 2012, **5**: 4952–5001.

[17] Önal PE, Uzun BB, Pütün EA. Steam pyrolysis of an industrial waste for bio-oil production. Fuel Process Technol. 2011, **92**: 879–885.

[18] Duman G, Okutucu C, Ucar S, Stahl R, Yanik J. The slow and fast pyrolysis of cherry seed. Bioresour Technol. 2011, **102**: 1869–1878.

[19] Martins AF, Diniz J, Stahl JA, Cardoso AL. Characterization of liquid products and char from the pyrolysis of eucalyptus sawdust. Quim Nova. 2007, **30**: 873–878.

[20] Weerachanchai P, Tangsathitkulchai C, Tangsathitkulchai M. Characterization of products from slow pyrolysis of palm kernel cake and cassava pulp residue. Korean J Chem Eng. 2011, **28**: 2262–2274.

[21] Pütüm AE, Özbay N, Pönal N, Pütüm E. Fixed-bed pyrolysis of cotton stalk for liquid and solid products. Fuel Process Technol. 2005, **86**: 1207–1219.

[22] Demiral I, Ayan EA. Pyrolysis of grape bagasse: effect of pyrolysis conditions on the product yields and characterization of the liquid product. Bioresour Technol. 2011, **102**: 3946–3951.

[23] Bridgwater AV. Review of fast pyrolysis of biomass and product upgrading. Biomass Bioenergy. 2012, **38**: 68–94.

[24] Butler E, Devlin G, Meier D, McDonnell K. A review of recent laboratory research and commercial developments in fast pyrolysis and upgrading. Renew Sustain Energy Rev. 2011, **15**: 4171–4186.

[25] Bridgwater AV, Meier D, Radlein D. A review of recent laboratory research and commercial developments in fast pyrolysis and upgrading. Org Geochem. 1999, **30**: 1479–1486.

[26] Bridgwater AV. Renewable fuels and chemicals by thermal processing of biomass. Chem Eng J. 2003, **91**: 87–102.

[27] Meier D. Renew sustain. State-of-the-art of fast pyrolysis in IEA bioenergy member countries. Energy Rev. 2013, **20**: 619–641.

[28] Kunii P, Levenspiel O. Fluidization engineering. In: Entrainment and Elutriation from Fluidized Beds, Chapter 7, 2nd edition. Robert E Krieger Publishing Company Inc, Huntington, New York, pp. 165–193. 1977. ISBN 0-409-90233-0.

[29] Bridgwater AV. Biomass fast pyrolysis. Thermal Sci. 2004, **8**: 21–49.

[30] Rozainee M. Effect of fluidising velocity on the combustion of rice husk in a bench-scale fluidised bed combustor for the production of amorphous rice husk ash. Bioresour Technol. 2008, **99**: 703–713.

[31] Penlidis A, McGregor JF, Hamielec AE. Continuous emulsion polymerization: design and control of CSTR trains. Chem Eng Sci. 1989, **44**: 273–281.

[32] Nogueira AL, Lona LMF, Machado RAF. Continuous polymerization in tubular reactors with prepolymerization: analysis using two-dimensional phenomenological model and hybrid model with neural networks. J Appl Polym Sci. 2004, **91**: 871–882.

[33] Palma M, Giudici R. Analysis of axial dispersion in an oscillatory-flow continuous reactor. Chem Eng J. 2003, **94**: 189–198.

[34] Almeida TM, Bispo MD, Cardoso ART, Migliorini MV, Schena T, de Campos MCV, Machado ME, López JA, Krause LC, Caramão EB. Analysis of products from pyrolysis of Brazilian sugar cane straw. J Agric Food Chem. 2013, **61**: 6812–6821.

[35] Lazzari E, Schena T, Primaz CT, Maciel GPS, Machado ME, Cardoso CAL, Jacques RA, Caramão EB. Production and chromatographic characterization of bio-oil from the pyrolysis of mango seed waste. Ind Crop Prod. 2016, **83**: 529–536.

[36] Torri IDV, Paasikallio V, Faccini CS, Huff R, Caramão EB, Sacon V, Oasmaa A, Zini CA. Bio-oil production of softwood and hardwood forest industry residues through fast and intermediate pyrolysis and its chromatographic characterization. Bioresour Technol. 2016, **200**: 680–690.

[37] Onorevoli B, Machado ME, Dariva C, Franceschi E, Krause LC, Jacques RA, Caramão EB. A one-dimensional and comprehensive two-dimensional gas chromatography study of the oil and the bio-oil of the residualcakes from the seeds of *Crambe abyssinica*. Ind Crop Prod. 2014, **52**: 8–16.

[38] Doumer ME, Arízaga GGC, da Silva DA, Yamamoto CI, Novotny EH, Santos JM, dos Santos LO, Wisniewski Jr A, de Andrade JB, Mangrich AS. Slow pyrolysis of different Brazilian waste biomasses as sources of soil conditioners and energy, and for environmental protection. J Anal Appl Pyrolysis. 2015, **113**: 434–443.

[39] Santos RM, Santos AO, Sussuchi EM, Nascimento JS, Lima ÁS, Freitas LS. Pyrolysis of mangaba seed: production and characterization of bio-oil. Bioresour Technol. 2015, **196**: 43–48.

[40] Figueiredo MKK, Romeiro GA, d'Avila LA, Damasceno RN, Franco AP. The isolation of pyrolysis oil from castor seeds via a low temperature conversion (LTC) process and its use in a pyrolysis oil-diesel blend. Fuel. 2009, **88**: 2193–2198.

[41] Czernik S, Bridgwater AV. Overview of applications of biomass fast pyrolysis oil. Energy Fuel. 2004, **18**: 590–598.

[42] Tanneru SK, Steele PH. Pretreating bio-oil to increase yield and reduce char during hydrodeoxygenation to produce hydrocarbons. Fuel. 2014, **133**: 326–331.

[43] Ospina V, Buitrago-Sierra R, López D. HDO of guaiacol over NiMo catalyst supported on activated carbon derived from castor de-oiled cake. Ingeniería InvestIgación. 2015, **35**: 49–55. doi:10.15446/ing. investig.v35n2.49838

[44] Mortensen PM, Grunwaldt JD, Jensen PA, Jensen AD. Influence on nickel particle size on the hydrodeoxygenation of phenol over Ni/SiO_2. Catal Today. 2016, **259**: 277–284.

[45] Vitolo S, Seggiani M, Frediani P, Ambrosini G, Politi L. Catalytic upgrading of pyrolytic oils to fuel over different zeolites. Fuel. 1999, **78**: 1147–1159.

[46] Tessarolo NS, Silva RVS, Vanini G, Casilli A, Ximenes VL, Mendes FL, Pinho AR, Romão W, de Castro EVR, Kaiser CR, Azevedo DA. Characterization of thermal and catalytic pyrolysis bio-oils by high-resolution techniques: 1 H NMR, GC× GC-TOFMS and FT-ICR MS. J Anal Appl Pyrolysis. 2016, **117**: 257–267.

[47] Shetty M, Murugappan K, Prasomsri T, Green WH, Leshkov YR. Reactivity and stability investigation of supported molybdenum oxide catalysts for the hydrodeoxygenation (HDO) of m-cresol. J Catal. 2015, **331**: 86–97.

[48] Persson NE, Blass SD, Rosenthal C, Bhan A, Schmidt LD. On-line deoxygenation of cellulose pyrolysis vapors in a staged autothermal reactor. RSC Adv. 2013, **3**: 20163–20170.

[49] Furimsky E. Hydroprocessing challenges in biofuels production. Catal Today. 2013, **217**: 13–56.

[50] Huber GW, Iborra S, Corma A. Synthesis of transportation fuels from biomass: chemistry, catalysts, and engineering. Chem Rev. 2006, **106**: 4044–4098.

[51] Wang K, Johnston PA, Brown RC. Comparison of *in-situ* and *ex-situ* catalytic pyrolysis in a micro-reactor system. Bioresour Technol. 2014, **173**: 124–131.

[52] Dorado C, Mullen CA, Boateng AA. Origin of carbon in aromatic and olefin products derived from HZSM-5 catalyzed co-pyrolysis of cellulose and plastics via isotopic labeling. Appl Catal B: Environ. 2015, **162**: 338–345.

[53] Stefanidis SD, Kalogiannis KG, Iliopoulou EF, Lappas AA, Pilavachi PA. In-situ upgrading of biomass pyrolysis vapors: catalyst screening on a fixed bed reactor. Bioresour Technol. 2011, **102**: 8261–8267.

[54] Patel MA, Baldanza MAS, Da Silva VT, Bridgwater AV. In situ catalytic upgrading of bio-oil using supported molybdenum carbide. Appl Catal A: General. 2013, **458**: 48–54.

[55] Wang L, Lei H, Bu Q, Ren S, Wei Y, Zhu L, Zhang X, Liu Y, Yadavalli G, Lee J, Chen S, Tang J. Aromatic hydrocarbons production from ex situ catalysis of pyrolysis vapor over Zinc modified ZSM-5 in a packed-bed catalysis coupled with microwave pyrolysis reactor. Fuel. 2014, **129**: 78–85.

[56] Ludlow-Palafox C, Chase HA. Microware pyrolysis of plastic wastes. In: Scheirs J, Kaminsky W, editors. Feedstock Recycling and Pyrolysis of Waste Plastics, Chapter 21. John Wiley & Sons Ltd, pp. 569–594. 2006. ISBN 9780470021545.

[57] Holland KM, US pat 5084141, 1992.

[58] Holland KM, US pat 5387321, 1995.

[59] Fernández Y, Arenillas A, Menéndez JA. Microwave heating applied to pyrolysis. In: Grundas S, editor. Advances in Induction and Microwave Heating of Mineral and Organic Materials, Chapter 31, p. 723. 2011. doi:10.5772/562

[60] Luque R, Menéndez JA, Arenillas A, Cot J. Microwave-assisted pyrolysis of biomass feedstocks: the way forward? Energy Environ Sci. 2012, **5**: 5481–5488.

[61] Budarin VL, Clark JH, Lanigan BA, Shuttleworth P, Breeden SW, Wilson AJ, Macquarrie DJ, Milkowski K, Jones J, Bridgeman T, Ross A. In-situ upgrading of biomass pyrolysis vapors: catalyst screening on a fixed bed reactor. Bioresour Technol. 2009, **100**: 6064–6068.

[62] Mehdizadeh M. Microwave/RF Applicators and Probes for Material Heating Sensing and Plasma Generation. William Andrew Publishing, Boston, p. 400. 2010. ISBN-10 0815515928.

[63] Javadi HHS. Microwave materials. In: Ishii T, editor. Handbook of Microwave Technology: Components and Devices, Chapter 19. Elsevier, Milwaukee, Wisconsin, pp. 605–643. 1995. ISBN 9780123746955.

[64] Thostenson ET, Chou TW. Microwave processing: fundamentals and applications. Compos Part A: App Sci Manuf. 1999, **30**: 1055–1071.

[65] Makibar J, Akarregi ARF, Amutio M, Lopez G, Olazar M. Performance of a conical spouted bed pilot plant for bio-oil production by poplar flash pyrolysis. Fuel Process Technol. 2015, **137**: 283–289.

[66] http://wwwpyrolysisoilnet/PRODUCTS_SOLUTIONS/Tire_Pyrolysis_Plant/ (accessed September 10, 2016).

[67] Pichini S, Rotolo MC, Bellotti P, Minutilloa A, Mastrobattistaa L, Pacifici R. Quali-quantitative analysis of best selling drugs from pharmacy, street market and traditional herbal medicine: a pilot study of market surveillance in Senegal. J Pharm Biomed. 2015, **104**: 62–66.

[68] Amutio M, Lopez G, Alvarez J, Olazar M, Bilba J. In-situ upgrading of biomass pyrolysis vapors: catalyst screening on a fixed bed reactor. Bioresour Technol. 2015, **194**: 225–232.

[69] Toor SS, Rosendahl L, Nielsen MP, Glasius M, Rudolf A, Iversen SB. Continuous production of bio-oil by catalytic liquefaction from wet distiller's grain with solubles (WDGS) from bio-ethanol production. Biomass Bioenerg. 2012, **36**: 327–332.

[70] Pérez JMM, Rocha JD, Cortez LAB, Medina MP, Luengo CA, Cascarosa E. Fast oxidative pyrolysis of sugar cane straw in a fluidized bed reactor. Appl Therm Eng. 2013, **56**: 167–175.

[71] Shemfe MB, Gu S, Ranganathan P. Techno-economic performance analysis of biofuel production and miniature electric power generation from biomass fast pyrolysis and bio-oil upgrading. Fuel. 2015, **143**: 361–372.

[72] https://patentesgooglecom (accessed September 10, 2016).

[73] https://patentscopewipoint (accessed September 10, 2016).

[74] Gómez N, Rosas JG, Cara J, Martínez O, Alburquerque JA, Sánchez ME. Slow pyrolysis of relevant biomasses in the Mediterranean basin. Part 1. Effect of temperature on process performance on a pilot scale. J Clean Prod. 2016, **120**: 181–190.

[75] Moncada JB, Aristizábal VM, Cardona A. Design strategies for sustainable biorefineries. J Biochem Eng. 2016. doi:101016/jbej201606009 (in press)

[76] Sadhukhan J, Lloyd JR, Scott K, Premier GC, Yu HE, Curtis T, Head IM. A critical review of integration analysis of microbial electrosynthesis (MES) systems with waste biorefineries for the production of biofuel and chemical from reuse of CO_2. Renew Sust Energ Rev. 2016, **56**: 116–132.

[77] Kokossis AC, Tsakalova M, Pyrgakis K. Design of integrated biorefineries. Comput Chem Eng. 2015, **81**: 40–56.

[78] Mota FAS, Viegas RAA, Lima AS, Santos FP, Caselli FTR. Pyrolysis of lignocellulose biomass: a review. Revista GEINTEC. 2015, **5**: 2511–2525. ISSN: 2237-0722. doi: 107198/S2237-0722201500040003

[79] Roy C, Blanchette D, Korving L, Yang J, Caumia B. Development of a novel vacuum pyrolysis reactor with improved heat transfer potential. In: Bridgewater AV, Boocock DGB, editors. Thermochem Biomass Convers. Blackie Academic and Professional, London, UK, p. 351. 1997.

[80] Alvarez J, Lopez G, Amutio M, Bilbao J. Bio-oil production from rice husk fast pyrolysis in a conical spouted bed reactor. Fuel. 2014, **128**: 162–169.

[81] Tang L, Huang H. Plasma pyrolysis of biomass for production of syngas and carbon adsorbent. Energy Fuels. 2005, **19**: 1174–1178.

[82] Huang X, Cheng D, Chen F, Zhan X. Reaction pathways of hemicellulose and mechanism of biomass pyrolysis in hydrogen plasma: a density functional theory study. Renew Energy. 2016, **96**: 490–497.

[83] Tang L, Huang H, Hao H, Zhao K. Development of plasma pyrolysis/gasification systems for energy efficient and environmentally sound waste disposal. J Electrostat. 2013, **71**: 839–847.

[84] Yan B, Xu P, Guo CYYJ, Cheng Y. Experimental study on coal pyrolysis to acetylene in thermal plasma reactors. J Chem Eng. 2012, **207**: 109–116.

[85] Yan B, Cheng Y, Jin Y, Guo CY. Performance of a conical spouted bed pilot plant for bio-oil production by poplar flash pyrolysis. Fuel Process Technol. 2012, **100**: 1–10.

[86] Punčochář M, Ruj B, Chatter PK. Development of process for disposal of plastic waste using plasma pyrolysis technology and option for energy recovery. Procedia Eng. 2012, **42**: 420–430.

[87] Fernández Y, Menéndez JA. Influence of feed characteristics on the microwave-assisted pyrolysis used to produce syngas from biomass wastes. J Anal App Pyrolysis. 2011, **91**: 316–322.

[88] Lam SS, Russell AD, Chase HA. Microwave pyrolysis, a novel process for recycling waste automotive engine oil. Energy. 2010, **35**: 2985–2991.

[89] Domínguez A, Fernández Y, Fidalgo B, Pis JJ, Menéndez JA. Bio-syngas production with low concentrations of CO_2 and CH_4 from microwave-induced pyrolysis of wet and dried sewage sludge. Chemosphere. 2008, **70**: 397–403.

[90] Domínguez A, Menéndez JA, Fernández Y, Pis JJ, Valente Nabais JM, Carrott PJM, Ribeiro Carrott MML. Conventional and microwave induced pyrolysis of coffee hulls for the production of a hydrogen rich fuel gas. J Anal App Pyrolysis. 2007, **79**: 128–135.

[91] Huang YF, Kuan WH, Lo SL, Lin CF. Total recovery of resources and energy from rice straw using microwave-induced pyrolysis. Bioresour Technol. 2008, **99**: 8252–8258.

[92] Zhao X, Song Z, Liu H, Li Z, Li L, Ma C. Microwave pyrolysis of corn stalk bale: a promising method for direct utilization of large-sized biomass and syngas production. J Anal App Pyrolysis. 2010, **89**: 87–94.

[93] Hofmann L, Antal MJ. Numerical simulations of the performance of solar fired flash pyrolysis reactors. Sol Energy. 1984, **33**: 427–440.

[94] Shakya BD. PhD Thesis School of Chemical and Biomolecular Engineering. University of Sydney, Sydney, Australia, p. 98. 2007.

[95] Li R, Zeng K, Soria JGM, Gauthier D, Rodriguez R, Flamant G. Product distribution from solar pyrolysis of agricultural and forestry biomass residues. Renew Energy. 2016, **89**: 27–35.

[96] Zeng K, Gauthier D, Li R, Flamant G. Solar pyrolysis of beech wood: effects of pyrolysis parameters on the product distribution and gas product composition. Energy. 2015, **93**: 1648–1657.

[97] Zeng K, Gauthier D, Lu J, Flamant G. Parametric study and process optimization for solar pyrolysis of beech wood. Energy Convers Manage. 2015, **106**: 987–998.

[98] Zeng K, Flamant G, Gauthier D, Guillot E. Solar pyrolysis of wood in a lab-scale solar reactor: influence of temperature and sweep gas flow rate on products distribution. Energy Procedia. 2015, **69**: 1849–1858.

[99] Zeng K, Minh DP, Gauthier D, Hortala EW, Nzihou A, Flamant G. The effect of temperature and heating rate on char properties obtained from solar pyrolysis of beech wood. Bioresour Technol. 2015, **182**: 114–119. doi:10.1016/j.biortech.2015.01.112

[100] Chopra S, Jain AK. A review of fixed bed gasification systems for biomass. Agric Eng Int CIGR J. 2007, **5**: 1–23.

[101] Rao MS, Singha SP, Sodha MS, Dubey AK, Shyam M. Stoichiometric, mass, energy and exergy balance analysis of countercurrent fixed-bed gasification of post-consumer residues. Biomass Bioenery. 2004, **27**: 155–171. doi:10.1016/j.biombioe.2003.11.003

[102] Lv PM, Xiong ZH, Chang J, Wu CZ, Chen Y, Zhu JX. An experimental study on biomass air–steam gasification in a fluidized bed. Bioresour Technol. 2004, **95**: 95–101. doi:10.1016/j.biortech.2004.02.003

[103] Li XT, Grace R, Lim CJ, Watkinson AP, Chen HP, Kim JR. Biomass gasification in a circulating fluidized bed. Biomass Bioenergy. 2004, **26**: 171–193. doi:10. 1016/S0961-9534(03)00084-9

[104] Joubert JE, Carrier M, Dahmen N, Stahl R, Knoetze JH. Inherent process variations between fast pyrolysis technologies: a case study on *Eucalyptus grandis*. Fuel Process Technol. 2015, **131**: 389–395. doi:10.1016/j.fuproc.2014.12.012

[105] Kook JW, Choi HM, Kim BH, Ra HW, Yoon SJ, Mun TY, Kim JH, Kim YK, Lee JG, Seo MW. Gasification and tar removal characteristics of rice husk in a bubbling fluidized bed reactor. Fuel. 2016, **181**: 942–950. doi:10. 1016/j. fuel. 2016. 05. 027

[106] Wagenaar BM, Venderbosch RH, Carrasco J, Strenziok R, Van der BJA. Rotating cone bio-oil production and applications. In: Bridgwater AV, editor. Progress in Thermochemical Biomass Conversion. Blackwell Science, Oxford, pp. 4952–5001. 2001. doi:10.1002/9780470694954.ch105

[107] Garcia-Perez M, Chaala A, Pakdel H, Kretschmer D, Roy C. Characterization of bio-oils in chemical families. Biomass Bioenergy. 2007, **31**: 222–242. doi:10.1016/j.biombioe.2006.02. 006

[108] Kim J-S. Production, separation and applications of phenolic-rich bio-oil—a review. Bioresour Technol. 2015, **178**: 90–98. doi:10.1016/j.biortech.2014.08.121

[109] Kim T-S, Kim J-Y, Kim KH, Lee S, Choi D, Choi I-G, Choi JW. The effect of storage duration on bio-oil properties. J Anal Appl Pyrolysis. 2012, **95**: 118–125. doi:10.1016/j.jaap.2012.01.015

[110] Shen DK, Gu S. The mechanism for thermal decomposition of cellulose and its main products. Bioresour Technol. 2009, **100**: 6496–6504. doi:10.1016/j.biortech.2009.06.095

[111] Shen DK, Gu S, Luo KH, Wang SR, Fang MX. The pyrolytic degradation of wood-derived lignin from pulping process. Bioresour Technol. 2010, **101**: 6136–6146. doi:10.1016/j.biortech.2010.02.078

[112] Kanaujia PK, Sharma YK, Garg MO, Tripathi D, Singh RJ. Review of analytical strategies in the production and upgrading of bio-oils derived from lignocellulosic biomass. J Anal Appl Pyrolysis. 2014, **105**: 55–74. doi:10.1016/j.jaap.2013.10.004

[113] Mortensen PM, Grunwaldt JD, Jensen PA, Knudsen KG, Jensen DA. A review of catalytic upgrading of bio-oil to engine fuels. Appl Catal A General. 2011, **407**: 1–19. doi:10.1016/j. apcata.2011.08.046

[114] Zhang L, Liu R, Yin R, Mei Y. Upgrading of bio-oil from biomass fast pyrolysis in China: a review. Renew Sustain Energy Rev. 2013, **24**: 66–72. doi:10.1016/j.rser.2013.03.027

[115] Lehto J, Oasmaa A, Solantausta Y, Kytö M, Chiaramonti D. Review of fuel oil quality and combustion of fast pyrolysis bio-oils from lignocellulosic biomass. Appl Energy. 2014, **116**: 178–190. doi:10.1016/j.apenergy.2013.11.040

[116] Chiaramonti D, Oasmaa A, Solantausta Y. Power generation using fast pyrolysis liquids from biomass. Renew Sustain Energy Rev. 2007, **11**: 1056–1086. doi:10.1016/j.rser.2005.07.008

[117] Yang Z, Kumar A, Huhnke RL. Review of recent developments to improve storage and transportation stability of bio-oil. Renew Sustain Energy Rev. 2015, **50**: 859–870. doi:10.1016/j.rser.2015.05.025

[118] Cao JP, Zhao XY, Morishita K, Wei XY, Takarada T. Fractionation and identification of organic nitrogen species from bio-oil produced by fast pyrolysis of sewage sludge. Bioresour Technol. 2010, **101**: 7648–7652. doi:10.1016/j.biortech.2010.04.073

[119] Cunha ME, Schneider JK, Brasil MC, Cardoso CA, Monteiro LR, Mendes FL, Pinho A, Jacques RA, Machado ME, Freitas LS, Caramão EB. Analysis of fractions and bio-oil of sugar cane straw by one-dimensional and two-dimensional gas chromatography with quadrupole mass spectrometry (GC × GC/qMS). Microchem J. 2013, **110**: 113–119. doi:10.1016/j.microc.2013.03.004

[120] Sukhbaatar B, Steele PH, Ingram LL, Kim MG. An exploratory study on the removal of acetic and formic acids from bio-oil. Bioresources. 2009, **4**: 1319–1329.

[121] Bertero M, Gorostegui HA, Orrabalis CJ, Guzmán CA, Calandri EL, Sedran U. Characterization of the liquid products in the pyrolysis of residual chañar and palm fruit biomasses. Fuel. 2014, **116**: 409–414. doi:10.1016/j.fuel.2013.08.027

[122] Sukhbaatar B, Steele PH, Kim MG. Use of lignin separated from bio-oil in oriented strand board binder phenol-formaldehyde resins. Bioresources. 2009, **4**: 789–804.

[123] Mourant D, Wang Z, He M, Wang XS, Garcia-Perez M, Ling K, Li C-Z. Mallee wood fast pyrolysis: effects of alkali and alkaline earth metallic species on the yield and composition of bio-oil. Fuel. 2011, **90**: 2915–2922. doi:10.1016/j.fuel.2011.04.033

[124] Isahak WNRW, Hisham MWM, Yarmo MA, Hin TY. A review on bio-oil production from biomass by using pyrolysis method. Renew Sustain Energy Rev. 2012, **16**: 5910–5923. doi:10.1016/j.rser.2012.05.039

[125] Vitasari CR, Meindersma GW, de Haan AB. Water extraction of pyrolysis oil: the first step for the recovery of renewable chemicals. Bioresour Technol. 2011, **102**: 7204–7210. doi:10.1016/j.biortech.2011.04.079

[126] Abou-Yousef H, Hassan EB. Efficient utilization of aqueous phase bio-oil to furan derivatives through extraction and sugars conversion in acid-catalyzed biphasic system. Fuel. 2014, **137**: 115–121. doi:10.1016/j.fuel.2014.07.067

[127] Teella A, Huber GW, Ford DMJ. Separation of acetic acid from the aqueous fraction of fast pyrolysis bio-oils using nanofiltration and reverse osmosis membranes. J Membrane Sci. 2011, **378**: 495–502. doi:10.1016/j.memsci.2011.05.036

[128] Bharti A, Banerjee T. Enhancement of bio-oil derived chemicals in aqueous phase using ionic liquids: experimental and COSMO-SAC predictions using a modified hydrogen bonding expression. Fluid Phase Equil. 2015, **400**: 27–37. doi:10.1016/j.fluid.2015.04.029

[129] Das DD, Schnitzer MI, Monreal CM, Mayer P. Chemical composition of acid-base fractions separated from biooil derived by fast pyrolysis of chicken manure. Bioresour Technol. 2009, **100**: 6524–6532. doi:10.1016/j.biortech.2009.06.104

[130] Wang S, Wang Y, Cai Q, Wang X, Jin H, Luo Z. Multi-step separation of monophenols and pyrolytic lignins from the water-insoluble phase of bio-oil. Sep Pur Technol. 2014, **122**: 248–255. doi:10.1016/j.seppur.2013.11.017

[131] de Oliveira AM, Biajoli AFP, Fidelis CHV, Silva RGC, Augusto F. Trends in sample preparation. In: Arruda MAZ, editor. Nova Science, New York, p. 142. 2006.

[132] Xu R, Ferrante L, Briens C, Berruti FJ. Flash pyrolysis of grape residues into biofuel in a bubbling fluid bed. J Anal Appl Pyrolysis. 2009, **86**: 58–65. doi:10.1016/j.jaap.2009.04.005

[133] Xu R, Ferrante L, Briens C, Berruti FJ. Bio-oil production by flash pyrolysis of sugarcane residues and post treatments of the aqueous phase. J Anal Appl Pyrolysis. 2011, **91**: 263–272. doi:10.1016/j.jaap.2011.03.001

[134] Xie J, Su D, Yin X, Wu C, Zhu J. Thermodynamic analysis of aqueous phase reforming of three model compounds in bio-oil for hydrogen production. Int J Hydro Energy. 2011, **36**: 15561–15572. doi:10.1016/j.ijhydene.2011.08.103

[135] Sanna A, Vispute TP, Huber GW. Hydrodeoxygenation of the aqueous fraction of bio-oil with Ru/C and Pt/C catalysts. Appl Catal B Environ. 2015, **165**: 446–456. doi:10.1016/j.apcatb.2014.10.013

[136] Vispute TP, Huber GW. Production of hydrogen, alkanes and polyols by aqueous phase processing of wood-derived pyrolysis oils. Green Chem. 2009, **11**: 1433–1445. doi:10.1039/B912522C

[137] Huber GW, Dumesic JA. An overview of aqueous-phase catalytic processes for production of hydrogen and alkanes in a biorefinery. Catal Today. 2006, **111**: 119–132. doi:10.1016/j.cattod.2005.10.010

[138] Gerdes C, Simon CM, Ollesch T, Kaminsky W. Design, construction, and operation of a fast pyrolysis plant for biomass. Eng Life Sci. 2002, **2**: 167–174. doi:10.1002/1618-2863(200206)2:6<167::AID-ELSC167>3. 0. CO, 2-F

[139] Bayerbach R, Meier D. Characterization of the water-insoluble fraction from fast pyrolysis liquids (pyrolytic lignin). Part IV: structure elucidation of oligomeric molecules. J Anal Appl Pyrolysis. 2009, **85**: 98–107. doi:10.1016/j.jaap.2008.10.021

[140] Djokic MR, Dijkmans T, Yildiz G, Prins W, Van Geem KM. Quantitative analysis of crude and stabilized bio-oils by comprehensive two-dimensional gas-chromatography. J Chromatogr A. 2012, **1257**: 131–140. doi:10.1016/j.chroma.2012.07.035

[141] von Muhlen C, Zini CA, Caramao EB, Marriott PJ. Applications of comprehensive two-dimensional gas chromatography to the characterization of petrochemical and related samples. J Chromatogr A. 2006, **1105**: 39–50. doi:10.1016/j.chroma.2005.09.036

[142] Marriott PJ, Shellie R. Principles and applications of comprehensive two-dimensional gas chromatography. TrAC Trends Anal Chem. 2002, **21**: 573–583. doi:10.1016/S0165-9936(02)00814-2

[143] Tessini C, Müller N, Mardones C, Meier D, Berg A, von Baer D. Chromatographic approaches for determination of low-molecular mass aldehydes in bio-oil. J Chromatogr A. 2012, **1219**: 154–160. doi:10.1016/j.chroma.2011.10.093

[144] Andersson T, Hyotylainen T, Riekkola M-L. Analysis of phenols in pyrolysis oils by gel permeation chromatography and multidimensional liquid chromatography. J Chromatogr A. 2000, **896**: 343–349. doi:10.1016/S0021-9673(00)00678-6

[145] Tienpont B, David F, Pereira A, Sandra P. In-vial pyrolysis (PyroVial) with pre- and post-sample treatment combined with different chromatographic techniques. J Chromatogr A. 2011, **1218**: 8303–8312. doi:10.1016/j.chroma.2011.09.053

[146] Yanik J, Stahl R, Troeger N, Sinag A. Pyrolysis of algal biomass. J Anal Appl Pyrolysis. 2013, **103**: 134–141. doi:10.1016/j. jaap.2012.08.016

[147] Johnston PA, Brown RC. Quantitation of sugar content in pyrolysis liquids after acid hydrolysis using high-performance liquid chromatography without neutralization. J Agric Food Chem. 2014, **62**: 8129–8133. doi:10. 1021/jf502250n

[148] Choi YS, Johnston PA, Brown RC, Shanks BH, Lee K-H. Detailed characterization of red oak-derived pyrolysis oil: integrated use of GC, HPLC, IC, GPC and Karl-Fischer. J Anal Appl Pyrolysis. 2014, **110**: 147–154. doi:10.1016/j.jaap.2014.08.016

[149] Undri A, Abou-Zaid M, Briens C, Berruti F, Rosi L, Bartoli M, Frediani M, Frediani P. J Anal Appl Pyrolysis. 2015, **114**: 208–221. doi:10.1016/j.jaap.2015.05.019

[150] Jacques RA, Caramão EB. Two dimensional liquid chromatography. Sci Chromatogr. 2009, **1**: 73–82.

[151] Le Masle A, Angot D, Gouin C, D'Attoma A, Ponthus J, Quignard A, Heinisch S. Development of on-line comprehensive two-dimensional liquid chromatography method for the separation of biomass compounds. J Chromatogr A. 2014, **1340**: 90–98.

[152] Sarrut M, Corgier A, Crétier G, Le Masle A, Dubant S, Heinisch S. Potential and limitations of on-line comprehensive reversed phase liquid chromatography × supercritical fluid chromatography for the separation of neutral compounds: an approach to separate an aqueous extract of bio-oil. J Chromatogr A. 2015, **1402**: 124–133.

[153] Tomasini D, Cacciola F, Rigano F, Sciarrone D, Donato P, Beccaria M, Caramão EB, Dugo P, Mondello L. Complementary analytical liquid chromatography methods for the characterization of aqueous phase from pyrolysis of lignocellulosic biomasses. Anal Chem. 2014, **86**: 11255–11262.

[154] Zhang Q, Chang J, Wang T, Xu Y. Review of biomass pyrolysis oil properties and upgrading research. Energy Convers Manag. 2007, **48**: 87–92.

[155] Kanaujia PK, Sharma YK, Agrawal UC, Garg MO. Comprehensive two dimensional gas chromatography review. Trends Anal Chem. 2013, **42**: 125–136.

[156] Pighinelli ALMT, Boateng AA, Mullen CA, Elkasabi Y. Evaluation of Brazilian biomasses as feedstocks for fuel production via fast pyrolysis. Energy Sustain Dev. 2014, **21**: 42–50.

[157] Lira CS, Berruti FM, Palmisano P, Berruti F, Briens C, Pécora AAB. Fast pyrolysis of Amazon *tucumã* (*Astrocaryum aculeatum*) seeds in a bubbling fluidized bed reactor. J Anal Appl Pyrolysis. 2013, **99**: 23–31.

[158] Tessarolo NS, dos Santos LRM, Silva RSF, Azevedo DA. Chemical characterization of bio-oils using comprehensive two-dimensional gas chromatography with time-of-flight mass spectrometry. J Cromatogr A. 2013, **1279**: 68–75.

[159] Spilã K, Kuoppala E, Fagernãs L, Oasmaa A. Characterization of biomass-based flash pyrolysis oils. Biomass Bioenergy. 1998, **14**: 103–113.

[160] Oasmaa A, Kuoppala E, Solantausta Y. Fast pyrolysis of forestry residue. 2. Physicochemical composition of product liquid. Energy Fuels. 2003, **17**: 433–443.

[161] Fagernäs L, Kuoppala E, Tiilikkala K, Oasmaa A. Chemical composition of birch wood slow pyrolysis products. Energy Fuels. 2012, **26**: 1275–1283.

[162] Wiggers VR, Meier HF, Wisniewski AJ, Barros AAC, Maciel MRW. In-situ upgrading of biomass pyrolysis vapors: catalyst screening on a fixed bed reactor. Bioresour Technol. 2009, **100**: 6570–6577.

[163] Kuddusi M, Karagöz S. Analytical pyrolysis of biomass using gas chromatography coupled to mass spectrometry. Trends Anal Chem. 2014, **61**: 11–16.

[164] Mullen CA, Boateng AA. Catalytic pyrolysis-GC/MS of lignin from several sources. Fuel Process Technol. 2010, **91**: 1446–1458.

[165] Liaw SS, Perez VH, Zhou S, Justo OR, Perez MG. Py-GC/MS studies and principal component analysis to evaluate the impact of feedstock and temperature on the distribution of products during fast pyrolysis. J Anal Appl Pyrolysis. 2014, **109**: 140–151.

[166] Ralph J, Hatfield RD. Pyrolysis-GC-MS characterization of forage materials. J Agric Food Chem. 1991, **39**: 1426–1437.

[167] Del Río JC, Gutiérrez A, Hernando M, Landín P, Romero J, Martínez AT. Determining the influence of eucalypt lignin composition in paper pulp yield using Py-GC/MS. J Anal Appl Pyrolysis. 2005, **74**: 110–115.

[168] Barbosa LCA, Maltha CRA, Silva VL, Colodette JL. Determination of the siringyl/guaiacyl ratio in eucalyptus wood by pyrolysis-gas chromatography/mass spectrometry (PY-GC/MS). Quim Nova. 2008, **31**: 2035–2041.

[169] Fortes ICP, Baugh PJ. Study of analytical on-line pyrolysis of oils from macauba fruit (Acrocomia sclerocarpa M) via GC/MS. J Braz Chem Soc. 1999, **10**: 469–477.

[170] Vinhal JO, Lima CF, Barbosa LCA. Analytical pyrolysis of the kernel and oil of babassu palm (*Orbignya phalerata*). J Anal Appl Pyrol. 2014, **107**: 73–81.

[171] Lu Q, Dong C-K, Zhang X-M, Tian H-Y, Yang Y-P, Zhu Xi-F. Selective fast pyrolysis of biomass impregnated with $ZnCl_2$ to produce furfural: analytical Py-GC/MS study. J Anal Appl Pyrol. 2011, **90**: 204–212.

[172] Carvalho WS, Cunha IF, Pereira MS, Ataíde CH. Thermal decomposition profile and product selectivity of analytical pyrolysis of sweet sorghum bagasse: effect of addition of inorganic salts. Ind Crop Prod. 2015, **74**: 372–380.

[173] Cardoso CR, Ataíde CH. Analytical pyrolysis of tobacco residue: effect of temperature and inorganic additives. J Anal Appl Pyrolysis. 2013, **99**: 49–57.

[174] Sfetsas T, Michailof C, Lappas A, Li Q, Kneale B. Qualitative and quantitative analysis of pyrolysis oil by gas chromatography with flame ionization detection and comprehensive two-dimensional gas chromatography with time-of-flight mass spectrometry. J Chromatogr A. 2011, **1218**: 3317–3325. doi:10.1016/j.chroma.2010.10.034

[175] Case PA, Truong C, Wheeler MC, DeSisto WJ. Calcium-catalyzed pyrolysis of lignocellulosic biomass components. Bioresour Technol. 2015, **192**: 247–252.

[176] Liu Z, Phillips JB. Comprehensive two-dimensional gas chromatography using an on-column thermal modulator interface. J Chromatogr Sci. 1991, **29**: 227–231.

[177] Cortes HJ, Winniford B, Luong J, Pursch M. Comprehensive two dimensional gas chromatography review. J Sep Sci. 2009, **32**: 883–904.

[178] Özbay N, Apaydın-Varol E, Uzun EB, Pütün AE. Characterization of bio-oil obtained from fruit pulp pyrolysis. Energy. 2008, **33**: 1233–1240.

[179] Busetto L, Fabbri D, Mazzoni R, Salmi M, Torri C, Zanotti V. Techno-economic performance analysis of biofuel production and miniature electric power generation from biomass fast pyrolysis and bio-oil upgrading. Fuel. 2011, **90**: 1197–1207.

[180] Wannapeera J, Fungtammasan B, Worasuwannarak N. Qualitative and quantitative analysis of pyrolysis oil by gas chromatography with flame ionization detection and comprehensive two-dimensional gas chromatography with time-of-flight mass spectrometry. J Anal Appl Pyrol. 2011, **92**: 99–105.

[181] Wang Y, Chen Q, Norwood DL, McCaffrey J. Recent development in the applications of comprehensive two-dimensional gas chromatograph. J Liquid Chromatogr Related Technol. 2010, **33**: 1082–1115.

[182] Dallüge J, Beens J, Brinkman UAT. Comprehensive two-dimensional gas chromatography: a powerful and versatile analytical tool. J Chromatogr A. 2003, **1000**: 69–108.

[183] Von Mühlen C, Zini CA, Caramão EB, Marriott PJ. Nomenclature for comprehensive multidimensional chromatography in portuguese language. Quím Nova. 2007, **30**: 682–687.

[184] Adahchour M, Beens J, Vreuls RJJ, Brinkman UAT. Recent developments in comprehensive two-dimensional gas chromatography (GC×GC): IV. Further applications, conclusions and perspectives. Trends Anal Chem. 2006, **25**: 821–840.

[185] Adam F, Bertoncini F, Coupard V, Charon N, Thiébaut D, Espinat D, Hennion M-C. Using comprehensive two-dimensional gas chromatography for the analysis of oxygenates in middle distillates: I. Determination of the nature of biodiesels blend in diesel fuel. J Chromatogr A. 2008, **1186**: 236–244.

[186] Flego C, Gigantiello N, Parker Jr WO, Calemma V. A comprehensive two-dimensional gas chromatography coupled with quadrupole mass spectrometry approach for identification of C_{10} derivatives from decalin. J Chromatogr A. 2009, **1216**: 2891–2899.

[187] Hua R, Li Y, Liu W, Zheng J, Wei H, Wang J, Lu X, Kong H, Xu G. Determination of sulfur-containing compounds in diesel oils by comprehensive two-dimensional gas chromatography with a sulfur chemiluminescence detector. J Chromatogr A. 2003, **1019**: 101–109.

[188] Moraes MSA, Zini CA, Gomes CB, Bortoluzzi JH, von Mühlen C, Caramão EB. Application of comprehensive two-dimensional gas chromatography (GC×GC) in the characterization of biodiesel/diesel blends: application to beef tallow biodiesel. Quím Nova. 2011, **34**: 1188–1192.

[189] Migliorini MV, Moraes MSA, Machado ME, Caramão EB. Caracterization of phenols in bio-oil from the pyrolysis of peach core by GC/MS e GC×GC/TOFMS. Sci Chromatogr. 2013, **5**: 47–65.

[190] Hoh E, Matovska K, Lehotay SJ. Optimization of separation and detection conditions for comprehensive two-dimensional gas chromatography–time-of-flight mass spectrometry analysis of polychlorinated dibenzo-*p*-dioxins and dibenzofurans. J Chromatogr A. 2007, **1145**: 210–221.

[191] Pedroso MP, de Godoy LAF, Fidélis CHV, Ferreira EC, Poppi RJ, Augusto F. Comprehensive two-dimensional gas chromatography (GC × GC). Quím Nova. 2009, **32**: 421–430.

[192] Von Mühlen C, Khummueng W, Zini CA, Caramão EB, Marriott PJ. Detector technologies for comprehensive two-dimensional gas chromatography. J Sep Sci. 2006, **29**: 1909–1921.

[193] Kallio M, Hyotyläianen T, Lehtonen M, Jussila M, Hartonen K, Shimmo M, Riekkola ML. Comprehensive two-dimensional gas chromatography in the analysis of urban aerosols. J Chromatogr A. 2003, **1019**: 251–260.

[194] Adahchour M, Beens J, Vreuls RJJ, Brinkman UAT. Recent developments in comprehensive two-dimensional gas chromatography (GC×GC): II. Modulation and detection. Trends Anal Chem. 2006, **25**: 540–553.

[195] Beens J, Adahchour M, Vreuls RJJ, Van Altena K, Brinkman UAT. Simple, non-moving modulation interface for comprehensive two-dimensional gas chromatography. J Chromatogr A. 2001, **919**: 127–132.

[196] Gaines RB, Frysinger GS. Temperature requirements for thermal modulation in comprehensive two-dimensional gas chromatography. J Sep Sci. 2004, **27**: 380–388.

[197] Adahchour M, Beens J, Brinkman UAT. Recent developments in the application of comprehensive two-dimensional gas chromatography. J Chromatogr A. 2008, **1186**: 67–108.

[198] Barceló D. Comprehensive two dimensional gas chromatography. In: Ramos L, editor. Comprehensive Analytical Chemistry Series, volume 55. Elsevier, Amsterdam. 2009. eBook ISBN 9780080932699.

[199] Bertsch W. Recent developments in the application of comprehensive two-dimensional gas chromatography. J High Resol Chromatogr. 2000, **23**: 167–181.

[200] Godoy LAF, Hantao LW, Pedroso MP, Poppi RJ, Augusto F. Quantitative analysis of essential oils in perfume using multivariate curve resolution combined with comprehensive two-dimensional gas chromatography. Anal Chim Acta. 2011, **699**: 120–125. doi:10.1016/j.aca.2011.05.003

[201] Striebich RC, Contreras J, Balster LM, West Z, Shafer LM, Zabarnick S. Identification of polar species in aviation fuels using multidimensional gas chromatography-time of flight mass spectrometry. Energy Fuels. 2009, **23**: 5474–5482. doi:10.1021/ef900386x

[202] Vendeuvre C, Ruiz-Guerrero R, Bertoncini F, Duval L, Thiebaut D. Comprehensive two-dimensional gas chromatography for detailed characterization of petroleum products. Oil and Gas Sci Technol. 2007, **62**: 43–55. doi:10.2516/ogst:2007004

[203] Von Mühlen C, de Oliveira EC, Zini CA, Caramão EB, Marriott PJ. Characterization of nitrogen-containing compounds in heavy gas oil petroleum fractions using comprehensive two-dimensional gas chromatography coupled to time-of-flight mass spectrometry. Energy Fuels. 2010, **24**: 3572–3580. doi:10.1021/ef1002364

[204] Khummueng W, Trenerry C, Rose G, Marriot P. Application of comprehensive two-dimensional gas chromatography with nitrogen-selective detection for the analysis of fungicide residues in vegetable samples. J Chromatogr A. 2006, **1131**: 203–214. doi:10.1016/j.chroma.2006.07.035

[205] Marsman JH, Wildschut J, Mahfud F, Heeres HJ. Identification of components in fast pyrolysis oil and upgraded products by comprehensive two-dimensional gas chromatography and flame ionization detection. J Chromatogr A. 2007, **1150**: 21–27. doi:10.1016/j.chroma.2006.11.047

[206] Marsman JH, Wildschut J, Evers P, de Koning S, Heeres HJ. Identification and classification of components in flash pyrolysis oil and hydrodeoxygenated oils by two-dimensional gas chromatography and time-of-flight mass spectrometry. J Chromatogr A. 2008, **1188**: 17–25. doi:10.1016/j.chroma.2008.02.034

[207] Windt M, Meier D, Marsman JH, Heeres HJ, de Koning S. Micro-pyrolysis of technical lignins in a new modular rig and product analysis by GC–MS/FID and GC×GC–TOFMS/FID. J Anal Appl Pyr. 2009, **85**: 38–46. doi:10.1016/j.jaap.2008.11.011

[208] Kallio MK. Comprehensive Two-Dimensional Gas Chromatography: Instrumental and Methodological Development [Academic Dissertation]. University of Helsinki, Helsinki. 2008.

[209] Purcaro G, Tranchida PQ, Ragonese C, Conte L, Dugo P, Dugo G, Mondello L. Evaluation of a rapid-scanning quadrupole mass spectrometer in an apolar × ionic-liquid comprehensive two-dimensional gas chromatography system. Anal Chem. 2010, **82**: 8583–8590. doi:10.1021/ac101678r

[210] Korytar P, Haglund P, Deboer J, Brinkman UAT. Comprehensive two-dimensional gas chromatography for the analysis of organohalogenated micro-contaminants. Trends Anal Chem. 2006, **25**: 373–396. doi:10.1016/j.trac.2005.12.003

[211] Mondello L, Casilli A, Tranchida PQ, Dugo G, Dugo P. Comprehensive two-dimensional gas chromatography in combination with rapid scanning quadrupole mass spectrometry in perfume analysis. J Chromatogr A. 2005, **1067**: 235–243. doi:10.1016/j.chroma.2004.09.040

[212] Schmarr H-G, Bernhardt J, Fischer U, Stephan A, Müller P, Durner D. Two-dimensional gas chromatographic profiling as a tool for a rapid screening of the changes in volatile composition occurring due to microoxygenation of red wines. Anal Chim Acta. 2010, **672**: 114–123. doi:10.1016/j.aca.2010.05.002

[213] Song SM, Marriott P, Wynne P. Comprehensive two-dimensional gas chromatography—quadrupole mass spectrometric analysis of drugs. J Chromatogr A. 2004, **1058**: 223–232. doi:10.1016/j.chroma.2004.08.087

[214] Mondello L, Casilli A, Tranchida PQ, Lo Presti M, Dugo P, Dugo G. Comprehensive gas chromatography coupled to mass spectrometry for the separation of pesticides in a very complex matrix. Anal Bioanal Chem. 2007, **389**: 1755–1763. doi:10.1007/s00216-007-1412-y

[215] Frysinger GS, Gaines RB. Comprehensive two-dimensional gas chromatography with mass spectrometric detection (GC × GC/MS) applied to the analysis of petroleum. J Sep Sci. 1999, **22**: 251–255. doi:10.1002/(SICI)1521-4168(19990501)22:5<251::AID-JHRC251>3.0.CO,2-V

[216] Kochman M, Gordin A, Alon T, Amirav A. Flow modulation comprehensive two-dimensional gas chromatography–mass spectrometry with a supersonic molecular beam. J Chromatogr A. 2006, **1129**: 95–104. doi:10.1016/j.chroma.2006.06.079

[217] Moraes MAS, Bortoluzzi JH, Migliorini MV, Zini CA, Caramão EB. Cromatografia gasosa bidimensional abrangente aplicada à análise qualitativa dos componentes majoritários do bio-óleo da pirólise de bagaço de laranja. Sci Chromatogr. 2011, **3**: 301–314. doi:10.4322/sc.2011.018

[218] Maciel GPS, Machado ME, Barbara JA, Dal Molin D, Caramão EB, Jacques RA. GC × GC/TOFMS analysis concerning the identification of organic compounds extracted from the aqueous phase of sugarcane straw fast pyrolysis oil. Biomass Bioenergy. 2016, **85**: 198–206. doi:10.1016/j.biombioe.2015.11.009

[219] Schneider JK, Cunha ME, Santos AL, Maciel GPS, Brasil MC, Pinho AR, Mendes FL, Jacques RA, Caramão EB. Comprehensive two dimensional gas chromatography with fast-quadrupole mass spectrometry detector analysis of polar compounds extracted from the bio-oil from the pyrolysis of sawdust. J Chromatogr A. 2014, **1356**: 236–240. doi:10.1016/j.chroma.2014.06.053

[220] Araujo RC, Pasa VDM, Marriot PJ, Cardeal ZL. Analysis of volatile organic compounds in polyurethane coatings based on Eucalyptus sp. bio-oil pitch using comprehensive two-dimensional gas chromatography (GC×GC). J Anal Appl Pyrolysis. 2010, **88**: 91–97. doi:10.1016/j.jaap.2010.02.012

[221] Almeida TM, Bispo MD, Cardoso ART, Migliorini MV, Schena T, Campos MCV, Machado ME, López JA, Krause LC, Caramão EB. Preliminary studies of bio-oil from fast pyrolysis of coconut fibers. 2013, **61**: 6812–6821. doi:10.1021/jf401379s

[222] Silva RVS, Casilli A, Sampaio AL, Ávila BMF, Veloso MCC, Azevedo DA, Romeiro GA. The analytical characterization of castor seed cake pyrolysis bio-oils by using comprehensive GC coupled to time of flight mass Spectrometry. J Anal Appl Pyrolysis. 2014, **106**: 152–159. doi:10.1016/j.jaap.2014.01.013

[223] Tessarolo NS, Silva RC, Vanini G, Pinho A, Romão W, Castro EVR, Azevedo DA. Assessing the chemical composition of bio-oils using FT-ICR mass spectrometry and comprehensive two-dimensional gas chromatography with time-of-flight mass Spectrometry. Microchem J. 2014, **117**: 68–76. doi:10.1016/j.microc.2014.06.006

[224] Salehi E, Abedi J, Harding T. Bio-oil from sawdust: pyrolysis of sawdust in a fixed-bed system. Energy Fuels. 2009, **23**: 3767–3772. doi:10.1021/ef900112b

[225] Martinez A, Rodriguez ME, York SW, Preston JF, Ingram LO. Use of UV absorbance to monitor furans in dilute acid hydrolysates of biomass. Biotechnol Progr. 2000, **16**: 637–641. doi:10.1021/bp0000508

[226] Lanças FM. A cromatografia líquida moderna e a espectrometria de massas: finalmente "compatíveis"? Sci Chromatogr. 2009, **1**: 35–61.

[227] Chiaradia MC, Collins CH, Jardim ICSF. O estado da arte da cromatografia acoplada à espectrometria de massas na análise de compostos tóxicos em alimentos. Quim Nova. 2008, **31**: 623–636. doi:10.1590/S0100-40422008000300030

[228] van der Hage ERE, Boon JJ. On-line Curie-point pyrolysis-high-performance liquid chromatographic-mass spectrometric analysis of lignin polymers. J Chromatogr A. 1996, **736**: 61–75. doi:10.1016/0021-9673(95)01345-8

[229] Kano H, Okamoto T, Kitagawa S, Iiguni Y, Ohtani H, Ito H, Iwaia K, Kuno M. A novel analytical pyrolysis device applicable for measurements of less volatile pyrolyzates. J Anal Appl Pyrolysis. 2015, **113**: 165–173. doi:10.1016/j.jaap.2014.12.008

[230] Dugo P, Cacciola F, Kumm T, Dugo G, Mondello L. Comprehensive multidimensional liquid chromatography: theory and applications. J Chromatogr A. 2008, **1184**: 353–368. doi:10.1016/j.chroma.2007.06.074

[231] Francois I, Sandra K, Sandra P. Comprehensive liquid chromatography: fundamental aspects and practical considerations—a review. Anal Chim Acta. 2009, **641**: 14–31. doi:10.1016/j.aca.2009.03.041

[232] Guiochon G, Marchetti N, Mriziq K, Shalliker R. Implementations of two-dimensional liquid chromatography. J Chromatogr A. 2008, **1189**: 109–168. doi:10.1016/j.chroma.2008.01.086

[233] Bedani F, Schoenmakers PJ, Janssen H. Theories to support method development in comprehensive two-dimensional liquid chromatography—a review. Sep Sci. 2012, **35**: 1697–1711. doi:10.1002/jssc.201200070

[234] Jandera P, Hájek T, Česla P. Comparison of various second-dimension gradient types in comprehensive two-dimensional liquid chromatography. Sep Sci. 2010, **33**: 1382–1397. doi:10.1002/jssc.200900808

[235] Camenzuli M, Schoenmakers PJ. A new measure of orthogonality for multi-dimensional chromatography. Anal Chim Acta. 2014, **838**: 93–101. doi:10.1016/j.aca.2014.05.048

[236] Elsner V, Laun S, Melchior D, Koehler M, Schmitz OJ. Analysis of fatty alcohol derivatives with comprehensive two-dimensional liquid chromatography coupled with mass spectrometry. J Chromatogr A. 2012, **1268**: 22–28. doi:10.1016/j.chroma.2012.09.072

[237] Blumberg L, Klee MS. A critical look at the definition of multidimensional separations. J Chromatogr A. 2010, **1217**: 99–103. doi:10.1016/j.chroma.2009.08.081

[238] Carr PW, Stoll DR. Two-Dimensional Liquid Chromatography—Principles, Practical Implementation and Applications. Agilent Technologies Inc, Germany, p. 182. 2015. ISBN 5991-2359EN.

[239] François I, de Villiers A, Tienpont B, David F, Sandra P. Comprehensive two-dimensional liquid chromatography applying two parallel columns in the second dimension. J Chromatogr A. 2008, **1178**: 33–42. doi:10.1016/j.chroma.2007.11.032

[240] Rieux L, Lubda D, Niederlander HAG, Verpoorte E, Bischoff R. Fast, high-efficiency peptide separations on a 50-μm reversed-phase silica monolith in a nanoLC–MS set-up. J Chromatogr A. 2006, **1120**: 165–172. doi:10.1016/j.chroma.2006.01.038

[241] Rigano F, Albergamo A, Sciarrone D, Beccaria M, Purcaro G, Mondello L. Nano liquid chromatography directly coupled to electron ionization mass spectrometry for free fatty acid elucidation in mussel. Anal Chem. 2016, **88**: 4021–4028. doi:10.1021/acs.analchem.6b00328

[242] Lanças FM. Vantagens e limitações da miniaturização em cromatografia líquida. Sci Chromatogr. 2009, **1**: 51–60.

[243] Silva RGC, Collins CH, Bottoli CBG. Capillary liquid chromatography: state of the art and current applications. Quim Nova. 2011, **34**: 841–849. doi:10.1590/S0100-40422011000500020

[244] Fanali C, Dugo L, Dugo P, Mondello L. Capillary-liquid chromatography (CLC) and nano-LC in food analysis. Trends Anal Chem. 2013, **52**: 226–238. doi:10.1016/j.trac.2013.05.021

[245] Cappiello A, Famiglini G, Palma P, Trufelli H. Advanced liquid chromatography-mass spectrometry interface based on electron ionization. Anal Chem. 2007, **79**: 5364–5372. doi:10.1021/ac070468l

Chapter 6

Perceptions on Internal and External Factors Impacting the U.S. Nonfood Advanced Biofuel Industry

Henry Jose Quesada-Pineda, Jeremy Withers and Robert Smith

Additional information is available at the end of the chapter

http://dx.doi.org/10.5772/64611

Abstract

The goal of this chapter is to introduce and discuss internal and external barriers impacting the nonfood advanced biofuel industry in the United States. Since 2005 when the EPAct was created, 59 cellulosic biofuel projects have been attempted in the U.S. with little commercial success. An initial list of internal and external barriers was extracted from secondary sources using qualitative analysis techniques such as grounded theory. Once the list was validated, a survey was sent to the biofuel industry members to gain more knowledge and clarification on the initial list of barriers. Statistical analysis revealed differences in perceptions from industry members when barriers were compared by project status, technology, and type of project. In addition, barriers for marketing and distribution of advanced biofuel's coproducts and by-products were identified and ranked by industry members, academicians, and other stakeholders.

Keywords: biofuel, cellulosic biofuel, internal and external barriers, coproducts, by-products

1. Introduction

The development of an environmental bioeconomy is necessary in the U.S. to reduce fossil fuel energy dependency. The term energy is classified into three main categories: fossil, nuclear, and renewable. The main fossil fuels are petroleum, coal, natural gas, and nuclear material. They are currently nonrenewable and contribute to the accumulation of greenhouse gases (GHGs), one of the causes of climate change. Fossil fuels, namely, petroleum for transportation fuel, are being consumed at an increasing rate from diminishing finite reserves. One model estimates that, at the current usage rates, fossil fuel reserves of oil, coal, and gas will last approximately 35, 107, and 137 years, respectively [1]. Other researchers have estimated that fossil fuel depletion will occur between the years 2100 and 2200 [2].

There are three primary methods to create liquid advanced biofuel (AB) and its coproducts: direct microbial conversion (DMC-biochemical), simultaneous saccharification and fermentation (SSF-thermochemical), or a hybrid of these techniques [3]. These two main approaches are further broken down into six secondary options for developing cellulosic biofuel: (1) catalytic pyrolysis and hydrotreating to hydrocarbons; (2) gasification and Fischer-Tropsch synthesis to hydrocarbons; (3) gasification and methanol-to-gasoline synthesis; (4) dilute acid hydrolysis, fermentation to acetic acid, and chemical synthesis to ethanol; (5) enzymatic hydrolysis to ethanol; and (6) consolidated bioprocessing (single-step enzyme production, hydrolysis, and fermentation) to biofuel [3].

Liquid biofuel is one such renewable energy source. Biofuel is a fuel additive capable of increasing octane levels by blending it into the U.S. fuel supply, or can be used as a fuel in internal combustion engines [4]. The total renewable biofuel sector is currently diversified into first (1G)-, second (2G)-, and third (3G)-generation lignocellulosic biomass forms of energy. For example, 1G is derived from corn and sugarcane, 2G advanced biofuel is derived from wood, grasses, municipal wastes, and crop residues, and 3G is derived from algae. Biomass is considered as living or nonliving agricultural vegetation such as wood and grass crops. In this case, biomass is typically differentiated by dedicated wood and grass energy crops, and unmerchantable timber and forest waste. Lignocellulosic feedstock's price currently ranges from \$50 to 80/ton of biomass [5]. These feedstocks could be from unmerchantable timber, forest thinnings (slash), sawdust, waste paper, mill residues, paper mill sludge, grasses, and grass variety residues. All biomass feedstock differs in moisture content and may have different costs. Dedicated energy crops are considered for energy use only. In this study, dedicated energy crops are categorized and differentiated as herbaceous crops (grasses) and wood-based crops. Herbaceous grass crops are harvested annually, with only the roots surviving the nongrowth cold seasons (e.g., switchgrass, *Miscanthus*). Wood-based crops, including fast-growing trees such as poplar, are harvested on a 3- to 12-year rotation cycle; harvest rotation cycles for slower growing trees may be as long as 25 years.

For this study, nonfood lignocellulosic biomass consisted specifically of biomass from wood and from grass varieties for the current purpose of substituting fossil petroleum-based fuels with renewable biofuel. Advanced biofuel is a contemporary liquid fuel for transportation produced primarily from cellulose and hemicellulose of renewable lignocellulosic biomass. It is derived from lignocellulose, which consists of three major components: cellulose, hemicellulose, and lignin. The cellulose and hemicellulose portions are the desired components for producing the highest value-added biofuel coproducts. Lignocellulosic biofuel currently has the greatest potential for energy, being the most abundant and rapidly renewable resource produced by photosynthesis [6]. The lignin portion typically becomes a process by-product, but recently was considered a coproduct when blended as filler for wood products.

http://dx.doi.org/10.5772/64611

This study presents results on an investigation conducted between 2014 and 2016 related to the status of AB projects in the U.S. market. It was found that the majority of AB projects never achieved a commercialization stage. Therefore, the research team was interested in learning more about the barriers and factors that have prevented the AB industry to reach commercial state, including impact not only on the AB production itself but also in coproducts and by-products of the AB industry.

2. Factors affecting the advanced biofuel industry

2.1. Biofuel policy

There are a multitude of government policies using a push-type strategy to bring the bioeconomy technology to the marketplace. The Environmental Protection Agency (EPA), U.S. Department of Agriculture (USDA), Energy Information Administration (EIA), Department of Energy (DOE), and Department of Defense (DOD) have jointly developed these policies to drive the bioeconomy. According to Reidy [7], the major goals and policy incentive's objectives driving the bioeconomy marketplace are the following:

1. **To reduce GHG emissions and sequester carbon**
   - Advanced carbon capture and storage (DOE Grants for R+D)
   - Federal Transit Administration (FTA) investments in GHG and energy reduction (Tigger) (DOT Grants)
2. **To achieve greater energy efficiency**
   - Efficient clean fossil energy systems (DOE Grants)
   - Integrated biorefineries grants program (DOE Grants)
   - Advanced marine and hydrokinetic grant program (DOE Grants)
   - Clean energy fund (DOE Grants)
   - Clean diesel grant program (EPA Grants)
3. **To integrate rural programs into efforts to increase energy security**
   - Transportation fuel and biofuels: Rural Energy for America Program (REAP)
4. **To stimulate economic growth and development**
   - Federal Transit Administration (FTA) Clean Fuels (DOT Grants)
5. **To obtain economically feasible conversion technologies**
   - Clean coal-to-liquid or gaseous fuel technologies grant program (NSF Grants)

Six main policies were created in the United States to bolster, develop, and implement the four incentives driving the bioeconomy. Sequentially, they are: (1) Clean Air Act 1970—through current amendments [8], (2) Energy Policy Act of 2005 (EPAct) [9, 10], (3) Advanced Energy Initiative 2006 [11], (4) Renewable Fuels Standards (RFS) of Energy Independence and Security Act of 2007 (EISA) [12–14]), (5) California Low Carbon Fuel Standard (LCFS) [15], and (6) Food, Conservation, and Energy Act of 2008 [16].

As of 2015, there were six policies driving the inception of advanced biofuels, and EISA carried the most focus toward developing biofuel projects while removing market share from the fossil industry. There are a host of incentives for industry development of advanced biofuels (AB), such as the 2005 EPAct creating the Renewable Fuel Standard, and its modification with 2007 EISA, and new components of RFS2: Renewable Volume Obligations (RVO), Renewable Identification Number (RIN), and Code of Federal Regulations (CFR). These policies provided production tax credits and research and development (R+D) funding to promote the RFS concept of replacing 35 billion gallons of fossil fuel with drop-in biofuel blends. The policy subsidies and incentives were the drivers leading to advanced biofuel (AB) project attempts from 2005 to 2015.

Biofuel projects are divided into three generations by feedstock type: first generation is ethanol —corn and sugarcane; second generation (2G) is advanced biofuel—wood, grass, and crop residues; and third generation (3G) is algae and butanol. Those feedstocks are in the $50–80 p/ton range. This chapter is focused on 2G wood and grass advanced biofuel. Wood and grass feedstock (lignocellulose) is typically separated by its major components in order of value: cellulose, hemicellulose, and lignin.

2.2. Advanced biofuel project status

The U.S. total renewable biofuels (TRFs) projects are classified as pilot with costs ranging $9 million or less, demonstration project costs ranging $100 million or less, or commercial projects costs ranging $100–500 million [17–19]. These three project types are further divided into five operational status categories: cancelled, shutdown, under construction, planning, and operating. Cancelled projects are considered terminal. Shutdown projects were stopped and put on hold, but potentially could be restarted at a later time. Under construction projects are currently being built, and planning projects are in the research and development phase, prior to construction. For operating projects, construction was completed and attempts at biofuel production have begun. References [18] and [19] provided the only accessible publication covering a large portion of wood-based biofuel projects, separated by location, type, and status, from their Forisk-Wood Bioenergy U.S. (WBUS) database. They indicated 36 cancelled projects, 4 shutdown projects, and 12 projects in planning or construction stages, stating that 75% have failed to advance [18, 19].

Currently, few advanced biofuel projects are producing biofuel, with none reaching sustainable commercial production economies of scale where biofuel project size to produce commercial-level biofuel was greater than costs. Some documents in the literature identified barriers, but the authors only focused on broad categories. The most inclusive documents provided a partial list of wood-based biofuel projects by type and status [18, 19]. In examining literature on barriers to advanced biofuel projects, the following 10 main barriers were determined: (1) high capital risks, (2) Organization of the Petroleum Exporting Countries

(OPEC)-based price distortions, (3) constrained blending markets, (4) policy fluctuations, (5) financing, (6) production costs, (7) global financial situation, (8) economic hurdles, (9) efficiency, effectiveness, and scaling technology, and (10) too many technology paths.

2.3. Factors impacting the advanced biofuel industry

Prior to 2005 EPAct, the corn ethanol industry was preestablished to close in 40 years, moving away from utilizing government subsidizes and close to achieving commercial production economies of scale. This subsidized preestablishment was the first barrier to advanced biofuel and 3G biofuel technologies. The EPAct led to a second barrier: different subsidy and expectation levels among the renewable fuel types. The EPAct created the RFS that forced the fossil fuel industry to relinquish approximately 10% yearly of the production output over the next 17 years until 2022. This created another barrier: a line drawn in the sand between OPEC-backed fossil fuel companies and government support of the emerging bioeconomy. Additionally, methyl tertiary butyl ether (MTBE) was increasingly being banned for environmental and health-related concerns, but fossil fuel companies needed the MTBE to increase the octane content of diesel and gasoline. MTBE was able to be transported in fossil fuel's current infrastructure, but biofuel has to be transported separately to the refinery and was more expensive. This was a third blow to the fossil fuel industry: reduction of their monopoly with market share percentage loss over time, MTBE could become banned with potential lawsuits, and unable to maximize delivery economies of scale without expensive upgrades to infrastructure for ethanol. These led to initial fossil infrastructure upgrades and supporting biofuel as a lubricant and octane enhancer with the 2005 EPAct.

The 2007 Energy Independence and Security Act and its modified RFS (EISA-RFS2) brought more specificity, policy incentive type drivers, and, subsequently, more barriers. The fossil fuel industry opposed the new RFS-2 and, to date, mounts continual media attacks to repeal the RFS. By 2007, the steady decline of fossil fuel consumption should have triggered more concern with the near-term potential for constrained blending markets. In 2012, the blend wall arrived; the advanced biofuel projects saturated market demand, with nowhere to put their fuel for blending above their mandate since D6 (RIN code for renewable fuel based on corn ethanol) by itself was filling more fuel capacity than available. The blend wall led to the next major barrier: political involvement in an attempt to create demand. The government was forced to balance the fallout of subsidizing and building an industry with diminishing room to put their products as they strive to meet mandated production economies of scale.

Lack of infrastructure and lack of factual knowledge are the main barriers to the public not having enough flex fuel vehicles and ethanol pumps to maintain low gas prices. The main barrier to all groups is time. Transportation fuel stations are willing to upgrade infrastructure [20] when the vehicles have upgraded technology. Republicans will not budge until the demand increases. Democrats cannot increase the infrastructure demand until they have control of the House and Senate. The vehicle demand will not increase until the vehicle infrastructure for higher blends is affordable. Advanced biofuel projects will have to receive subsidies until that happens. The public would not support another tax (i.e., carbon tax), while petroleum and gas prices are low [21]. Therefore, time is the overarching barrier with certainty, in an uncertain climate.

The knowledge gaps from the broad barrier categories are not precise enough to fully aid in developing an industry. Furthermore, 75% of AB projects have been lost since inception [18, 19]. No articles were found analyzing if AB location, status, or technology type was a barrier. The Renewable Fuel Standard (RFS) appears to work for some and not for others. Examining the barriers across multiple bioeconomy groups, such as academia, government, biofuel publishers, advanced biofuel projects, and the remainder of the bioeconomy, was pivotal to determine a progression of barriers and how the level of understanding changes when moving outwards from the proprietary inner workings of companies to the broader bioeconomy. No consolidated lists were found of coproducts and by-products from 2G AB companies. The focus was mainly placed on their funding and technology issues, as if they are not utilizing their secondary products.

Therefore, this study was deemed necessary due to the perceived advanced biofuel investment risk, investment potential in the bioeconomy, infrastructure need, and 75% loss of projects in less than 8 years. Additionally, a simplified understanding of internal and external barriers across and within industry stakeholders groups and market and distribution barriers of their products was needed to drive faster return on investment from reducing risk, as conditioned bioeconomy reinforcement. Determination of these knowledge gaps in a singular document will more quickly aid in bioeconomy collaboration maximizing the RFS-2 potential.

3. Methods

This research was conducted in two phases. Phase one identified all wood and grass (nonfood) AB projects that have been attempted by their status, location, feedstock, and technology type in the U.S. During phase two, a survey was conducted requesting industry members to rank internal and external barriers for the AB industry. In addition, industry members, academicians, government representatives, and other stakeholders were also asked to rank marketability and distribution barriers of biofuel's coproducts and by-products. After compiling survey responses, interviews with a selected group of industry members were conducted to discuss and gain more insights on the specific barriers.

The geographical location, operational status, and demographics information for each project were determined by examining secondary sources of information such as technical reports, peer-reviewed papers, trade journals, and newspapers. These were based on the biofuel industry terminology used in the Wood Bioenergy U.S. database according to Forisk Consulting [22] along with acquired secondary sourced data from the literature review. The data were used to individually classify and code categories directly associated with advanced biofuel projects as follows: type (pilot, demonstration, and commercial), operational status (cancelled, shutdown, operating, planning, and under construction), demographic (project, name, and location), feedstock type used, and contact information.

http://dx.doi.org/10.5772/64611

Grounded theory was used to examine peer-reviewed papers, industry reports, technical reports, trade journals, and newspapers to detect barriers impacting the AB industry. The goal of the grounded theory analytical technique is to classify and categorize information based on higher level categories. The technique starts with an initial open coding involving labeling, data segmentation, conceptualizing, and developing categories. Higher level grouping and categorization includes axial coding to analyze the most significant and frequent data from the initial coding, thus relating categories to subcategories [23]. Following the extraction of barriers, a list of the most common by-products, and coproducts were also extracted from secondary sources.

The outputs of grounded theory (list of barriers) were used to design a questionnaire to have biofuel industry members provide their perceptions on the list of barriers impacting the AB industry separated by internal, external, and marketing and distribution of coproducts and by-products. In addition, discussions with a sample of the biofuel industry experts were conducted to clarify survey results and gain additional insights. Industry members were chosen by direct requests from the projects identified in the first phase. The survey included Likert-type questions, open-ended questions, and close-ended questions. The Likert-scale questions were developed for nine different constructs that were identified during the literature review. A scale from 1 to 5 was used, where 1 was strongly disagree and 5 was strongly agree.

4. Results

4.1. Project status

A total of 59 AB projects were identified and classified by project status (**Figure 1**). The geographical distribution visually indicated that there was a relationship by region and project status for the Eastern part of the U.S. and in Mississippi. The geographic location analysis indicated that most of the advanced biofuel projects are located in the Eastern region, but the proportion rates of projects when comparing the Eastern and the Western regions does not show any significant difference between regions. Mississippi seems to have state policies designed to attract the industry. Other projects seem to be uniformly scattered in the Eastern region. In total, 19 projects were cancelled or shutdown. Of the 59 projects started since 2007, only 13 are operating in 2015.

A contingency table analysis indicated that the majority of projects have been started in the Eastern region ($n = 41$, 82%). Given that there could be a relationship between the regions and the status of projects, a test was conducted to test if the proportion of status of projects was the same for both regions. The results of the Chi-square test indicated that there was no significant relationship between regions and status of projects ($p = 0.3260$).

There are five stages of technology development for advanced biofuel projects (**Figure 2**). Each stage is representative of the feasibility of planning, financial constraints, proving conceptual design, and intellectual rights. Finally, repeat the success. The average pilot plant typically costs $10 million or less, the average demonstration plant cost is less than $100 million, and a commercial plant cost varies from $100 to $500 million. **Figure 2** shows the number of individual projects by technology status achieved from 2005 to current.

Figure 1. Map of all advanced biofuel projects since 2005 (Withers [23]).

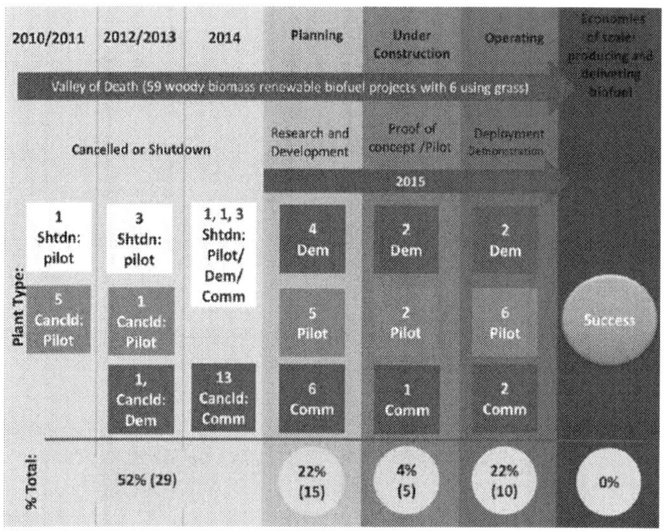

Figure 2. Project stages of technology development and percentage status where Shtdn = shutdown, Cancld = canceled, Dem = demonstration, and Comm = commercial (Withers 2016 [23]).

4.2. Perception of industry members on internal and external barriers

4.2.1. Internal barriers

A total of 16 industry members participated in the initial survey. Participants generally agreed that internal barriers include technology yield per ton (56%), technology conversion (50%), and lack of continuous project growth (44%). Participants did not view the following categories as barriers: coproducts marketing (69%), coproducts distribution (56%), by-products marketing (63%), by-products distribution (63%), strategy (56%), management (50%), and product development (44%).

Table 1 shows the median responses on internal barriers by project type, project status, and project technology. All participants had to indicate project type, project status, and technology type. Each of these categories was further divided in subcategories as shown in **Table 1**. Responses across project status are very similar and do not show a clear distinction between the subcategories. In the case of the category project type, it seems that industry members classified as pilot have a higher perception on barriers than the ones identified as open and closed. Also, in the technology type category, industry members classified as biochemical seem to have a stronger perception of internal barriers than the other technology types.

Median	TYPE			STATUS		TECHNOLOGY			
INTERNAL BARRIERS	Commercial	Demonstration	Pilot	Closed	Open	Planning	Biochemical	Hybrid	Thermochemical
Product development	2	2.5	4	2.5	3.5	2.5	4	3	2
Byproducts marketing	2	2	3	2.5	2	2	4	3	2
Byproducts distribution	2	2.5	3	2.5	2	2	4	2	2
Co-products marketing	2	2	3	2.5	2	2	4	2	2
Co-products distribution	2	2.5	3	2.5	2.5	2	4	2	2
Continuous project growth	3	3.5	4	3.5	3.5	3	4.5	3	3
Management	2	2.5	3	2	2.5	2.5	2.5	3	2
Strategy	2	2.5	4	3	2.5	2	3.5	2	2
Technology conversion rate	2	3.5	4	4	3.5	2.5	4	2	4
Technology high titer and yield per ton	3	3.5	4	4	3.5	3.5	4	4	4

Table 1. Median values of internal barriers by type, status, and technology.

A contingency analysis was conducted to compare the differences within each category or group. It was found that there were no differences within type (commercial, demonstration, and pilot) and status (closed, open, and planning). However, the contingency analysis by the technology group (biochemical, hybrid, and thermochemical) yielded a significant difference on internal barriers by-products distribution and coproducts marketing on the biochemical technology type. Given that the number of counts by cells was less than five in some cases, a Fisher's exact test was then performed on these categories; the Fisher's test determined that by-products distribution (p = 0.074) and coproducts marketing (p = 0.028) were significant barriers for a significance level of 0.1.

4.2.2. External barriers

In the case of external barriers, biofuel industry members agreed that funding (100%), renewable volume obligation (75%), EPA pathway process (75%), and RFS and RINs (56%) were external barriers. Noticeable uncertainty was placed in DOE pathway process and waiver credits. The categories of competitors, energy costs, suppliers, and third-party relationships yielded fairly similar disagreement.

The sample was also divided in categories, similar to the internal barriers analysis. The median responses on external barriers by project type, project status, and technology type were also examined (**Table 2**). Overall, the data show that industry members in all categories have a higher perception of external barriers than internal barriers. A contingency analysis was performed to compare the subcategories by project type, project status, and technology type to determine if the differences within each subcategory were significant. It was found that there were no differences in the category project status (closed, open, and planning are the same) on the perception of barriers. However, significant differences were found in the project type and technology type categories. By project type, differences were found on the perception of barriers competitors (demonstration and pilot different than commercial) and energy costs (pilot different than commercial and demonstration). And differences were found on the perceptions of barriers competitors (biochemical is different), energy costs (biochemical and hybrid different), and third-party relationships (biochemical is different to the other two). In all cases, an exact Fisher's test was conducted with a significance level of 0.1.

4.3. Marketability and distribution barriers for coproducts and by-products from advanced biofuel industries

In this part of the study, a ranking and classification of barriers impacting the marketability and distribution of lignocellulosic biofuel's coproducts and by-products were conducted. Coproducts and by-products are an important component of the AB industry business model. Without the proper marketing and commercialization strategies, coproducts and by-products cannot be commercialized. As it is today, AB industry needs to have revenue from its coproducts and by-products in order to remain competitive.

The advanced biofuel production process yields by-products and further processing generates subsequent coproducts. The list showed in **Table 3** was obtained through research from secondary sources. Combining or improving by-products can lead to desired coproducts. Unused by-products increase expenses [25], since they require disposal; as a result, increasing the value from by-products and coproducts could help sustain a biofuel project [26]. Vivekanandhan [26] suggests that many of the biofuel industry small-scale projects do not generally collect coproducts due to high opex (ongoing) costs foregoing added profit potential, while the opposite is true for commercial scale projects. The coproducts and by-products are more valuable to reduce energy costs when burned for biofuel projects are placed in landfill as waste [28]. Therefore, understanding harmful by-product waste streams is economically and environmentally beneficial when planning scaling projects to reduce harmful impact [25, 29]. According to Doherty et al. [29] and Gellerstedt et al. [30] providing value-added coproducts may lead to improved biorefinery financial success, and some coproducts could actually be more valuable than the biofuel itself [25].

Median	TYPE			STATUS			TECHNOLOGY		
EXTERNAL BARRIERS	Commercial	Demonstration	Pilot	Closed	Open	Planning	Biochemical	Hybrid	Thermochemical
Competitors	2	3	3	2	2	2	4.5	1	2
Funding	5	5	5	5	5	5	5	5	5
Suppliers	2	2.5	4	2	3	2.5	4	3	2
DOE pathway process	3	4	4	4.5	3.5	3	4	3	4
EPA pathway process	4	5	4	3.5	4.5	4.5	5	4	4
USDA pathway processes	4	3.5	4	4	3.5	3.5	4	3	4
Production tax credits	4	4	4	3.5	4	3.5	4	4	4
Renewable fuel policy standards	3	5	4	3	4	5	5	2	5
Waiver credits	3	4	4	3.5	3.5	4	4.5	4	4
Renewable volume obligation	5	5	4	3.5	5	5	5	4	5
Renewable identification numbers	3	4.5	5	3.5	4	4	5	2	4
Energy costs	2	2	5	4	4	2	3.5	4	2
3rd party relationships	3	3	3	3.5	2.5	3	4	3	2

Table 2. External median quantiles by type, status, and technology.

Product	Source	Process	Market	Examples of producing companies
		Gases and fuels		
Syngas	Biomass of lignin	Gasification	Production of ethanol, methanol, dimethyl ether, olefins, propanol and butanol [25, 34–36]	
Hydrogen	Lignin	Gasification	Fuel cells, industrial uses [25]	
Carbon dioxide	Sugars	Fermentation	Industrial uses, beverage, dry ice [25]	Lanza Tech
Carbon monoxide				Lanza Tech
Synthetic gasoline and diesel		Biochemical/ thermochemical/ hybrid	Liquid fuels	Joule, Sundrop, Envergent, Abengoa, Fiberight, Ensyn

Product	Source	Process	Market	Examples of producing companies
Jet fuel		Biochemical/ thermochemical/ hybrid		Envergent, Frontline, GEVO, Fulcrum, Byogy, Vertimass, Virent, Lanza Tech
Methane		Biochemical		Enerkem, Intrexon, Calysta, Siluria, Oberon, Kiverdi, Mango materials, Industrial microbes
Lignin	Lignin	Hydrolysis	Fuel for heat and electricity, fertilizer, wood adhesive, color additive, reinforcing filler, animal feed, yeast production [25, 37, 27]	Renmatix
Naphtha		Distillation	Fuel source solvent	Joule
		Organic acids		
Succinic acid	Glucose	Fermentation in high CO_2	Food additive, plasticism surfactants, detergents, solvents, textiles, and pharmaceuticals [29]	Myriant, Riverdia, BioAmber, Novozymes, DSM
Lactic acid polylactic acid	Glucose	Fermentation	Food and beverages, textiles [25]	Invista, Plaxica, Lanza Tech, IOC, Nature Works, Calysta, Direvo, Purac, Leaf Technologies, Myriant
Acetic acid	Glucose	Fermentation	Food additive and industrial chemicals, resins, and alcohols [25]	Zeachem, American Process
Fumaric acid	Glucose	Fermentation	Food additive, production of resins and alcohols [25]	Novozymes, Myriant
Oleic acid				
Acrylic acid				Myriant
Adipic acid				Renovia, Verdezyne
Levulinic acid				GFB Biochemical, Mercurious
		Alcohols		
n-buterol	Glucose	Fermentation	Liquid fuel, food additive, solvent [25]	
Xylitol	Xylose	Hydrogenation	Sweetener [25]	ZuChem, Xylitol, Taurus

http://dx.doi.org/10.5772/64611

Product	Source	Process	Market	Examples of producing companies
Sorbitol				joule
Arabinitol				
Aromatic compounds				
Xylose, arabinose		Dehydration	Solvent, pesticides, resins, liquid fuel [25]	Taurus, DuPont
Benzene, toluene, xylene	Lignin	Catalysis	Solvents, pesticides, resins, liquid fuel [25]	Virent, GEVO, Avantium
Olefins		Pyrolysis	Production of polyethylene [25]	SABIC, Byogy, INEOS
Biobenzene		Catalytic	Food and beverage packaging, textiles, automobiles, detergents, construction materials, and paints and coatings [38]	Virent, Anellotech's
Macromolecules				
Cellulose nanofibers	Cellulose	Chemical-mech treatment	Structural composites, plastics, films [25]	
Polyhydroxyalkanoate	Lignin	Fermentation	Biodegradable plastic use in films, packaging, fibers, coatings, foams, and medical [25]	
Lignosulfonates	Lignin	Sulfonation	Dispersants, emulsifiers, binders, sequestrants, adhesives, fillers, dust prevention [25]	
Carbon fiber	Lignin	Melt spinning	Reinforcement for automotive plastics [25]	BETO
High purity lignin	Lignin		Coatings, emulsifiers, gels, antimicrobial products [25]	
Other products				
Cellulose nanofibers	Cellulose	Hydrolysis	Animal feed [25]	
Protein	Protein		Animal feed [25]	Cargill, Calysta, Valicor
Biochar	Lignin	Combustion	Fuel, soil additive and carbon sequestration [25]	Cool planet, Mercurious
Betulinol	Forest residues		Antioxidant [29]	
Propanediol (PDO)	Sugars	Fermentation	Deicing fluids, engine coolants, heat transfer fluids, polyurethanes, solar thermal,	DuPont, Joule

Product	Source	Process	Market	Examples of producing companies
			unsaturated polyester resins, [39]	
Butanediol, biobutadiene	Dextrose or sucrose	Fermentation	Plastics, solvents, electronic chemicals, and elastic fibers [40]	Joule, Myriant, Genomatica
N butanol	Sugars	Fermentation	solvents, glycol ethers, acetate, acrylate [41]	Green Biologics, DuPont, GEVO
Polyethylene terephthalate (PET)	Isobutanol	biochemical	Films and bottles for packaging, fibers for nonwovens, textiles, automotive resin.	Anellotech's, GEVO, Joule
Farnesene	plant sugars	Fermentation	Solvents, emollients, vitamins [42]	Amyris, Intrexon, Chromatin
Polyamides	Syngas	Fermentation	Precursor for specialty plastics [43]	Arkema, Avantium, Genomatica, DuPont, Terryl
5c and 6c sugars				GeoSyn fuels, Sweetwater Energy, Kakira, San Martinho, Cascades, Buriram, Applied Biorefinery
Omega 3's and 7's				Solarvest, Nature Works, Lanza Tech, IOC, Calysta, KD-Pharma, BioProcess Algae, Cellana
		Waxes		
Furfural	Pentose and hexoses	Hydrolysis	Food additive in vanilla, resins [44, 45]	Chempolis, DuPont, Glucan Biorenewables, Mercurious
Suberin	Forest residues		Fatty acid [29]	

Table 3. List of potential coproducts and by-products from AB industry.

Many initial biofuel projects as in early in 2005, did not focus on these secondary products, but instead focused on more pressing technology and funding issues. Forty-two percent of all projects included in this study are pilot and demonstration plants designed for testing purposes, with reduced focus on secondary outputs. The commercial facilities are realizing the value of their coproducts and are restrategizing. For example, Virent Biogasoline, a commercial biofuel company impacted by the blend wall, changed its website to list available quantities of various coproducts they produce. Discussing survey results with the industry indicated there are at least 44 coproducts produced, nearly twice the number identified from the literature. This increase was based on companies currently stymied by blend-wall limitations that reduce demand to fund production economies of scale. These limitations drive stakeholders to consider new markets beyond biofuel to meet shareholder financial expectations. Advanced biofuel companies are currently focused on shifting to platform technologies, targeting higher value coproducts and the available funding arena [32, 33].

In addition to the perception of AB industry members, perceptions of other AB industry stakeholders such as academicians, government representatives, and journalists are included to rank AB's barriers for by-product and coproducts. Altogether, a total of 44 responses were obtained from all stakeholders. Out of the 44 responses, 28 respondents provided usable data to this section, identifying barriers to coproduct marketability (N = 27) and distribution (N = 28), as well as by-product marketability (N = 28) and distribution (N = 22), see **Tables 4** and **5**. Cost, financing, and public awareness were the main barriers across the four classifications. There are many similarities of response between the four categories of coproducts and by-products marketability and distribution barriers, such as infrastructure, fossil industry control, public perception, and policy. Some responses are very similar to the internal and external barriers analyzed in the previous section; however, many are unique to this study, such as sole source risk, heated rail car shortage, and flooding a niche market.

The perceived need of coproducts and by-products' infrastructure to support the already subsidized industry was not expected. Nor did the industry expect to be stymied by the blend wall, the fossil fuel industry buying cellulosic waiver credits (CWCs) and lobbying against them, politics, or a slowly developing infrastructure. It would seem the advanced biofuel industry initially did not examine the end-user market demand and capabilities for additional by-products and coproducts. The survey results indicated that by-product and coproducts infrastructure are a niche market and saturated in the short term, since the industry was already shifting toward platform technologies. According to, there was a 9% growth in premium renewable biochemicals in 2015, which implies that the shift to platform technology would potentially become a barrier, as well, in a niche market. Reidy [32] stated the industry is moving to produce and sell premium products. Selling premium products would imply the niche market barrier may only affect those in competition with advanced biofuels that already produce nonrenewable premium chemicals, such as the fossil fuel industry. The shift in this industry to compete at a multiproduct platform level other than biofuel in new markets was an attempt to avoid sole source risk and maximize by-products potential and funding. Rural economic development was one of the three primary objectives established by the government. The survey results indicated that some projects face lack of heated distribution channels from declining rural rail systems. In the short term, premium coproducts, such as waxes, will have to be developed to offset the cost of changing perceived risk to increase demand for the revitalization of the heated rural rail infrastructure.

Coproducts marketability		Coproducts distribution	
Main barrier	**Secondary barrier**	**Main barrier**	**Secondary barrier**
Biointegrity of supply chain	Access to capital	Competition and distribution restriction	Access to capital
Consumer awareness of larger societal benefits	Available volume	Cost	Available volume
Cost	Lack of benefit to producers	Financial support	Competition
Finding high credit-worthy third party for off-take	Limited market and competition from non-renewable sources	Flooding a niche market	Consumer demand
GMO	Not being focused	GMO isolation	Controlled by oil companies
Government uncertainty	Obligated parties	Government uncertainty	Misinformation about the need for the industry as a whole
I did not know that enough co-products produced to be impeded	Perception of cost and efficacy	Immature supply chain infrastructure	Obligated parties
Lack of clear end user demand	Poor policy	Lack of clear end user demand	Product purity
Lack of incentives	Public ignorance	Limited volume	Requires heated tankers or rail cars for shipment
Low identified uses	Quality	Market fragmentation	Small markets
Oversupply	Separation of water	No infrastructure	Market fragmentation
Process economics	Sole source risk	Oil industry	
Public awareness		Poor policy	
Public perception		Requires additional fractionation—no local fractionators	
Quality of F-T wax for use as a wax		Scale match or biointegrity of chemicals	
Specifications		Unavailability	
Technology		Unclear markets breeds unclear distribution channels	

Table 4. Coproducts marketability and distribution barriers.

By-products marketability		By-products distribution	
Main barrier	**Secondary barrier**	**Main barrier**	**Secondary barrier**
Cost	Conditioning and transportation to markets	Controlled by oil companies	Consumer demand
Financing	Controlled by oil companies	Cost	Lack of balance sheets
GMO	Distance to market	Financing	Lack of benefit to producers

http://dx.doi.org/10.5772/64611

By-products marketability		By-products distribution	
Main barrier	Secondary barrier	Main barrier	Secondary barrier
High value product development	Investment	Flooding a niche market	Lack of product knowledge by customers
I did not know that enough by-products produced to be impeded	Lack of balance sheets	GMO isolation	Lack of true public education
Lack of awareness among public	Lack of benefit to producers	Lack of awareness among public	Low volume
Lack of clear end user demand	Limited markets	Logistics	Unfamiliarity
Lack of incentives	No perceived need	Market demand	
Low identified uses	Not being Focused	Marketing	
Low volume	Poor Policy	No infrastructure	
Low value wood ash		No local markets for ash	
No infrastructure		Oil industry	
Oversupply		Production technology	
Price		Transport cost	
Quality		Unclear markets breeds unclear distribution channels	
Specifications			
Technology			
Value proposition			

Table 5. By-products marketability and distribution barriers.

5. Conclusions

The barrier analysis indicated the perspectives on barriers to production of advanced biofuel are different by project type, status, and technology. The barrier impact changed across time and type of project. The closed projects faced the same barriers; however, fewer barriers than the current projects now that the blend wall is a permanent factor. Discussions with bioeconomy industry representatives about the implications of the blend wall led to an improved RFS model and improved understanding of the system.

Overall, timing is the main barrier to advanced biofuel projects. If the decline in fuel consumption was realized by all parties, the advanced biofuel group may not currently exist. However, the outcome of timing has created the realization that the remaining advanced biofuel projects are now rapidly moving to become advanced biochemical platform technology companies, quickly and annually claiming market share of global premium coproducts. They are well poised to either blend higher levels of biofuel and/or premium coproducts, dependent upon the full spectrum of petroleum barrel price and demand. Additionally, they are unifying their efforts to become a household lifestyle premium brand. Will the petroleum industry realize its marketing myopia and grow with the bioeconomy global brand, or will it inadvertently continue as the increasingly undesired environmentally unfriendly brand? A review of the literature did not distinguish any lists of barriers to the marketability and distribution of coproducts and by-products. However, through the survey and interviews in this study, an extensive list of barriers was developed, including 27 coproducts marketability and 28 coproducts distribution barriers, and 28 by-products marketability and 22 by-products distribution barriers. The main barriers were cost, funding, fossil industry control of market, and public awareness

To move the bioeconomy forward faster, developing an incremental greenhouse gas (GHG) carbon tax is needed on an incremental level to fund the developing infrastructure, public education, and factual perception to bolster the demand for biofuel and biochemicals. The funding is privately earmarked, ready, and in bearish stance, awaiting public demand. The information compiled in this study can aid the biofuel industry and the bioeconomy in future pursuits; it can provide guidance to inform R+D to reduce costs and improve perceived risk, increasing investment viability.

Author details

Henry Jose Quesada-Pineda*, Jeremy Withers and Robert Smith

*Address all correspondence to: quesada@vt.edu

Virginia Tech, Blacksburg, United States

References

[1] Shafiee, S., & Topal, E. (2008).When will Fossil Fuel Reserves be Diminished. Energy Policy 37, Science Direct (181–189). Retrieved from http://www.academia.edu/4853226/When_will_ fossil fuel_reserves be_diminished

[2] Chiari, L., & Zecca, A. (2011). Constraints of fossil fuels depletion on global warming projections. Energy Policy, 39(9) 5026–5034.

[3] Brown, T., & Brown, R. (2012). A review of lignocellulosic biofuel commercial-scale projects in the United States. Biofuels Byproducts and Biorefining, 7(3):235–245. DOI: 10.1002/bbb.1387

[4] Szczodrak, J., and Fiedurek, J. (1996). Technology for conversion of lignocellulosic biomass to ethanol. Biomass and Bioenergy, 10(5–6):367–375.

[5] Fueling Growth (2013). 2013 E2 Advanced Biofuel Market Report. Retrieved from http://www.fuelinggrowth.org/e2-advanced-biofuel-market-report-2013/

[6] Moxley, G., & Zhang, P. (2007). More accurate determination of acid-labile carbohydrates in lignocellulose by modified quantitative saccharification. Energy & Fuels, 21(6):3684–3688. DOI: 10.1021/ef7003893

[7] Riedy, M. J. 2013. Financing Advanced Biofuels, Biochemicals and Biopower in Integrated Biorefineries: Status of available domestic and International Financing Mechanisms. Biomass 2013: How the advanced bioindustry is reshaping American energy. July 31-August 1, 2013. Washington, DC Convention Center.

[8] National Highway Traffic Safety Administration (2016 March). CAFÉ–Fuel Economy. Retrieved from http://www.nhtsa.gov/fuel-economy

[9] U.S. Department of Energy-Energy.gov. (2015b). The Energy Policy Act of 2005 (EPAct 2005). Retrieved from http://energy.gov/eere/femp/articles/energy-policy-act-2005

[10] U.S. Environmental Protection Agency (2009). Renewable Fuel Standard (RFS). Retrieved from http://www.epa.gov/otaq/fuels/renewablefuels/index.htm

[11] The White House (2006). The Advanced Energy Initiative. Retrieved from https://georgewbush-whitehouse.archives.gov/ceq/advanced-energy.html

[12] U.S. Department of Energy (2013). 3rd. Annual MSW to Biofuels Summit, Orlando, FL. February. Retrieved from http://energy.gov/sites/prod/files/2014/04/f14/duff msw_to_biofuels_ summit.pdf

[13] Sorda, G. et al. (2010, November). An Overview of Biofuel Policies Across the World. Retrieved from http://www.sciencedirect.com/science/article/pii/S0301421510005434?np=y

[14] U.S. Environmental Protection Agency (2007). Summary of the Energy Independence and Security act. Retrieved from https://www.epa.gov/laws-regulations/summary-energy-independence-and-security-act

[15] California Energy Commission (2016). Low Carbon Fuel Standard. Retrieved from http://www.energy.ca.gov/low_carbon_fuel_standard/

[16] U.S. DOE (2010, April). Annual Energy Outlook 2010. Retrieved from http://www.eia.gov/oiaf/aeo/pdf/0383(2010).pdf

[17] ABLC 2015. (2015, March 11–13). Advanced Bioeconomy Leadership Conference. Capital Hilton, Washington D.C. Retrieved from http://www.biofuelsdigest.com/bdigest/2015/01/05/ablc-2015-initial-line-up-announced-for-the-advanced-bioeconomy-leadership-conference/

[18] Mendell, B., & Lang, A. (2012). The Rise and Fall of Wood-Based Biofuels, Part 1. Forisk Consulting, Wood BioEnergy U.S. 5(2), page 1. Retrieved from http://forisk.com/wordpress// wpcontent/assets/WBUS_Free_201304.pdf

[19] Mendell, B., & Lang, A. (2013). The Rise and Fall of Wood-Based Biofuels, Part 2. Forisk Consulting, Wood BioEnergy U.S. Retrieved from http://forisk.com/wordpress// wpcontent /assets/WBUS_Free_201307.pdf

[20] Love's Biofuel representative (2015, July). Personal discussion in Washington DC.

[21] Coleman, B. (2016, February). The Advanced Bioeconomy Leadership Conference. Grand Hyatt, Washington D.C. Forisk Consulting (2014). Project development. Wood Bioenergy U.S., 6(2):12.

[22] Charmaz, K. (2006). Constructing Grounded Theory: A Practical Guide Through Qualitative Analysis. London: SAGE.

[23] Withers, J. (2015). Barriers Impacting United States Advanced Biofuel Projects. MS Thesis. Department of Sustainable Biomaterials. Blacksburg, VA: Virginia Tech. .

[24] Patton, J. (2010). Value-added coproducts from the production of cellulosic ethanol. Central Grasslands Research Center, North Dakota State University.

[25] Closset, G., Raymond D., & Thorp B. (2005). The Integrated Forest Products Biorefinery. An Agenda 2020 Program. Retrieved from http://www.pyne.co.uk/Resources/user/ glasgow/ Biorefinery%20Business%20Case-June1%20_2_.pdf

[26] Vivekanandhan, S., Zarrinbakshs, N, Misra, M, and Mohanty, A.K. 2013. Coproduces of Biofuel Industries in Value-added Biomaterials Uses: A Move towards a Sustainable Bioeconomy. Chapter in Liquid, Gaseous and Solid Biofuels-Conversion Techniques. Editor Zhen Fang, ISBN 978-953-51-1050-7. InTech Publishers.

[27] Poursorkhabi, V., Misra, M., & Mohanty, A. K. (2013). Extraction of lignin from a coproduct of the cellulosic ethanol industry and its thermal characterization. BioResources, 8(4):5083–5101.

[28] Soderholm, P., & Lundmark, R. (2009). The development of forest-based biorefineries: implications for market behavior and policy. Forest Products Journal, 59(1/2):6–16.

[29] Doherty, W. O. S., et al., (2011). Value-adding to cellulosic ethanol: lignin polymers. Industrial Crops and Products, 33(2):259–276.

[30] Gellerstedt, G., Sjoholm, E., & Brodin, I. (2010). The wood-based biorefinery: a source of carbon fiber. The Open Agriculture Journal, 3:119–124. Retrieved from www.benthamscience.com

[31] Berven, D. (2016, Feb. 12). Personal discussion at the Advance Bioeconomy Leadership Conference.

[32] Reidy, M. (2015, March 12). Personal discussion at the Advanced Bioeconomy Leadership Conference.

[33] Petrus, L., & Noordermeer M. (2006, July 27). Biomass to biofuels, a chemical perspective. Green Chemistry, 8(10):861–867.

[34] Stewart, N., et al. (2008). Plants to power: bioenergy to fuel the future. Trends in Plant Science, 13(8):421–429.

[35] Buaban, B., et al. (2010). Bioethanol production from ball milled bagasse using an on-site produced fungal enzyme cocktail and xylose-fermenting *Pichia stipitis*. Journal of Bioscience Bioengineering, 110:18–25.

[36] Singh, P., et al. (2010). Biopulping of lignocelluose material using different fungal species: a review. Review of Environmental Science and Biotechnology, 9:141–151.

[37] Virent (2014, Dec.). Virent Announces World's First Demonstration of Full Range Bio-Aromatics Production. Retrieved from http://www.virent.com/news/virent-announces-worlds-first-demonstration-of-full-range-bio-aromatics-production/

[38] Dupont (2016). Susterra 1, 3-propanediol is an Innovative, Specialty Idol that Provides both High Performance and Renewable Content. Retrieved from http://www.dupont-tateandlyle.com/susterra

[39] Genomatica, & BASF (2015, Sept.). BASF and Genomatica Expand License Agreement for 1,4-butanediol (BDO) from Renewable Feedstock. Retrieved from https://www.basf.com/en/company/news-and-media/news-releases/2015/09/p-15-347.html

[40] Green Biologics (2016). n-butanol. Retrieved from http://www.greenbiologics.com/n-butanol.php

[41] Amyris (2015, Oct.) Amyris Achieves Record Low Cost Farnesene Production. Retrieved from https://amyris.com/amyris-achieves-record-low-cost-farnesene-production/

[42] Lanza Tech, & Lane, J. (2013, December). Evonik and Lanza Tech Sign 3-year Development Deal for Biobased Plastics. Retrieved from http://www.biofuelsdigest. com/bdigest/2013/12/09/evonik-and-lanzatech-sign-3-year-development-deal-for-bio-based-plastics/

[43] Larsson S, et al. (1998). Commercial harvest of willow wood chips in Sweden. Proceedings of the 10th European Conference, Biomass for Energy and Industry, Wurzburg C.A.R.M.E.N. 200–203.

[44] Goncalves et al. (2013). Industrial-scale steam explosion pretreatment of sugarcane straw for enzymatic hydrolysis of cellulose for production of second generation ethanol and value-added products. Bioresource Technology, 130:168–173.

[45] ABLC 2016. (2016, March 16–19). Advanced Bioeconomy Leadership Conference. Grand Hyatt, Washington D.C. Retrieved from http://www.biofuelsdigest.com/bdigest/2015/11/19/ ablc-2016-initial-line-up-announced-for-advanced-bioeconomy-leadership-week/

Chapter 7

The Biofuel Crops in Global Warming Challenge: Carbon Capture by Corn, Sweet Sorghum and Switchgrass Biomass Grown for Biofuel Production in the USA

Roland Ahouélété Yaovi Holou and
Valentin Missiakô Kindomihou

Additional information is available at the end of the chapter

http://dx.doi.org/10.5772/65690

Abstract

This research evaluates potential carbon capture of sweet sorghum, switchgrass, and corn grown in Portageville, Missouri, from 2007 to 2009. Our results showed that corn grain C content averaged 43%, whereas C grain captured was 1.3–4.7 Mg C ha^{-1} depending on year and N rate. N fertilization significantly increased C capture, but not C content of grain. C capture by switchgrass depended on cultivars and harvest date. Switchgrass cv. Alamo biomass contained 46% C compared to 44% C for Blackwell's. Alamo maximum C capture depended on year, being 9.8 Mg C ha^{-1} in 2008 and 13.4 Mg C ha^{-1} in 2009. C is equivalent to 32.3–49.6 Mg CO_2 ha^{-1}, while Blackwell captured 3.7–4.4 Mg C ha^{-1}. C in sweet sorghum biomass ranged from 42 to 45%, whereas total C capture ranged from 3.2 to 13.8 Mg ha^{-1} according to year, soil, and N rate. The highest C capture appeared in loam. Sweet sorghum aboveground biomass showed 82% C captured in the stalk. When converted into CO_2, C captured by sweet sorghum was equivalent to 12–51 Mg CO_2 ha^{-1}. In addition to their biofuel potential, corn, switchgrass, and sweet sorghum can substantially contribute to environmental cleaning by capturing a significant amount of CO_2.

Keywords: carbon, corn, sweet sorghum, switchgrass, global warming, CO_2

1. Introduction

For many decades, reports have increasingly proven that the climate of our planet is changing mostly because of anthropogenic effects that are increasing the global warming [1–3]. The increase of the concentration of certain molecules such as carbon dioxide (CO_2), methane, nitrous oxide, chlorofluorocarbons, aerosol, and sulfur in the atmosphere is one of the leading causes of the temperature increase. Actions are being taken to reduce the concentration of greenhouse gases in the atmosphere and to better protect the ozone layer [4]; one of these actions is the use of biofuel in engines [5–7]; [3].

Interestingly, plants have a unique ability to uptake CO_2 from the air through their stomata and use it for photosynthesis [8]. Therefore, plants help clean the environment by capturing the CO_2 [9, 11], and integrating it into the metabolic systems that make up carbohydrates and other carbon-containing compounds. When they die, plants return the carbon they sequestered in their roots and leaves to the soil [4]. The mechanism by which plants capture the atmospheric carbon and move it to a soil has been reviewed [12]. The previous author showed that grasses transfer more carbon to the soil than trees. However, because of the biological activity of microorganisms and some anthropogenic reasons, not all the carbon captured by plants is sequestered in the soil [12, 13]. Some carbon captured by plants returns to the atmosphere in the form of CO_2 [14, 8]. Although there is controversy about the impact of biofuel crops on the climate, the majority of the papers that have evaluated the environmental impact of these crops showed that each crop can significantly decrease the emission of greenhouse gases [15]. For instance, the greenhouse gas emitted by the cellulosic ethanol produced from switchgrass is 94% smaller than that from gasoline, so the production of ethanol from switchgrass is much cleaner than the production of gasoline [16].

Roots are the main carbon depositor in the soil, and they decompose very slowly [17, 8]. Unfortunately, it is difficult to study carbon sequestration in roots [8]. Nevertheless, Andress [18] proved that the amount of CO_2 sequestered by switchgrass in the soil is 138.1 kg of CO_2/Mg of aboveground biomass. In the northern plains of the US, switchgrass grown for biofuel production returned to the soil up to 4.42 Mg C $ha^{-1}year^{-1}$ [19]. Sorghum can capture 3.4–7.2 Mg of CO_2 per hectare [20]. In addition, these previous authors found that sorghum roots can accumulate up to 14% of the total carbon capture in the above and underground biomass. In general, roots can contribute from 7 to 43% of the total C sequestered by a plant [21]. In the case of corn, the amount of C sequestered in the roots is 60% more than that of the stover [22]. Residues from crops are a significant way to improve C sequestration into a soil [20]. Usually, plants that produce a lot of biomass capture the most C. One of the strategies currently used is to pay farmers a carbon credit to entice them to grow better crops that can help reduce global warming. In general, carbon dioxide costs $100 per metric ton [23, 24].

Our research proved that sweet sorghum and switchgrass can produce a significant amount of biomass, suggesting that their C capture will be high. Unfortunately, very little is known about the carbon capture potential of these crops in Missouri. Nevertheless, this kind of information is needed to make a complete economic and environmental assessment of these biofuel crops. The objectives of this research have been (1) to determine the carbon capture by corn grain, switchgrass, and sweet sorghum biomass, and (2) to determine the impact of nitrogen (N) application and the soil type on carbon capture. For sweet sorghum, we also studied the partitioning of the carbon capture between the leaves and the stalk. Our general

http://dx.doi.org/10.5772/65690

goal is to show that in addition to producing biofuel, sweet sorghum, switchgrass, and corn have additional advantages in their ability to clean the environment and therefore help resolve the global warming challenge.

2. Methods

The research was carried out in South-eastern Missouri over a 2-year period (2008–2009) on Tiptonville silt loam soil (fine-silty, mixed, superactive, thermic Oxyaquic Argiudolls). The test was located at Portageville (36°24′N, 89°41′W) in 2008 and Hayward (36°23′N, 89°39′W) in 2009. Weather data were obtained from Missouri Historical Agricultural Weather Database (www.agebb.missouri.edu/weather/history). Electronic weather stations (Campbell Scientific Inc., Logan, UT) were established at each test site to measure hourly air temperature, relative humidity, wind direction and speed, solar radiation, and rainfall.

The experimental design [25–27] was a four-replicate, randomized complete block. Each block consisted of seven N treatments. Each N treatment corresponded to a plot. Each plot was 8.30 m long and 3.05 m wide (four rows spaced 76.2 cm apart).

Corn (*Zea mays* cv. P33N58) was planted mid- to late April at 79,071 seeds ha^{-1} on a silt loam soil type and the sweet sorghum (*Sorghum bicolor* cv. M81E), in May at 296,516 seeds ha^{-1} on loamy, clayed, and sandy soils. The nitrogen rates applied were 0, 45, 90, 134, 179, 224, and 269 kg N ha^{-1} on the corn while 0, 22, 45, 67, 90, 112, and 135 kg N ha^{-1} on the sweet sorghum. Less than 2 weeks after planting, the field received two applications of atrazine (2-chloro-4-ethylamine-6-isopropyl amino-S-triazine) at 1.1 kg ha^{-1} active ingredient to control weeds. Additionally, the field was regularly hoed as needed to reduce weeds not controlled by the herbicides. Three weeks after planting, the appropriate N rate was broadcast by hand on each plot using ammonium nitrate (17% nitrate-N, 17% ammonium-N). Ammonium nitrate was chosen because it does not have urea's potential for ammonia volatilization, simplifying the test by minimizing and removing the uncertainty of ammonia losses. The field was irrigated as needed. The corn field was five furrow irrigated with 76-mm water applications per year. At maturity, the two middle rows in each plot were harvested using a plot combine. The grain yield was calculated per plot based on 15% moisture content. By the sorghum side, five furrow irrigations (76-mm water applications) per year were made on the loam and clay compared with six to eight sprinklers (25-mm applications) per year on the sand. The sandy soil was irrigated with linear move sprinkler irrigation, whereas the loam and the clay soils were furrow irrigated. The sweet sorghum heads were removed a month before the harvest of the stalk. This was done to maximize sugar in the stalk. Four and a half months after germination (**Table 1**), sweet sorghum was harvested using a hay sickle mower. The fields used in 2007 and 2008 had been previously planted in cotton and soybean, respectively. In 2009, the sorghum planted on the clay soil followed soybean, the sorghum planted on the loam followed corn, and the sorghum planted on the sand followed cotton.

N rate (kg ha^{-1})	Carbon capture in corn grain (kg ha^{-1})†	
	2008	2009
0	2536.9 b	546.1 c
45	2513.1 b	1318. b
90	3591.6 ab	2471.4 a
134	4727.1 a	2690.2 a
179	4558.2 a	2886.2 a
224	3588.4 ab	2948.8 a
269	3884.5 ab	2917.0 a

†: Numbers followed by different letters are statistically different within the same column at $p \leq 0.05$.

Table 1. Mean separation of the carbon capture in corn grain.

Switchgrass (*Panicum virgatum* cv. Alamo) was drill planted in 6-m wide strips (east to west), spaced parallel 100 m apart and 450 m long by a cotton field near Portageville (MO, USA; 36.4253°N, 89.6994°W) [28, 29]. The main soil in the field was a Bosket fine sandy loam (bosket, fine loamy, mixed active, thermic, mollic, and Typic *Hapludales*) soil. The field was located in the upper Mississippi River Delta region where the topography is nearly flat and southwest winds in May and June sometimes cause damage to young crops. The farmer planted the switchgrass as a wind break to minimize blowing sand injury to cotton seedlings. An added benefit of the strips is habitat for birds and rabbits. The field is burned every 4 years in April and mowed annually in September or October, and switchgrass strips have not received any lime, pesticides, N, P, or K since establishment in 1990. This crop required fewer nutrients probably because it fixed the atmospheric N and had mycorrhizae activity. The optimal N fertilization rate applied in the US is not clear but ranged from 120 to 224 kg N ha^{-1} [29].

In 2008 and 2009, switchgrass biomass was monthly harvested from May to November. For each sampling date, four plots in the field were evaluated. The biomass was harvested from the center of the strips by hand in a floristically homogeneous subplot of 1.67 m^2 using a hay sickle mower. Total fresh biomass weight was separated into leaves, stem, and head and oven dried for constant weight. Switchgrass biomass was determined by extrapolation from the biomass obtained in the 1.67-m^2 subplot.

The biomass of four samples of each N treatment was dried in an oven and analyzed for carbon using the LECO SC 44 Carbon Analyser (Leco Corp, St Joseph, MI) adapted from NRCS [30]. The carbon content (%) was directly read from the machine. The carbon capture in the biomass was calculated as follows:

$$\text{Carbon capture (kg ha}^{-1}) = \text{dried Biomass (kg ha}^{-1}) \times \text{Carbon content (\%)} \quad (1)$$

The equivalent captured CO_2 was calculated based on the oxidation reaction of carbon:

$$C + O_2 \rightarrow CO_2 \tag{2}$$

Based on the molecular weight of carbon (14 g) and that of CO_2 (44 g), each gram of carbon sequestered by a plant is equivalent to 3.14 g of CO_2 uptake. The equivalent amounts of CO_2 captured by the plants were extrapolated considering that ratio.

The equivalent CO_2-sequestered switchgrass soil was extrapolated based on a previous study done by Andress [18] who proved that switchgrass sequestered 138.1 kg of CO_2/Mg of aboveground biomass. The data were analyzed using the Proc mixed model in SAS 9.2 (SAS Institute Inc., Cary, NC). Significant differences were assumed for $p \leq 0.05$. The year of the study, the soil type, and the N rate were considered the main fixed factors, whereas the block (repeat) was classified as a random variable. For the Proc mixed model, the estimation method was the restricted maximum likelihood (REML). Means were separated and grouped by letter by using the macro developed by Saxton [31]. Significant differences are assumed for $p < 0.05$.

3. Results

3.1. Carbon capture in corn grain

In general, the carbon content (%) and the amount of carbon captured (kg ha^{-1}) in corn grain depended on the year ($p < 0.0001$). In 2008, 43.8% of the grain was carbon compared to 42.9% in 2009. Additionally, N fertilization significantly affected the amount of C sequestered in the grain ($p < 0.0001$), but not the carbon content of the grain ($p = 0.4051$). Usually, the carbon capture in the grain increased as the N rate went up (**Table 1**).

Moreover, the impact of the N rate on the C capture in the grain was more pronounced in 2009 ($p < 0.0001$) than in 2008 ($p = 0.03$) (**Figure 1**).

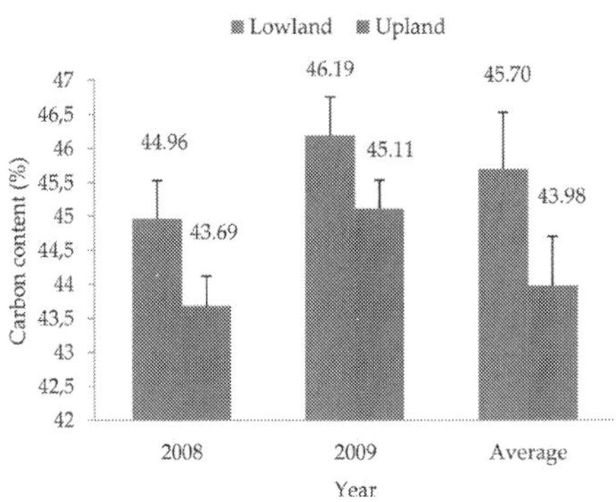

Figure 1. Carbon content of switchgrass biomass.

This confirmed the N need that the corn had on the loam in 2009, which was planted after corn in the rotation system. The corn was then unable to capture as much carbon from the air when compared to the previous year.

3.2. Carbon capture in switchgrass biomass

The carbon content in switchgrass biomass depended on the year of the study and the variety. Switchgrass var. Alamo contained more C (45.73%) than the Blackwell variety (44%). The carbon content of the underground biomass was similar to that of the aboveground. The amount of carbon sequestered in the aboveground biomass depended on the switchgrass variety, the year of the study ($p = 0.002$), and the date ($p < 0.0001$). The amount of C captured by the Alamo variety was more than three times that of the Blackwell one (**Table 2**).

Usually, the maximum carbon captured by the Alamo variety was reached in October (**Table 2**). In 2008, the Alamo variety captured a maximum of 9.8 Mg ha^{-1} compared to 13.4 Mg ha^{-1} in 2009. By contrast, the Blackwell variety captured 3.7 Mg ha^{-1} in 2008 as opposed to 4.4 Mg ha^{-1} in 2009. When the maximum carbon captured by the Alamo variety was converted into CO_2, it was equivalent to 32.3 and 49.6 Mg ha^{-1} in 2008 and 2009, respectively (**Table 2**). By contrast, the Blackwell variety captured 13.6 Mg CO_2 ha^{-1} in 2008 compared to 16.2 Mg CO_2 ha^{-1} in 2009. The amount of CO_2 sequestered to the soil by Alamo ranged from 3.03 to 4.02 Mg ha^{-1} according to the year (**Table 2**) compared to 1.17–1.34 Mg ha^{-1} for the Blackwell variety (**Table 2**).

Variables	Year	Alamo					Blackwell
		May	June	July	October	November	July
				kg ha^{-1} year^{-1}			
Carbon capture in aboveground biomass	2008	1571.6	4933.5	7823.7	9807.2	9276.3	3692.2
	2009	3816.5	8737.5	11074.3	13400.6	9560.8	4382.5
CO_2 capture by plant	2008	5815.0	18254.1	28947.8	36286.7	34322.2	13661.0
	2009	14121.0	32328.6	40974.9	49582.2	35375.0	16215.1
CO_2 sequestered to soil	2008	499.6	1556.6	2438.9	3031.2	2810.3	1174.1
	2009	1148.3	2628.9	3331.9	4024.7	2826.4	1342.8

Table 2. Carbon capture in switchgrass cv. Alamo.

3.3. Sweet sorghum carbon capture

In addition to its biofuel potential, sweet sorghum captured 3.2–13.8 Mg ha^{-1} according to the year, the soil, and the N rate (**Table 3**). The highest carbon capture was recorded in the loam. Usually, the carbon content of the aboveground biomass ranged from 41.9 to 44.6% (**Table 3**). Unlike in the loam and the sand, the carbon in the clay was not affected by the N rate (**Table 4**). This suggested that the lack of available N in the soil reduced the ability of sweet sorghum to assimilate atmospheric CO_2 into its metabolic systems to build organic compounds. The fact that in 2009 (year where sweet sorghum was planted after corn in the loam) the impact of the N rate on the C content was significant only in the loam confirmed that sweet sorghum had difficulty taking up the amount of N it needed from the soil to perform photosynthesis. In general, the application of N improved the C content of the biomass.

Variables	Year	Soil type	N rate (kg ha^{-1})†							
			0	22	45	67	90	112	134	Mean
Whole biomass carbon content (%)	2007	Loam	42.8	42.9	42.9	42.9	42.9	42.9	42.5	42.8
	2008	Clay	42.9	43.1	43.2	43.5	43.2	43.2	43.3	43.2
		Loam	43.2	43.3	43.3	43.4	43.3	33.8	43.3	41.9
		Sand	44	43.4	43.4	44.5	43.4	43.4	44.3	43.8
	2009	Clay	42.9	42.9	43	43.1	43.2	43	43.1	43
		Loam	44.3	44.8	44.4	44.6	44.8	43.8	45.2	44.6
		Sand	44.4	44.4	44.5	44.6	44.4	44.5	44.5	44.5
Total carbon yield (Mg ha^{-1})	2007	Loam	6.8	5.8	7.5	8.6	6.6	7.9	6.1	7
	2008	Clay	4.9 b	5.1 b	5.9 ab	8.4 a	7.8 a	7.1 ab	8.0 a	6.7
		Loam	11.4	10.6	13.8	11.5	11	10	11.7	11.4
		Sand	4.5	3.7	5.1	4	5.1	6.2	5	4.8
	2009	Clay	3.3 c	5.6 abc	5.6 abc	6.9 ab	7.2 ab	7.8 a	7.7 ab	6.3
		Loam	3.2 c	4.7 bc	4.4 bc	6.3 ab	5.3 abc	6.4 ab	7.7 a	5.6
		Sand	4.8 c	5.1 c	7.6 a	7.1 ab	5.0 c	6.2 abc	5.4 bc	5.8
Equivalent CO_2 captured by biomass (Mg ha^{-1})	2007	Loam	25.3	21.5	27.6	31.8	24.2	29.2	22.5	26
	2008	Clay	18.1	19	21.8	31	29	26.1	29.6	24.9
		Loam	42.2	39.3	51.1	42.5	40.7	36.9	43.4	42.3
		Sand	16.8	13.8	18.8	14.9	19	23	18.7	17.9
	2009	Clay	12.2	20.8	20.6	25.7	26.8	29.1	29.6	23.4
		Loam	11.9	17.6	16.5	23.4	19.7	23.7	28.3	20.9
		Sand	17.8	18.9	28.1	26.2	18.5	23	20.1	21.6

†Numbers followed by different letters are statistically different within the same column at $p \leq 0.05$.

Table 3. Carbon content and capture by sweet sorghum aboveground biomass.

Variable†	2007	2008			2009		
	Loam	Clay	Loam	Sand	Clay	Loam	Sand
Carbon content whole biomass	*	ns	ns	***	ns	**	ns
Leaves C content	*	ns	ns	ns	ns	ns	ns
Stalk content stalk	ns	ns	ns	ns	ns	*	ns
Total capture yield	ns	*	ns	ns	**	*	*
Carbon capture leaves (kg ha^{-1})	ns	ns	ns	ns	ns	ns	*
Carbon capture stalk (kg ha^{-1})	ns	*	ns	ns	**	*	ns

†The values (i.e., symbols) in the table are the probability associated with the test of the impact of *N*; $^{*}p < 0.05$; $^{**}p < 0.01$, and $^{***}p < 0.001$.

Table 4. Impact of *N* on the carbon content and capture in sweet sorghum aboveground biomass.

However, the impact of N on the biomass C content depended on the organ. Indeed, in both years, the C content of the leaves did not depend on the N rate (**Table 4**). However, in 2009, the impact of the N rate on the carbon content of the stalk was significant ($p = 0.0135$). These results suggested that the accumulation of organic compounds in the stalk was affected by the lack of N. The predominant organic molecules in sweet sorghum stalks are sugars. Therefore, the decrease of the sugar content in the stalk may explain why its carbon content decreased as N is lacking. In other words, the lower C content means less C available to make the sugars so less sugar.

Nitrogen fertilization affected the C capture in sweet sorghum biomass depending on the soil and the year. Unlike the carbon content, the carbon capture in the clay was significantly affected by the N rate (**Table 4**). These results showed that the application of N is required in the clay if an increase of N capture by sweet sorghum is persuaded. Similarly, on the loam, when sweet sorghum is grown after corn, its ability to sequester the atmospheric carbon is limited by N ($p = 0.03$) (**Table 4**). Except in the loam 2009 ($p = 0.016$), the N rate did not affect the amount of carbon captured in the leaves (**Table 2**). By contrast, the N fertilization improved the sequestration of the C in the stalk in the clay in both years and in the loam in 2009. The C accumulation in the stalk and in the sand was never affected by the N rate (**Figure 2**). These results suggested that the significant impact of the N rate on the total C capture in the sand was due to its effect on the leaves. Therefore, in sand, the leaves are more sensitive to the C capture than the stalk in cases when the N is lacking. By contrast, in the clay and in the loam, when N is deficient, the carbon accumulation in the stalk is highly affected. In general, 82% of the carbon captured in the biomass is found in the stalk (**Figure 2**). When converted into equivalent CO_2, the amount of C captured by sweet sorghum was 11.9–51.1 Mg ha^{-1} according to the soil type and the N rate (**Table 3**).

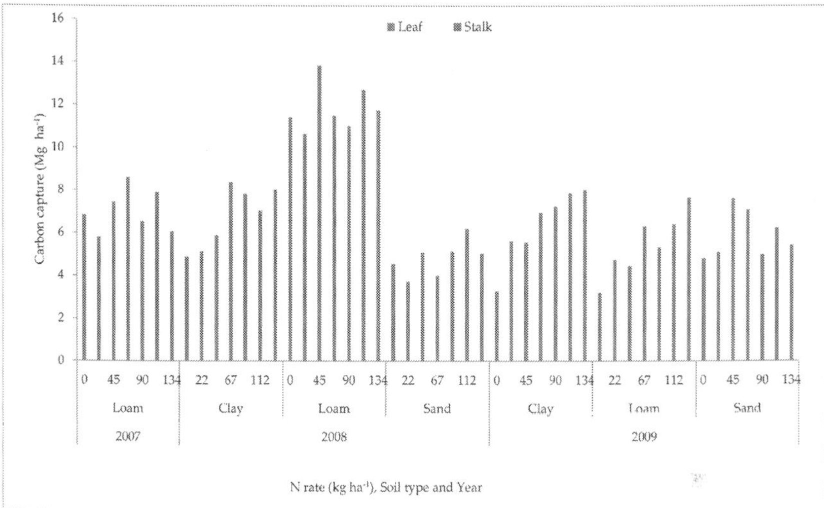

Figure 2. Impact of the soil type, the year, and the N rate on the partitioning of the C capture in sweet sorghum leaves and stalk.

4. Discussion

Our results proved that the carbon content ranged from 42.9 to 43.8% in corn grain, 44–45.7% in switchgrass and 41.9–44.6% for sweet sorghum biomass. Generally, the carbon capture by corn depended on the N rate. The maximum C capture by corn grain was recorded with the application of 134 kg N ha^{-1} in 2008 (4.7 Mg C kg ha^{-1}) compared to 2.9 Mg C ha^{-1} in 2009 with 224 kg N ha^{-1}.

We observed that the maximum C capture by switchgrass cv. Alamo was reached in October, ranging from 9.8 Mg C ha^{-1} (in 2008) to 14.4 Mg C ha^{-1} (in 2009). That is three to four times higher than what corn put in its grain. However, the C capture is smaller with switchgrass var. Blackwell (3.7 Mg ha^{-1} in 2008 compared to 4.4 Mg C ha^{-1} in 2009) which is the same range as what corn had in its grain. Furthermore, sweet sorghum captured 3.2–13.8 Mg C ha^{-1} (**Table 3**), suggesting that switchgrass can sequester more C than sweet sorghum. The amount of C sequestered by sweet sorghum in our study was higher than the 3–7 Mg of CO_2 per hectare recorded by [20].

The high ability of switchgrass to capture atmospheric carbon was also recorded [32]. In most cases, sweet sorghum captured two to three times the amount of C that corn put in its grain. We found that switchgrass captured more carbon than sweet sorghum and corn grain. These results showed the additional environmental impact that switchgrass and sorghum may have over corn grain. Because of their high C capture, sweet sorghum and switchgrass can clean the environment from CO_2 as has been shown feasible with other crops [9–11]. The C sequestered in the soil will also significantly increase the positive environmental effects of these crops. Other authors also pointed to the potential C sequestration in crop roots as an important component of the fight against climate change. The theoretical maximum C sequestered in the soil by switchgrass in our study was about 4 Mg ha^{-1}, a little less than the 4.42 Mg C $ha^{-1}year^{-1}$ observed by previous authors in the northern plains of the US [19].

Finally, farmers that grow these crops should be compensated with carbon credit. However, if not well managed, the captured carbon can return to the air. To avoid that scenario, farming techniques that disturb the soil less should be encouraged [12, 13]. Farming techniques that minimize soil disturbance (e.g., non-tillage and use of cover crops) can help sequester C in the soil, and consequently reduce the effects of global warming [20]. While non-tillage is ideal, it is also impractical. Still, it is important to point out that in contrast to tillage that increases the rate of soil C mineralization, non-tillage improves the storage of soil C [33, 34, 21].

5. Conclusion

Our results showed that since the carbon capture by sweet sorghum, switchgrass, and corn are statistically significant, these crops can help reduce the concentration of CO_2 in the environment and therefore contribute to the reduction of global warming.

Acknowledgements

Research for this chapter was funded by financial support from the Missouri Life Science Research Board, and Missouri Fertilizer and Ag. Lime Board (USA).

Author details

Roland Ahouélété Yaovi Holou[1,3] and Valentin Missiakô Kindomihou[2*]

*Address all correspondence to: vkindomihou@yahoo.fr

1 University of Missouri-Delta Center, Portageville, MO, USA

2 Laboratory of Applied Ecology, Faculty of Agronomic Sciences, University of Abomey-Calavi, Benin Republic, West Africa

3 DiasporaEngager, Augusta, GA, USA

References

[1] Weaver A.J., Green C. Global climate change: Lessons from the past—Policy for the future. Ocean and Coastal Management. 1998;39(1-2):73–86. DOI: 10.1016/S0964-5691(98)00015-5

[2] Kaufmann R.K., Kauppi H., Stock J.H. Emissions, concentrations, & temperature: A time series analysis. Climatic Change. 2006;77(3):249–278. DOI: 10.1007/s10584-006-9062-1

[3] Tamang P.L., Bronson K.F., Malapati A., Schwartz R., Johnson J., Moore-Kucera J. Nitrogen requirements for ethanol production from sweet and photoperiod sensitive sorghums in the Southern High Plains. Agronomy Journal. 2011;103:431–440. DOI: 10.2134/agronj2010.0288

[4] Ma Z., Wood C.W., Bransby D.I. Impacts of soil management on root characteristics of switchgrass. Biomass and Bioenergy. 2000;18:105–112. DOI: 10.1016/S0961-9534(99)00076-8

[5] Ney R.A., Schnoor J.L. Incremental life cycle analysis: Using uncertainty analysis to frame greenhouse gas balances from bioenergy systems for emission trading. Biomass and Bioenergy. 2002;22:257–269. DOI: 10.1016/S0961-9534(02)00004-1

[6] Zhang C., Han W., Jing X., Pu G., Wang C. Life cycle economic analysis of fuel ethanol derived from cassava in southwest China. Renewable and Sustainable Energy Reviews. 2003;7(4):353–366. DOI: 10.1016/S1364-0321(03)00057-1

[7] Petrou E.C., Pappis C.P. Biofuels: A survey on pros and cons. Energy and Fuels. 2009;23(2):1055–1066. DOI: 10.1021/ef800806g

[8] Wilhelm W.W., Johnson J.M.F., Hatfield J.L., Voorhees W.B., Linden D.R. Crop and soil productivity response to corn residue removal: A literature review. Agronomy Journal. 2004;96:1–17.

[9] Herzog H. Peer reviewed: what future for carbon capture and sequestration? Environmental Science & Technology. 2001;35(7):148–153. DOI: 10.1021/es012307j

[10] Liu R., Li J., Shen F. Refining bioethanol from stalk juice of sweet sorghum by immobilized yeast fermentation. Renewable Energy. 2008;33:1130–1135. DOI: 10.1016/j.renene.2007.05.046

[11] Xiong S., Zhang Q.G., Zhang D.Y., Olsson R. Influence of harvest time on fuel characteristics of five potential energy crops in northern China. Bioresource Technology. 2008;99(3):479–485. DOI: 10.1016/j.biortech.2007.01.034

[12] Kuzyakov Y.V. Tracer studies of carbon translocation by plants from the atmosphere into the soil (a review). Eurasian Soil Science. 2001;34:28–42.

[13] Parrish D.J., Fike J.H. The biology and agronomy of switchgrass for biofuels. Critical Reviews in Plant Sciences. 2005;24:423–459. DOI: 10.1080/07352680500316433

[14] Gale W.J., Cambardella C.A. Erratum: Carbon dynamics of surface residue- and root-derived organic matter under simulated no-till (Soil Science Society of America Journal, 64: (190–195). Soil Science Society of America Journal. 2002;66(6):2040. DOI: 10.2136/sssaj2000.641190x

[15] Searchinger T., Heimlich R., Houghton R.A., Dong F., Elobeid A., Fabiosa J., Tokgoz S., Hayes D., Yu T.H. Use of U.S. croplands for biofuels increases greenhouse gases through emissions from land-use change. Science. 2008;319:1238–1240. DOI: 10.1126/science.1151861

[16] Schmer M.R., Vogel K.P., Mitchell R.B., Perrin R.K. Net energy of cellulosic ethanol from switchgrass. Proceedings of the National Academy of Sciences of the United States of America. 2008;105:464–469. DOI: 10.1073/pnas.0704767105

[17] Molina J.A.E., Clapp C.E., Linden D.R., Allmaras R.R., Layese M.F., Dowdy R.H., Cheng H.H. Modeling the incorporation of corn (Zea mays L.) carbon from roots and rhizodeposition into soil organic matter. Soil Biology and Biochemistry. 2001;33:83–92. DOI: 10.1016/S0038-0717(00)00117-6

[18] Andress D. Soil carbon changes for bioenergy crops. ANL/ES/RP-113201, Argonne National Lab., Argonne, IL (US). 2004.

[19] Frank A.B., Berdahl J.D., Hanson J.D., Liebig M.A., Johnson H.A. Biomass and carbon partitioning in switchgrass. Crop Science. 2004;44:1391–1396.

[20] Sainju U., Whitehead W., Singh B. Carbon accumulation in cotton, sorghum, and underlying soil as influenced by tillage, cover crops, and nitrogen fertilization. Plant and Soil. 2005;273(1):219–234. DOI: 10.1007/s11104-004-7611-9

[21] Kuo S., Sainju U., Jellum E. Kuo S., Sainju U., Jellum E. Winter cover crop effects on soil organic carbon and carbohydrate in soil. Soil Science Society of America Journal. 1997;61(1):145–152. DOI: 10.2136/sssaj1997.03615995006100010022x

[22] Balesdent J., Balabane M. Major contribution of roots to soil carbon storage inferred from maize cultivated soils. Soil Biology & Biochemistry. 1996;28(9):1261–1263. DOI: 0.1016/0038-0717(96)00112-5

[23] Baker M., Babcock B. Value maximization from corn fractionation: feed, greenhouse gas reductions, and cointegration of ethanol and livestock. Integration of agricultural and energy systems. Proceedings of a Conference February 12–13, 2008, Atlanta, GA. 2008;67–80.

[24] Szulczyk K., McCarl B., Cornforth G. Market penetration of ethanol. Renewable and Sustainable Energy Reviews. 2010;14(1):394–403. DOI: 10.1016/j.rser.2009.07.007

[25] Holou R.A.Y., Kindomihou V. Impact of nitrogen fertilization on the oil, protein, starch, and ethanol yield of Corn [*Zea mays* (L)] grown for Biofuel production. Journal of Life Sciences. 2011;5:1013–1021.

[26] Holou R.A.Y., Stevens G. Juice, sugar, and bagasse response of sweet sorghum (*Sorghum bicolor* (L.) Moench cv. M81E) to N fertilization and soil type. Global Change Biology Bioenergy. 2011;4:302–310. DOI: 10.1111/j.1757-1707.2011.01126.x.

[27] Holou R.A.Y., Stevens G., Kindomihou V. Impact of nitrogen fertilization on the nutrient removal by corn grain. Crop Management. 2011. DOI: 10.1094/CM-2011-1223-02-RS.

[28] Holou R.A.Y., Stevens G., Sinsin B. Biomass production and soil nutrient removal by switchgrass grown for biofuel production. Future Science, Biofuels. 2011;2:629–637. DOI: 10.4155/bfs.11.140.

[29] Holou RAY, Stevens G, Kindomihou V. Return of aboveground nutrients by switchgrass into the surrounding soil during senescence. Biofuels. 2013;4(2):169–183. DOI: 10.4155/BFS.12.79

[30] NRCS. Soil survey laboratory methods manual. Soil Survey Investigations Report. 1996; N° 42. Version 3.0. USDA. National Soil Survey Center.

[31] Saxton A.M. A macro for converting mean separation output to letter groupings in Proc Mixed. Proceeding of 23rd Annual SAS User Group International Conference, Nashville, TN, 22–25 March, 1998; SAS Inst., Cary, NC:1243-1246.

[32] Liebig M.A., Johnson H.A., Hanson J.D., Frank A.B. Soil carbon under switchgrass stands and cultivated cropland. Biomass and Bioenergy. 2005;28(4):347–354. DOI: 10.1016/j.biombioe.2004.11.004

[33] Jastrow J. Soil aggregate formation and the accrual of particulate and mineral-associated organic matter. Soil Biology and Biochemistry. 1996;28(4–5):665–676. DOI: 10.1016/0038-0717(95)00159-X

[34] Allmaras R., Schomberg H., Douglas C., Dao T.. Soil organic carbon sequestration potential of adopting conservation tillage in US croplands. Journal of Soil and Water Conservation. 2000;55(3):365–373.

Chapter 8

Theoretical Considerations for Economics of Second- and Third-Generation Biofuels

Fouzia Tabssum and Javed Iqbal Qazi

Additional information is available at the end of the chapter

http://dx.doi.org/10.5772/65735

Abstract

Ground is being prepared all over the world for installation of biofuel plants which can govern the sustainable supply of cleaner fuels at affordable prices and predictable amounts. At the dawn of this century biofuels identified low cost feedstocks, their diverse pretreatments, different methods of saccharifications and fermentations and those for cultivation of biodiesel yielding organisms. Bioalcohols, biohydrogen and biogas represent the biofuels which are derived from microbial work on the biowaste-resources. Extensions in this sector have focused the solar energy captured by the microalgae from which oils can be extracted for biodiesel. Undoubtedly, all forms of available energies on this planet earth had/have been derived, directly or indirectly, from the solar inputs. In this chapter pivotal role of solar insolation will be discussed albeit for regeneration as well as processing of lignocellulosic biomass for obtaining biofuels. Conclusively, biofuels' sustainable supplies, role of solar energy has been dreamt at various steps of the process; from the collection of biowaste resources through steps of pretreatment, saccharification / fermentation and purification of the product. This chapter discusses the subject matter into two major sub-headings: 1) Biofuels from lignocellulosic / food industrial wastes and 2) Cultivation of microbes for biodiesel.

Keywords: biodiesel, bioethanol, biohydrogen, economizing biofuel production, lignofuels, second-generation bioenergy, third-generation bioenergy

1. Introduction

The form of life we humans know and understand to some extend is impossible without the sun. The biosphere will not be sustainable without continuous rain of sun energy. Recalling our basic information on energy conversion in and between different systems, efficiencies can never attain 100% levels. Same is true for biotic components of the biosphere. Food transfer through different trophic levels and metabolites' circulation and their transfer through biochemical pathways all pass through this limitation. Thus, sun energy is pivotal to the existence and sustenance of all forms of energy in the biosphere. The fossils fuels on whom, at present we are dependent to "no return" level had been derived from ancient organisms whose presence at that time was also dependent on the sun energy. Now when the fossil fuels' reservoirs are depleting and further they have polluted our air, soil, and water, we must look for alternatives, which must be sustainable and environment friendly too. Energy from biomass can be condensed into low volume high efficiency fuels through microbial fermentation and/or other routes of bio conversion of abundantly available matter. However, food competing sources of carbohydrates cannot be diverted to biofuels' refineries by ignoring the ever increasing human population. Whereas feedstock for second- and third-generations' biofuels are abundantly available at little or no cost. Biowastes have a great potential in this regard. For instance, it has been estimated that up to 442 billion liters per year of bioethanol could be produced from lignocellulosic materials [1]. But their pretreatments, saccharification, and then fermentation still need a lot of work to be done. Low-cost and environment compatible strategies to render lignocellulosic biomass accessible for enzymatic saccharifications have to be developed. Further for lignocellulosic wastes of different plant origins, different methods may be required for optimum outputs. Liberation of undesired compounds following pretreatments and saccharification is another area demanding more inputs. Inhibitors' resistant saccharifying and/or fermenting microbes and enzymes have to be isolated, developed, characterized, and optimized for select feedstocks and locations. Jumping to third-generation biofuels, biowastes have to be identified to support rapid growth of microalgae and fast fixation of CO_2. This chapter describes some theoretical considerations which need to be considered through experimental verification to economize the biofuels' production. Although the science of bioenergy, especially for second- and third-generation biofuels is still at its infancy, but taking into account the drastically exhausted fossil fuels' reservoirs, the already polluted biosphere, ever increasing human population, and its elevating demands for more comforts, which need more supply of energy, we are left with limited choices including the fascinating future of biofuels.

2. Biofuels from lignocellulosic/agri/food industrial wastes

Fossil sources of energy are depleting at a very rapid rate and alternative sources of renewable energy are being searched. Agricultural wastes are produced daily and get accumulated into different environments. They pose special problems of solid waste management in urban environments and industrial locations. On the other hand, they are potential source of fermentable carbohydrates. These wastes could be identified as low-cost feedstocks for biofuels' production. For example, bioethanol can be fermented from diverse feedstocks including corn, sugarcane, wood, and fruit and vegetable wastes such as pawpaw and sweet potatoes, etc. In addition to the production of biofuels, utilization of lignocellulosic agriwastes will concomitantly solve the solid waste disposal problem, as the fermentation residues can be used as solid fertilizers. A number of physical and chemical treatments methods have been reported for the production of bioethanol from diverse categories of plant biomass. However, researchers are continuously describing new methods and techniques for economic yields of bioethanol [2–7]. From the biofuels headings, bioethanol is becoming increasingly popular as fossil fuel additive and for reducing the stress of decline in crude oil availability. Bioethanol productions require sustainable supplies of fermentable sugars, efficient fermenting microbes, a few nutrients (depending upon the nature of feedstock as well as microbes), and optimized culture conditions. Resultantly, bioethanol productions have been described from diverse waste resources such as market vegetable waste, carrot discard, hydrolyzed agricultural wastes, banana peels, and pulp and peels of mango. [8–13].

Interests in the area of bioethanol production from organic waste materials emerged in the late 1980s. Since then lignocellulosic material extraction and enzymatic hydrolysis have been reported extensively, however, for development of technically feasible and economically viable large-scale enzyme-based biomass to ethanol conversion processes addressing diverse waste resources and fermentation conditions a lot of work still has to be done to cover different regions of the world. The success of cellulose to ethanol conversion processes has been described as a function of cellulose fiber pretreatment, enzyme selection, and operating conditions [14]. Nigam and Singh [15] have discussed that researchers have been redirecting their interests in biomass-based fuels for sustainable development in the content of economical and environmental considerations. These researchers further stressed that renewable bioresources are available globally in the form of residual agricultural biomass and wastes. However, much research has to be done for the development of an effective, economical, and efficient conversion process.

Estimates for bioconversion of globally wasted crop appear promising. For example, Kim and Dale [16] have described that by employing the wasted crops for bioenergy production conflict between human food and industrial use can be avoided. Crops' residues and industrial lignocellulosic wastes can be considered feedstocks for bioethanol productions. Wasted crops have a potential to produce up to 49.1 GL year^{-1} of bioethanol. These authors further discussed that from the residues and wasted crops potentially 49.1 GL year^{-1}, ethanol production is possible. Accordingly, bioethanol could replace 353 GL of gasoline when used in E85 fuel.

Following unprecedented growth of human population and industrialization, ethanol demand is increasing continuously. Conventional crops of bioethanol production are unable to meet the demand due to their primary value for food and feed. Lignocellulosic substances including agricultural wastes have therefore become attractive feedstocks for bioethanol production. They are cost effective, renewable, and abundant. The promising technology of bioethanol from waste resources has several challenges and limitations such as biomass transport and handling, and efficient pretreatment methods for total delignification of lignocellulosics. Novel pretreatment methods may increase yields of fermentable sugars after enzymatic saccharification [17].

Voluminous work regarding the diversity of lignocellulosic resources dictate for development of diverse methods of pretreatment and to economize further recognition of different categories of saccharifying and fermenting microbes, which might be required for different sorts of substrates to be processed at different locations in a country. Local sociocultural and economic situations can influence the overall efficiency of biofuels' productions from waste biomass. In this context, cost of transport of raw material from its natural origin or industrial processing units must be considered and if it is profitable then the pretreatment and subsequent concentration/detoxification processes be accomplished at the first place. Alternatively, the small pretreatment as well as fermentation plants might be installed at or near locations of the biowaste resources. Heat for the pretreatment should be secured directly/indirectly from sun rays. Efficient acidic or alkaline pretreatments for different lignocellulosic substrates have to be identified. To reduce the expenditure of neutralizing a substrate pretreated with acid/alkali, respective category of acidophilic or alkaliphilic cellulolytic, and/or ethanologenic microbes must be searched and recruited accordingly. Similarly, a lignocellulosic or a food industrial waste known for higher amounts of certain inhibitors' yield during the initial processes might be attacked with the inhibitor(s)' resistant and/degrading microbes, which must be cellulolytic/ethanologenic too or their enzymes thereof. The points highlighted above will definitely add to the economics of biofuels production from biowaste resources. A layout of schematic thoughts in this regard has been depicted in **Figure 1**, while consideration of different lignocellulosic substrates is summarized in **Table 1**.

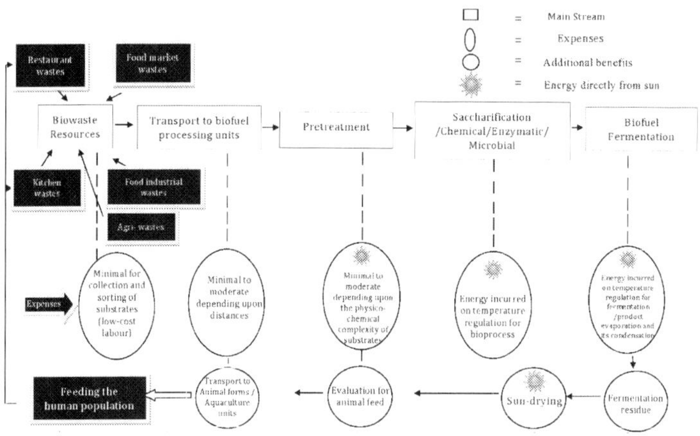

Figure 1. An overview of different steps of theoretical considerations to promote economics of second-generation biofuels deriving from solid wastes.

As can be seen from **Figure 1** that overall conversion of lignocellulosic substrates to low volume high-energy content biofuels involves expenditure of energy at various steps. The raw material processing including washing with hot water, decontaminating or sterilizing, fermentative, and product purification steps need thermal inputs of energy. At least four steps of overall process depicted in **Figure 1** can be accomplished by solar energy inputs. Some other steps to improve economics of lignofuels and their application at present time when the fossil fuels still represent efficient fuels for transport sector, at least, are given below.

Lignocellulosic/ biowaste substrate	Specificity of pretreatment	Microbial requirement	Benefit of specific microbe(s)	Possible locations, e.g., industrial vicinity/ natural habitat	Biofuel/Potential additional benefit(s)	Reference(s)
Sugarcane bagasse/lingo-cellulosic biomass	Ball milling, dilute sulfuric acid, steam explosion, delignification	*Pichia stipitis* BCC 15191, recombinant *Escherichia coli* KO11, *Paecilomyces variotii*	Consumption of both xylose and glucose	Sugar mills/ agro-industrial derivatives	Bioethanol/ Consumption of surplus lignocellulosic material, animal feed from fermentation residue	[17, 19, 20]
Wheat Straw	Knife milling with 0.7–1.0 mm rejection screen, washed with water and dried	*Pichia stipitis* NRRL Y-7124 *Pichia stipitis* A	Adapted at increased concentration of hydrolysate	Agri-waste	Bioethanol/value added product from abundant agriwastes/ sustainable solid waste management	[17]
Rice straw	Chopped to 5–6 mm size range	*Candida shehatae* NCL-3501	Co-ferment glucose and xylose	Agriwaste	Bioethanol/value added product from abundant agricultural wastes/ sustainable solid waste management	[17]
Corn straw/corn stover	Chopped, steam explosion (3.5 MPa, 275°C, 2 min) 2% NaOH, 80°C, 1 h, ammonia fiber expansion	*Saccharomyces cerevisiae* ATCC 26603 Pichia stipitis NRRL Y-7124 *S. cerevisiae*	Ferment only glucose Ferment glucose first and then xylose from the mixture	Agriwaste	Bioethanol/value added product from abundant agricultural wastes/ sustainable solid waste management	[21, 17]
Hard wood	Mechanical communication Steam explosion	Diversity of bacteria and yeasts capable	Vary from substrate to substrate	Forests	Bioethanol/value added product	[22]

Lignocellulosic/ biowaste substrate	Specificity of pretreatment	Microbial requirement	Benefit of specific microbe(s)	Possible locations, e.g., industrial vicinity/ natural habitat	Biofuel/Potential additional benefit(s)	Reference(s)
		of saccharifying fermenting different substrates			from forest residues.	
Kitchen wastes (rich in glucose, starch and cellulose)	Hot water, acid solutions, liquefaction/ saccharification	Enzymatic hydrolysis, *Saccharomyces cerevisiae*	Easily available	Cafeterias, restaurants, dining halls, food plants and household kitchens	Bioethanol/ reduced ethanol costs/no requirement of fermentation nutrients	[23–25]
Wastewater, agricultural and industrial wastes, animal by-products	Anaerobic digestion	Microbial community	Easy availability from herbivores' dung	Animal and agricultural farms	Biogas/prevent pollution; stabilized biofertilizer	[26, 27]
Crop residues, livestock waste, food waste Starch-cellulosic-based wastes, dairy wastes, palm oil mill effluent and glycerol	1. Dark fermentation 2. Photofermentation	Bacterial consortia	No need of illumination Light energy inputs required	Agricultural farms	H_2/pollution-free fuel; waste treatment; mitigating global warming	[28, 29]
CO_2	Concentration	Microalgal culture		Cement sector	Biodiesel/ transformation of polluting gas to valuable product; waste water treatment	[30]

Table 1. Compatibility of different pretreatments and fermentative microbes for economic production of biofuels from diverse and abundantly available lignocellulosic and other biowastes.

2.1. Transport of feedstocks

Lignocellulosic feedstocks from agricultural fields and relevant industrial units are required to be transported to pretreatment and/or biofuel's production units. This involves an obvious expenditure; and at present will necessitate consumption of fossil fuels to run the transporting trucks. Important to consider is that biofuels are being attempted/generated to lessen the present burden of fossil fuels' consumption. To reduce the cost of biofuels and render them compatible at present with the fossil fuels' supplies, every step is required to be optimized. In this regard it might prove quite fruitful if the lignocellulosic biowaste feedstocks are pretreated at or near to their generation locations, so that relatively low-volume/mass pretreated material ready for saccharification or biofermentation can be transported at affordable expenses.

To further reduce the feedstock transportation costs, processes of pretreatments and/or saccharification can be accomplished near to the lignocellulosic biowaste origins and the sugar streams can be transported dynamically employing pipe system. Alternatively, sugars syrups can be concentrated into low-volume thick fluids to reduce the cost by usual transport systems. In a country like Pakistan, processes of pretreatments and sugars syrups' concentration can be accomplished by direct solar energy. Another strategy would be installation of those industrial units in one locality whose wastes can be coutilized cost effectively.

2.2. Concept of cluster(s) of industrial units having wastes of mutual interest

Man can achieve maximum efficiency and sustainability of his efforts addressing construction of production units approaching a level and design not superseding natural system(s). In this regard man has been imitating the nature subconsciously in the past. While conferring to the cumulative maturity time has come to recognize/realize that the nature tailored models are best to follow. Exploration of biogeochemical cycles within and among different ecosystems has taught us that sustainability relies upon the utilization of waste of one unit as resource by the other and vice versa. Thus, to save cost presently we have to pay to treat industrial effluents/solid wastes or to transport them to other locations for treatment to utilize them as biowaste resource, industrial units can be managed in the form of different consortia in which for a given set of industrial units, waste(s) from one unit can serve feedstock for another. For instance, sugar industries in Pakistan produce millions of tons of sugarcane bagasse (SCB). Chemical pretreatment of SCB requires the application of dilute acid(s) or alkali. Many other industrial effluents are characterized with considerably high or low pH and are not accompanied with further process interfering chemicals. Coinstallation of such relevant industrial units can solve many problems including low cost pretreatment of SCB and value-added consumption of the otherwise negative-valued acidic/alkaline industrial effluents, for instance.

2.3. Tax relaxation for biofuels' consumers

New trends whose immediate impact on modern human minds reflects reduction in efficiency of desired task(s) are difficult to get acceleration. Their acceptance in the society must be escalated with some incentive. Bioethanol-driven motor cars, especially at the start, might not run with speed comparable to petrol-driven engines. Whereas the environmental friendly emission from biofuel-driven motor cars might be strengthen with reduction in the motor vehicle taxes.

Waste biomass for biofuels' generation will have to be utilized from different sources for sustainability of the process. The wastes, in general, represent solid wastes while organic content-rich food industrial effluents can also be utilized for the fermentation of different types of biofuels. An overview of different lignocellulosic wastes, their needs of pretreatments, and microbes along with potential biofuels and additional benefits is given in **Table 1**.

Regarding the fluid wastes, continuous and sustainable resources are represented by food, industrial, and domestic sewages, which may contain varying levels of fermentable organic loads. Schemes have been developed and shown in **Figures 2** and **3**, for upgrading the waste effluents into biofuels' generation (**Figures 2** and **3**). Responding to above theoretical notions for improving economics of second-generation liquid biofuels, it appears pertinent here refer to the work of Nigam and Singh [15]. These authors concluded that four challenges are imperative for sustainable biofuel productions:

1. Improvements in enzymatic hydrolysis through the provision of low cost crude enzymes characterized with higher specific activities.

2. The development of such microbial strains which are not only robust fermenting organisms, but must be resistant to inhibitors present in hydrolysates. Further, the developed strains must be able to consume all sugars present in the raw material yielding optimum productivity of alcohols and be resistant to higher concentration of the product too.

3. Development of integration strategies to decrease the number of steps of production process.

4. Useful consumption of any byproducts and wastes generated to reduce the energy demand and protect the environment.

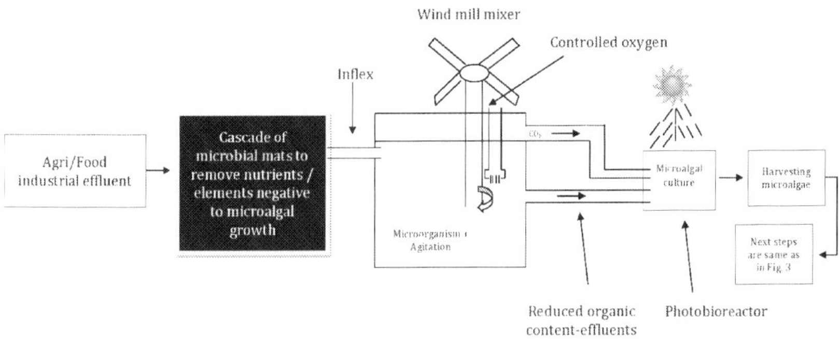

Figure 2. Rebust bioreactor for reducing agri/food industrial effluents' BOD with concomitant efficient yield of microalgae for biodiesel production.

http://dx.doi.org/10.5772/65735

Nigam and Singh [15] further added that there is much potential for biofuel market and now it is a matter of time before they compete with petroleum-based fuels. Technological developments, in future, will enhance energy balance and reduce emissions and production cost of biofuels.

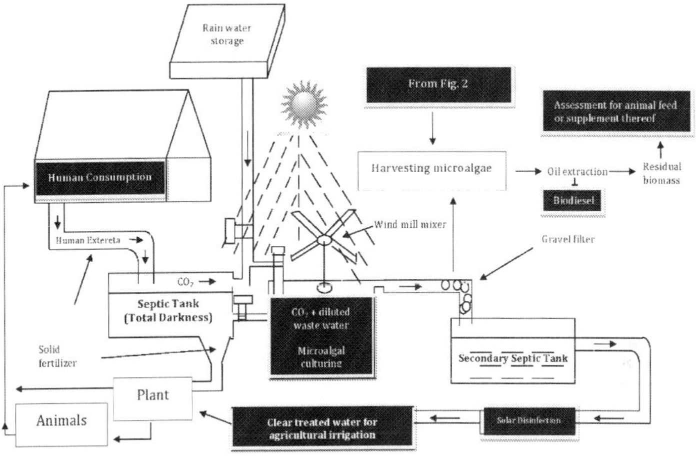

Figure 3. A workable multistep plane for sustainable treatment of domestic/organic content-rich wastewaters with concomitant microalgal yield.

3. Cultivations of microbes for biodiesel (third-generation biofuels)

First-generation biofuels, primarily produced from food crops and mostly oil seeds, will remain far behind to achieve targets of biofuel production, climate change mitigation, and economic growth. Whereas second-generation biofuels from lignocellulosic wastes are still in their infancy. Meanwhile, another bioenergy sector derives from nonfood feedstocks such as microalgae. Different types of photobioreactors and open ponds to cultivate microalgae with additional benefits of wastewater treatment and as food additive for human health and for aquaculture have been described [18].

In an agri-based country like Pakistan, a vast diversity of nutritionally enriched effluents from food industries like corn steep liquor, molasses, and whey are produced in abundant amounts. These can be semipretreated with solar insolation derived heat and fermented for the generation of CO_2 and less BOD effluents. Thus, processed effluents can then be employed for rapid growth of microalgae for biodiesel production. For this particular notion a robust bioreactor has been designed (**Figure 2**). The food/other industrial effluents pretreated in the bioreactor (**Figure 2**) and made optimum through dilutions and/or by incorporating essential nutrients for the cultivation of microalgae can then be routed to **Figure 3** for obtaining biofuel and ultimate treatment of the effluents.

Domestic sewage can be biotreated specifically designed for the generation of effluents suitable for cultivation of microalgae. Such practices would incur economy of the process as treatment of domestic/industrial effluents will be achieved in a profitable manner. Whereas in the case of water shortage, the fresh water sources will not be affordable for diverse biotechnological processes without involving additional costs. In this context, biofuels' generation from biowastes/effluents would be an appealing field of development for sustainable supplies of biofuels and water for irrigation/other processes (**Figure 3**).

Mata et al. [18] concluded that for algal biodiesel development of strategies for large-scale cultivation and harvesting would be required. Growth conditions and provision of affordable nutrients for large-scale algal cultivations have to be identified for optimum yields. For this oil extraction strategies, provision of light, CO_2, and nutrients and turbulence, temperature, and O_2 levels must be optimized for a given location for conserving efficient oil content and biomass yield. These workers further indicated that using sources of CO_2, nutrient-rich wastewaters, or inexpensive fertilizers are expected to increase algal yields. In addition to biofuel production, considering microalgae biomass for different applications, such as food, agriculture, and medicine, will contribute to the sustainability and market competitiveness of the microalgal industry [18].

Another stream of biodiesel might originate from the identification of such biowastes, whose pretreatments/hydrolyses can lead to the successful cultivation of oleaginous yeasts. Even CO_2 has been claimed as a waste resource for cultivating microalgae to extract biodiesel (**Table 1**). Many other waste resources will hopefully be soon imagined as resources for biotechnological productions of value-added products in the near future.

4. Conclusion

After excessive exploitation of the fossil fuels' reservoirs and rendering the biosphere highly polluted, humans are left to extract/ferment fuels from biomass as we had been obtaining foods directly and/or indirectly from first trophic level. Parts of the plant biomass we preferably consume as food cannot be afforded longer for biofuels' production. Although the lignocellulosic residues and agriwastes represent potential sources of bioenergy productions. Accordingly, low/no cost feedstocks have been identified properly in different countries. Now the challenges of processing the feedstocks in terms of their economical and environmentally sustainable pretreatments, saccharification, and energy productions have been accepted by scientists of different disciplines. It is hoped that collaborative efforts of mechanical engineers, chemical engineers, biologists, and other scientists will bring the human wisdom to the hub of biotechnology for sustainable production of biofuels in the near future. Time is not too far when the humans will intentionally be cultivating specific species of plants and harvesting them or their parts presently considered as wastes to supply the biorefineries. Another business sector will represent biofuel productions from the lignocellulosic residues and industrial wastes.

Author details

Fouzia Tabssum and Javed Iqbal Qazi*

*Address all correspondence to: qazi.zool@pu.edu.pk

Microbial Biotechnology Laboratory, Department of Zoology, University of the Punjab, Lahore, Pakistan

References

[1] Bohlmann GM. Process economic considerations for production of ethanol from biomass feedstocks. Indian Biotechnology. 2006; 2: 14–20.

[2] Khan MN, Dwivedi AK, Gupta B. Bioethanol production from agricultural wastes and its economical impacts: Review. Biotechnology. 2016; 5(3): 322–323.

[3] Chafle S, Parmar V, Biya S. Utilization of vegetable and fruit waste for bio-energy generation. Journal of Automation and Control Engineering. 2014; 2(2): 143–145.

[4] Vishwakarma HS, Kumar A, Singh J, Dwivedi S, Kumar M. Production of ethanol from fruit wastes by using *Saccharomyces cerevisiae*. International Journal of Renewable Energy Technology Research. 2014; 3(10): 1–5.

[5] Vazirzadeh M, Robati R. Investigation of bio-ethanol production from waste potatoes. Annals of Biological Research. 2013; 4(1): 104–106.

[6] Hossain ABMS, Fazliny AR. Creation of alternative energy by bio-ethanol production from pineapple waste and the usage of its properties for engine. African Journal of Microbiology Research. 2010; 4(9): 813–819.

[7] Akin-Osanaiye BC, Nzelibe HC, Agbaji AS. Production of ethanol from *Carica papaya* (pawpaw) agro waste: effect of saccharification and different treatments on ethanol yield. African Journal of Biotechnology. 2005; 4(7): 657–659.

[8] Babu S, Harinikumar KM, Singh RK, Pandey A. Optimization of bioethanol production from fruit wastes using isolated microbial strains. International Journal of Advanced Biotechnology and Research (IJBR). 2014; 5(4): 598–604.

[9] Chiranjeevi T, Uma A, Radhika K, Rani GB, Prakasham RS, Rao PS, Umakanth AV. Enzymatic hydrolysis of market vegetable waste and subsequent ethanol fermentation-Kinetic evaluation. Journal of Biochemical Technology. 2014; 5(4): 775–781.

[10] Aimaretti NR, Ybalo CV, Rojas ML, Plou FJ, Yori JC. Production of bioethanol from carrot discards. Bioresource Technology. 2012; 123: 727–732.

[11] Singh A, Kuila A, Adak S, Bishai M, Banerjee R. Utilization of vegetable wastes for bioenergy generation. Agricultural Research. 2012; 1(3): 213–222.

[12] Arumugam R, Manikandan M. Fermentation of pretreated hydrolyzates of banana and mango fruit wastes for ethanol production. Asian Journal of Experimental. Biological Science. 2011; 2(2): 246–256.

[13] Patle S, Lal B. Ethanol production from hydrolysed agricultural wastes using mixed culture of *Zymomonas mobilis* and *Candida tropicalis*. Biotechnology Letters. 2007; 29(12): 1839–1843.

[14] Champagne P. Bioethanol from agricultural waste residues. Environmental Progress. 2007; 27(1): 51–57

[15] Nigam PS, Singh A. Production of liquid biofuels from renewable resources. Progress in Energy and Combustion Science. 2011; 37: 52–68.

[16] Kim S, Dale BE. Global potential bioethanol production from wasted corps and crop residues. Biomass & Bioenergy. 2004; 26: 361–375.

[17] Sarkar N, Ghosh KS, Bannerjee S, Aikat K. Bioethanol production from agricultural wastes: An overview. Renewable Energy. 2012; 37: 19–27.

[18] Mata TM, Martins AA, Caetano NS. Microalgae for biodiesel production and other applications: A review. Renewable and Sustainable Energy Reviews. 2010; 14: 217–232.

[19] Dias MOS, Junqueira TL, Rossell CEV, Filho RM, Bonomi A. Evaluation of process configurations for second generation integrated with first generation bioethanol production from sugarcane. Fuel Processing Technology. 2013; 109: 84–89.

[20] Zerva A, Savvides AL, Katsifas EA, Karagouni AD, Hatzinikolaou DG. Evaluation of *Paecilomyces variotii* potential in bioethanol production from lignocelluloses through consolidated bioprocessing. Bioresource Technology. 2014; 162: 294–299.

[21] Lau MW, Dale BE. Cellulosic ethanol production from AFEX-treated corn stover using *Saccharomyces cerevisiae* 424 A(LNH-ST). Proceedings of the National Academy of Science of the U.S.A. 2009; 106: 1368–1373.

[22] Ruffell J. Pretreatment and hydrolysis of recovered fibre for ethanol production. Master of Applied Science, The University of British Columbia. 2008.

[23] Cheng CL, Lo YC, Lee KS, Lee DJ, Lin CY, Chang JS. Biohydrogen production from lignocellulosic feedstock. Bioresource Technology. 2011; 102: 8514–8523.

[24] Cekmecelioglu D, Uncu ON. Kinetic modeling of enzymatic hydrolysis of pretreated kitchen wastes for enhancing bioethanol production. Waste Management. 2013; 33: 735–739.

[25] Matsakas L, Kekos D, Loizidou M, Christakopoulos P. Utilization of household food waste for the production of ethanol at high dry material content. Biotechnology for Biofuels. 2014; 7(4): 1–9.

[26] Karellas S, Boukis I, Kontopoulos G. Development of an investment decision tool for biogas production from agricultural waste. Renewable and Sustainable Energy Reviews. 2010; 14: 1273–1282.

[27] Hamad TA, Agll AA, Hamad YM, Bapat S, Thomas M, Martin KB, Sheffield JW. Study of combined heat, hydrogen and power system based on a molten carbonate fuel cell fed by biogas produced by anaerobic digestion. Energy Conversion and Management. 2014; 81: 184–191.

[28] Chong ML, Sabaratnam V, Shirai Y, Hassan MA. Biohydrogen production from biomass and industrial wastes by dark fermentation. International Journal of Hydrogen Energy. 2009; 34: 3277–3287.

[29] Guo XM, Trably E, Latrille E, Carrère H, Steyer JP. Hydrogen production from agricultural waste by dark fermentation: A review. International Journal of Hydrogen Energy. 2010; 35: 10660–10673.

[30] Bermudez SPC, Perez JSG, Rittmann BE, Saldivar RP. Photosynthetic bioenergy utilizing CO_2: an approach on flue gases utilization for third generation bioguels. Journal of Cleaner Production. 2015; 98: 53–65.

Chapter 9

Emerging Green Technologies for Biodiesel Production

Hanifa Taher and Sulaiman Al-Zuhair

Additional information is available at the end of the chapter

http://dx.doi.org/10.5772/66187

Abstract

The current global energy demand is met by burning the non-renewable fossil fuels. As the demand is escalating, resources and reserves are diminishing. In addition, the environment is threatened by the continuous emission of greenhouse gases; mainly CO_2, which is worrying. Therefore, searching for alternatives is inevitable. Biodiesel received a considerable attention to potentially replace petroleum-based fuels. It can be produced from oil-rich feedstocks through several methods using different technologies, including transesterification. Although alkali catalyzed biodiesel process is commercially viable, several challenges were raised. In this chapter, an overview of the current status of biodiesel production approaches is discussed and the emerging technologies are highlighted. The chapter rewards the attention of using green processes, where the effectiveness of using; microalgae biomass as a green feedstock (compared to conventional crop-based seeds), lipases as green catalysts (compared to conventional chemical catalysts), and green and tunable solvents, such as neoteric solvents and supercritical fluids (compared to conventional volatile organic solvents) are addressed.

Keywords: biodiesel, microalgae, lipase, green solvents, microwave

1. Biodiesel

The continuous dwindling fossil fuels supply and increasing atmospheric carbon dioxide emission have put the pressure on finding and developing sustainable alternative fuels. Biodiesel, which is a mixture of fatty acids alkyl esters, is the proposed alternative that can replace to the conventional petroleum diesel. The physical properties of biodiesel are similar to petroleum diesel and can be used without any modifications in the engine [1, 2]. In addition, biodiesel is a renewable, non-toxic and biodegradable fuel that can decreases the reliance on fossil fuels and reduces harmful gasses emissions [3–8]. Furthermore, biodiesel has lower sulfur and aromatic contents. According to the U.S. Department of Energy statistics and analysis [9], in 2015, the production of biodiesel in the United States alone reached 1.2×10^9 gallons. The statistics also indicated that the current biodiesel production in first 6 months of 2016 is about 21% higher than that obtained in 2015, during same duration.

2. Feedstocks

2.1. Conventional feedstocks

Triglycerides from oil-rich feedstock's, such as soybean, rapeseed, canola, sunflower, and palm, have been commonly used due to their abundant availability [10–16]. The first initiation was by the diesel engine inventor, Rudolf Diesel, who tested the use of peanut oil. However, natural oils are viscose with inappropriate cetane number. Thus, the idea was not accepted and vegetable oils were replaced by petroleum oil [7, 17–19]. The recent concern about limited oil reservoirs and oil explosion increased activities recalled the attention to use oil-rich feedstocks. Oils dilution with solvent, thermal cracking, pyrolysis, micro emulsions and transesterification has been suggested to overcome this viscosity limitation [20, 21]. Among these, transesterification with short chain alcohols, such as methanol and ethanol, in the existence of proper catalyst is the preferred and commonly used approach.

Although vegetable oils are available in large quantities, biodiesel production from these vegetable oils competes with their use as food source, which results in increasing their prices and affect food market. In addition, cultivating oil-rich crops requires lands and freshwater. It was reported that vegetable oil accounts for more than 60% of biodiesel overall production cost [5, 22]. From that prospective, non-edible oils such as those from non-edible plants that are not used in nutrition and can grow in the unfertile lands were suggested. However, freshwater requirement still exist. The use of waste oils and fats have been recommended, where there use is a waste management process [23], but contains large amount of free fatty acids and water which increases the production cost. Furthermore, cannot satisfy the ever-increasing global demands of diesel [24–26].

2.2. Green feedstock

Choosing an inexpensive and more sustainable oil feedstock is the critical step to get cost-effective biodiesel. Currently, microalgae, which are micro-organisms, received a promising attention. Due to their high oil content and growth rate, they have been considered as potential feedstock that can replace the conventional diesel [27–31]. Furthermore, cultivation of micro-algae cells does not require land development neither freshwater. Several algae strains were found to grow in seawater and wastewater. The oil contents of such feedstocks are usually between 20% and 50% and reach be in some strains to 80% in dry basis. The strain may also change its composition by altering the growth conditions, such as light, nutrients, and temperature. The stressful environment usually results to higher oil productivity.

More interestingly, microalgae cells contain of protein, carbohydrates and lipids, which extend the application domains of produced biomass from food to biofuel. Microalgae cells are also used for CO_2 mitigation. However, to use microalgae biomass for biodiesel production, several steps have to be carried out, which are strain selection, biomass production and harvesting, and oil extraction and conversion.

3. Production technologies

Oils derivatization by transesterification is the most common approach used commercially. Typically, transesterification is a reaction between the oil and a short-chain alcohol that results to form an esters mixture and glycerol, as side product. The reaction is commonly take place in the presence of catalyst that can speed up the reaction, where three moles of alcohols are needed to react with 1 mol of the oil. Higher alcohol to oil molar ratios than the stoichiometry, however, is usually employed to produce more biodiesel.

3.1. Conventional catalysts

As mentioned earlier, transesterification reactions are chemically catalyzed, which can be either base or acid catalysts, depending on the oil quality free fatty acids (FFAs) and water contents). Alkali catalysts, such as sodium hydroxide (NaOH) and potassium hydroxide (KOH), are the commonly used, due to their low cost and high achievable yields of more than 98% within a hour at reasonable temperature of 60°C [7, 32]. The reaction starts by preparing the alkoxide solution and charging it to the reactor with oils. The reaction is then heated to the reaction temperature for few hours. Products are then separated by gravity and crude biodiesel is obtained, which needs further washing to recover unreacted oils and alcohols. In addition to neutralization to receive the catalyst used.

Although the process is simple and commercially used, it is not practical with feedstocks containing high free fatty acids (FFAs) and water contents such as those from non-edible and waste oil due to soap formation that lowers the overall production yield and require large amount of catalysts [33, 34]. Pretreatment of oil prior transesterification by acid esterification have been suggested, where sulfuric acid (H_2SO_4) is commonly adopted. Although it could be beneficial in enhancing oil quality, the process very slow and requires large amount of alcohols. In addition, acids are corrosive [5, 17, 35].

3.2. Green catalysts

The use of enzymes, which are green catalysts, has been suggested. Among the several available enzymes, lipases (EC 3.1.1.3), which are of hydrolytic enzymes received an increasing attention in biodiesel production. Lipase-catalyzed biodiesel production due to their ability to act on ester bonds, at mild temperatures with less energy needs [35].

Lipases, namely the non-specific one, can convert oils from different sources, including from hose cooking, without any pre-treatment needs with easy product separation and no soap formation. Among the several studied, lipases from *Candida antartica* [36–44], *Pseudomonas fluorescens* [45, 46], *Pseudomonas cepacia* [47], *Candida rugosa* [48–50] and *Rizhomucor miehei* [51, 52] are commonly used.

Although lipases are superior compared to chemical catalysts, accumulation of glycerol which is a by-product negatively affect the enzyme activity and reaction yield. Glycerol accumulation increases reaction mixture viscosity and forms hydrophilic layer around rhe enzyme, preventing the reaction substrate to reach enzyme active site [53, 54]. The highest glycerol inhibition effect was found when silica, which has the highest micro-pores structure, was used in the immobilization protocol of the lipase. Continuous removal of produced glycerol from reaction mixture and/or using *tert*-butanol as solvent was proposed [55, 56]. The used of silica gel that can absorb glycerol is also advantageous in such case [57].

The activity of the enzyme was also found to decrease when more than 1.5 molar equivalents of alcohol is used. This is because at certain concentration, alcohols which are hydrophilic becomes insoluble in oils and tends to strip-off the hydration layer of water from the lipase. Therefore, inhibition and lose in activity [2, 58–60]. Numerous solutions have been proposed to overcome short-chain alcohols inhibition limitation. These include step-wise alcohols addition [38, 61], use of acetates as acceptors [57, 58, 62], lipase pretreatment and activity enhancement [39], use of genetically modified methanol-tolerant lipase and improving the polarity of the reaction medium using organic solvents. The latter is commonly adopted method.

On the other hand, enzymatic biodiesel production is not yet commercialized due to enzymes high costs. Immobilization of the lipase is usually considered to re-use the enzyme in several cycles. Immobilization can also enhance the stability. For example, Novozym®435, which is an immobilized enzyme form of *Candida antartica,* was reused for 12 continuous cycles without any detectable loss in the activity [57] when the non-edible oil from Jatropha was transesterified with methanol. Whereas, when *tert*-butanol was used as reaction media Novozym®435 activity was maintained for 200 cycles [63].

4. Reaction medium

The solvent-free reaction systems are always the preferable one in enzyme catalyzed processes, however when the reaction is catalyzed by a lipase the use of solvents is essential to prevent the inhibition. By introducing hydrophobic solvents to the reaction, the solubility of reaction substrates increases resulting in reduced inhibition effect of hydrophilic substrates/products. In addition, the viscosity and transport limitations of reaction mixture to enzyme active sites decreases, which results in increased reaction yield [64, 65].

4.1. Conventional organic solvents

Numerous organic solvents have been used in biodiesel production, where the hydrophobicity was considered as main factor in selecting the proper solvent [66]. It was found that the biodiesel production rate increases with the increase in the hydrophobicity of the solvent used, and hydrophilic solvents resulted tend to strip-off the bound water from the enzyme surface is used [39, 67–70]. Generally, the stripping was reported to take place when an organic solvent with Log P (hydrophobicity) <2 used. *n*-Hexane, which has log $P = 3.5$, has been commonly used, where its effect on enhancing the production yield, compared to solvent-free system, was observed in several studies. These includes the work of Nelson et al. [36], who tested the effect of using *n*-hexane in tallow fats transesterification with methanol at 3:1 methanol to oil molar ratio when catalyzed by *Mucor miehei* lipase. High yield reaching 95%, compared to 19% in solvent free, was obtained.

tert-Butanol is another solvent used in lipase catalyzed biodiesel processes. It has been selected as a capable alternative to *n*-hexane that cannot dissolve glycerol and minimize its inhibition effect [1, 13, 71–74]. A high methanol to oil molar ratio of 6:1 could be reached in soybean oils transesterification with Novozym®435, resulting in 60% yield, compared to only 10% in solvent-free system.

4.2. Green solvents

Although organic solvents enhance the production yield, an additional downstream unit is required to separate the solvent from the products, resulting in an additional production cost. Moreover, organic solvents are toxic and volatile and their use could pose several environmental issues that should be minimized. Efforts have been made to find alternative non-toxic and environmental benign solvents. In this regard, supercritical CO_2 and ionic liquids (ILs) have been suggested.

4.2.1. *Supercritical carbon dioxide*

Supercritical fluids are fluids at temperatures and pressures above their critical points. They have been used in several applications. Among the different fluids, supercritical carbon dioxide ($SC-CO_2$) and is been the most commonly used. Supercritical CO_2 is a non-toxic and cheap fluid that appear in abundant with moderate critical parameters [75]. Compared to organic solvents, $SC-CO_2$ has liquid solubilization capacity and gas diffusivity and viscosity, where small changes in the conditions can lead to a significant increase in the properties. These unique physiochemical properties allow it to be used in several applications, including separation and reaction [76–78]. Moreover, easy products separation can be achieved using $SC-CO_2$.

Although $SC-CO_2$ has been commonly used in esters transesterification in the presence of lipase, its employment in biodiesel production is still new [79]. The compatibility of $SC-CO_2$ with lipases is well recognized, and by using it in biodiesel production, the mass transfer of reaction substrates into enzyme active sites would be enhanced. In spite of the high pressure uses, it was clearly verified that it has minimal effect on enzyme inhibition at pressure less than 200 bars [80, 81]. Comparable yields to organic solvent were achieved when palm kernel and Jatropha oils were transesterified in the presence of Novozym®435 in $SC-CO_2$ [75, 82–84].

Higher yield of 80% was obtained when $SC-CO_2$ was used in microalgae lipids transesterification in the presence of same lipase [42].

Supercritical CO_2 has been also used to extract oils for biodiesel production, such as those oils from vegetable crops [76, 77], microalgae cells [85–90] and fats from animal meat. The use of $SC-CO_2$ is adopted to minimize the use of toxic solvents and utilize the leftover, after extractions, in other applications such as in food and pharmaceutical industries, unlike *n*-hexane which is toxic and its use is an energy intensive process. Its effectiveness depends on the selected extraction conditions; namely the temperature, pressure and flow rate, where increasing the pressure increases $SC-CO_2$ density and the extraction yield whereas the temperature has two opposite effect that become equal at crossover pressure. By increasing the extraction temperature, $SC-CO_2$ density decreases and reduces its capability to solubilize the desired solute, while solute vapor pressure increases resulting in more solutes extraction. Several studies had considered the effectiveness of using $SC-CO_2$. For example, similar yields performance of *n*-hexane were reported for oils extraction from *Spirulina platensis* [86], *Spirulina maxima* [87] and *Pavlova* sp. [89] microalgae cells. A higher efficiency was reported in extracting oils from *Chlorococum* sp. and *Nannochloropsis* sp. [90, 91].

As mentioned earlier, $SC-CO_2$ has many advantages. However, high pressure is needed for pumping and reaching the supercritical state of CO_2, making the process costly. Depressurization to separate the biodiesel from enriched $SC-CO_2$ could negatively affect lipase structural confirmation, therefore its stability. To minimize the effect, continuous operation has been also considered. It has been successfully employed for soybean [92], corn oil [93, 94], microalgae and sunflower oils [95]. Taher et al. [55, 56] had stated that the feasibility of using $SC-CO_2$ for energy production is not evident, but combining oil extraction conversions to biodiesel in $SC-CO_2$ in one integrated system would be feasible and the additional pumping cost for energy production could be justified and make the overall process more feasible [42, 44, 96]. On the other hand, the presence of water could results to carbonic acid formation that change the reaction pH and denaturant the lipase. CO_2 may also react with the amine groups on the surface of lipase to form carbamates [97].

4.2.2. Ionic liquids

Ionic liquids (ILs) are liquids of low crystallization tendency. They are composed of cations and anions and distinguished from conventional solvents in their non-vapor pressure feature. Thus, known to as "designer solvents." They have developed as green alternative solvents to replace the conventional volatile solvents in several process, including biodiesel production. The first attempt to use them in lipase catalyzed reactions was with [bmim][PF_6] and [bmim][BF_4], which were used in several reactions, including transesterification [98]. The selection the proper IL depends on its effect to enhance the enzyme activity reaction substrates/products solubility [99–103].

Typically, by judicious selection of the alkali chain on the cation and anion group, the physiochemical properties of designed IL can be tuned. For example, symmetric and shorter alkyl chains cations in the IL result in a higher melting temperature than those with asymmetric cations [104, 105], and increasing chain branching results in an increased the melting point

[106, 107]. However, it decreases with the increase in anion size. On the other hand, ILs with symmetric and fluorinated anions, have high viscosity, which is not preferable in enzyme catalyzed reactions. Ionic liquids based on cations with aromatic phenyl ring also have high viscosity as well [108].

The miscibility of the reaction substrates with the IL and IL hydrophobicity are main factor affecting the overall reaction yield, where the high solubility of reaction substrates to enhance reaction rate and low solubility of the biodiesel in the IL are the desired features in biodiesel production. The hydrophobicity of the IL depends mainly on the anions used. For example, [PF_6^-] and [Tf_2N^-] anions, which are hydrophobic results in making hydrophobic ILs, where those with hydrophilic anions, such as [Cl^-], [Br^-], [I^-], [NO_3^-], [CH_3COO^-] and [CF_3COO^-] for hydrophilic ILs. The hydrophobicity of ILs can also be affected by the length of the alkyl chain on the cation, in which longer alkyl chain results in a more hydrophobic IL [109–111]. Similar to of organic solvents, hydrophobic ILs are preferable in enzyme catalyzed biodiesel process, wherein hydrophilic ILs may strip-off the essential hydration layer and deactivate the lipase. Moreover, the nucleophilicity of the anion used in the IL combination affect lipase activity and stability, where high nucleophilicity of an IL may affect lipase structure activity interacting with the positively charged sites in lipase [112].

Among the several tested ILs in biodiesel production, [PF_6^-] and [NTf_2^-] based ILs are commonly used. For example, [bmim][PF_6] and [emim][PF_6] where used in sunflower oil transesterification Novozyme®435. High yield, reaching 98%, was achieved in [emim][PF_6] due to its higher hydrophobicity, however, insignificant products were obtained when [BF_4^-]-based ILs were tested [113]. The high yield obtained in [emim][PF_6] compared to [bmim][PF_6] is due to the ability of long-chain cation-based ILs to dissolve reaction substrates, thus the reaction take place in a two-phase system resulting in moderate efficiency.

In addition to ILs uses in lipase catalyzed reaction, they have been employed as green catalyst to overcome the reaction complication and product purification issues in chemical catalyzed reactions. The brønsted acidic ILs [$PY(CH_2)_4SO_3H$][HSO_4] and [$CyN_{1,1}PrSO_3H$][Tos] was found to be effective in transesterifying cottonseed (92% yield) and coconut (98% yield) oils, respectively, where comparable to that using concentrated sulfuric acid were obtained [114, 115]. Similar yield was also obtained from esterification of long-chain free fatty acids in [NMP][CH_3SO_3] [116]. In addition, biodiesel yield of 87 and 97% were also achieved using the basic IL [bmim][OH] [117] and [hmm][OH], respectively [118].

Ionic liquids have been used to extract oils, commonly from microalgae cells as they can be used with wet cells without the need cell walls disruption. In such processes, hydrophilic ILs are used where they have the capability to dissolve algal cell components leaving the oils insoluble and float. The extraction from wet cells of *Chlorella vulgaris* was tested using IL [emim][DEP], where 40% higher than *n*-hexane-methanol (7:3 v/v) mixture was obtained [119]. The effect of adding a polar solvent with the IL was also evaluated [120]. For example, a mixture of [emim][CH_3SO_4] and methanol was tested with *Chlorella sp.* cells containing 70% water, and an yield of 75% was achieved at 1:1.2 (w/w) solvents ratio.

The main challenge of employing ILs at industrial scale is in their high costs. Therefore, the recycling step is important. In addition, when long alkyl chain on cation-based ILs is used, the separation step is not easy and continuous recovery of biodiesel from reaction mixture as they produced is vital. Combination ILs with SC$-CO_2$ (IL-SC$-CO_2$) has been recently suggested, where biodiesel can be recovered using SC$-CO_2$ in an effective manner. Such system was tested for biodiesel production from triolein using Novozym®435 in different ILs and high yields reaching 98% was obtained after 6 h [121].

5. Conclusions

Biodiesel production using chemical catalysts and solvents from received the attention to replace conventional diesel fuel. However, the process is not commercialized due to many shortcomings raised. The employment of green catalysts and solvents, either by SC$-CO_2$ or by ILs has been suggested to several technical restrictions. The use of integrated processes that combine the use of different green catalysts and solvents in a one process to enhance product separation and solvent recover is discussed.

Author details

Hanifa Taher[1*] and Sulaiman Al-Zuhair[2]

*Address all correspondence to: healblooshi@masdar.ac.ae

1 Chemical and Environmental Engineering Department, Masdar Institute, Abu Dhabi, UAE

2 Chemical and Petroleum Engineering Department, UAE University, AlAin, UAE

References

[1] Demirbas, A., Progress and recent trends in biodiesel fuels. Energy Conversion and Management 2009, 50, 14–34.

[2] Fjerbaek, L.; Christensen, K. V.; Norddahl, B., A review of the current state of biodiesel production using enzymatic transesterification. Biotechnology and Bioengineering 2009, 102(5), 1298–1315.

[3] Bajpai, D.; Tyagi, V. K., Biodiesel: source, production, composition, properties and its benefits. Journal of Oleo Science 2006, 55(10), 487–502.

[4] Fellows, P., Food Processing Technology: Principles and Practice. Woodhead Publishing Limited and CRC Press LLC: USA, 2000.

[5] Al-Zuhair, S., Production of biodiesel:possibilities and challenges. Biofuels, Bioproducts and Biorefining 2007, 1(1), 57–66.

[6] Demirbas, A., Importance of biodiesel as transportation fuel. Energy Policy 2007, 35(9), 4661–4670

[7] Fukuda, H.; Kondo, A.; Noda, H., Biodiesel fuel production by transesterification of oils. Journal of Bioscience and Bioengineering 2001, 92(5), 405–416.

[8] Ranganathan, S. V.; Narasimhan, S. L.; Muthukumar, K., An overview of enzymatic production of biodiesel. Bioresource Technology 2008, 99(10), 3975–3981

[9] Monthly Biodiesel Production Report with Data for August 2016; U.S. Department of Energy, 2016; 1–10.

[10] Antunes, W. M.; Veloso, C. d. O.; Henriques, C. A., Transesterification of soybean oil with methanol catalyzed by basic solids. Catalysis Today 2008, 133–135, 548–554.

[11] Liu, X.; He, H.; Wang, Y.; Zhu, S.; Piao, X., Transesterification of soybean oil to biodiesel using CaO as a solid base catalyst. Fuel 2008, 87(2), 216–221.

[12] Dubé, M. A.; Tremblay, A. Y.; Liu, J., Biodiesel production using a membrane reactor. Bioresource Technology 2007, 98(3), 639–647

[13] Al-Zuhair, S.; Ling, F. W.; Jun, L. S., Proposed kinetic mechanism of the production of biodiesel from palm oil using lipase. Process Biochemistry 2007, 42(6), 951–960.

[14] Kalam, M. A.; Masjuki, H. H., Biodiesel from palmoil-an analysis of its properties and potential. Biomass Bioenergy 2002, 23(6), 471–479.

[15] Saka, S.; Kusdiana, D., Biodiesel fuel from rapeseed oil as prepared in supercritical methanol. Fuel 2001, 80(2), 225–231.

[16] Mekhilef, S.; Siga, S.; Saidur, R., A review on palm oil biodiesel as a source of renewable fuel. Renewable and Sustainable Energy Reviews 2011, 15(4), 1937–1949.

[17] Akoh, C. C.; Chang, S.-W.; Lee, G.-C.; Shaw, J.-F., Enzymatic approach to biodiesel production. Journal of Agricultural and Food Chemistry 2007, 55(22), 8995–9005.

[18] Basha, S. A.; Gopal, K. R.; Jebaraj, S., A review on biodiesel production, combustion, emissions and performance. Renewable Sustainable Energy Reviews. 2009, 13(6–7), 1628–1634.

[19] Sharma, Y. C.; Singh, B.; Upadhyay, S. N., Advancements in development and characterization of biodiesel: a review. Fuel 2008, 87(12), 2355–2373.

[20] Fan, X.; Burton, R., Recent development of biodiesel feedstocks and the applications of glycerol: a review. The Open Fuels and Energy Science Journal 2009, 1, 100–109.

[21] Helwani, Z.; Othman, M. R.; Aziz, N.; Fernando, W. J. N.; Kim, J., Technologies for production of biodiesel focusing on green catalytic techniques: a review. Fuel Process Technology. 2009, 90(12), 1502–1514.

[22] Lai, C.-C.; Zullaikah, S.; Vali, S. R.; Ju, Y.-H., Lipase-catalyzed production of biodiesel from rice bran oil. Journal of Chemical Technology and Biotechnology 2005, 80(3), 331–337.

[23] Phan, A. N.; Phan, T. M., Biodiesel production from waste cooking oils. Fuel 2008, 87(17–18), 3490–3496.

[24] Predojevi?, Z. J., The production of biodiesel from waste frying oils: a comparison of different purification steps. Fuel 2008, 87(17–18), 3522–3528.

[25] Taufiqurrahmi, N.; Mohamed, A. R.; Bhatia, S., Production of biofuel from waste cooking palm oil using nanocrystalline zeolite as catalyst: process optimization studies. Bioresource Technology 2011, 102(22), 10686–10694.

[26] Zhang, Y.; Dube, M. A.; McLean, D. D.; Kates, M., Biodiesel production from waste cooking oil: 1. process design and technological assessment. Bioresource Technology 2003, 89(1), 1–16.

[27] Adamczak, M.; Bornscheuer, U. T.; Bednarski, W., The application of biotechnological methods for the synthesis of biodiesel. European Journal of Lipid Science and Technology 2009, 111, 800–813.

[28] Chisti, Y., Biodiesel from microalgae. Biotechnology Advances 2007, 25(3), 294–306.

[29] Spolaore, P.; Joannis-Cassan, C.; Duran, E.; Isambert, A., Commercial applications of microalgae. Journal of Bioscience and Bioengineering 2006, 101(2), 87–96.

[30] Vijayaraghavan, K.; Hemanathan, K., Biodiesel production from freshwater algae. Energy and Fuels 2009, 23(11), 5448–5453.

[31] Sheehan, J.; Dunahay, T.; Benemann, R.; Roessler, G.; Weissman, C., A look back at the U.S. department of energy- aquatic species program: biodiesel from algae; National Renewable Energy Laboratory, USA, 1998.

[32] Atadashi, I. M.; Aroua, M. K.; Abdul Aziz, A. R.; Sulaiman, N. M. N., The effects of water on biodiesel production and refining technologies: a review. Renewable and Sustainable Energy Reviews 2012, 16(5), 3456–3470.

[33] Ma, F.; Clements, L. D.; Ha, M. A., Biodiesel fuel from animal fat. Industrial and Engineering Chemistry Research 1998, 37(9), 3768–3771.

[34] Sivasamy, A.; Cheah, K. Y.; Fornasiero, P.; Kemausuor, F.; Zinoviev, S.; Miertus, S., Catalytic applications in the production of biodiesel from vegetable oils. ChemSuSChem 2009, 2(4), 278–300.

[35] Marchetti, J. M.; Miguel, V. U.; Errazu, A. F., Possible methods for biodiesel production. Renewable & Sustainable Energy Reviews. 2007, 11, 1300–1311.

[36] Nelson, L. A.; Foglia, T. A.; Marmer, W. N., Lipase-catalyzed production of biodiesel. Journal of American Oil Chemists' Society 1996, 73(9), 1191–1195.

[37] Shimada, Y.; Watanabe, Y.; Samukawa, T.; Sugihara, A.; Noda, H.; Fukuda, H.; Tominaga, Y., Conversion of vegetable oil to biodiesel using immobilized *Candida antarctica* lipase. Journal of the American Oil Chemists' Society 1999, 76 (7), 789–793.

[38] Watanabe, Y.; Shimada, Y.; Sugihara, A.; Noda, H.; Fukuda, H.; Tominaga, Y., Continuous production of biodiesel fuel from vegetable oil using immobilized *Candida antarctica* lipase. Journal of the American Oil Chemists' Society 2000, 77(4), 355–360.

[39] Samukawa, T.; Kaieda, M.; Matsumoto, T.; Ban, K.; Kondo, A.; Shimada, Y.; Noda, H.; Fukuda, H., Pretreatment of immobilized *Candida antarctica* lipase for biodiesel fuel production from plant oil. Journal of Bioscience and Bioengineering 2000, 90(2), 180–183.

[40] Fedosov, S. N.; Brask, J.; Pedersen, A. K.; Nordblad, M.; Woodley, J. M.; Xu, X., Kinetic model of biodiesel production using immobilized lipase *Candida antarctica* lipase b. Journal of Molecular Catalysis B: Enzymatic 2013, 85–86, 156–168.

[41] Watanabe, Y.; Shimada, Y.; Sugihara, A.; Tominaga, Y., Conversion of degummed soybean oil to biodiesel fuel with immobilized *Candida antarctica* lipase. Journal of Molecular Catalysis B: Enzymatic 2002, 17(3–5), 151–155.

[42] Taher, H.; Al-Zuhair, S.; Al-Marzouqi, A. H.; Haik, Y.; Farid, M., Enzymatic biodiesel production of microalgae lipids under supercritical carbon dioxide: process optimization and integration. Biochemical Engineering Journal 2014, 90, 103–113.

[43] Taher, H.; Al-Zuhair, S.; AlMarzouqui, A.; Hashim, I., Extracted fat from lamb meat by supercritical CO_2 as feedstock for biodiesel production. Biochemical Engineering Journal 2011, 55(1), 23–31.

[44] Al-Zuhair, S.; Hussein, A.; Al-Marzouqi, A. H.; Hashim, I., Continuous production of biodiesel from fat extracted from lamb meat in supercritical CO_2 media. Biochemical Engineering Journal 2012, 60, 106–110.

[45] Guldhe, A.; Singh, B.; Rawat, I.; Permaul, K.; Bux, F., Biocatalytic conversion of lipids from microalgae *Scenedesmus obliquus* to biodiesel using *Pseudomonas fluorescens* lipase. Fuel 2015, 147, 117–124.

[46] Devanesan, M. G.; Viruthagiri, T.; Sugumar, N., Transesterification of jatropha oil using immobilized *Pseudomonas fluorescens*. African Journal of Biotechnology 2007, 6(21), 2497–2501.

[47] Noureddini, H.; Gao, X.; Philkana, R. S., Immobilized *Pseudomonas cepacia* lipase for biodiesel fuel production from soybean oil. Bioresource Technology 2005, 96(7), 769–777.

[48] Moreno-Pirajàn, J. C.; Giraldo, L., Study of immobilized *Candida rugosa* lipase for biodiesel fuel production from palm oil by flow Microcalorimetry. Arabian Journal of Chemistry 2011, 4(1), 55–62.

[49] Tan, Z.; Fang, M.; Du, H.; Song, L.; Ren, D.; Tang, X.; Han, X.; Liang, X., Production of biodiesel catalyzed by *Candida rugosa* lipase at interface of W/O microemulsion system. Journal of the Brazilian Chemical Society 2014, 25, 1704–1711.

[50] Lee, J. H.; Kim, S. B.; Kang, S. W.; Song, Y. S.; Park, C.; Han, S. O.; Kim, S. W., Biodiesel production by a mixture of *Candida rugosa* and *Rhizopus oryzae* lipases using a supercritical carbon dioxide process. Bioresource Technology 2011, 102(2), 2105–2108.

[51] Huang, D.; Han, S.; Han, Z.; Lin, Y., Biodiesel production catalyzed by *Rhizomucor miehei* lipase-displaying *Pichia pastoris* whole cells in an isooctane system. Biochemical Engineering Journal 2012, 63, 10–14.

[52] Huang, J.; Xia, J.; Yang, Z.; Guan, F.; Cui, D.; Guan, G.; Jiang, W.; Li, Y., Improved production of a recombinant *Rhizomucor miehei* lipase expressed in *Pichia pastoris* and its application for conversion of microalgae oil to biodiesel. Biotechnology for Biofuels 2014, 7(1), 1–11.

[53] Dossat, V.; Combes, D.; Marty, A., Continuous enzymatic transesterification of high oleic sunflower oil in a packed bed reactor: influence of the glycerol production. Enzyme and Microbial Technology 1999, 25(3–5), 194–200.

[54] Xu, Y.; Nordblad, M.; Nielsen, P. M.; Brask, J.; Woodley, J. M., In Situ visualization and effect of glycerol in lipase-catalyzed ethanolysis of rapeseed oil. Journal of Molecular Catalysis B: Enzymatic 2011, 72 (3–4), 213–219.

[55] Chen, H.-C.; Ju, H.-Y.; Wu, T.-T.; Liu, Y.-C.; Lee, C.-C.; Chang, C.; Chung, Y.-L.; Shieh, C.-J., Continuous production of lipase-catalyzed biodiesel in a packed-bed reactor: optimization and enzyme reuse study. Journal of Biomedicine and Biotechnology, 2011, 1-6.

[56] Azócar, L.; Navia, R.; Beroiz, L.; Jeison, D.; Ciudad, G., Enzymatic biodiesel production kinetics using co-solvent and an anhydrous medium: a strategy to improve lipase performance in a semi-continuous reactor. New Biotechnology 2014, 31(5), 422–429.

[57] Modi, M. K.; Reddy, J. R. C.; Rao, B. V. S. K.; Prasad, R. B. N., Lipase-mediated conversion of vegetable oils into biodiesel using ethyl acetate as acyl acceptor. Bioresource Technology 2007, 98(6), 1260–1264

[58] Du, W.; Xu, Y.; Liu, D.; Zeng, J., Comparative study on lipase-catalyzed transformation of soybean oil for biodiesel production with different acyl acceptors. Journal of Molecular Catalysis B - Enzymatic 2004, 30(3–4), 125–129.

[59] Li, L.; Du, W.; Liu, D.; Wang, L.; Li, Z., Lipase-catalyzed transesterification of rapeseed oils for biodiesel production with a novel organic solvent as the reaction medium. Journal of Molecular Catalysis B: Enzymatic. 2006, 43(1–4), 58–62

[60] Zheng, Y.; Quan, J.; Ning, X.; Zhu, L.-M.; Jiang, B.; He, Z.-Y., Lipase-catalyzed transesterification of soybean oil for biodiesel production in tert-amyl alcohol. World Journal of Microbiology and Biotechnology 2009, 25(1), 41–46.

[61] Shimada, Y.; Watanabe, Y.; Sugihara, A.; Tominaga, Y., Enzymatic alcoholysis for biodiesel fuel production and application of the reaction to oil processing. Journal of Molecular Catalysis B: Enzymatic 2002, 17(3–5), 133–142.

[62] Modi, M. K.; Reddy, J. R. C.; Rao, B. V. S. K.; Prasad, R. B. N., Lipase-mediated transformation of vegetable oils into biodiesel using propan-2-ol as acyl acceptor. Biotechnology Letters 2006, 28(9), 637–640.

[63] Lu, J.; Chen, Y.; Wang, F.; Tan, T., Effect of water on methanolysis of glycerol trioleate catalyzed by immobilized lipase *Candida* sp. 99–125 in organic solvent system. Journal of Molecular Catalysis B: Enzymatic 2009, 56(2–3), 122–125.

[64] Klibanov, A. M., Improving enzymes by using them in organic solvents. Nature 2001, 209, 241–246.

[65] Zaks, A.; Klibanov, A., Enzymatic catalysis in organic media at 100 degrees C. Science 1984, 224, 1249–1251.

[66] Adamczak, M.; Krishna, S. H., Strategies for improving enzymes for efficient biocatalysis. Food Technology and Biotechnology 2004, 42(4), 251–264.

[67] Klibanov, A. M., Why are enzymes less active in organic solvents than in water? Trends in Biotechnology 1997, 15(3), 97–101.

[68] Doukyu, N.; Ogino, H., Organic solvent-tolerant enzymes. Biochemical Engineering Journal 2010, 48(3), 270–282.

[69] Yang, L.; Dordick, J. S.; Garde, S., Hydration of enzyme in nonaqueous media is consistent with solvent dependence of its activity. Biophysical Journal 2004, 87(2), 812–821.

[70] Gorman, L. A. S.; Dordick, J. S., Organic solvents strip water off enzymes. Biotechnology and Bioengineering 1992, 39(4), 392–397.

[71] Yang, Z.; Zhang, K.-P.; Huang, Y.; Wang, Z., Both hydrolytic and transesterification activities of *Penicillium expansum* lipase are significantly enhanced in ionic liquid [Bmim][Pf6]. Journal of Molecular Catalysis B: Enzymatic 2010, 63(1–2), 23–30.

[72] Peng, W.; Wu, Q.; Tu, P., Pyrolytic Characteristics of heterotrophic chlorella protothecoides for renewable bio-fuel production. Journal of Applied Phycology 2001, 13(1), 5–12.

[73] Lai, J.-Q.; Hu, Z.-L.; Wang, P.-W.; Yang, Z., Enzymatic production of microalgal biodiesel in ionic liquid [Bmim][Pf6]. Fuel 2012, 95, 329–333.

[74] Royon, D.; Daz, M.; Ellenrieder, G.; Locatelli, S., Enzymatic production of biodiesel from cotton seed oil using t-butanol as a solvent. Bioresource Technology 2007, 98(3), 648–653.

[75] Rathore, V.; Madras, G., Synthesis of biodiesel from edible and non-edible oils in supercritical alcohols and enzymatic synthesis in supercritical carbon dioxide. Fuel 2007, 86(17–18), 2650–2659.

[76] Del Valle, J. M.; Rivera, O.; Mattea, M.; Ruetsch, L.; Daghero, J.; Flores, A. S., Supercritical CO_2 processing of pretreated rosehip seeds: effect of process scale on oil extraction kinetics. The Journal of Supercritical Fluids 2004, 31(2), 159–174.

[77] Reverchon, E.; Marrone, C., Modeling and simulation of the supercritical CO_2 extraction of vegetable oils. Journal of Supercritical Fluids 2001, 19(2), 161–175.

[78] Sovova, H.; Zarevucka, M.; Vacek, M.; Stransky, K., Solubility of two vegetable oils in supercritical CO_2. Journal of Supercritical Fluids 2001, 20(1), 15–28.

[79] Romero, M. D.; Calvo, L.; Alba, C.; Habulin, M.; Primoži?, M.; Knez, Ž., Enzymatic synthesis of isoamyl acetate with immobilized *Candida antarctica* lipase in supercritical carbon dioxide. The Journal of Supercritical Fluids 2005, 33(1), 77–84.

[80] Celia, E.; Cernia, E.; Palocci, C.; Soro, S.; Turchet, T., Tuning *Pseudomonas cepacea* lipase (PCL) activity in supercritical fluids. The Journal of Supercritical Fluids 2005, 33(2), 193–199.

[81] Novak, Z.; Habulin, M.; Krmelj, V.; Knez, Å. E., Silica aerogels as supports for lipase catalyzed esterifications at sub- and supercritical conditions. The Journal of Supercritical Fluids 2003, 27(2), 169–178.

[82] Oliveira, D.; Oliveira, J. V., Enzymatic alcoholysis of palm kernel oil in n-hexane and $SCCO_2$. The Journal of Supercritical Fluids 2001, 19(2), 141–148.

[83] Madras, G.; Kolluru, C.; Kumar, R., Synthesis of biodiesel in supercritical fluids. Fuel 2004, 83(14–15), 2029–2033.

[84] Varma, M. N.; Madras, G., Synthesis of biodiesel from castor oil and linseed oil in supercritical fluids. Industrial and Engineering Chemistry Research 2007, 46(1), 1–6.

[85] Taher, H.; Al-Zuhair, S.; Al-Marzouqi, A. H.; Haik, Y.; Farid, M.; Tariq, S., Supercritical carbon dioxide extraction of microalgae lipid: process optimization and laboratory scale-up. The Journal of Supercritical Fluids 2014, 86, 57–66.

[86] Andrich, G.; Zinnai, A.; Nesti, U.; Venturi, F., Supercritical fluid extraction of oil from Microalga Spirulina (Arthrospira) platensis. Acta Alimentaria 2006, 35(2), 195–203.

[87] Mendes, R. L.; Nobre, B. P.; Cardoso, M. T.; Pereira, A. P.; Palavra, A. N. F., Supercritical carbon dioxide extraction of compounds with pharmaceutical importance from microalgae. Inorganica Chimica Acta 2003, 356, 328–334.

[88] Mendes, R. L.; Fernandes, H. L.; Coelho, J. P.; Reis, E. C.; Cabral, J. M. S.; Novais, J. L. M.; Palavra, A. N. F., Supercritical CO_2 extraction of carotenoids and other lipids from Chlorella vulgaris. Food Chemistry 1995, 53(1), 99–103.

[89] Cheng, C.-H.; Du, T.-B.; Pi, H.-C.; Jang, S.-M.; Lin, Y.-H.; Lee, H.-T., Comparative study of lipid extraction from microalgae by organic solvent and supercritical CO_2. Bioresource Technology 2011, 102(21), 10151–10153.

[90] Halim, R.; Gladman, B.; Danquah, M. K.; Webley, P. A., Oil extraction from microalgae for biodiesel production. Bioresource Technology 2011, 102, 178–185.

[91] Andrich, G.; Nesti, U.; Venturi, F.; Zinnai, A.; Fiorentini, R., Supercritical fluid extraction of bioactive lipids from the microalga *Nannochloropsis* sp. European Journal of Lipid Science and Technology 2005, 107(6), 381–386.

[92] Dalla Rosa, C.; Morandim, M. B.; Ninow, J. L.; Oliveira, D.; Treichel, H.; Oliveira, J. V., Continuous lipase-catalyzed production of fatty acid ethyl esters from soybean oil in compressed fluids. Bioresource Technology 2009, 100(23), 5818–5826.

[93] Ciftci, O. N.; Temelli, F., Continuous production of fatty acid methyl esters from corn oil in a supercritical carbon dioxide bioreactor. The Journal of Supercritical Fluids 2011, 58(1), 79–87.

[94] Ciftci, O. N.; Temelli, F., Continuous biocatalytic conversion of the oil of corn distiller's dried grains with solubles to fatty acid methyl esters in supercritical carbon dioxide. Biomass and Bioenergy 2013, 54, 140–146.

[95] Rodrigues, A. R.; Paiva, A.; da Silva, M. G.; Sim??es, P.; Barreiros, S., Continuous enzymatic production of biodiesel from virgin and waste sunflower oil in supercritical carbon dioxide. The Journal of Supercritical Fluids 2011, 56(3), 259–264.

[96] Al-Zuhair, S.; Taher, H., Supercritical Fluids Technology in Lipase Catalyzed Processes. CRC Press, USA, 2016.

[97] Wright, H. B.; Moore, M. B., Reactions of aralkyl amines with carbon dioxide. Journal of the American Chemical Society 1948, 70(11), 3865–3866.

[98] Madeira Lau, R.; Van Rantwijk, F.; Seddon, K. R.; Sheldon, R. A., Lipase-catalyzed reactions in ionic liquids. Organic Letters 2000, 2(26), 4189–4191.

[99] Lozano, P.; de Diego, T.; Carrié, D.; Vaultier, M.; Iborra, J. L., Lipase catalysis in ionic liquids and supercritical carbon dioxide at 150°C. Biotechnology Progress 2003, 19(2), 380–382.

[100] Lozano, P.; De Diego, T.; Carrie, D.; Vaultier, M.; Iborra, J. L., Over-stabilization of *Candida antarctica* lipase B by ionic liquids in ester synthesis. Biotechnology Letters 2001, 23(18), 1529–1533.

[101] Dang, D. T.; Ha, S. H.; Lee, S.-M.; Chang, W.-J.; Koo, Y.-M., Enhanced activity and stability of ionic liquid-pretreated lipase. Journal of Molecular Catalysis B: Enzymatic 2007, 45 (3–4), 118–121.

[102] Kaar, J. L.; Jesionowski, A. M.; Berberich, J. A.; Moulton, R.; Russell, A. J., Impact of ionic liquid physical properties on lipase activity and stability. Journal of the American Chemical Society 2003, 125(14), 4125–4131.

[103] Klahn, M.; Lim, G. S.; Wu, P., How ion properties determine the stability of a lipase enzyme in ionic liquids: a molecular dynamics study. Physical Chemistry Chemical Physics 2011, 13(41), 18647–18660.

[104] Ohno, H.; Yoshizawa, M., Ion conductive characteristics of ionic liquids prepared by neutralization of alkylimidazoles. Solid State Ionics 2002, 154, 303–309.

[105] Endres, F.; Zein El Abedin, S., Air and water stable ionic liquids in physical chemistry. Physical Chemistry Chemical Physics 2006, 8(18), 2101–2116.

[106] Xue, L.; Gurung, E.; Tamas, G.; Koh, Y. P.; Shadeck, M.; Simon, S. L.; Maroncelli, M.; Quitevis, E. L., Effect of alkyl chain branching on physicochemical properties of imidazolium-based ionic liquids. Journal of Chemical and Engineering Data 2016, 61(3), 1078–1091.

[107] Zhang, S.; Lu, X.; Zhou, Q.; Li, X.; Zhang, X.; Li, S., Ionic Liquids: Physicochemical Properties. Elsevier Science, Netherlands, 2009.

[108] Ochędzan-Siodłak, W.; Dziubek, K.; Siod?ak, D., Densities and viscosities of imidazolium and pyridinium chloroaluminate ionic liquids. Journal of Molecular Liquids 2013, 177, 85–93.

[109] Hernández-Fernández, F. J.; Ríos, A. P. D. L.; Tomás-Alonso, F.; Gómez, D.; Víllora, G., Stability of hydrolase enzymes in ionic liquids. The Canadian Journal of Chemical Engineering 2009, 87(6), 910–914.

[110] Aki, S. N. V. K.; Brennecke, J. F.; Samanta, A., How polar are room-temperature ionic liquids? Chemical Communications 2001, 5, 413–414.

[111] Huddleston, J. G.; Visser, A. E.; Reichert, W. M.; Willauer, H. D.; Broker, G. A.; Rogers, R. D., Characterization and comparison of hydrophilic and hydrophobic room temperature ionic liquids incorporating the imidazolium cation. Green Chemistry 2001, 3(4), 156–164.

[112] Lozano, P.; de Diego, T.; Guegan, J.-P.; Vaultier, M.; Iborra, J. L., Stabilization of a-chymotrypsin by ionic liquids in transesterification reactions. Biotechnology and Bioengineering 2001, 75(5), 563–569.

[113] Sunitha, S.; Kanjilal, S.; Reddy, P. S.; Prasad, R. B. N., Ionic liquids as a reaction medium for lipase-catalyzed methanolysis of sunflower oil. Biotechnology Letters 2007, 29(12), 1881–1885.

[114] Han, M.; Yi, W.; Wu, Q.; Liu, Y.; Hong, Y.; Wang, D., Preparation of biodiesel from waste oils catalyzed by a brønsted acidic ionic liquid. Bioresource Technology 2009, 100(7), 2308–2310.

[115] Wu, Q.; Chen, H.; Han, M.; Wang, D.; Wang, J., Transesterification of cottonseed oil catalyzed by brønsted acidic ionic liquids. Industrial and Engineering Chemistry Research 2007, 46(24), 7955–7960.

http://dx.doi.org/10.5772/66187

[116] Zhang, L.; Xian, M.; He, Y.; Li, L.; Yang, J.; Yu, S.; Xu, X., A brønsted acidic ionic liquid as an efficient and environmentally benign catalyst for biodiesel synthesis from free fatty acids and alcohols. Bioresource Technology 2009, 100(19), 4368–4373.

[117] Zhou, S.; Liu, L.; Wang, B.; Xu, F.; Sun, R. C., Biodiesel preparation from transesterification of glycerol trioleate catalyzed by basic ionic liquids. Chinese Chemical Letters 2012, 23(4), 379–382.

[118] Ren, Q.; Zuo, T.; Pan, J.; Chen, C.; Li, W., Preparation of biodiesel from soybean catalyzed by basic ionic liquids [Hnmm]oh. Materials 2014, 7(12), 8012.

[119] Choi, S.-A.; Oh, Y.-K.; Jeong, M.-J.; Kim, S. W.; Lee, J.-S.; Park, J.-Y., Effects of ionic liquid mixtures on lipid extraction from *Chlorella vulgaris*. Renewable Energy 2014, 65, 169–174.

[120] Young, G.; Nippgen, F.; Titterbrandt, S.; Cooney, M. J., Lipid extraction from biomass using co-solvent mixtures of ionic liquids and polar covalent molecules. Separation and Purification Technology 2010, 72(1), 118–121.

[121] Lozano, P.; Bernal, J. M.; Vaultier, M., Towards continuous sustainable processes for enzymatic synthesis of biodiesel in hydrophobic ionic liquids/supercritical carbon dioxide biphasic systems. Fuel 2011, 90(11), 3461–3467.

Chapter 10

Biogas, Biodiesel and Bioethanol as Multifunctional Renewable Fuels and Raw Materials

Venko Beschkov

Additional information is available at the end of the chapter

http://dx.doi.org/10.5772/65734

Abstract

Nowadays the world economy is based mainly on petrol as an energy source and raw material for chemical products. The global economic growth in the past century has led to high energy consumption, mainly from fossil fuels, such as coal, oil, and natural gas. The extensive use of fossil fuels formed and stored underground for millions of years has made impossible for the present vegetation on Earth to treat the emitted carbon dioxide by photosynthesis, leading to strong emissions of carbon dioxide and greenhouse effect with the consequent climate changes. One of the ways to cope with this global problem is to close the carbon cycle in nature by the use of renewable biofuels enabling recycling the sources of biological origin by energy production and consumption of the resulting carbon by photosynthesis. Some of these biofuels are biogas (a mixture of methane and carbon dioxide) generated from organic waste; ethanol, produced by fermentation of carbohydrates; and biodiesel, produced by transesterification of lipids. Another feature of this approach is the utilization of organic waste as energy, thus leading to multiple benefits for the environment: waste treatment with energy production, closing the natural carbon cycle, and saving of fossil fuels. Biofuels with their feedstock also serve as raw materials for new technologies for chemicals being now produced from petrol, natural gas, and coal.

Keywords: biomass, renewable fuels, biogas, biodiesel, ethanol, carbon dioxide recycling

1. Introduction

During the twentieth century, the world economy was mainly based on petrol as an energy source and raw material for chemical products. Energy consumption has increased steadily over the last century due to the world population growth and the technical progress development. The enormous economic growth on a global scale in the past century has led to the extensive use of fossil fuels, such as coal, oil, and natural gas.

A reason of concern was the extensive use of fossil fuels formed and stored underground for millions of years. It has made impossible for the present vegetation on Earth to treat the emitted carbon dioxide by photosynthesis. The result was greenhouse effect with the consequent climate changes. The climate changes assigned to the increased emissions of greenhouse gases forced humanity to develop alternative energy sources, one of them being biomass, either fresh or residual.

Another reason for humanity to turn to the renewable energy resources is the concern of depletion of the overall oil reserves [1], because the eight great economies (except Brazil) and many other nations depend on oil, the consequences of inadequate oil availability could be severe. Therefore, there are great incentives in exploring alternative energy sources.

One of the ways to cope with this global problem is to close the carbon cycle in nature by the use of renewable fuels enabling recycling the sources of biological origin by energy production and consumption of the resulting carbon by photosynthesis. Such biofuels are biogas (a mixture of methane and carbon dioxide) generated by anaerobic digestion of organic waste, ethanol, produced by fermentation of carbohydrates, and biodiesel, produced by transesterification of lipids.

The main feature of these approaches is the utilization of organic waste as energy, thus leading to multiple benefits for the environment: waste treatment with energy production, closing the natural carbon cycle, and saving of fossil fuels. Moreover, there are options to utilize the biofuels and their derivatives and residues as sources for chemical production.

2. Biogas

The anaerobic digestion of organic waste is a well spread process in nature. The huge amounts of natural gas collected underground are formed by this process during millions of years. The result is gas, containing about 95% methane with some contaminations. Nowadays, this process is used for agricultural waste treatment producing biogas with satisfactory heating capacity. Biogas is a mixture of methane and carbon dioxide with some contaminations of hydrogen sulfide, mercaptans, ethane, etc. The methane content varies from 55 to 90% volume depending on the substrate nature and content, the method of digestion, etc. The gas containing less than 50% methane is not combustible.

Biogas is broadly distributed in countries with developed agriculture (like India, China, Brazil, etc.), being a cheap and environmentally friendly option for the simultaneous solution of waste treatment problems and energy demand. Anaerobic digestion is also a convenient technology for activated sludge utilization and waste treatment in the food industry, pulp and paper industry, in household waste treatment, etc.

http://dx.doi.org/10.5772/65734

The anaerobic digestion with biogas production is a complicated process of consequent hydrolysis of organic macromolecules (carbohydrates and proteins) to oligosaccharides and peptides, acidogenesis to volatile fatty acids (mainly formic, acetic, and propionic), acetogenesis, and methanogenesis [2]. The overall process is shown in the scheme in **Figure 1**.

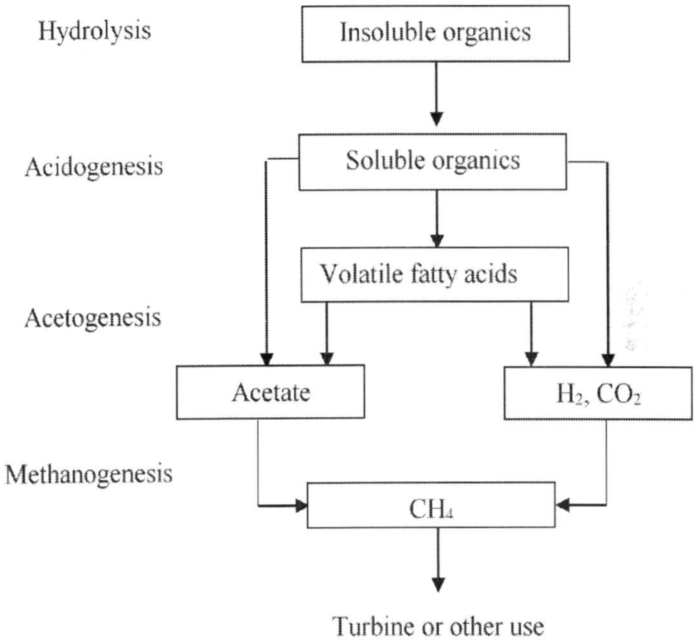

Figure 1. Four steps in biogas production [2].

The gross chemical reaction describing the first step, i.e., hydrolysis is:

$$(C_6H_{10}O_5)n + nH_2O = nC_6H_{12}O_6 \tag{1}$$

In the second step, acidogenic bacteria convert the products of hydrolysis into simple organic compounds, mostly short-chain carboxylic acids, ketones and alcohols.

The chemical reactions are shown below. Glucose is parallely converted into ethanol and propionic acid:

$$C_6H_{12}O_6 \leftrightarrow 2CH_3CH_2OH + 2CO_2 \tag{2}$$

$$C_6H_{12}O_6 + 2H_2 \leftrightarrow 2CH_3CH_2COOH + 2H_2O \quad (3)$$

The acetogenic reactions are

$$CH_3CH_2COO^- + 3H_2O \leftrightarrow CH_3COO^- + H^+ + HCO_3^- + 3H_2 \quad (4)$$

$$C_6H_{12}O_6 + 2H_2O{=}2CH_3COOH + 2CO_2 + 4H_2 \quad (5)$$

$$CH_3CH_2OH + O_2{=}CH_3COOH + H_2O \quad (6)$$

$$2HCO_{3^-} + 4H_2 + H^+{=}CH_3COO^- + 4H_2O \quad (7)$$

In the last step, methanogenic bacteria convert acetic acid into CH_4 and CO_2:

$$CH_3COOH = CH_4 + CO_2 \quad (8)$$

There is another parallel pathway to produce methane by reduction of carbon dioxide resulting from formic acid degradation:

$$HCOOH \leftrightarrow CO_2 + H_2 \quad (9)$$

$$CO_2 + 4H_2 \leftrightarrow CH_4 + 2H_2O \quad (10)$$

The activity of the digesting bacteria and biogas production of gas is most rapid in two temperature ranges: between 29°C and 41°C (fermentation is known as mesophilic) or between 49°C and 60°C (the thermophilic range). The mesophilic regime between 32°C and 35°C is more reliable for stable and continuous production of methane. Biogas produced outside this temperature range is rich of carbon dioxide, it is not combustible and that is why it has no calorific value. The thermophilic regime gives higher yield of biogas, but with less net energy efficiency because of the energy losses for high temperature maintenance.

Different methanogenic strains are responsible for these parallel and competitive processes. The bacteria from the genus *Methanosarcina* are capable to grow during the catabolism of acetate to CO_2 and CH_4 [3], whereas the strains *Methanobacterium* and *Methanobrevibacter* convert carbon dioxide by reduction of hydrogen to methane [4–8].

The balance between decarboxylation of acetic acid and carbon dioxide reduction is important for the methane content in the resulting biogas. If only decarboxylation of acetic acid occurs,

http://dx.doi.org/10.5772/65734

methane content will be 50% only. The high methane content in the biogas means that carbon dioxide reduction prevails.

Anyway, all methanogenic strain are vital in neutral media, i.e., for pH values between 6 and 8. Big deviations either in the acid domain or in the alkaline one lead to strong inhibition and even to death. Acidogenesis is one of the inevitable steps in biogas production. On one hand, methanogenesis is favorized by fatty acid formation, but on the other hand, it could be strongly inhibited by their accumulation due to the pH drop. In such cases, the produced gas is very rich in carbon dioxide and it is not combustible. That is why, one must be very careful in the feeding strategy by substrate and in the selection of bioreactor and flow organization.

One suitable way to minimize the effect of acid accumulation on the biogas formation is to distribute spatially the consecutive processes of biogas formation and to carry them out simultaneously. Such a construction is the baffled bioreactor separated into consecutive compartments fed from the one end with outflow at the other one, cf. **Figure 2**. It is known that such reactors are stable toward disturbances in feed, pH oscillations, temperature variations, etc. [9].

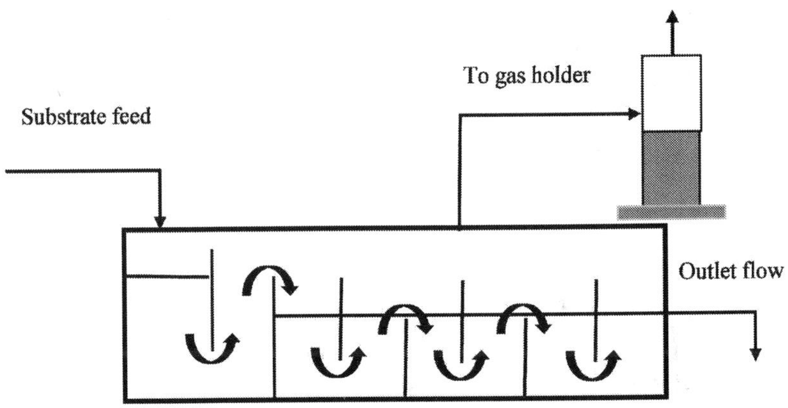

Figure 2. Multistage bioreactor for biogas production.

The main advantage of this type of reactor in the considered case is the distribution of the different consecutive processes (hydrolysis, acidogenesis, acetogenesis, and methanogenesis) in different reactor compartments. Due to its one-way feeding, the intermediate products in one compartment passed as substrates to the next one. Because of this feeding organization, different bacteria are spontaneously cultivated, specialized to transform different intermediates of the overall methanogenesis.

There are successful applications of such bioreactor for biogas production by residual stillage from ethanol distillation as a feed [10]. The intermediate profile and the microbial distribution in eight compartment bioreactor for this process are illustrated in **Figure 3** by author's experimental data. Acetic and propionic acids prevail in the first three compartments where reducing sugars are present, due to the hydrolysis of carbohydrates. Obviously, the first three steps of biogas production, i.e., hydrolysis, acidogenesis, and acetogenesis predominantly take place in the first three compartments where sugars are present and the concentration of methanogenic bacteria is very low or negligible. Methanogenics prevails in the next compartments 5–7, which corresponds to the very low acid concentration. There are few methanogenic bacteria in compartment 8 which corresponds to the negligible acid concentration. The biogas productivity rate for those experiments was ca. 4 vol.biogas/vol.reactor/day.

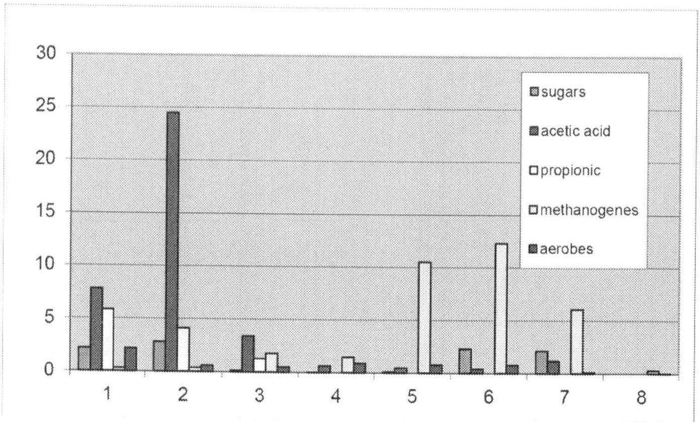

Figure 3. Acid and microbial profiles along the compartments in the bioreactor (own data).

The biogas produced by anaerobic digestion finds applications in different area. First, it could be used directly for heating purposes. Next, after some processing to remove carbon dioxide and sulfur-containing compounds the biogas could replace partially the natural gas for local applications. This biogas could be supplied directly in the pipelines, it could be used for the public transport and for electricity production by cogeneration.

Another promising application is the direct electricity production in fuel cells [11]. For this purpose, biogas should be scrubbed for carbon dioxide and sulfur compounds removal and then the purified methane could be fed to solid oxide fuel cells (SOFC) and molten carbonate fuel cells (MCFC). In this case, methane is directly converted to hydrogen and carbon monoxide by a catalyst in the anodic space. Another approach is to convert methane into carbon monoxide by steam reforming (SR) or partial oxidation reforming (POX) and consequent water-gas shift reaction to isolate hydrogen, supplied to a fuel cell:

$$CH_4 + H_2O = CO + 3H_2(SR) \quad (11)$$

$$CH_4 + \frac{1}{2}O_2 = CO + 2H_2(POX) \quad (12)$$

2.1. Biogas as a source of organic fuels

There are some papers claiming to utilize biogas as a source for other organic fuel production by catalytic auto-thermal reforming [12, 13]. Another approach is to use biogas being a mixture of methane and carbon dioxide to produce synthesis gas (mixture of carbon monoxide and hydrogen) [14]:

$$CH_4 + CO_2 = 2CO + 2H_2 \tag{13}$$

Furthermore, the synthesis gas could be converted into light hydrocarbons by the Fischer-Tropsch process.

3. Biodiesel

Biodiesel consists of methyl or ethyl esters of fatty acids produced by transesterification of natural lipids. Different natural fats are used as raw materials, namely, rapeseed, soybean, processed residual sunflower oil, animal fats, and some kinds of algae. The latter are attractive because they can utilize the carbon dioxide from flue gases by photosynthesis thus reducing the emissions of greenhouse gases [15].

The energy content of biodiesel is within 37 and 40 MJ/L compared to 46 MJ/L of the traditional diesel fuel. Biodiesel does not contain sulfur compounds.

The idea for the use of vegetable oils as fuel for diesel engines is more than 100 years old [16, 17]. Just in the 1970s, the petrol crises and the enhanced environmental conscience in the modern societies have led to the secondary discovery of this possible alternative to the hydrocarbon-based fossil fuels. However, the direct use of vegetable oil as a fuel is not convenient, because of its very high viscosity, high flame point, trend to polymerization, etc., all leading to engine damage [18].

Transesterification with low alcohols is the best modification of natural oil for the biodiesel purposes.

Today, biodiesel is in commercial use throughout the world. It is used as a single fuel or blended with traditional diesel (with 30–36%).

Biodiesel is produced in the European Union since 1992. The world production attained 3.8 mln tons in 2005 to reach 3.7 mln tons only in the USA in 2007. The total world production for 2016 is about 15 mln tons.

3.1. Benefits of biodiesel use

Biodiesel does not contain sulfur and aromatic compounds and its use in the conventional engines leads to reduction of emissions of noncombusted hydrocarbons and carbon monoxide. Comparison of the emissions resulting by the use of biodiesel and traditional one is shown in **Table 1**.

Emission type	Biodiesel	Traditional diesel
	(% of EU standards from 1993)	
Carbon monoxide	60	180
Hydrocarbon total	90	150
Nitrogen oxides	65	60
Fine particulates	95	190

Table 1. Comparison of emissions released by biodiesel and conventional diesel fuel [19].

3.2. Problems in biodiesel production and use

The main disadvantage at biodiesel production and use is the uncertain standardization depending on the source of lipids. It reflects the different cetane number and the variable temperature of gelatinization depending on the esters and the raw material type.

Another severe problem is crude glycerol, released as byproduct after transesterification of lipids. Its amount is about 10% of the substrate and it is almost equal of the methanol used. This residual glycerol is contaminated by potassium hydroxide, water, some nonreacted lipids, some soaps, and monoglycerides and diglycerides. The low quality of this product makes it impossible for direct practical application.

Provided the annual world production of biodiesel is about 15 mln metric tons, one could expect that 1.5 mln metric tons of crude glycerol would be released. It is an enormous amount and it poses the necessity for its application and processing.

Pure glycerol has various practical applications but it could be hardly replaced by the residual crude glycerol after biodiesel production. That is why new application should be sought.

Recent studies show the opportunity for crude glycerol utilization as syngas by steam reforming [20–22], cf. Eq. (9). Other applications are proposed, for example, hydrogen production by photo-fermentation [23, 24], or as a fuel in fuel cells and microbial fuel cells [25, 26]. However, in these cases the contamination by methanol is not recommended [27, 28].

3.3. Glycerol utilization for hydrogen and other chemicals production

The large amounts of residual crude glycerol prompted to the search of simultaneous waste treatment and for new applications as a raw material, alternative to the petrol for the traditional organic synthesis [29–31]. Such efforts are directed toward production of chemicals of broad industrial importance, e.g., polyols as precursors of plastics (2, 3-butanediol, 1, 3-propandiol) [32–35], propionic acid [36, 37], succinic acid [38], or hydrocarbons by catalytic reforming [39], for epichlorohydrin, some ethers [40], polyesters, etc.

Among the potential applications of waste glycerol are the production of biodegradable polymers for packaging [31, 32, 41, 42], as antifreezing agents [43] as substrate for microbial syntheses, etc. Crude glycerol has been used as carbon source in the nutrition media for biopolymer production by the species *Bacillus* and *Pseudomonas* [44–46].

http://dx.doi.org/10.5772/65734

3.4. Microbial conversion of glycerol into chemical products

The bacterium *Rhodopseudomonas palustris* is capable for photo-fermentative conversion of crude glycerol to hydrogen [23]. The conversion rates and the yields depend on the concentrations of the added nitrogen containing compounds. Higher yields of hydrogen and also ethanol are registered at the use of *Enterobacter aerogenes* HU-101 [47]. There are works on the production of different chemicals from glycerol in microbial processes. Different bacteria (from the genera *Klebsiella, Clostridium*, and *Enterobacter*) are capable to convert glycerol, producing basic chemicals, differing by the intermediate reactions and products [30]. The metabolic pathway of glycerol conversion by bacteria from the genus *Klebsiella* was proposed and discussed by Saxena et al. [48] and Zhang et al. [49]. It is shown in **Figure 4**. It is seen that two diols (1,3-propandiol and 2,3-butanediol) are produced by two competitive mechanisms. Those two diols are interesting as precursors for polymer production, e.g., polypropylene and butadiene. Besides succinic and lactic acids are produced, ethanol too.

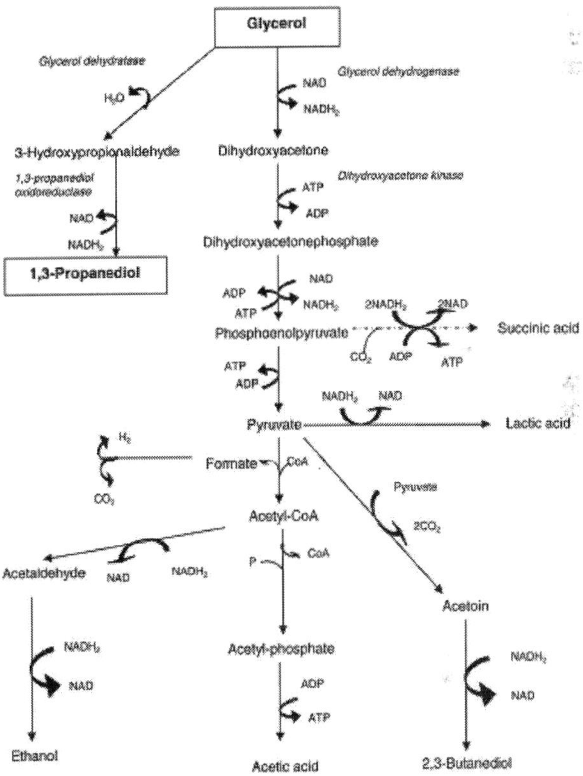

Figure 4. Metabolic pathway for glycerol digestion by bacteria from the genus *Klebsiella* [48, 49].

The studies of glycerol conversion at the metabolism of bacteria from the genus *Clostridium* show similar processes like in the previous case [50–52]. At *Clostridium* mainly 1,3-propanediol, organic acids (formic, acetic, butyric, and lactic) as well as n-butanol are produced.

Additionally, formic and acetic acids are also produced. These two carboxylic acids are very important for the consequent production of biogas being a mixture of methane and carbon dioxide.

The studies of the metabolism of *Enterobacter* bacteria show predominant formation of ethanol and hydrogen [53–55].

3.5. Glycerol for biogas production

The microbial production of acetic and formic acids from glycerol is interesting with the relationship of biogas production by anaerobic digestion. The two main pathways for biogas production by methanogenic bacteria are based on acetate decarboxylation, or carbon dioxide reduction by hydrogen, both produced from formic acid decomposition, cf. Eqs. (8) and (10).

Conversion of glycerol into biogas by anaerobic fermentation is an interesting option to produce renewable energy together with waste glycerol treatment [56–58]. It is reported that glycerol considerably enhances biogas formation by properly selected microbial population [59]. There are also many studies for the glycerol impact on biogas yield from various substrates, such as cattle dung [60–62], pig manure [57, 62, 63], activated sludge [64–66], as well as at more complicated mixtures of cellulose and household waste [57, 58, 67].

In any case, the results are considerable enhancement of biogas yield from 180 to 400% with respect to the reference substrate. It is typical, however, that the amounts of the added glycerol are restricted to 1–4% wt. from the main substrate. Addition of bigger amounts of glycerol leads to strong acidification of the broth and inhibition of methanogenesis [39, 56]. It means that no considerable amounts of crude glycerol could be utilized as biogas.

However, it was reported recently that crude glycerol may serve as a single substrate for biogas production with pretty high yield, i.e., 0.345 L biogas/g COD [68].

For attainment of maximum efficient biogas production different schemes of bioreactor feed are studied, as well as choice of the reactor construction, flow organization, etc.

4. Ethanol

Ethanol is a renewable energy source produced through fermentation of sugars. Ethanol is widely used as a partial gasoline replacement worldwide. Fuel ethanol that is produced from corn has been used in gasohol or oxygenated fuels since the 1980s. These gasoline fuels contain up to 10% ethanol by volume. As a result, the US transportation sector now consumes about 4540 million liters of ethanol annually, about 1% of the total consumption of gasoline. Recently, US automobile manufacturers have announced plans to produce significant numbers of flexible-fueled vehicles that can use an ethanol blend with 85% ethanol and 15% gasoline by volume—alone or in combination with gasoline. Using ethanol-blended fuel for automobiles can significantly reduce petroleum use and exhaust greenhouse gas emission.

Ethanol is also a safer alternative to methyl tertiary butyl ether (MTBE), the most common additive to gasoline used to provide cleaner combustion.

However, the cost of ethanol as an energy source is relatively high compared to fossil fuels. A dramatic increase in ethanol production using the current corn starch-based technology (or other cereals) may not be practical for small countries because corn production for ethanol will compete for the limited agricultural land needed for food and feed production. Additional drawback is the increasing prices of cereals used extensively as substrate for ethanol production by fermentation due to the enhanced demand and thus putting the third-world countries in disadvantaged position.

An alternative potential source for low-cost ethanol production is to utilize lignocellulosic biomass (LCB) such as straw, stems, cobs, grass, sawdust, wood chips, and forestry waste. This approach is known as "second-generation" ethanol production. Extensive research has been completed on conversion of lingo-cellulose to ethanol in the past two decades [69–74].

The conversion includes two main steps: pretreatment with hydrolysis of cellulose in the LCB to fermentable reducing sugars and fermentation of the sugars to ethanol [75].

The purpose of pretreatment is to remove lignin and hemicellulose, reduce cellulose crystallinity, and increase the porosity of the materials. Physical, physical-chemical, chemical, and biological processes have been used for pretreatment of LCB. The hydrolysis is usually catalyzed by cellulolytic enzymes, and the fermentation is carried out by yeasts or bacteria. The factors that have been identified to affect the hydrolysis of cellulose include porosity (accessible surface area) of the waste materials, cellulose fiber crystallinity, and lignin and hemicellulose content [2, 75]. The presence of lignin and hemicellulose impedes the access of cellulases to cellulose, thus reducing the efficiency of the hydrolysis. Removal of lignin and hemicellulose, reduction of cellulose crystallinity, and increase of porosity in pretreatment processes can significantly improve the hydrolysis [76, 77].

Another disadvantage of the enzymatic hydrolysis of cellulose is the strong product inhibition of glucose and therefore the low-product concentrations and the low-process rate. The easier but not environmentally friendly way is to use acid hydrolysis by sulfuric acid [78]. In this case, higher concentrations of fermentable sugars and oligosaccharides are produced. The usual approach is to employ a two-step dilute acid hydrolysis, where the hemicellulose is hydrolyzed to xylose and recovered in the first stage and a more vigorous second-stage hydrolysis is employed for conversion of cellulose to glucose [79].

Several different organisms have been proposed for convert fermenting sugars into ethanol. The mostly spread ones are the yeast, *Saccharomyces cerevisiae,* due to its robust growth rate and high ethanol tolerance, up to 23% [80, 81]. The effort to use thermostable yeast has shown that they suffer from low ethanol tolerance [82].

The bacterium, *Zymomonas mobilis,* has been shown to produce higher ethanol yields but with lower ethanol tolerance [82].

After fermentation ends the "beer" containing 2–12% ethanol is subjected to distillation to produce the azeotropic mixture of 96% (vol.) ethanol and 4% water. This mixture is not appropriate for blending with gasoline for the water separation, particularly at low temperatures. That is why additional drying is required to reach water content of less than 1%.

The classical methods are extractive distillation by adding solvents like benzene, cyclohexane, or ether to break the azeotrope.

Most advantageous is the molecular sieve drying technology, where the azeotrope is passed through a bed of synthetic zeolite with uniform pore sizes which preferentially adsorb water molecules. After the bed becomes saturated, it is regenerated by heating or evacuating the bed to remove the adsorbed water. The most efficient technology is the vapor-phase "pressure swing" adsorption molecular sieve process [83]. Nowadays, this process is preferred to the classical extractive distillation due to the clean process and the lack of side chemical products due to extraction and distillation.

The problem to be solved is the stillage processing. Stillage is the waste after ethanol distillation and it contains a lot of cellulose residues, nonfermented oligosaccharides, proteins, etc., with COD reaching 70 kg/m^3. The stillage amounts are between 1 L/kg feedstock for cereals and 20 L/kg for cellulosic substrate from coniferous origin.

There are different ways to treat this waste. One of them is to use it as animal feed after evaporation and concentration. Another option is to use it as substrate for single cell protein production with the subsequent use as animal fed of the residue. The simplest and the straightforward method is to use stillage for biogas production by anaerobic digestion [84, 85]. According to our experience, the produced biogas has over 70% (vol.) methane content. The COD was decreased from 70 to 1 g/L (over 98% efficiency) and after some additional treatment the wastewater could safely discharged or used for irrigation.

There is a relatively new proposal for consolidated bioprocessing (CBP) of lingo-cellulosic materials consisting in cellulase production, substrate hydrolysis, and fermentation accomplished in a single process by cellulolytic microbes [86].

CBP offers the potential for lower biofuel production costs due to simpler substrate processing, low energy inputs, and higher conversion efficiencies compared to separate hydrolysis and fermentation processes. It is an economically attractive goal for "third-generation" biofuel production.

4.1. Ethanol applications as fuel and raw materials

The use of ethanol as fuel depends on the oil prices on the global market and the local regulations in the different states. However, they are other options to use the "bioethanol" as a substrate and a raw material for chemical purposes. Besides the well-known applications as commodity product, chemical reactant, and solvent, ethanol may serve as a source of hydrogen production by steam reforming [87, 88] or chemical products, like ethylacetate [89, 90].

http://dx.doi.org/10.5772/65734

4.2. Problems and drawbacks

The main problems associated to ethanol production for fuel purposes are either from economical or environmental point of view. The economic problems are related to the prices of cereals competing its application as food. Therefore, the extensive use of cereals for industrial or fuel purposes may be unfavorable for their alimental needs. Next, the demand of new area for crop growing may lead to deforestation and disturbing the biodiversity and environmental balance.

The use of second-generation raw materials for ethanol production (cellulose-based waste) is restricted due to environmental reasons. Not all of the waste lingo-cellulosic biomass could be safely converted into ethanol without disturbing the natural ecological processes. The extensive use of lingo-cellulose for ethanol production could lead to deforestation and threat on biodiversity in large area of land. That is why decision making on the size and the rate of lingo-cellulosic waste use for this and for any other purpose should be very carefully, after precise and thorough environmental analysis.

5. Carbon dioxide utilization

Unfortunately, each of the described processes of biomass utilization ends with the inevitable release of carbon dioxide, resulting of the fuel combustion. Next, crops growing in industrial scale require considerable input of energy, most frequently taken from oil-based fuels. That is why abiotic carbon dioxide utilization and conversion is a major task for the future research and technology of biomass-based renewable fuels. There are two trends for this challenge: utilization photosynthesis by vegetation and recycling the biomass by chemical or electrochemical reduction to organic fuels, like methane.

The first one is to pass flue gases containing carbon dioxide through greenhouse area containing algae capable to produce lipids and other organics being biofuels. Algae can be grown in open ponds, closed-loop systems, and photo-bioreactors. Algae are capable of much higher yields with lower resource inputs than other feedstock and that is why they are moved to an own category. The following biofuels could be produced by algae: biodiesel, butanol, gasoline, methane, ethanol, and kerosene [91, 92].

There are some problems associated with the efficiency of photosynthesis for industrial purposes, the utilization rate of carbon dioxide, etc. Another drawback regarding algae is that biofuel produced from them are less chemically stable than biodiesel produced from other sources because the biofuels have unsaturated bonds in their molecules and they are subjected to spontaneous polymerization.

Another approach is to utilize carbon dioxide as a raw material for various chemical products, like ethers, dimethylcarbonate, as antiknocking additive, monomers for plastics production, acyl carbonates [93]. All of these products are currently produced from petrol and that is why carbon dioxide recycling is important to greenhouse emissions but also to the reduced use of oil as a whole.

On the other hand, carbon dioxide is irreplaceable tool for supercritical extraction of biologically active and thermo-instable substances from natural products.

An attractive approach is to reduce electrochemically carbon dioxide to methane in presence of methanogenic bacteria [94–97] or to ethanol and acetic acid [98].

6. Conclusions

The present review demonstrates different options for biomass utilization to replace, at least partially, the use of fossil fuels and thus to reduce the pressure of greenhouse gases emissions and to close the carbon cycle in nature within the present times. The biomass and the waste of biofuel production could be also utilized as raw material for various chemical manufacturing currently produced from oil. However, in some cases this option may additionally influence the environment and create secondary pollution. It could happen in cases of bioethanol and biodiesel production when large area of land is required for crop growth and much energy for crop production is requested. That is why local solutions about the use of renewable fuels based on biomass should be made very careful after thorough environmental analysis.

More attractive option is biogas production and utilization, because it is always related to simultaneous waste processing and energy production with closing the carbon cycle at local level.

Of course, the most promising research is dedicated to carbon dioxide recycling turning it to fuels or value-added chemicals.

Acknowledgements

This work was supported by the National Science Fund of Bulgaria under the grant DFNI-E02/16, 2014.

Author details

Venko Beschkov

Address all correspondence to: vbeschkov@yahoo.com

Institute of Chemical Engineering, Bulgarian Academy of Sciences, Sofia, Bulgaria

References

[1] Campbell, C. J. and Laherrere, H. 1998, The end of cheap oil, Scientific American, 3, 78–83.

[2] Garcia-Heras, J. L. Reactor sizing, process kinetics, and modelling of anaerobic digestion of complex wastes, In: J. Mata-Alvarez (editor), Biomethanization of the organic fractions in municipal solid wastes, IWA-Publishing, London, UK, 2003.

[3] Ferry, J. G. 1997, Enzymology of the fermentation of acetate to methane by Methanosarcina thermophila., Biofactors, 6, 25–35.

[4] Conrad R. 2005, Quantification of Methanogenic Pathways Using Stable Carbon Isotopic Signatures: A Review and a Proposal., Organic Geochemistry, 36, 739–752.

[5] Zyakun, A.M., Bondar, V.A., Laurinavichus, K.S., Shipin, S.S., Belyaev, S.S. and Ivanov, M.V. 1988, Fractionation of carbon isotopes under the growth of methane- 10 producing bacteria on various substrates (in Russian), Mikrobiologicheskii. Zhurnal, 50, 16–22.

[6] Gelwicks, J. T., Risatti, J. B. and Hayes, J. M. 1994, Carbon isotope effects associated with aceticlastic methanogenesis, Applied and Environmental Microbiology, 60, 467–472.

[7] Games, L. M., Hayes, J. M. and Gunsalus, R. P. 1978, Methane-producing bacteria: Natural fractionations of the stable carbon isotopes, Geochimica and Cosmochimica Acta, 42, 1295–1297.

[8] Krzycki J.A., Kenealy W.R., deNiro M.J. , Zeikus J.G., Stable Carbon Isotope Fractionation by *Methanosarcina barkeri* during Methanogenesis from Acetate, Methanol, or Carbon Dioxide-Hydrogen. 1987, Applied and Environmental Microbiology, 53, 2597–2599.

[9] Grobicki, A., Stuckey, D.C. 1992, Hydrodynamic characteristics of the anaerobic baffled reactor, Water Research, 26, 371–378.

[10] Wilkie, A. C., Riedesel, K. J. and Owens, J. 2000, Stillage characterization and anaerobic treatment of ethanol stillage from conventional and cellulosic feedstocks, Biomass and Bioenergy 19, 63–102.

[11] O'Hayre, R., Cha, S. W., Collela, W., Prinz, F. B. Fuel Cell Fundamentals (2nd Edition), John Wiley & Sons, N. Jersey, 2009, pp. 371–395.

[12] Cassinelli, W.H., Damyanova, S., Parizotto, N.V., Zanchet, D., Bueno, J.M.C. and Marques, C.M.P. 2014, Applied Catalysis A: General, 475, 256–269.

[13] Mortola, V.B., Damyanova, S., Zanchet, D., and Bueno, J.M.C. 2011, Surface and structural features of Pt/CeO2-La2O3-Al2O3 catalysts for partial oxidation and steam reforming of methane, Applied Catalysis B Environmental 107(3–4), 221–236.

[14] Damyanova S., Pawelec, B., Arishtirova, K. and Fierro, J.L.G. 2011, Biogas reforming over bimetallic PdNi catalysts supported on phosphorus-modified alumina, Fuel and Energy Abstracts 36(17),10635–10647.

[15] Clarens, A.F., Resurreccion, E.P., White, M.A. and Colosi, L.M. 2010, Environmental life cycle comparison of algae to other bioenergy feedstocks, Environmental Science Technology, 44, 1813–1819.

[16] Knothe, G. 2001, Historical perspectives on vegetable oil – based diesel fuels. Inform 12 (11), 1103-1107. http://biodiesel.org/reports/20011101_gen-346.pdf.

[17] Henriques, R. 1898, Über partielle Verseifung von Ölen und Fetten. Zeitschrift für Angewandte Chemie, 15, 338–345.

[18] Srivastava, A. and R. Prasssad. 2000, Triglycerides-based diesel fuels. Renewable and Sustainable Energy Reviews, 4, 111–133.

[19] van Dyne, D.L. and Raymer, P.L. 1992, Biodiesel production potential from industrial rapeseed in the southeastern U.S., Report for The Southeastern Regional Biomass Energy Program, pp. 1–20. http://biodiesel.org/reports/19921031_gen-239.pdf.

[20] Dou, B., Dupont, V. and Williams, P. T. 2008, Computational fluid dynamics simulation of gas – solid flow during steam reforming of glycerol in a fluidized bed reactor. Energy Fuels, 22, 4102–4108.

[21] Adhikari, S., Fernando, S. D., To, S. D. F., Bricka, R. M., Steele, P. H. and Haryanto, A. 2008, Conversion of glycerol to hydrogen via a steam reforming process over Nickel catalysis. Energy Fuels, 22, 1220–1226.

[22] Avasthi, K.S. R.N. Reddy, R.N., Patel, S. 2013, Challenges in the Production of Hydrogen from Glycerol – A Biodiesel Byproduct Via Steam Reforming Process, Procedia Engineering, 51, 423–429.

[23] Sabourin-Provost, G. and Hallenbeck, P. C. 2009, High yield conversion of a crude glycerol fraction from biodiesel production to hydrogen by photofermentation. Bioresource Technology, 100, 3513–3517.

[24] S. Adhikari, S. D. Fernando, A. Haryanto, 2009, Hydrogen production from glycerol: An update, Energy Conversion and Management, 50 (10), 2600–2604.

[25] Ragsdale, S. R. and Ashfield, C. B. 2008, Direct–glycerin fuel cell for mobile applications. ECS Transactions, 16, 1847–1854.

[26] Arechederra, R. L. and Minteer, S. D. 2009, Complete oxidation of glycerol in an enzymatic biofuel cell. Fuel Cells, 9, 63–69.

[27] Lammers, P.J., Kerr, B.J., Weber, T.E., Bregendahl, K., Lonergan, S. M., Prusa, K. J., Ahn, D. U., Stoffregen, W. C., Dozier, W. A. and Honeyman, M. S. 2008, Growth performance, carcass characteristics, meat quality and tissue histology of growing pigs fed crude glycerin-supplemented diets. Journal of Animal Science, 86, 2962–2970.

[28] Yang, F. X., Hanna, M. A. and Sun, R. C. 2012, Value-added uses for crude glycerol - A byproduct of biodiesel production. Biotechnology and Biofuels, 5, 5–13.

[29] Pagliaro, M. and Michele Rossi, M. The Future of Glycerol: New usages for a versatile raw material, RSC Publishing, Cambridge. ISBN: 0854041249, pp 32–81.

[30] Da Silva, G. P., Mack, M. and Contiero, J. 2009, Glycerol: A promising and abundant carbon source for industrial microbiology, Biotechnology Advances, 27, 30–39.

[31] Ashby, R. D., Wyatt, V. T., Foglia, T. A. and Solaiman, D. K. Y. Industrial products from biodiesel glycerol, in: Biocatalysis & Bioenergy, Hou, C. T. and Shaw, J. J. (Eds.), John Wiley & Sons, Inc., Hoboken, 2009, pp. 131–154.

[32] Biebl H., Menzel K., Zeng A.P. and Deckwer W.D., 1999, Microbial production of 1,3-propanediol, Applied Microbiology and Biotechnology, 52, 289–297.

[33] Wilkens, E., Ringel, A.K., Hortig, D., Willke, T. and Vorlop, K.D. 2012, High-level production of 1, 3-propanediol from crude glycerol by *Clostridium butyricum* AKR102a, Applied Microbiology and Biotechnology, 93,1057–1063.

[34] Oh, B. R., Seo, J. W., Heo, S. Y., Hong, W. Y., Luo, L.H., Son, J. H., Park, D. H. and Kim, C. H. 2012, Bioprocess and Biosystems Engineering, 35, 159–165.

[35] Fermentive production of 1,3-propanediol; United States Patent 5254467; http://www.freepatentsonline.com/5254467.html.

[36] Himmi, E. H., Bories, A., Boussaid, A., Hassani, L. 2000, Propionic acid fermentation of glycerol and glucose by Propionibacterium acidipropionici and Propionibacterium freudenreichii ssp. Shermanii, Applied Microbiology and Biotechnology, 53, 435–440.

[37] Boyaval, P. and Corre, C. 1995, Production of propionic acid, Lait, 75, 453–461.

[38] Lee, P.C., Lee, W.G., Lee, S.Y., Chang, H.N., 2001, Succinic acid production with reduced by-product formation in the fermentation of *Anaerobiospirillum succiniciproducens* using glycerol as a carbon source, Biotechnology and Bioengineering, 72, 41–48.

[39] López, J.Á.S., De los Ángeles Martín Santos, M., Pérez, A.F.C. and Martín, A.M. 2009, Bioresource Technology, 100, 5609–5615.

[40] Klepácová, K., Mravec, D., Hájeková, E. and Bajus, M. 2003, Petroleum and Coal, 45(1-2), 54–57.

[41] Braunegg, G., Genser, K., Bona, R., Haage, G., Schellauf, F., Winkler, E. 1999, Production of PHAs from agricultural waste material. Macromololecular Symposia, 144, 375–383.

[42] Tohoku Electric turns glycerin into bioplastics, http://www.greencarcongress.com/2007/08/tohoku-elec-dev.html

[43] Todorova, S. Utilization of waste glycerol from biodiesel production by *Bacillus subtilis,* science & technologies, nautical & environmental studies, Patent BG 110823: Liquid Substance for Defreezing of Roads and Highways, 2, 2015, pp. 52–56.

[44] Sathianachiyar, S. and Devaraj, A. 2013, Biopolymer producing by bacterial species using glycerol, a byproduct of biodisel. International Journal of Scientific and Research Publications, 3, 8, 1–5.

[45] Saimmai, A., Rukadee, O., Sbhon, V. and Maneerat, S. 2012, Biosurfactant production by bacillus subtilis TD4 and pseudomonas aeruginosa SU7 grown on crude glycerol

obtained from biodisel production plant as sole carbon source. Journal of Scientific and Industrial Research (India), 71, 06, 396–406.

[46] Anastacio, G.S., Santos, K. O., Suarez, P. A., Torres, F. A., De Marco, J. L. and Parachin, N. S. 2014, Utilization of glycein byproduct derived from soybean oil biodiesel as a carbon source for heterologous protein production in pichia pastoris. Bioresource Technology, 152, 505–510.

[47] Ito, T., Nakashimada, Y., Senba, K., Matsui, T. and Nishio, N. 2005, Hydrogen and ethanol production from glycerol-containing wastes discharged after biodiesel manufacturing process. Journal of Bioscience and Bioengineering, 100, 260–265.

[48] Saxena, R., Anand, P., Saran, S. and Isar, J. 2009, Biotechnology Advances, 27, 895.

[49] Zhang, Y., Huang, Z., Dua, C., Li, Y. and Cao, Z. 2009, Metabolic Engineering, 11, 101–106.

[50] González-Pajuelo, M., Meynial-Salles, I., Mendes, F., Soucaille, P. and Vasconcelos, I. 2006, Applied Environmental Microbiology, 72(1), 96–101.

[51] Biebl H. 2001, Fermentation of glycerol by Clostridium pasteurianum-batch and continuous culture studies, Journal of Industrial Microbiology and Biotechnology, 27(1), 18–26.

[52] Szymanowska-Powałowska, D. 2014, 1,3-Propanediol production from crude glycerol by *Clostridium butyricum* DSP1 in repeated batch, Electronic Journal of Biotechnology, 17 (6), 322–328.

[53] Reungsang, S., Sittijunda, Angelidaki, I. 2013, Simultaneous production of hydrogen and ethanol from waste glycerol by *Enterobacter aerogenes* KKU-S1, International Journal of Hydrogen Energy, 38(4), 1813–1825.

[54] Raymond E., Nwachukwu, S., Shahbazi, A., Wang, L., Worku, M., Ibrahim, S. and Schimmel, K. 2013, Optimization of cultural conditions for conversion of glycerol to ethanol by *Enterobacter aerogenes* S012, AMB Express, 3, 12. doi: 10.1186/2191-0855-3-12

[55] Jitrwung, R. and Yargeau, V. 2015, Biohydrogen and bioethanol production from biodiesel-based glycerol by Enterobacter aerogenes in a continuous stirred tank reactor, International Journal of Molecular Sciences, 16, 10650–10664. doi:10.3390/ijms160510650

[56] Kolesárová, N., Hutňan, M., Špalková, V., Lazor, M. (2010). In: Proceedings from the 37th International Conference of SSCHE, Tatranske´ Matliare, Slovakia, 1126–1139.

[57] Yang, Y., Tsukahara, K. and Sawayama, S. 2008, Biodegradation and methane production from glycerol-containing synthetic wastes with fixed-bed bioreactor under mesophilic and thermophilic anaerobic conditions. Process Biochemistry, 43(4), 362–7.

[58] Amon, T., Kryvoruchko, V., Amon, B. and Schreiner, M. Investigations on the effect of crude glycerol from biodiesel production as addition enhancing the biogas generation

http://dx.doi.org/10.5772/65734

from silomais, grain mare, rapeseed cake and pig manure (in German). Institut für Landtechnik, Universität für Bodenkultur Wien. Source: http://www.nas.boku.ac.at/fileadmin/_/H93/H931/AmonPublikationen/SEEGEndbericht.pdf; 2004 [accessed 25.05.09].

[59] Glycerin as a biogas feedstock. http://news.mongabay.com/bioenergy/2006/12/glycer-in-as-biogas-feedstock.html

[60] Castrillón, L., Fernández-Nava, Y., Ormaechea, P. and Marañón, E. 2001, Optimization of biogas production from cattle manure by pre-treatment with ultrasound and co-digestion with crude glycerin, Bioresource Technology, 102(17), 7845–7849.

[61] Robra, S., Serpa da Cruz, R., Oliveira, A.M., Almeida Neto, J.A. and Santos J.V. 2010, Generation of biogas using crude glycerin from biodiesel production as a supplement to cattle slurry. Biomass and Bioenergy, 34(9), 804–806.

[62] Astals, S., Nolla-Ardèvol, V., Mata-Alvarez, J. 2012, Anaerobic co-digestion of pig manure and crude glycerol at mesophilic conditions: Biogas and digestate, Bioresource Technology, 110, 63–70.

[63] Wohlgemut, O. 2009, Co-Digestion of Hog Manure with Glycerol to Boost Biogas and Methane Production, MSc. Thesis, University of Manitoba, Winnipeg, Manitoba, p. 51.

[64] Siles, J.A., Martín, M.A., Chica, A.F. and Martín, A. 2010, Anaerobic co-digestion of glycerol and wastewater derived from biodiesel manufacturing. Bioresource Technology, 101(16), 6315e21.

[65] Fountoulakis, M.S., Petousi, I, Manios, T. 2010, Co-digestion of sewage sludge with glycerol to boost biogas production. Waste Management, 30(10), 1849e53.

[66] Fountoulakis, M.S. and Manios, T. 2009, Enhanced methane and hydrogen production from municipal solid waste and agro-industrial by-products co-digested with crude glycerol, Bioresource Technology, 100(12), 3043–3047.

[67] Ostrem, K. Greening waste: Anaerobic Digestion for Treating the Organic Fraction of Municipal Solid Wastes, M.S. Thesis, Columbia University, May 2004, p. 6.

[68] Hutnan, M., Kolesárová, N. and Bodik, I. 2013, Anaerobic digestion of crude glycerol as sole substrate in mixed reactor. Environmental Technology, 34(13–16), 2179–87.

[69] Sassner, P., Galbe, M. and Zacchi, G. 2008, Techno-economic evaluation of bioethanol production from three different lignocellulosic materials, Biomass and Bioenergy, 32(5), 422–430.

[70] Schneider, H., Maleszka, R., Neirinck, L., Veliky, I. A., Wang, P. Y. and Chan, Y. K. 1983, Ethanol production from D-xylose and several other carbohydrates by *Pachysolen tannophilus* and other yeasts, Pentoses and Lignin, 27, 57–71.

[71] Shafizadeh, F. and Lai, Y.Z. 1975, Thermal degradation of 2-deoxy-D-arabino-hexonic acid and 3-deoxy-D-ribo-hexono-1,4-lactone, Carbohydrate Research, 42(1), 39–53.

[72] Singh, A., Pant, D., Korres, N. E., Nizami, A.-S., Prasad, S. and Murphy, J. D. 2010, Key issues in life cycle assessment of ethanol production from lignocellulosic biomass: Challenges and perspectives, Bioresource Technology, 101(13), 5003–5012.

[73] Velásquez-Arredondo, H. I., Ruiz-Colorado, A. A. and De Oliveira, S. 2010, Ethanol production process from banana fruit and its lignocellulosic residues: Energy analysis, Energy, 35(7), 3081–3087.

[74] Sukumaran, R. K., Surender, V. J., Sindhu, R., Binod, P., Janu, K. U., Sajna, K. V., Rajasree, K. P. and Pandey, A. 2010, Lignocellulosic ethanol in India: Prospects, challenges and feedstock availability, Bioresource Technology, 101(13), 4826–4833.

[75] Yang, B. and Wyman, C. E. 2008, Pretreatment: The key to unlocking low-cost cellulosic ethanol, Biofuels, Bioproducts, and Biorefining, 2,26–40 .

[76] Kerr, R. A. 2005, Bumpy road ahead for world's oil. Science, 310, 1106–1108.

[77] Tassinari, T., Macy, C. and Spano, L. 1980, Energy-requirements and process design considerations in compression-milling pretreatment of cellulosic wastes for enzymatic-hydrolysis. Biotechnology and Bioengineering Symposium, 22, 1689–1705.

[78] Mackie, K., Deverell, K. and Callander, I. Aspects of wood hydrolysis via the dilute sulphuric acid process. In: Duckworth HE, Thompson EA, editors. International Symposium on Ethanol from Biomass, 1982 Oct 13-15; Winnipeg, Canada. Ottawa: The Royal Society of Canada, 1983. pp. 271–308.

[79] Beck, M. J. and Strickland, R. C. 1984, Production of ethanol by bio-conversion of wood sugars derived from two-stage dilute acid hydrolysis of hardwood. Biomass, 6, 101–110.

[80] Ingledew, W. M. The biochemistry of alcohol production In: Lyons TP, Kelsall DR, Murtagh JE, editors. The Alcohol Textbook. Nottingham, UK: Nottingham University Press, 1995. pp. 55–79.

[81] Lynd, L. R., Ahn, H. J., Anderson, G., Hill, P., Kersey, D. S. and Klapatch, T. 1991, Thermophilic ethanol production: investigation of ethanol yield and tolerance in continuous culture. Applied Biochemistry and Biotechnology,28/29, 549–570.

[82] Esser, K., Karsch, T. 1984, Bacterial ethanol production: Advantages and disadvantages. Process Biochemistry, 19(3), 116–121.

[83] Swain, R. L. B. Molecular sieve dehydrators. How they became the industry standard and how they work. In: Jacques KA, Lyons TP, Kelsall DR, editors. The Alcohol Textbook, 3rd ed. Nottingham, UK: Nottingham University Press, 1999. pp. 289–293.

[84] Kinoshita, C. M. 1991, Cogeneration in the Hawaiian sugar industry. Bioresource Technology, 35, 231–237.

[85] Perea, P. 1981, The merits of excess bagasse as fuel for generating electricity. Sugar y Azucar, 76(5), 42–46.

[86] Carere, C. R., Sparling, R. Cicek, R and Levin, D. B. 2008, Third Generation Biofuels via Direct Cellulose Fermentation, International Journal of Molecular Sciences, *9*(7), 1342–1360.

[87] Andonova, S., Naves de Ávila Neto, C., Arishtirova, K. andDamyanova, S. 2011, Structure and redox properties of Co promoted Ni/Al 2O 3 catalysts for oxidative steam reforming of ethanol, Applied Catalysis B Environmental, 105(3), 346–360.

[88] Zanchet, D., Santos, J. B. O., Damyanova, S., Gallo, J. M. R. and Bueno, J. M. C. 2015, Toward Understanding Metal-Catalyzed Ethanol Reforming, ACS Catalysis, 5, 3841–3863.

[89] Sato, A. G., Volanti, D. P., Meira, D. M., Damyanova, S., Longo, E. and Bueno, J. M. C. 2013, Effect of the ZrO2 phase on the structure and behavior of supported Cu catalysts for ethanol conversion, Journal of Catalysis, 307, 1–17.

[90] Freitas, I. C., Damyanova, S., Oliveira, D.C., Marques, C. M. P. and Bueno, J. M. C. 2014, Effect of Cu content on the surface and catalytic properties of Cu/ZrO2 catalyst for ethanol dehydrogenation, Journal of Molecular Catalysis A Chemical, 381, 26–37.

[91] Optimizing biofuel production from algae using carbon dioxide emissions, February 24, 2016, American Chemical Society, https://www.sciencedaily.com/releases/2016/02/160224151017.htm

[92] Demirbas, A. 2011, Biodiesel from algae, biofixation of carbon dioxide by microalgae: A solution to pollution problems, Applied Energy, 88, 3541–3547.

[93] Sankaranarayanan, S., Srivasan, K. 2012, Carbon dioxide – A potential raw material for the production of fuel, fuel additives and bio-derived chemicals, Indian Journal of Chemistry, 51A, 1252–1262.

[94] Hara, M., Onaka, Y., Kobayashi, H., Fu, Q., Kawaguchi, H., Vilcaez, J. and Sato, K. 2013, Mechanism of electromethanogenic reduction of CO_2 by a thermophilic methanogen, Energy Procedia, 37, 7021–7028.

[95] Kuhl, K. P., Hatsukade, T., Cave, E. R.,, D. N., Kibsgaard, J. and Jaramillo, T. F. 2014, Electrocatalytic conversion of carbon dioxide to methane and methanol on transition metal surfaces, Journal of the American Chemical Society, *136*(40), 14107–14113.

[96] Zhu, Y.,, S., Ye, Y., Zhang, X., Wang, L., Zhu, W., Cheng, F. and Tao, F. 2012, Catalytic conversion of carbon dioxide to Methane on Ruthenium–Cobalt Bimetallic Nanocatalysts and Correlation between Surface Chemistry of Catalysts under Reaction Conditions and Catalytic Performances, *ACS Catalysis, 2*(11), 2403–2408.

[97] Cheng, S. and Logan, B. E. Electromethanogenic reactor and processes for methane generation. US Patent No. 8,440,438, issued May 14, 2013.

[98] Min, S. U., Yong, J., Yao, Z., Ping G. and Daping, L. 2013, Coupled bioelectrochemical system for reducing CO_2 to simple organic compounds in the presence of H2, Chinese Journal of Applied Environmental Biology, 19(5), 827–832.

Chapter 11

Potential of Cellulosic Ethanol to Overcome Energy Crisis in Pakistan

Saima Mirza, Habib ur Rehman,
Waqar Mahmood and Javed Iqbal Qazi

Additional information is available at the end of the chapter

http://dx.doi.org/10.5772/66534

Abstract

Liquid biofuel industry in Pakistan may become a promising source for saving our foreign exchange and environment. Currently, bioethanol production is dependent on cane molasses, a product of sugar industry. Harnessing of more bioethanol from lignocellulosic waste crop residue has potential to respond to the fuel scarcity. Lignocellulose exists in nature as a polymer and serves as the largest sink for fixed global carbon and could be used both as a carbon source for microbial growth-assisted bioethanol production and for fabricating enzymes for more energetic simultaneous production to represent an important segment of the renewable energy sector. An exciting aspect of this research is the development of new biorefining techniques that facilitate the extraction of sugar-derived biofuel by processing of waste crop residues by employing novel nature inspired lignolytic enzyme. Further research will explore more avenues for stabilization of system in terms of process parameters for optimum bioethanol yield from enzymatically hydrolyzed lignin waste streams. The chapter can be considered as an anticipatory work and exploration of new dimensions for promotion of nature-inspired enzyme-assisted lignocellulose-based bioethanol production industry, which maximizes sustainable development opportunities especially in energy sector.

Keywords: crop residue conversion into biofuel, agriculture waste bioethanol, enzymatic ethanol, lignin biofuel, sustainable ethanol

1. Introduction

There has been a universal consensus that greenhouse gases (GHG) such as methane (CH), carbon dioxide (CO) and nitrous oxide (NO) are the main cause of global warming. This extreme apprehension forced many nations of the world to reach treaty on Japanese Protocol.

Pakistan signed the Kyoto Protocol on Climate Change in 2004 and accredited the scientific invention's potential as a possible way to control the emissions of GHG [1]. Transportation consumes approximately 27% of primary energy [2]. In EU25 countries, transportation consumes about 28% of the energy and more than 80% is for lane transport [3]. The current oil requirement is around 12 million tons a day and there is forecasting that it may increase to about 16 million tons per day in 2030. The 30% of the world oil production is used to fulfill the fuel requirements for transport. In the current energy blend in Pakistan, the share of petroleum products is about 40%. Its use has grown up suddenly, mainly by fuel oil and the gasoline [4]. The immediate focus of this chapter is to review the current developments in the dimensions of waste crop residue-based bioethanol production and applications of nature-inspired enzymes efficient production of the liquid fuel in Pakistan.

Pakistan is facing severe energy crisis in these days which is resulting in many social and economic problems. Solution of this crisis might come from alternative(s), addressing cheap and eco-friendly fuel sources. This utmost and urgent requirement is likely to be achieved by biomass resources that have been mainly ignored previously, while they are accessible in enough quantities to solve the energy crisis in the country. Agriculture has remained the basis of Pakistan's economy as it provides employment to 45% of the population and provides feedstocks to agro-based industry. Clean energy supply is critical in agricultural areas in Pakistan where biofuels are currently not an option because of the lack of cost-effective and efficient biofuel production technologies, although villagers depend on conventional diesel for powering different agriculture machinery and gasoline for transporting agricultural goods from farm/market to end consumers. Fuel shortage has also led to a cut in electricity production. It is thus clear that the major limiting factor is energy which creates barriers for developing economies. Careful estimates show that by 2050, Pakistan's energy needs will increase three times without a concurrent increase in supply. Pakistan plans to cut natural gas supply by around 4,247,527 m^3/d to fertilizer plants and compressed natural gas (CNG) pumps to increase electricity supply to cities facing daily rolling blackouts. In 2012, Pakistan's natural gas supply had about 15 billion m^3/year deficit with increasing tendency. Large biofuel production plants that can contribute significant amounts of sustainable fuel are the only solution to supplement the power shortage in the country. Rise in conventional fuel price and its continuous depletion naturally brings great demand for the innovative biofuel production technologies as a Clean Energy Solution in Pakistan. The valuable progress starts with the beneficial use of the waste material and crop residues as feedstocks which otherwise represent environmental liabilities. The development in this sector will further provide opportunities to create multiple symbiotic partnerships among the farmers and the business community.

Ethanol is an important energy source which has the huge potential to lessen the sole dependency on fossil-derived fuels and alleviate hazards to our environment. Additionally, it is an ideal precursor molecule because of its promising potential as fuel, beverage, feedstock, antiseptic and industrial solvent. Currently, it is replacing around 3% of the gasoline that is derived from fossils throughout the world which is compatible with petroleum and recommended for transportation in both blended and pure forms. The consumption of ethanol is around 1.6 million tons and consumption of fuel ethanol can be increased up to 160,000 tons by 10% blending of bioethanol in petrol. According to Chris Somerville, director of the

Energy Biosciences Institute, USA, annual production of ethanol from corn is around 13–14 billion gallons in the United States, which is equal to 10% of gasoline use. In Brazil, 40% of liquid transportation fuel is bioethanol and 15% of the nation's electricity is deriving from it. Therefore, it is assumed that current technology is healthy enough to produce ethanol as an alternative fuel to some extent for immediate partial replacement of oil. However, corn ethanol which is already in use has several drawbacks as corn is a food crop. Furthermore, when cost-to-benefit ratio regarding equipment and processing involved in planting, harvesting and transporting corn ethanol are considered, it becomes incomparable with gasoline. Therefore, hardy, fast-growing plants, like switchgrass, elephant grass and miscanthus, are more favorable feedstocks options. These grasses can grow up to 10 feet height, thrive on marginal land and can survive even with little or no fertilizer [5]. Moreover, cellulose-rich waste material of paper and other industries and waste crop residues can also be considered attractive options for cost-effective and even zero cost bioethanol productions by using nature inspired enzymatic processes.

2. Biomass: a cheap and sustainable biofuel resource

Pakistan is an agricultural country and huge capacity of biomass in the form of waste crop residues such as rice straw, wheat straw, cotton stalks, maize stalks, sugarcane tops and so on are available for bioethanol production. Pakistan annually generates around 69 million tons of field-based crop residues. Field-based crop residues are generally considered useless and it has been estimated that 50 million tons of residue/waste is produced every year from major crops (including 6.88 million tons of sugarcane bagasse). These are either burned in the fields/homes or buried in the land. Direct burning of this field base left over emits carbon dioxide and smoke which are hazardous for health and a source of ozone with risks to the atmosphere. Excluding domestic consumption and commercial usage, the net available resource potential of four crops, that is (wheat, cotton, rice and corn) for biomass power generation, is estimated to be about 10.942 million tons [6]. These estimations showed that Pakistan is endowed with abundant availability of biomass resources, which can be economically exploited for developing a sustainable biomass energy system, because the country has been perennially facing power demand-supply gap, which is currently estimated at 4500–5500 MW [7]. The system is being maintained by resorting to load shedding, often extending up to 12–16 h. Pakistan has strategies to add 9700 MW of electricity generation capacity by 2030 as per the Medium Term Development Framework [8], which would partially take care of current fuel scarcities. In this context, power generation through biomass can also play an important role in bridging the overall demand-supply gap. It would be essential to expand and diversify the resource base, particularly in the context of continuous access to electricity in all regions of the country. Large numbers of industries in Pakistan are currently dependent on liquid fuels for meeting their captive demand for electricity and heat. The situation is therefore ideally suited for promoting biomass-based liquid biofuel production as a sustainable and renewable alternative for the industrial sector as well. If only field left over crop residues are used for production of bioethanol, even then a sufficient amount of bioethanol can be produced to cut oil import and improve the profitability of the farming sector.

It is known that some developed countries like the United States and Brazil are largest bioethanol producers and ethanol production in these countries is achieved by fermentation of corn glucose [9]. The production of ethanol from molasses is not new, but some areas need to be researched for enhanced yield. Until now, in Pakistan, sugar mills distilleries are operational for ethanol production using molasses, but in order to utilize molasses fully and get maximum benefit, it is important to increase the number of distillery units on one hand and assurance of possible involvement of efficient enzymes and engineered microbes on the other hand. Furthermore, application of mathematical imitations would be used to explore efficient way for optimized yield without intervention of any pilot plant [10, 11]. The major steps for large-scale microbial production of ethanol are fermentation of sugars, distillation and dehydration.

3. Microbial fermentation

Different ethanologenic microbes have been known to have promising qualities like limited growth requirements, genetically amenable, higher sugar and ethanol tolerance. Bioethanol production by two strains (mutant and wild) of yeast *Kluyveromyces marxianus* have been documented [12]. Wild strain showed maximum specific growth rate at 40°C while mutant showed maximum specific growth and ethanol formation rates at 45°C from fermentation of diluted molasses. Results of this study anticipated that large-scale production may also be economically feasible by employing these microbes. Yeast-assisted bioethanol production process is more common and commercially applicable method in Pakistan [13]. *Zymomonas mobilis* is also attracting more attention for bioethanol production due to less process limitations [14, 15]. Different experimental studies in this regard revealed that optimum ethanol production up to 55.8 g L^{-1} can be achieved by *Zymomonas mobilis* at 30°C after 48 h of retention time [16, 17].

Sugar beet, molasses and sugarcane juice are one of the most vital and easily accessible raw materials for the fermentative production of alcohol. The increased cost of molasses has triggered many distilleries to search alternate sources of feed stocks for the production of bioethanol in Pakistan. In starch industry, a by-product called enzose hydrol contains 5, 12, 56 and 5% of oligosaccharides, maltose, glucose and maltotriose, respectively, and is a cheap and good source of fermentable sugars. Mostly oligo-saccharides and maltotriose are not completely consumed by ethanologenic microbes and therefore need pretreatment [18]. Similarly, bioconversion of cellulose into ethanol can be accompanied by various microbes as well as by some filamentous fungi, including *Neurospora crassa* [19, 20] *Zymomonas* [21], *Trichoderma viride* [22], *Paecilomyces* sp. [23], *Zygosaccharomyces rouxii* [24] and *Aspergillus* sp. [25], termites' gut enzymes, genetically engineered bacteria such as *Escherichia coli* [26] and thermophilic, anaerobic bacteria, such as *Clostridium thermocellum* [27]. Thus, certain possible methods need to be designed for economical production of ethanol from agricultural farm residues by employing most effective microbes [28]. Among such agro-based wastes, wheat straw is one of the most plentiful crop residues which has broadly been studied and is abundantly available too [29].

Current investigations are focusing on pretreatment of the hard biomass, that is, lignocellulosic sugarcane bagasse, rice straw and wheat straw and subsequent production of ethanol from the pretreated biomass using ethanologenic microorganism. Different fungal species have promising potential for breaking down lignin and therefore may be applied for efficient ethanol fermentation. Hypothetical yield of ethanol is 0.511 g per gram of glucose consumed. Practically, this yield cannot be achieved because part of fermentable glucose is consumed for cell maintenance, for synthesis of by-products like glycerol and lactic acid and therefore is not completely converted into ethanol. Nevertheless, at the manufacturing level, under ideal conditions, it remains 90–95% of the hypothetical yield [30]. Ethanol formation represents a specific loop of the general cellular metabolism; however, its general production route is shown in **Figure 1** [31].

Generalized bioethanol production is as follows [31].

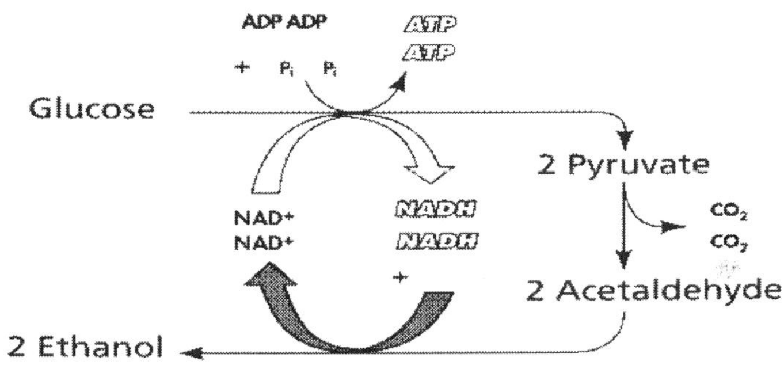

Figure 1. Part of the fermentative metabolism directly involved in the ethanol production.

4. Bioethanol production potential of industrial sector in Pakistan

Few industries in Pakistan are already involved in bioethanol production from by-products or industrial effluents, but it is necessary to develop nature inspired bioethanol production on a large scale that may not only provide a solution to Pakistan's power shortages but can also be profitable enough to render their viability in local conditions. The biomass like rice straw, sugarcane molasses, bagasse and wheat stubble are the chief resources of lignocellulose feedstocks worldwide [32]. One of the largest available biomass is rice straw which is about 7.31×10^{14} rice stubbles per year in world and 90% of its annual global production comes from the Asia [33]. Another abundantly available biomass, a by-product of sugarcane processing, is the sugarcane bagasse which represents important source for fuel generation systems and ethanol production due to its high easily accessible sugar contents for fermentation [34].

Sugar industry is the biggest agro-industry in Pakistan after textile and has been playing key role in the production of ethanol. There are about 76 sugar mills in Pakistan already which are producing seasonal ethanol from around 2.5 million metric tons (MMT) of molasses. However, being an agricultural country, the best option is second-generation ethanol. However, for this, the complex lignin-cellulose-hemicelluloses matrix of the biomass has to be broken and the carbohydrate polymers need to undergo hydrolyses to yield fermentable sugars. The important source of the livelihood of farmers is sugar industry and their 70% population is dependent on it. The yield of sugar in Pakistan is about 85.95 kg per 100 kg of sugarcane. The molasses production from sugarcane is approximately 40 kg per ton of cane from which ethanol production is approximately10 L. There is 270,000 tonnes per annum current production capacity for ethanol of fuel grade in our country which can readily be increased up to 400,000 tonnes per annum through the rise in employments and feedstock like waste crop residue [18, 35]. This molasses-to-bioethanol conversion process is conducted in distilleries. But most of the distilleries are located onsite in sugar mills which make the production cycle an integrated one. The mills, after processing sugarcane, store the molasses in storage tanks on-site and then pass it on to the distilleries for bioethanol production. Simple molecular sieve technology is used for bioethanol production in most of these mills which requires 1.5 million USD capital expenditure and can be completed in 5–6 months.

AL-Abbas sugar mill production plant is situated exactly in the center of one of the huge sugarcane growing areas of province Sindh at Mirwah Gorchani. This area is also known to be the most fruitful regarding sugarcane cultivation in Pakistan, assuring the supply of good crop of sugarcane throughout the entire season for the sugarcane plant. The plant is linked to the national highway by means of a mile of metal led road and is also accessible by a web of many other roads from different directions which facilitate transport of sugarcane from the plantation sites to the sugarcane plant. Total crushing potential of this sugarcane crushing plant is about 7500 M ton per day. AL-Abbas sugar mill established the largest ethanol distillery plant in 1999. The plant design is equipped with highly advanced French technology using multieffect vacuum distillation. The ethanol production capacity of unit-I is approximately 87,500 L per day. The growing demand of ethanol has urged the management to set up unit-II with the same capacity of ethanol production. The bioethanol-based power plant of AL-Abbas sugar mill has 15 MW electricity production capacity.

Shakarganj sugar mill is located in Jhang, Pakistan. They are producing three different types of ethanol, that is, concentrated ethyl alcohol, denatured spirit and methylated spirit for industrial and alternate source of energy usage. The mill is exporting approximately 90% of its total ethyl alcohol and is a four-time award winner for the highest export of ethyl alcohol. The unit produces anhydrous alcohol employing eco-friendly dry dehydration technology. The denatured and methylated spirit is in high demand in local wood product and paint industries.

Another bioethanol-producing sugar mill is located in Nankana, Sheikhupura, Pakistan. The ethanol production potential of this distillery is 125,000 L/day. Besides this capacity, the distillery also produces ethanol of fuel grade with 99.8% purity from the mill molasses. State-of-the-art distributed control system (DCS) which not only promises for increased steadiness and but also approachability of plant is used. The distillery system is established with fewer number of devices and lesson wiring. The distillery can cut in half the costs related to applying and sustaining the loops by incorporating the transmitter controllers into the process and by opting not to tie any critical loops back into the DCS. The distillery is equipped with ultramodern machinery and is working on International Standard Operating processes to carry them to produce high-quality products and is meeting the demands of end users.

Crystalline Chemical Industries (Pvt.) Ltd (CCI) also practices sugarcane molasses fermentation for ethanol production, located in Sargodha, Pakistan. This unit of distillation exports about 90% of its ethanol produce.

Habib sugar mills Ltd. has industrial alcohol production capacity up to 142,500 L/day. Pinnacle distilleries (Pvt.) Ltd. is producing rectified spirit for portable applications, technical alcohol, anhydrous ethanol 99.7% minimum for manufacturing use. The fuel grade alcohol is produced up to 30,000 tons per year.

Almost all sugar industries in Pakistan are producing bioethanol mainly from molasses containing feasible level of fermentable sugars. Presently, the biomass proportions which can be economically converted into ethanol are sugar (sugarcane) and starch (e.g. corn). In future, there will be plentiful industrial scale progress in the subject of lignoethanol where the hard part of a plant (cellulose) will be converted into fermentable sugars and consequently converted to bioethanol. After microbial fermentation, the produce is subjected to distillation, dehydration and then is condensed for quality improvement and water and other impurities removal. However, due to high cost in the form of energy input, this traditional process is replaced with some energy saving processes (molecular sieve) mainly to avoid distillation completely for dehydration. This process involves the use of ethanol vapors under pressure and allows these vapors to pass through molecular sieve beads bed. The energy saving by this technology of dehydration accounts for 3000 btus/gallon (840 kJ/L) than that of azeotropic distillation.

If all raw sugarcane molasses is converted to bioethanol, then it has the potential to substitute 5–7% consumption of gasoline. This will be a very important contribution in future to lessen the burdens on Pakistan economy. The government of Pakistan should make policy to endorse the blending of ethanol in transportation fuels as early as it becomes conceivable [18]. With the production of bioethanol from Pakistan's own raw molasses, about 600 million of precious foreign exchange can be saved [36]. Besides this, other advantages of ethanol usage are good engine performance and better yields; it burns more efficiently and keeps our environment clean and more easily biodegradable, as well as consistent with the global focus on biofuel. No doubt this is a most effective way for production of bioethanol from raw/waste material; however, involvement of a variety of waste biomass or crop residue will be more optimistic for solution of energy issues. The main factor in ethanol production is the content of lignocellulose present in substrates which will be hydrolyzed by different hydrolyzing agents to provide fermentable glucose [37, 38]. The nature-inspired enzymes from wood fungus and termite may be used as an extra bonus in the presence of exiting bioethanol production technologies, which can convert the long chains of polysaccharides into monosacharaides. Different industries like forestry, pulp and paper, agriculture and food processing including municipal solid waste (MSW) and animal wastes are major producers of lignocellulosic waste materials [39, 40].

5. Present challenges for bioethanol production from lignocellulosic feedstocks

Currently, lignocellulolytic enzymes are derived from fungus, gut of termite and certain bacteria [41]. Established technology for bioethanol in Pakistan is relatively of low-tech approach to meet some needs by employing molasses and some selective biomass. Such limitations with biomass make the process and yield profit limited. At the same time, the farmers and agribusinesses cannot access recent technologies that may greatly expand the use of bioethanol to meet the demand for power in many applications.

The current energy scenario warrants the demand for research and development of biomass-based biofuel production systems. Biomass, due to its renewable nature and abundance, is becoming an increasingly attractive fuel source. Lignin, the second most abundant biomass constituting aromatic biopolymer on Earth, is highly recalcitrant to depolymerization. Lignin serves as bonding for hemicellulose and cellulose and creates an obstacle for penetration of any solution or enzyme to lignocellulosic structure which is the major structural component of all plants and can be depolymerized to fermentable sugars. Microbes enhance the conversion of lignin into fermentable sugars but there are some hurdles which need to be removed first. Recalcitrant nature of lignin could be tackled through different biocatalysts due to their non-hazardous and eco-friendly nature. Therefore, lignocellulolytic microorganisms like fungi and some bacteria are considered as promising biomass degraders especially for large-scale applications due to their potential yields of extracellular synergistically acting enzymes into the environment. These enzymes can contribute significantly in degradation of lignocellulosic material by converting long chain polysaccharides into their 5- and 6-carbon sugar components [42, 43]. Although lignin resists attack by most microorganisms, basidiomycetes, white-rot fungi are able to degrade lignin efficiently [44, 45]. Lignocellulolytic enzymes-producing fungi are widespread and include species from the ascomycetes (e.g. *Trichoderma reesei)* and basidiomycetes phyla such as white-rot (e.g. *Phanerochaete chrysosporium)* and brown-rot fungi (e.g. *Fomitopsis palustris)* [46, 47]. Few basidiomycetes, for example, *P. eryngii, P. chrysosporium* and *T. versicolor* can act as biocatalysts for ethanol production by having potential for lignin degradation/depolymerization. Ethanol fermentation requires high concentration of sugar solutions; therefore, biocatalytic conversion of lignocellulosic material into hydrolysate containing high concentration of sugar will be incentive for decreasing production cost. Therefore, variety of lignocellulytic material (wheat straw, rice straw and rice husk) could be degraded by basidiomycetes and subject to ethanologenic fermentation for ethanol production cost-effectively. However, some strains of white-rot fungi have promising potential to degrade lignin by simultaneous attack on lignin, hemicellulose and cellulose, whereas few can selectively work just on lignin. It is pertinent here to note that synergistic biocatalytic ability of white rot fungi would be source of efficient depolymerization method and will be helpful in proving that the heteropolymer lignin represents an untapped resource of renewable aromatic chemicals [48, 49]. Lignocellulosic biofuel production is not yet economically competitive with fossil fuels; therefore, to ensure successful utilization of all sugars is important for improving the overall economy especially in terms of maximum theoretical yield. Xylose is one of the most abundant sugars in lignocellulosic hydrolysate. Therefore, over expression of xylose isomerase will facilitate complete utilization of xylose present in hydrolysate which otherwise remains to varying extent in spent culture [18, 50]. Another matter of concerns regarding lignin depolymerization and its conversion into biofuels/bioethanol is repolymerization of lignin-derived low molecular weight sp. into high molecular weight molecules which are not easy to be degraded by microbes. Repolymerization is observed to occur within few hours after onset of lignin volarization. For this purpose, organization of most effective microbial sink for immediate utilization of low molecular species for bioethanol production is the most appealing option [51, 52].

For overcoming this bottleneck, microbial sink/consortium of different microbes with xylose overexpression is an offered strategy. Preventing repolymerization of low molecular weight lignin species into high molecular weight lignin compounds and ensuring the complete utilization and conversion of available sugars into bioethanol can make the bioprocess cost-effective. The description in this chapter will lead to development of technologies that can be helpful in efficient depolymerization of lignin and its simultaneous conversion into high-valued microbial-assisted advanced biofuel. The chapter represents need for development of road map for advanced level of biofuel production from waste crop residues. Nature-inspired enzymes' involvement is the most effective way for enhanced bioethanol production from biomass. The enzymes convert the long chains of polysaccharides into monosaccharaides. Currently, lignocellulytic enzymes used for ethanol production from cellulosic biomass are obtained from fungus, gut of termite and certain bacteria [40]. Present restrictions of enzymatic breakdown of lignocellulose-based biomass are mostly due to concern of enzymatic steadiness and vulnerability to inhibitors or by-products [53, 54]. Continuous bioengineering efforts and prospecting should provide novel enzymes with lower susceptibility to inhibitors and relatively higher specific activity [55]. Few insects such as termites have very efficient approaches to break the lignocellulose-based substrates as potential mean of bioenergy [56]. In case of lower termites, activity (cellulolytic) is normally dependent on enzymes produced by endosymbiotic, flagellated protists [57], while in case of higher termites, their guts contain lignocellulytic enzymes which combine with cellulases secreted by certain endosymbiotic gut bacteria [58, 59].

Hence, establishment of large-scale bioethanol production plant by treating waste crop residues with such novel enzyme will enhance the production and can successfully provide support to deteriorating economy of Pakistan (**Figure 2**). The resulting cleaner environment is another benefit that has monetary values that the government may be financially and ethically interested in. Such multiple positive benefits will attract different interested parties to involve in replication of process, each with resources and benefits to sustain and multiply. Additionally, the greenhouse gas emission will be reduced by burning bioethanol, as the net CO_2 emission is zero because the amount of CO_2 emitted on burning is equal to the amount of it, which is absorbed from the atmosphere by the process of plant photosynthesis which will be used for production of bioethanol [60]. In Pakistan, the current domestic production of raw oil presently satisfies only almost 25% of the country's consumption and remaining demands are met by importing fuels from abroad. This make Pakistan's economy vulnerable to different social and economic issues; however, incorporation of biofuel/bioethanol will reduce burden on country's economy significantly.

Federal Cabinet, Economic Coordination Committee (ECC) of Pakistan has decided to permit marketing of Ethanol 10 as motor vehicle fuel on the trial basis. Anhydrous ethanol can be blended with gasoline in different proportions having less than 1% water content. Many of the motor vehicles having gasoline engines operate well with ethanol blend of 10% in their fuels (E 10). The Government of Pakistan enacted a 15% duty on export of molasses to prefer the use of molasses for ethanol production rather than export [61]. The government of Pakistan should make policy to enforce the blending of ethanol in transportation fuels as early as it becomes conceivable [18]. Successful implementation of large-scale waste crop residue-based bioethanol production concept will attract private sector investment and company-farm partnership to accelerate the development and commercialization of new bioenergy solution to improve emerging economies and transform the lives of at least small farmers. The concept is readily adoptable by different agricultural regions as the essential supply of the feedstock is available in the form of agricultural residues that are sustainable and typically available abundantly and locally.

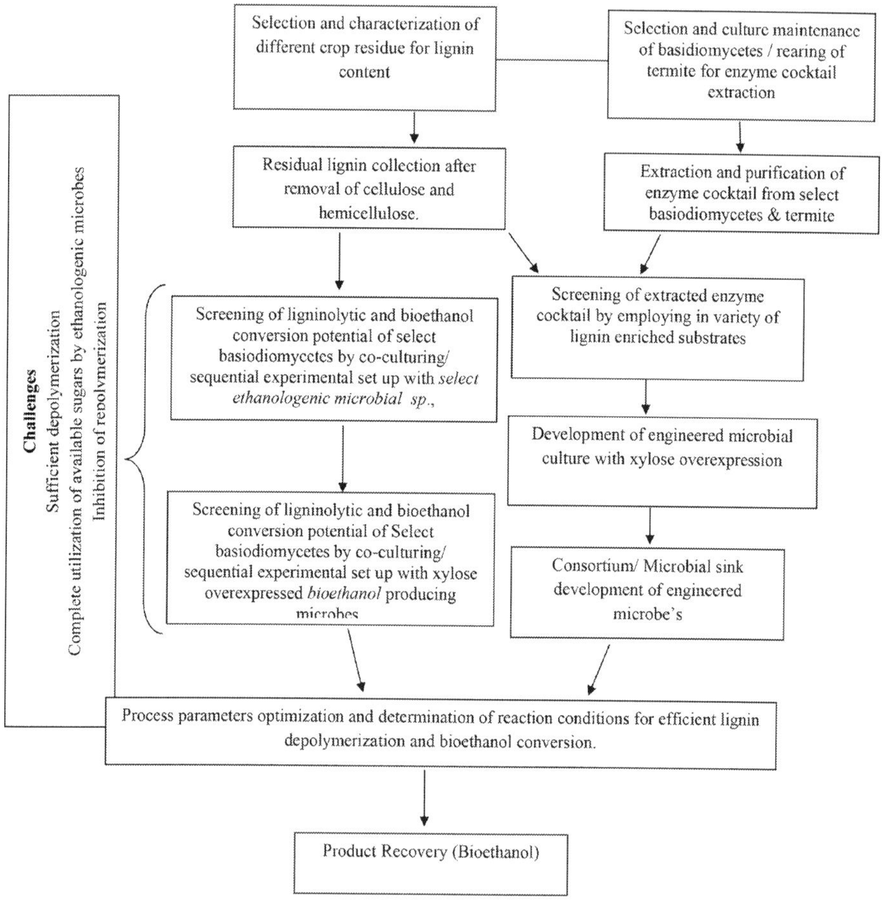

Figure 2. Schematic representation of nature inspired enzyme assisted bioethanol production process.

6. Concluding remarks

To import conventional energy resources, Pakistan is spending around 7 US$ billion equivalent to 40% of total imports. Careful estimates show that by 2050 Pakistan's energy needs will be increased three times while the supplies are not very inspiring. In 2012, Pakistan natural gas supply had about 15 billion-m^3/year deficit with increasing tendency. Rise in natural gas price (0.51 $/$m^3$) brings great potential to promote biomass-based biofuel production in Pakistan.

Pakistan being an agricultural country produces wheat, sugarcane and potatoes as one of its biggest crops [62–64]. Consequently, large amounts of wheat straw and sugarcane bagasse are obtained as by waste products. Due to high ambient temperature in most part of the year, poor post-harvest processing and storage of thousands of tons of biomass are wasted each year. In short access of promising process for the initiation of sustainable strategies for waste crop residue-based bioethanol production while consuming starch, cellulosic and lignin loads of effluents of respective origins in the country, bestowed with suitable biomass and temperature optima for successful cultivation of ethanologenic microbes, is expected to provide sustainable supplies of biofuels. Additionally, more than three billion acres worldwide which are not suitable for agriculture purpose due to dryness could be utilized for growing drought-hardy plants for biofuels production. The only disadvantage of using these crops is that they contain lignocellulose, a hard plant material, that needs more treatment than either corn or sugarcane to be converted into alcohol. Therefore, search for ways to make the overall process more efficient by reusing materials, changing the fermenting agent and searching for better and nature-inspired enzymes will be milestone in this regard. The process development of nature inspired enzyme-assisted conversion of agricultural and food waste into bioethanol that can be used as clean biofuel is demand of time. Adoption of these new dimensions for bioethanol production will definitely reduce pressure on energy and transportation sector, entire dependence on conventional fuels and can triumph fight against climate change.

Author details

Saima Mirza[1], Habib ur Rehman[2], Waqar Mahmood[3] and Javed Iqbal Qazi[4*]

*Address all correspondence to: qazi.zool@pu.edu.pk

1 Energy System Engineering, Punjab Bioenergy Institute, University of Agriculture, Faisalabad, Pakistan

2 Punjab Bioenergy Institute, University of Agriculture, Faisalabad, Pakistan

3 Centre for Energy Research and Development, University of Engineering and Technology, Lahore, Pakistan

4 Microbial Biotechnology Laboratory, Department of Zoology, University of the Punjab, Lahore, Pakistan

References

[1] Daily Times. Pakistan ratifies Kyoto Protocol, 2004. Available at: http://www.dailytimes.com.pk/default.asp?page, story 17.12.2004. p. 77.

[2] EIA. Office of Integrated Analysis and Forecasting. International Energy Outlook DOE/EIA-0484; US Department of Energy, Washington, U.S.A. 2006.

[3] Eurostat. Online database of the European Union, 2007. Available at: http://epp.eurostat.ec. europa.eu, Eurostat. Accessed on 08.05.2007.

[4] Qin D, et al., editors. Changes in atmospheric constituents and in radiative forcing, Contribution of Working Group I to the Fourth Assessment Report of the Intergovernmental Panel on Climate Change. C.U.P. United Kingdom and New York, USA, 2007, p. 996.

[5] Somerville C. New sources near for biofuels. *Havard Gazette*, USA, 2011. Available at: http://environment.harvard.edu/news/huce-headlines/new-sources-near-biofuels.

[6] Abbas S. Biomass power generation feasibility study [Internet]. in Agribusiness Development,Agrib. 2014. Available at: http://www.agribusiness.com.pk/biomass-power-generation-feasibility-study/.

[7] National Power Policy. Government of Pakistan, 2013.

[8] Policy for development of renewable energy for power generation. Government of Pakistan, 2006.

[9] MacDonald T, Yowell G, McCormack M. Staff report. US ethanol industry production capacity outlook. California Energy Commission. 2001. Available at: http//www.energy.ca.gov/reports/2001-08-29_600-01-017.

[10] Akpa J. Modeling of a bioreactor for the fermentation of palm wine by *Saccaharomyce cerevisiae* (yeast) and lactobacillus (bacteria). Biores. Technol. 2012; 3: 231–240.

[11] Ana M, Viorel T. Calin M. Simulation of fermentation bioreactor control for ethanol production. In 11th International Conference on Development and Application Systems, Romania, 2012, pp. 17–20.

[12] Aziz S, Hafeez ur Rehman M, Farman AS, Rajoka MI, Suhail A. Production of ethanol by indigenous wild and mutant strain of thermotolerant *Kluyveromyces Marxianus* under optimized fermentation conditions. Pak. J. Anal. Environ. Chem. 2009; 10: 25–33.

[13] Prasad S, Anoop S, Joshi H. Ethanol as an alternative fuel from agricultural, industrial and urban residues resources. Conserv. Recyc. 2007; 50: 1–39.

[14] Carlos A, Cardona O. Fuel ethanol production: Process design trends and integration opportunities. Biores. Technol. 2007; 98: 2415–2457.

[15] Bai F, Anderson W, Young M. Ethanol fermentation technologies from sugar and starch feedstock's. Biotechnol. Adv. 2010; 26: 89–105.

[16] David R, Dochain R, Mouret A, Sablayrolles M. Dynamical modeling of alcoholic fermentation and its link with nitrogen Consumption. Proceedings of the 11th International Symposium on Computer Applications in Biotechnology, Belgium, 2010, pp. 496–501.

[17] Dai W, Word D, Hahn J. Modeling and dynamic optimization of fuel-grade ethanol fermentation using fed-batch process. Control Eng. Pract. 2014; 22: 231–241.

[18] Arshad M. Khan, ZM, Khalil-ur-Rehman, Shah FA, Rajoka, FA. Optimization of process variables for minimization of byproduct formation during fermentation of blackstrap molasses to ethanol at industrial scale. Lett. Appl. Microbiol. 2008; 47: 410–414.

[19] Gong CS, Maun CM, Tsao GT. Direct fermentation of cellulose to ethanol by a cellulolytic filamentous fungus Monilia sp. Biotechnol. Lett. 1981; 3: 77–82.

[20] Yamauchi H, Akita O, Obata T, Amachi T, Hara S. Production and application of a fruity odor in a solid state culture of *Neurospora* sp. using pregelatinized polish rice. Agric. Biol. Chem. 1989; 53: 2881–2888.

[21] Matthew H, Ashley O, Brian K, Alisa E, Benjamin JS. *Wine making 101*, 2005. Available at: http//www.arches.uga.edu/~matthaas/strains.htm.

[22] Ito K, Yoshida K, Ishikawa T, Kobayashi S. Volatile compounds produced by fungus *Aspergillus oryzae* in rice koji and their changes during cultivation. J. Ferment. Bioengin. 1990; 70: 169–172.

[23] Gervais P, Sarrette M. Influence of age of mycelia and water activity on aroma production by *Trichoderma viride*. J. Ferment. Bioengin. 1990; 69: 46–50.

[24] Pastore GM, Park YK, Min DB. Production of a fruity aroma by *Neurospora* from beiju. Mycol. Res. 1994; 98: 25–35.

[25] Sugawara E, Hashimoto S, Sakurai Y, Kobayashi A. Formation by yeast of the HEMF 4-hydrpxy-2 or 5.-ethyl-5 or 2.-methyl-3 2H.-furanone. aroma components in Miso with aging. Biosci. Biotechnol. Biochem. 1994; 58: 1134–1135.

[26] Millichip RJ, Doelle HW. Large-scale ethanol production from Milo Sorghum using *Zymomonas mobilis*. Proc. Biochem. 1989; 24: 141–145.

[27] Ingram LO, Conway T, Clark DP, Sewell GW, Preston JF. Genetic engineering of ethanol production in *Escherichia coli*. Appl. Environ. Microbiol. 1987; 53: 2420–2425

[28] Krishna HS, Chowdary GV, Reddy SD, Ayyanna C. Simultaneous saccharification and fermentation of pretreated *Antigonum leptopus* Linn. leaves to ethanol. J. Chem. Technol. Biotechnol. 1999; 74: 1055–1060.

[29] Ballesteros M, Oliva JM, Negro MJ, Manzanares P, Ballesteros I. Ethanol from lignocellulosic materials by a simultaneous saccharification and fermentation process SFS with *Kluyveromyces marxianus* CECT 10875. Proc. Biochem. 2004; 39: 1843–1848.

[30] Hugot E, Jenkins GH. Handbook of Cane Sugar Engineering. Elsevier, Amsterdam, New York, USA, 1986.

[31] Arshad M. Bioethanol: A sustainable and environment friendly solution for Pakistan. A Scientific J. COMSATS – Sci. Vision. 2010–2011; 16–17: 21–26.

[32] Gruno M, Vaeljamaee P, Pettersson G, Johansson G. Inhibition of the *Trichoderma reesei* cellulases by cellobiose is strongly dependent on the nature of the substrate. Biotechnol. Bioengin. 2004; 86: 503–511.

[33] Kim S, and Dale BE. Global potential bioethanol production from wasted crops and crop residues. Biomass. Bioengin. 2004; 26: 361–375.

[34] Parkash, A. Modeling of ethanol production from molasses: A review. Ind. Chem. 2015; 3: 108.

[35] Arshad M. Employment generation through bio-ethanol production industry as renewable energy. In: COMSATS, Policies and Strategies for Successful Implementation of Employment Generating Programmes in Renewable Energies, Biotechnology, Agriculture, Environment and ICTs, Islamabad, Pakistan, 11–12 August, 2009.

[36] Sibtain RK. An Eco Friendly Resource. The News, Sunday, 2009; 25

[37] Curreli N, Agelli M, Pisu B, Rescigno A, Sanjust E, Rinaldi A. Complete and efficient enzymic hydrolysis of pretreated wheat straw. Proc. Biochem. 2002; 37: 937–941.

[38] Mosier N, Wyman C, Dale B, Elander R, Lee YY, Holtzapple M, Ladisch M. Features of promising technologies for pretreatment of lignocellulosic biomass. Biores. Technol. 2005; 96: 673–686.

[39] Champagne P. Feasibility of producing bioethanol from waste residues: A Canadian perspective feasibility of producing bio-ethanol from waste residues in Canada. Res. Conserv. Recyc. 2007; 50: 211–230.

[40] Kalogo, Y. Habibi S, MacLean HL, Joshi SV. Environmental implications of municipal solid waste-derived ethanol. Environ. Sci. Technol. 2007; 41: 35–41.

[41] Lynd LR. Overview and evaluation of fuel ethanol from cellulosic biomass: Technology, economics, the environment, and policy. Ann. Rev. Energ. Environ. 1996; 21: 403–465.

[42] Zhou S, Ingram LO. Synergistic hydrolysis of carboxymethyl cellulose and acid-swollen cellulose by two endoglucanases (CelZ and CelY) from *Erwinia chrysanthemi*. J. Bacteriol. 2000; 182: 5676–5682.

[43] Dashtban M, Schraft H, Qin W. Fungal bioconversion of lignocellulosic residues; opportunities & perspectives. Int. J. Biol. Sci. 2009; 5: 578–595.

[44] Abbas A, Koc H, Liu F, Tien M. Fungal degradation of wood: Initial proteomic analysis of extracellular proteins of *Phanerochaete chrysosporium* grown on oak substrate. Curr. Genet. 2005; 47: 49–56.

[45] Wong DW. Structure and action mechanism of ligninolytic enzymes. Appl. J. Biochem. Biotechnol. 2009; 157: 174–209.

[46] Ljungdahl LG. The cellulase/hemicellulase system of the anaerobic fungus *Orpinomyces* PC-2 and aspects of its applied use. Anna. New York Acad. Sci. 2008; 1125: 308–321.

[47] Yoon JJ, Cha CJ, Kim YS, Son DW, Kim YK. The brown-rot basidiomycete *Fomitopsis palustris* has the endo-glucanases capable of degrading microcrystalline cellulose. J. Microbiol. Biotechnol. 2007; 17: 800–805.

[48] Bugg TD, Ahmad M, Hardiman EM, Rahmanpour R. Pathways for degradation of lignin in bacteria and fungi. Nat. Prod. Rep. 2011; 28: 1883–1896.

[49] Jenkins T, Bovi A, Edwards R. Plants: Biofactories for a sustainable future? Phil. Trans. R. Soc. A: Phy. Math. Engin. Sci. 2011; 369: 1826.

[50] Hallenbeck, PC, Ghosh D, and Skonieczny YV. Microbiological and engineering aspects of biohydrogen production. Ind. J. Microbiol. 2009; 49: 48–59.

[51] Zhou H, Cheng JS, Wang BL, Gerald RF, Stephanopoulos G. Xylose isomerase overexpression along with engineering of the pentose phosphate pathway and evolutionary engineering enable rapid xylose utilization and ethanol production by *Saccharomyces cerevisiae*. Metabol. Engin. 2012; 14: 611–622.

[52] Kurosawa K, Wewetzer SJ, and Sinskey AJ. Engineering xylose metabolism in triacylglycerol producing *Rhodococcus opacus* for lignocellulosic fuel production. Biotechnol. Biofuels. 2013; 6: 134.

[53] Mousdale DM. Biofuels: Biotechnology, Chemistry, and Sustainable Development. CRC Press, Boca Raton, FL, 2008; pp. 66–78.

[54] Kristensen JB, Felby C, Jorgensen H. Yield determining factors in high-solids enzymatic hydrolysis of lignocellulose. Biotechnol. Biofuels. 2009; 2: 11.

[55] Lynd LR, Laser MS, Bransby D, Dale BE, Davison B, Hamilton R, Himmel M, Keller M, McMillan JD, Sheehan J, Wyman CE. How biotech can transform biofuels. Nat. Biotechnol. 2008; 26: 169–172.

[56] Martin MM. Cellulose digestion in insects. Comp. Biochem. Physiol. 1983; 75A: 313–324.

[57] Ohkuma, M. Symbioses of flagellates and prokaryotes in the gut of lower termites. Trends in Microbiol. 2008; 16: 345–352.

[58] Tokuda G, Watanabe H, Matsumoto T, Noda H. Cellulose digestion in the wood-eating higher termite, *Nasutitermes takasagoensis* (Shiraki): Distribution of cellulases and properties of endo-beta-1, 4-glucanase. Zool. Sci. 1997; 14: 83–93.

[59] Warnecke F, Luginbuhl P, Ivanova N, Ghassemian M, Richardson TH, Stege JT, Cayouette M, McHardy AC, Djordjevic G, Aboushadi N, Sorek R, Tringe SG, Podar M, Martin HG, Kunin V, Dalevi D, Madejska J, Kirton E, Platt D, Szeto E, Salamov A, Barry K, Mikhailova N, Kyrpides NC, Matson EG, Ottesen EA, Zhang X, Hernandez M, Murillo C, Acosta LG, Rigoutsos I, Tamayo G, Green BD, Chang C, Rubin EM, Mathur EJ, Robertson DE, Hugenholtz P. Leadbetter JR. Metagenomic and functional analysis of hindgut microbiota of a wood-feeding higher termite. Nature. 2007; 450: 560–565.

[60] Oleveira, M.D. De. Sugarcane and ethanol production and carbon dioxide balances. In: Pimentel, D, editors. Biofuels, Solar and Wind as Renewable Cambridge University Press, Cambridge, Energy Systems. Springer Science, United Kingdom and NewYork, USA. 2008.

[61] Sajid C. Use of Bagasse with coal as fuel: Sugar industry allowed to co-generate power in lean season. Daily Times, April 06, 2010.

[62] Hussain MF, Anwar S, Hussain Z. Economics of sugarcane production in Pakistan: A price risk analysis. Int. Res. J. Finance and Econ. 2006; 4: 70–77.

[63] Arifullah SA, Chisti AF, Zulifqar M, Yasmeen G, Farid N, Ahmed I. Estimating yield potential of the major crops and its implications for Pakistan's crops sector. Sarhad J. Agric. 2009; 25: 611–615.

[64] Nawab K, Amanullah SP, Rab A, Arif M, Khan MA, Mateen A, Munsif F. Impact of integrated nutrient management on growth and grain yield of wheat under irrigated cropping system. Pak. J. Bot. 2011; 4: 1943–1947.

Chapter 12

Jatropha Biofuel Industry: The Challenges

M. Moniruzzaman, Zahira Yaakob,
M. Shahinuzzaman, Rahima Khatun and
A.K.M. Aminul Islam

Additional information is available at the end of the chapter

http://dx.doi.org/10.5772/64979

Abstract

Considering environmental issues and to reduce dependency on fossil fuel many countries have politicized to replenish fossil fuel demand from renewable sources. Citing the potential of Jatropha mostly without any scientific and technological backup, it is believed to be one of the most suitable biofuel candidates. Huge grants were released by many projects for huge plantation of Jatropha (millions of hectares). Unfortunately, there has been no significant progress, and Jatropha did not contribute much in the energy scenario. Unavailability of high-yielding cultivar, large-scale plantation without the evaluation of the planting materials, knowledge gap and basic research gap seem to be the main reasons for failure. Thus, the production of Jatropha as a biofuel has been confronted with various challenges such as production, oil extraction, conversion and also its use as a sustainable biofuel. In this chapter, we disclose the challenges and possible remedy for the contribution in the biofuel industry.

Keywords: *Jatropha curcas*, biofuel industry, renewable energy, challenges, solutions

1. Introduction

Jatropha belongs to the family Euphorbiaceae and has 175 species. It has originated from tropical America and has spread all over the tropics and subtropics of Asia and Africa [1]. Throughout the world, more than 1,000,000 ha of Jatropha have been propagated. Majority (85%) of them are in the Asian countries, i.e., India, China and Myanmar; the remaining, 12% in Africa and 2% in Latin America (Brazil and Mexico). India is the largest cultivator of

Jatropha [2]. In the ancient times, Jatropha has been used in various fields, such as storm protection, soil erosion control, firewood, hedges and traditional medicines [3–6]. The seed oil of Jatropha is also used as lamp fuel, soap manufacturing ingredient, paints and as a lubricant [4, 7, 8]. The characteristics of Jatropha seed oil match with characteristics of diesel [9–11], thus it is called a biodiesel plant [12]. Jatropha grows on diverse wasteland without any agricultural impute (irrigation and fertilization) and has 40–60% oil content [12, 13]. Easy propagation, rapid growth, drought tolerance, pest resistance, higher oil content than other oil crops, adaptation to a wide range of environmental conditions, small gestation period, and optimum plant size and architecture (which make the seed collection more convenient; actually inconvenient [14]) are some characteristics of Jatropha [15], which makes it a promising crop for biofuel [16]. Although Jatropha ranked behind palm (palm > *Calophylum inophyllum*> Cocus sp. > Jatropha) according to annual oil yield/hectare, it is favoured as a non-edible feed stock [17, 18]. A number of earlier reports, proceedings, expectations and assumptions predicted that the seed yield of Jatropha range from 2 to 5 Mg/ha and even 7.8 to 12 Mg/ha- without any scientific and technological backup [19].

There is a complete mismatch between theoretical expectation and actual seed production of Jatropha in field conditions [9, 20–22]. The research on Jatropha opened the floodgate to the scientific community to grab funds and publish papers in high impact journals because seed oil of Jatropha has characteristics of biodiesel and this crop was non-native of arid, semiarid and subtropical regions. Singh and co-authors [19] depict the expectation and contribution picture from Jatropha policy (**Figure 1**). The reported yields of Jatropha in field conditions in India, Belgium, South Africa and Tanzania, are 0.5–1.4 mg/ha/yr, 0.5 mg/ha/yr, 0.35 mg/ha/yr and 2 mg/ha/yr, respectively [23]. The less productivity is because of unavailability of suitable high yielding varieties, large-scale plantation without evaluating the genetic potential of planted materials, consideration of Jatropha as no/low impute crop and lack of knowledge on agronomy. In this chapter, we present the challenges confronted by Jatropha as a biofuel and recommended solutions.

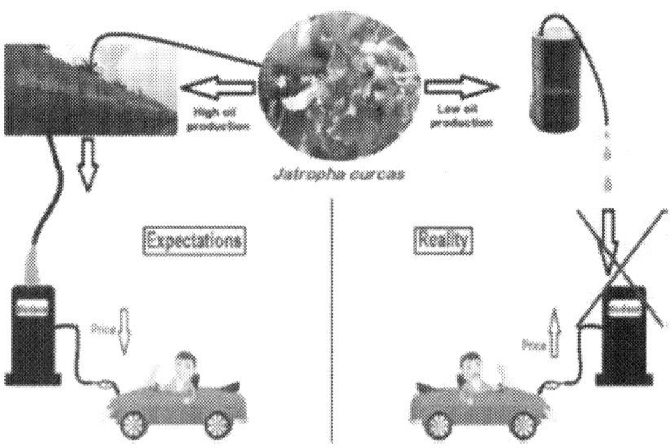

Figure 1. Expectation and contribution from Jatropha biofuel industry.

http://dx.doi.org/10.5772/64979

2. Potentials of Jatropha

Jatropha has multiple uses (**Figure 2**). Jatropha seed oil possesses biodiesel and jet fuel production potentials. Its wood, leaves and fruits have been using as firewood in rural areas. It also has industrial applications. Preparation of soap and cosmetics, and dyeing clothes and fishing nets are some of its common applications. Traditionally, Jatropha has been known as a medicinal plant. The therapeutic compounds from Jatropha can be used as anti-microbial, anti-inflammatory, healing, homeostatic, anti-cholinesterase, anti-diarrheal, anti-hypertensive and anti-cancer agents in modern pharmaceutical industry. As it contains toxins, before using Jatropha and/or its derivatives as a therapeutic agent, toxicological studies must be conducted. Jatropha seed cake can supplement animal feed and organic fertilisers as it bears higher percentage of protein and other nutrients. Soil erosion control and used as hedges are prehistoric uses of Jatropha.

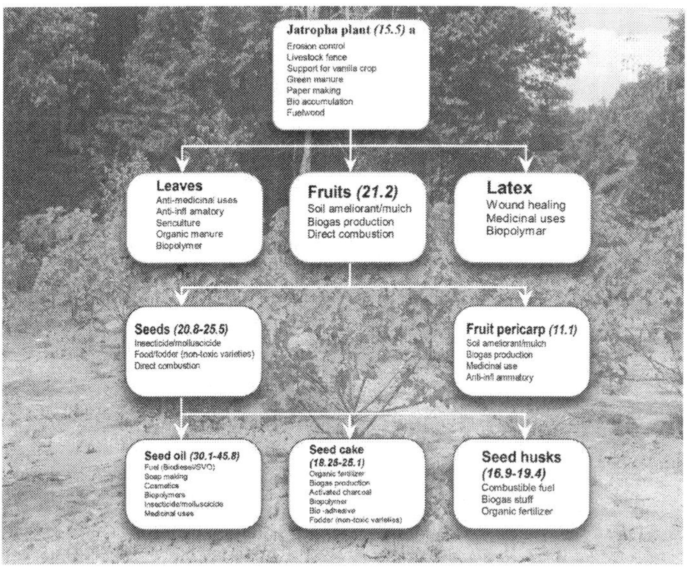

Figure 2. Multipurpose uses of *Jatropha curcas*.

2.1. Jatropha as energy source

Various features like, ease of production, sustainability and environmentally friendly nature of biomass draw attention as a potential renewable energy to replenish fossil fuel demand. Among the crops identified as energy crops for first generation biofuels, *Jatropha curcas* L. (JCL) has been acknowledged as one of the promising candidates [24].

Parts of Jatropha plant, like wood, fruit shells, seed husks and kernels [25], are used to produce energy. Raw oil is the major resource obtained from Jatropha. Depending on the variety/cultivars, decorticated seeds contain 40–60% oil [26–31]. The oil is utilised for many purposes, such as lighting, lubricating, making soap [32] and most importantly as biodiesel. Biodiesel from Jatropha comply with European biodiesel standards (**Table 1**).

Parameter	Jatropha oil
Density at 15 °C	0.920 g/cm^3
Viscosity at 30 °C	52 cSt
Flash point	240°C
Fire point	274 ± 3°C
Cloud point	9 ± 1°C
Pour point	4 ± 1°C
Cetane number	38
Caloric value	38.20 MJ/kg
Conradson carbon residue	0.8 ± 0.1 (%w/w)
Hydrogen	10.52 (%w/w)
Sulphur	0 (%w/w)
Oxygen	11.06 (%w/w)
Nitrogen	0
Carbon	76.11 (%w/w)
Ash content	0.03 (%w/w)
Neutralization number	0.92 mg KOH/g
Saponification value	198
Iodine number	94
Monoglycerides	Not detected
Diglycerides	2.7% m/m
Triglycerides	97.3% m/m
Water	0.07% m/m
Phosphorus	290 mg/kg
Calcium	56 mg/kg
Magnesium	103 mg/kg
Iron	2.4 mg/kg

Source: Giibitz et al. (1999).

Table 1. Chemical and physical properties of Jatropha oil.

There are approximately 24.60%, 47.25% and 5.54% of crude protein, crude fat and moisture, respectively, in Jatropha oil [33]. Both saturated and unsaturated fatty acids are present in the oil. The major saturated fatty acids are Palmitic acid (16:0) at 14.1% and stearic acid (18:0) at 6.7%, oleic acid (18:1) at 47.0% and linoleic acid (18:2) at 31.6%. The usefulness of Jatropha oil and its esters instead of petro-diesel has been reported [34]. The energy value of Jatropha seed oil (39MJ kg^{-1}) is higher than anthracite coal and is comparable to crude oil [35].

http://dx.doi.org/10.5772/64979

By mass, the shells bears about 35–40% of the dry fruit, that is, 60–65% of the seed weight [36]. There are approximately 42% husk and 58% kernel in a seed [25]. The gross energy value of Jatropha seed is 24 MJ kg^{-1} which is higher than lignite coal and cattle manure and is comparable to corn cobs [37].

The first step of oil extraction is the mechanical removal of the shell from the fruit. Jatropha seed shell contains cellulose (34%), hemicelluloses (10%) and lignin (12%) [25]. Approximately 11.1 MJ energy is driven from one (1) kg seed shell of Jatropha [35].Ash (4%), volatile matter (71%) and fixed carbon (25%) are the components of seed husk. Approximately, 16 MJ energy is driven from one (1) kg seed husk [36], which is comparable to wood.

Less energy expenditures and the prospect of using a cheap substrate make hydrogen (H_2) gas a lucrative source of future renewable energy. Lignocellulose biohydrogen can be produced by the fermentation of de-oiled Jatropha solid waste (DJSW) and Jatropha seed cake that contains lignocellulose [38–41]. Kuma and co-workers reported highest achievable cumulative hydrogen production (CHP) of 296 mL H_2 by the fermentation of de-oiled Jatropha waste under optimum conditions. The reported optimum conditions are; substrate concentration 211 g/L, pH 6.5 and temperature 55.4°C [40]. Lopes and co-workers produced 68.2 mL H_2/gVS_{iJSC} biohydrogen by dark fermentation of seed cake by a pure strain of the bacteria *Enterobacter aerogenes* without pretreatment of the substrate [41]. In the viewpoint of energy saving, it is significant.

2.2. Industrial use

The common use of viscous oil from Jatropha seed is for making soap [31]. The manufacturing of soft and durable soap from Jatropha oil becomes easy because of the high palmitic acid content and its hydrophobic nature. People in West Africa, Zambia, Tanzania and Zimbabwe are familiar with the use of Jatropha soap [42]. The presence of glycerine in Jatropha oil soap makes the white soap good for skin. It also have very good foaming [43] properties. Jatropha soap can be used for various skin diseases because of its medicinal properties [44]. The 36% linoleic acid (C18:2) content in Jatropha kernel oil is good for skin care [42, 45]. The oil is also an ingredient in hair conditioners [46].

Jatropha is rich in many phytochemical constituents. Alkaloids, coumarins, flavonoids, lignoids, phenols, saponins, steroids, tannins and terpenoids are found in different parts of this plant [47]. These components show anti-microbial [48], anti-inflammatory [49–51], healing, homeostatic [52], anti-cholinesterase [53, 54], anti-diarrheal [55–57], anti-hypertensive activities [58] and are anti-cancer agents [59, 60]. It is necessary to study the toxicity associated with these phytochemicals. The toxic effects could decrease its medicinal value.

The strong purgative activity of oil helps to cure skin diseases and to soothe rheumatic pain. It is also used in pesticides. The people in the Philippines have been using the dye obtained from Jatropha bark for colouring finishing nets, cloths and lines [61]. Another application of seed oil is in eczema treatment [62]. Manufacturing of soap and cosmetics from Jatropha derivatives (as an alternative to Karitee butter) is a non-energy application of Jatropha.

2.3. Other uses

From prehistoric days, Jatropha is used to make hedges. The advantage is that animals do not feed on it. Seed germinating plants bears taproots along with surface roots; Jatropha is a seed germinating plant, and it protects soil against erosion. It also act as a nutrient pump because the roots can uptake the leached down minerals and return them in the form of leaf fall, fruit debris and other organic remains.

The higher protein content (58.1% by weight) of Jatropha seed cake, after detoxification, when compared to that of soy meal (48%), makes it a valuable animal feed protein supplement. As it possesses most of the minerals nutrients— nitrogen, potassium, calcium, magnesium, sulphur, iron, phosphorus, zinc, copper and manganese, Jatropha seed cake is considered to be an excellent organic fertiliser [63, 64].

3. Limitation of Jatropha as a biofuel crop

– A good commercial variety with a higher yield and disease resistance is still lacking.

– High fluctuation of yield among trees.

– It requires proper irrigation and nutrients for fruiting, though it can survive on insufficient irrigation and nutrients.

– Relatively long gestation period: it requires 3–5 years to become commercially productive.

– The presence of toxic components limits its use as feed and therapeutic agents.

– Recent study reveals that Jatropha is susceptible to pests and diseases.

– Jatropha is sensitive to frost and water logging.

– Jatropha may be host for some diseases (cassava diseases).

– High viscosity of Jatropha seed oil limits its use in cool climate conditions.

– In certain environments, Jatropha may create weed problem.

4. Jatropha production challenges

4.1. Poor seed yield

Related experts suggest that the Jatropha seed yield of 4–5 Mg/ha/yr is needed for the commercial viability of the industry. If the usual seed yield of 3.75 Mg/ha with 30–35% oil content or 1.2 Mg/ha oil yield only then Jatropha would compete with soybeans (USA 0.38 Mg oil/ha) and rapeseed (Europe 1.0 Mg oil/ha) [65]. However, there is high flocculation of the unit seed yield and seed oil content of Jatropha. A number of authors reported that the low seed yield and the low seed oil content are the one of the most important barrier for commercial viability of Jatropha biodiesel industry [19, 66–69]. In India, a different location trial at diverse agro-climatic regions was conducted and the average seed yield was recorded as 0.5–1.4 Mg/ha/yr after 5 years of plantation [14]. A similar result was observed from plantation of 24 elite accessions with good plant architecture (height and branching pattern) in sodic soil [20]. In Belgium, the average seed yield was reported as 0.5 Mg seed/ha after 4 years of plantation, using the best known production techniques [70]. Recent assessment revealed that globally the average seed productivity of Jatropha is 1.6 Mg/ha which is equivalent to 0.475 Mg/ha/yr biodiesel productivity, which is not a safe position for the industry to be economically feasible. In South Africa, the highest seed yield was 0.35 Mg/ha after 5 years of plant growth [21]. A Jatropha silvi-pastoral production system in central-west Brazil where hybrid seeds were used, however, it could not ensure any significant seed yield, against the expectation of 2.4 kg/plant [22]. In Tanzania, a negligible gain at US$ 9 ha^{-1} with yields of 3 Mg/ha and a loss of US$ 65 ha^{-1} on lands with yields of 2 Mg/ha of seeds after 5-year investment [23] were obtained. In Panzhihua, China, Jatropha could not change local energy scenario and the industry has been confronted by a number of risk factors [71].

Developing of a higher yielding and more oil containing variety is one of the main effective solutions. However, a good commercial variety is still missing [72]. Zhang et al. [67] and Yu et al. [68] also point out that variety breeding is one of the main hurdles for Jatropha planting.

Actually, the current Jatropha breeding program is limited to conventional breeding and surveying of germplasm resources of wild Jatropha plants [67]. However, the study of modern biotechnology application on Jatropha improvement is limited [72]. Particularly, studies on cloning, expression and biological function annotation for Jatropha genes, which are responsible for economical traits, are largely absent.

The enhancement of unit seed yield of Jatropha for commercial use should be the main objective of cultivation. Therefore, the techniques of Jatropha cultivation refers to many field practices such as propagation, site preparation, tree density and canopy control, insects and diseases control, fertilization and irrigation management, cropping treatments [67, 68, 73]. Few studies on planting techniques and poor management for planting base limit large-scale plantation of Jatropha [68, 74]. However, there is limited research to demonstrate precisely and scientifically the impact of field operation on the seed yield of Jatropha. Moreover, there are no/a few detailed reports on field observation on the seed yield under different treatments of cultivation techniques. For example, data on tree density for Jatropha cultivation, canopy pruning intensity and frequency, insecticide effect as well as fertilization and irrigation efficiency are largely absent in the literature.

4.2. Consider as low impute crop

J. curcas is believed as a low input crop because of its ability to grow on barren land. However, it needs adequate nutrients as fertilizer and rainfall or irrigation for growing as a productive crop. On the other hand, excessive fertilization and irrigation may cause vegetative growth (biomass production) at the cost of fruit production. Moisture and nutrients have larger influence on the seed yield and oil productivity from the plantation on marginal lands. The plant growth and the seed yield of *J. curcas* were significantly increased under irrigated conditions as compared to non-irrigated conditions [9, 14]. It was observed that there was 750 kg/ha yield under irrigated conditions at the same time only 450 kg/ ha was recorded under rainfed conditions from 3-year-old plantations [75]. Application of nitrogen and phosphorus increased the growth, seed yield and oil yield of *J. curcas* [76]. Furthermore, another report by BAIF Development Research Foundation showed that there was about 500 kg/ha seed yield under rainfed conditions in the fifth year of plantation. However, after regular irrigation of the same plantation, in next year the seed yield was recorded about 1200 kg/ha [77].

The systematic studies for yield improvement, the agronomy (especially the irrigation and nutritional requirements) in different agro climatic conditions have not been adequately addressed, despite advocacy for large-scale plantation of *J. curcas* [78].

4.3. Pest and disease susceptibility

Control of insects and diseases is particularly one of the most important technical issues which could seriously shape Jatropha cultivation (**Figure 3**). Though it was claimed that Jatropha is free of pests and diseases, the current study do not support the claim. Recent studies reported that the plants were susceptible to viral infection (Cucumber mosaic virus), insect attack, rodents, powdery mildew, leaf spots, insect defoliations and fungal diseases of the soil [14, 21]. In Belgium, leaf miner *Stomphastis thraustica*, the leaf and stem miner *Pempelia morosalis* and the shield-backed bug *Calidea panaethiopica* are the major pests affecting Jatropha [79]. Fruit sap sucking predators *Scutellera perplexa* [80] and *Maconellicoccus hirsutus* have recently been investigated in India [81]. These infections caused approximately 60–80% damage to the standing Jatropha crop at different study sites [14, 79–81].

Figure 3. The photographs (A and B) show viral incidence in Jatropha.

Moreover, monocropping could result in the spreading of insects and diseases. In Panzhihua, China Yu and co-workers found 24 species of insects and diseases that are affecting Jatropha

http://dx.doi.org/10.5772/64979

[82]. Wu and co-authors reported eight diseases and seven species of insects on Jatropha in the dry-hot valley of Yunnan Province [83]. Jatropha monoculture expansion may spread insects and diseases.

4.4. Jatropha breeding objectives [72]

- Dry matter (increased fruit and carnal size) accumulation in fruits rather in cost of vegetative growth.

- Higher female flowers ratio per inflorescence for more fruits.
- Flowering and fruit maturity synchronizing for mechanisation of harvesting.
- Bigger seeds and more oil contents.
- More branching to produce more flowers, fruits and seeds.
- Oil quality improvement.
- Development of non-toxic variety for safe use.
- Disease and stress tolerant cultivar development.
- Improve plant architecture for deeper and smaller rooting.

5. Oil extraction challenges

Jatropha seeds contain 40–60% of oil depending on the variety [18, 84]. The first step of oil extraction is the removal of shells from the seeds after collecting the ripe fruits from trees. Seed oil can be extracted manually, mechanically, chemically and enzymatically. The oil extraction process is shown in **Figure 4**. Oil can be extracted by mechanical pressure, solvent extraction and enzymatic degradation of kernel. Mechanical extraction yields about 90% of total oil from the seed [85]. Solvent and enzymatic extraction yield almost 100% of oil from the seed. However, these are complex processes and take long time. Solvent extraction involves handling of large volume dangerous chemicals. Commercially suitable enzyme(s) is still not available for enzymatic extraction of oil from seed kernel [86, 87] till date.

In the mechanical process, a machine is used to exert pressure on seeds for the removal of oil. After cleaning and checking, the seeds are fed into the hopper of the machine. For Jatropha seed 0.41 L of oil is extracted from 1 kg. Mechanical parameters and pretreatment of seeds affect oil yields. The effects of treatment and physical parameters on the oil extraction are shown on **Figure 5**. The amount of oil that can be recovered from the seeds is affected by:

- *Throughput*: It is the amount of seed crashed per hour (kg/h). The higher throughput recovers less amount of oil per kg of seeds, because of short time exposure of seeds to pressure. It can be regulated by altering the turning pace of the screw throughput.

- *Oil point pressure*: It is the amount of pressure necessary to start oil flow from the seeds. If it is possible to reduce the oil point pressure, the oil extraction becomes easier.
- *Pressure*: The more the pressure, the more the oil recovered from the seeds. But oil recovery at high pressure brought more solid particles with oil. It makes the removal of solid particles more difficult. A pressure range of 50–150 bar is considered as the optimum operating pressure for engine-driven oil extraction.
- *Nozzle size*: A smaller pore causes higher pressure and therefore a higher oil yield. An ideal nozzle size is needed for smooth operation.
- *Hull content of the seeds*: Less energy should be used for pressing seeds so that there are no hull fibres in the crude oil. However, it appears like paste inside standard expellers, which sticks to the worm and keeps rotating along with it.

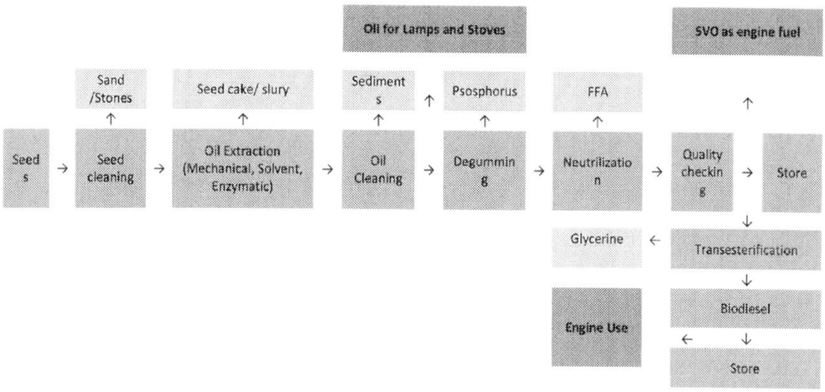

Figure 4. Oil extraction steps and use.

Pre-treatments		Oil Yield	Pressure	Temperature	Throughput	Energy/litter
heating	▲		▼	▲		▼
boiling		▼	▲	▲		▲
flaking		▼		▼		▼
moisture content	▼	▲	▲	▲	▼	▲
hull fraction	▼	▼	▼	▼	▼	▼
Mechanical factors						
RPM	▼	▲	▲		▼	▲
restriction size	▼	▲	▲	▲		▲

Figure 5. Pre-treatment and mechanical factors effect on seed oil recovery.

http://dx.doi.org/10.5772/64979

The mechanical method is easier and less expensive but produces less oil (8–9%). Heat is generated during the process that affects the quality of biodiesel. A high efficient oil recovery (90–98%) technique, solvent extraction, is the most widely used. However, high energy input and toxicity of solvent used are major disadvantage of this technique. Enzyme-based techniques may be the solution [88]. For extraction of oil from *Jatropha* seeds, aqueous enzymatic oil extraction (AEOE) is a promising technique. Plant cell walls are composed of a complex chemical structure. Enzymes that present in the system break cell walls and oil bodies and accelerate oil recovery. This eco-friendly process does not produce volatile organic compounds as atmospheric pollutants. Prolonged reaction time is the major disadvantage associated with AEOE. Moreover, suitable commercial enzyme is not available till date. Winkler et al. [87] studied enzyme supported oil extraction. They used alkaline protease, protease in combination with hemicelluloses and/or cellulose. Alkaline protease treatment produces 86% oil.

Ahmad et al. [89] isolated a bacteria marked MB4, which produces xylanases that enhanced the extraction yield of Jatropha oil. The advantages offered by this process are: protein in the residue can be further processed for other applications, no purified enzyme preparation is needed, and the resulting oil can be used for biodiesel production. Immobilization of lipase has gained immense potential in the biofuel industry mainly to reduce the production costs and to make the method more economical [90].

6. Conversion challenges

There are two types of methods, which are generally used for conversion of hydrocarbon fuels from renewable feedstock. One is the thermochemical process and another is the biochemical process. The thermochemical process is the conversion of biomass to hydrocarbons in the presence of high temperature and pressure. In the biochemical conversion process, biomass is converted into carbohydrates over some steps by the method of fermentation using enzymes or micro-organisms [91]. The thermochemical conversion can be carried out mainly by transesterification, pyrolysis, microemulsion, esterification, gasification, etc. Among these processes, pyrolysis and transesterification are the promising methods, which are used to produce Jatropha biofuel, mainly biodiesel and bio-jet-fuel. Details of these processes are discussed below.

6.1. Transesterification

Transesterification, also called alcoholysis, is the reaction where the oil converts into its corresponding fatty ester [92, 93]. This is a similar process to hydrolysis but here, alcohol is used instead of water. So, transesterification is the organic reaction where one ester transfer into another ester by interchanging the alkoxy moiety. The basic reaction involved in transesterification is shown in **Figure 6**. This reaction is used to decrease the high viscosity of triglyceride. Due to the reversible nature of this reaction, extra alcohol is used to move the equilibrium towards the product. A catalyst is used to promote the reaction rate and the product yield [94]. Two types of catalysts are used in the transesterification reaction. The acid catalyst makes the carbonyl group more reactive by donating a proton while the base catalyst remove a proton from alcohol to make it more reactive [93].

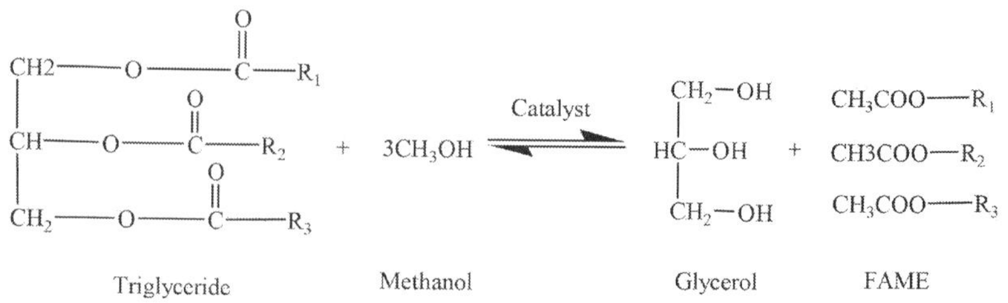

Figure 6. Catalytic transesterification of triglyceride.

The transesterification process of Jatropha oil produces mono fatty acid alkyl esters and glycerol as the by-product. In this process, methanol is the alcohol used due to its low price, low temperature reaction, minimum reaction time and high yield of fatty acid methyl esters [95]. This reaction is affected by several factors, such as molar ratio of glycerides and alcohol, reaction temperature, time, catalyst and also the free fatty acid content and moisture content in the Jatropha seed oil. Generally, the homogeneous base catalysts, NaOH and KOH, are used because of their higher yield and quality fatty acid methyl esters (FAMEs) [96]. However, homogeneous base catalyst for transesterification of Jatropha oil associates some problems. It is very difficult to separate the catalyst from the product and the purification step produces a large amount of alkaline wastewater. Treatment of this water also increases the production cost [97]. Because of the presence of the higher free fatty acid content and the moisture content in Jatropha oil, the base catalyst induce saponification reaction which decrease the production yield [98]. To overcome this problem, an acid catalyst can be used in transesterification of Jatropha vegetable oil, but with the acid catalyst, the reaction requires more oil-methanol molar ratios and the reaction will be very slow [99]. Another possible solution to overcome this problem is a two-step procedure for the treatment of Jatropha oil. First step is esterification of free fatty acid and the second step is transesterification of Jatropha oil triglyceride [100]. But this is also not cost effective. Instead of a homogeneous catalyst, a heterogeneous catalyst is a better option for transesterification of higher FFA containing vegetable oil because it can result in good conversion and a high yield of FAME with optimum reaction conditions [101]. Many researchers recommended the heterogeneous catalysts for transesterification of vegetable oil in their investigation [99, 102–111].

6.2. Pyrolysis/thermal cracking

Pyrolysis or cracking of vegetable oil is one of the promising routes to produce biofuel (biodiesel and bio-jet-fuel) because of the straight chain alkanes and high cetane number of the product [112–114]. Pyrolysis is defined as the thermal conversion of vegetable oils by heat in absence of air in favour of a catalyst into alkanes, alkenes, aromatics, carboxylic acids and little amounts of gaseous products [115]. When compared with transesterification of Jatropha vegetable oil for producing biofuel, the hydrocracking process needs a higher energy and temperature (280–300°C) [116] but the pyrolysed products have a higher cetane number and oxidation stability. Catalytic pyrolysis increases the yield of product by breaking large molecules, and also improves the quality of the product (biofuel).

Catalytic cracking of vegetable oil is a three-step mechanism. First one is the removal of oxygen by C=O bond hydrogenation, then C—O bond rupture and finally C—C bond breaking with the aid of a catalyst. The cracking reaction may occur by different routes such as: hydrodeoxygenation, decarboxylation and decarbonylation, which are shown in **Figure 7**. Each route produces shorter and straight chain hydrocarbons with the removal of water, CO, CO_2, etc. Catalytic cracking of Jatropha oil in the presence of different heterogeneous catalysts shows better result. The activities of different catalysts in cracking of Jatropha oil for producing biofuel are investigated under different conditions.

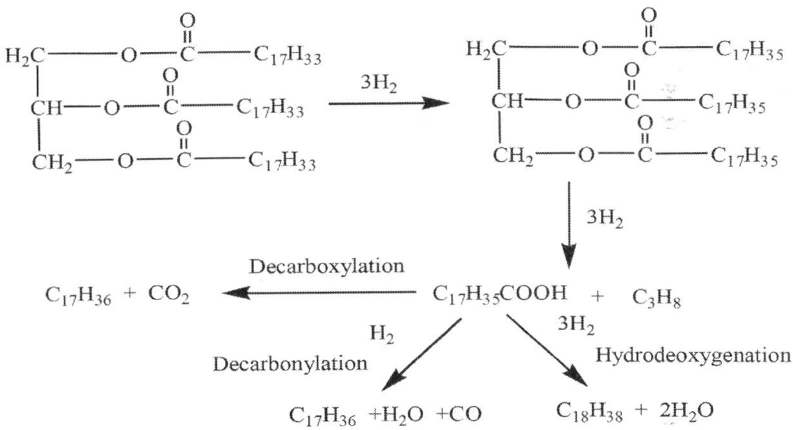

Figure 7. Catalytic cracking of triglyceride.

Today, the conventional catalysts which are used to crack vegetable oil are mainly sulphided silica, alumina-supported Ni-Mo, Co-Mo or Ni-W [117–122]. So using these catalysts needs the addition of sulphur containing compounds which are responsible for the production of sulphur residue with the end product and causes greenhouse emission like H_2S and also corrosion problem. Moreover, the noble metal catalysts are more active but the disadvantages with them are high price and rarity, as well as they are responsible for catalyst poisoning and impurities [123]. Recently some novel metal catalysts have been developed for hydrocracking of Jatropha oil. $PtPd/Al_2O_3$ and sulphided $NiMoP/Al_2O_3$ at 330–390°C temperature and 3 MPa pressure [124]; Ni/H-ZSM-5 [125]; sulphided form of $Co\text{-}Mo/Al_2O_3$, $Ni\text{-}W/SiO_2\text{–}Al_2O_3$; and $Ni\text{–}Mo/Al_2O_3$ have been developed for producing biofuel from Jatropha oil. Among the other catalytic systems, the homogeneous solid base catalyst is more beneficial for hydroprocessing Jatropha oil because of its reusability, low cost and high selectivity [126]. But the base catalyst produces soap with FFA and it needs a high purity Jatropha oil which is the main obstacle [127].

The major problems with Jatropha bio-jet-fuel are its freezing point and low yield. The freezing point of Jatropha hydrocarbon produced by catalytic cracking is higher than zero degrees whereas the freezing point of conventional jet fuel is less than −40°C [128–130]. To overcome this problem, a new catalyst system has to be designed for hydroprocessing Jatropha oil. There are many advantages of using metal supported on microporous zeolite catalysts for hydrocracking Jatropha oil due to the versatile characteristics of zeolite [131]. Zeolite catalysts have ion-exchange abilities with high porosity, broad surface area and concurrent-base character [132]. It can solve the diffusion limitation and increase the production yield due to its unique structure. For cracking reactions, high temperature (280–300°C) and pressure are necessary, which increase the production cost. So it is most important to select and improve the non-sulphided metal supported zeolite catalyst as well as the selection of optimum conditions (temperature, pressure and reaction time) for hydrocracking of Jatropha oil to produce diesel and jet-fuel range hydrocarbons.

The mostly used zeolites for cracking reaction are Zeolite Y, Meso-Y, H/ZSM-5, Na/ZSM-5, Ni/ZSM-5, Ru/ZSM-5, Zeolite β, SAPO-11, SAPO-34, Ni/SAPO-11, ultrastable-Y zeolite (USYZ), rare earth-Y zeolite (REY), Bentonite P-140, SBA-15, MCM-41, etc. [133–141]. The catalytic activity of several zeolites depends on its structure, shape and size of the substrate, polarity and the reaction parameters such as temperature, pressure, time, etc. A high reaction temperature shows the high activity of zeolite. Some investigations showed the cracking result of different vegetable oils with different supported zeolite catalysts. The cracking of sunflower oil with the CaO/SBA-14 catalyst under 160°C temperature and 5 hours reaction time showed 95% biodiesel yield [142], whereas SAPO-11 showed 83–90% yield on the treatment of palm oil under 7 wt% Ni loading, 493°C temperature and 2 MPa H_2 pressure for 6 hours [117]. On the other hand, Mesoporous-Y zeolite showed 40.5% bio-jet-fuel yield at 400°C temperature for 3 h cracking [138].

7. Use challenges

7.1. Crude oil use

Non-edible Jatropha oil is the promising alternative as bio-energy for diesel and jet engine. But, due to the high viscosity, large molecular mass and chemical structure of Jatropha oil, it cannot be used directly to the compression ignition (C.I.) engines for long time. From the study it is very clear that using Jatropha oil directly can cause some problems to the engine [143–145]. The main problems are pumping, burning and atomization with the injector system of compression ignition engine. Unburned Jatropha oil can distort the injector nozzle, stick to the ring and damage the cylinder of the diesel engine [146, 147]. It also makes the emission of particulate substance such as smoke, unburned hydrocarbon and carbon which affect human health and pollute the environment [146]. Hence, the better way to use Jatropha oil directly to the diesel engine is by the reduction of its viscosity by means of blending *Jatropha curcas* oil with diesel oil in different proportions.

There have been some investigations to the use of Jatropha oil blends in diesel engines. At different proportions of Jatropha oil and diesel blends, it shows different oil properties and engine performance. **Table 2** [143, 146] shows the different properties of diesel and diesel/oil blends in different proportions. From the data it is clear that the viscosity and density gradually decreased by decreasing the amount of crude Jatropha oil in the diesel/oil blend. It is observed that the oil blend containing more than 20% Jatropha oil have high viscosity compared to the maximum viscosity limit of the diesel engine. So, the viscosity needs to be reduced more to make the blend usable for diesel engine. Where the permeable limit of fuel viscosity for diesel engine is maximum 870 kg/m^3, the viscosity of oil/diesel blend for 2.6:97.4 proportion is 868 kg/m^3 which is very near to the maximum limit. But, addition of Jatropha oil with diesel decreases the exhaust gas temperature. So, only a small portion (about 2.6%) of Jatropha oil can be used with diesel fuel as the ignition-accelerator additives.

Jatropha oil:diesel	Pure Jatropha	70:30	50:50	30:70	20:80	2.6:97.4*	Pure diesel
Density (kg/m^3)	917	900	891.5	881	876	868	866
Viscosity (cSt) at 30°C	36.9	23.45	17.5	9.8	6.9	5.9	5.7
Exhaust gas temperature (°C)	210	230	245	252	260	275	280
Flash point (°C)	99	–	94	–	90	88	86
Pour point (°C)	−3	–	+6	–	+12	+15	+15
Specific gravity	0.918	–	0.892	–	0.877	0.869	0.867
Calorific value (Mg/kg)	42	–	43		44.2	45.2	45.9

Table 2. Different properties of Jatropha oil-diesel blends.

7.2. By-product: seed cake and glycerine

Jatropha curcas seed oil is the most offering alternative source of feedstock for biofuel industries. From Jatropha biofuel plant, some by-products are produced. The main by-products are seed cake and glycerine. To make sustainable and economically viable industry, it is necessary to use by-products properly. But there are some problems to recover and use these by-products directly. Jatropha seed oil and seed cake contain 58–64% protein with high nutritional value [148]. But, they also contain some toxic ingredients such as: phorbol esters, lectins, trypsin inhibitors, phytate, saponins, tannins, etc., which make them non-edible for human, fish, goat and mice. Also, the process to recover glycerine from biofuel is not easy. So, proper steps should be taken for using these by-products.

Seed cake: Due to the presence of toxic components, Jatropha seed cake cannot be used as a feed meal for human, fish, goat, chicken and rat. Phorbol ester is responsible for cancer, skin irritation, tumour promotion and purgation [149]. Lectins cause haemorrhagic spot and trypsin inhibitor causes adverse effect in monogastrics [150]. So, before using seed cake as an animal feed it needs to be detoxified. Lectins and trypsin inhibitors are heat sensitive and they decrease during biofuel processing at about 160°C. But phorbol ester decreases only 5% at high temperature. By increasing the digestible organic matter and metabolizable energy, heat treatment increases the nutritive value of *Jatropha curcas* seed meal [151]. So, the main issue is to neutralize the toxic phorbol ester from Jatropha seed meal before use. **Table 3** [151, 152] shows the different toxic components present in Jatropha and soybean meal.

Toxic component	Toxic variety	Non-toxic variety	Soybean meal	Result of heat treatment
Phorbolester (mg/g kernel)	2.70	0.11	–	5% decrease
Lectins (mg/ml assay)	102	51	0.32	Decrease
Trypsin inhibitors (mg/g meal)	21.1	26.5	3.9	Decrease
Phytate (% in meal)	10.1	8.9	1.5	No change
Saponins (%)	2.6	3.4	4.7	No change

Table 3. Toxic components present in Jatropha curcas oil and seed cake.

There have been several methods to detoxify the toxic phorbol ester from Jatropha oil and seed kernel. Some of them are fungal isolation, γ-irradiation, adsorption, plasma application, ozonation, etc. During the Jatropha oil refining and purification processes about 55% phorbol ester can be removed by bleaching and de-acidification step but degumming and deodorizing step cannot remove phorbol ester [151]. About 44% phorbol ester can be removed by chemical treatment by using $NaHCO_3$, whereas the combination of chemical treatment and heat treatment can remove 56% and the combination of chemical treatment and ozonation can remove up to 75% phorbol ester from *Jatropha curcas* seed oil and seed meal [153]. Gamma irradiation can remove 71.35% of phorbol esters at 50 kGy absorption dose. But this method takes long time, high temperature and high dose of gamma irradiation which is not economic. On the other hand, fungal isolation can remove 97.8% of phorbol esters from Jatropha seed and oil [154]. Among all the processes, the adsorption process is more effective to detoxify Jatropha oil phorbol ester and this process can remove up to 99.5% phorbol ester present in the Jatropha seed oil. Un-detoxified oil contains 2.70 mg/g phorbol ester. After detoxification by the adsorption process, Jatropha oil contains 0.02 mg/g phorbol ester, which is lower than permissible limit (0.09 mg/g) [152, 155].

In the adsorption process, the one-time adsorption carried out with 3.2% (w/v) bentonite 200 as the adsorbent at 32°C temperature, 100 rpm stirring rate and 15 min adsorption time can remove 98% phorbol ester. Two-time adsorption with 0.8% (w/v) bentonite 200 under same conditions can remove 99.50% phorbol ester [152].

Glycerine: Glycerine is the major by-product of Jatropha biofuel processing plant. So, the recovery and proper use of this product is more beneficial for the Jatropha biofuel project. But, the recovery of glycerine from biofuel is not an easy process. Traditionally, glycerine is recovered from biofuel by washing with water. In this process, water is mixed with the biodiesel, agitating the mixture gently, allowing the mixture to separate the several phases and finally glycerine is extracted from the water phase [156]. But, this process is not favourable because water causes many problems when used to wash biofuel. Washing of biofuel needs the use of deionized water, produce large amount of wastewater. This water can degrade the biofuel by hydrolysis and it can increase the processing time with multiple drying, multiple washes and water-biodiesel separation steps [157]. The suitable alternative to recovery glycerine from the biofuel is the adsorption with a bed of ion exchange resin [158]. But this process is not established yet. Recovery of glycerine from biofuel by ion exchange resin is the combination of four steps: filtration, physical adsorption, ion exchange and removal of soap by glycerine affinity.

8. Alternative of Jatropha

There are many potential non-edible oil-rich plants in almost every country (mostly tropical and subtropical). Mostly they are wild and naturally growing. They may or may not have yet been explored for oil producing potential. In India, 11 tree species (*Garcinia indica, Azadirachta indica, Hevea brasiliensis, Calophylluum inophyllum, Madhuca indica, Mesua ferrea, Mallotus philippines, Ricinus communis, Pongamia glabra, Salvadora* and *Shorea robusta*) are largely distributed that have biodiesel producing potentials [17]. *Camelina sativa, Gossypium hirsutum, Cynara cardunculus, Abutilon muticum, Simmondsia chinensis, Passiflora edulis, Aleurites moluccana, Carnegiea gigantea, Pachira glabra, Croton megalocarpus* and *Terminalia bellirica* have high content of non-edible oil in their seeds [159]. These plants may also be explored for their suitability to meet the blending requirements rather than focusing on a single candidate (Jatropha).

9. Lifecycle assessment (LCA) of the Jatropha biofuel

Environmental impact of biofuels is determined by lifecycle assessment [160, 161]. The environmental flows throughout the lifecycle stages of a product or services are evaluated by LCA [162]. The four interacting individual phases—scoping, inventory analysis, impact assessment and interpretation—are the basic of LCA study (ISO, 2006). International Organization for Standardization (ISO) regulates it as 14040:2006 and 14044:2006 standards.

Jatropha cultivation, oil extraction, conversion of seed oil into biodiesel and biodiesel use are four major phases of the Jatropha biodiesel system. **Figure 8** shows the system boundaries of the Jatropha biodiesel. The flow processes, inputs and outputs in the Jatropha biodiesel system are summarized in **Table 4**. Energy balance, global warming potential (GWP), and land-use impact, net energy gain (NEG), net energy ratio (NER), ecosystem structural quality (ESQ) and ecosystem functional quality (EFQ) are most relevant impact categories of LCA systems [161, 163]. The data on LCA of Jatropha biodiesel is not sufficient though there are some reports on the LCA methodology [163–167] and LCA [168, 169] for Jatropha biodiesel.

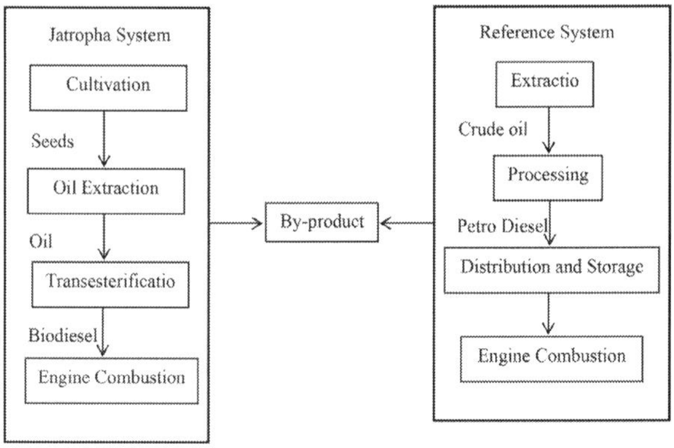

Figure 8. System boundaries of the Jatropha biodiesel system.

Phase	Processes	Inputs	Products	By-products
(a) Cultivation				
Seedling production	– Seeding in nurseries	– Energy, machines, infrastructure	– Air emissions	–
Plantation establishment	– Planting cuttings – Transplantation – Direct seeding – Land preparation	– Energy, machines, auxiliaries	– Air emissions – Standing biomass – Seeds	–
Plant management	– Pruning, canopy management, fertilising, irrigation, harvesting	– Energy, machines, infrastructure auxiliaries	– Air emissions	– Woody cuttings
(b) Oil extraction	– Mechanical – Solvent-based	– Energy, machines, infrastructure, auxiliaries	– Air emissions – Raw oil – Wastewater	– Seed cake – Fruit shells
(c) Transesterification	– Catalysis – Transport of biodiesel	– Energy, infrastructure	– Air emissions – Biodiesel – Wastewater	–
(d) Use	– Combustion		–Air emissions	–

Table 4. Flow processes, inputs and outputs of the Jatropha biodiesel system.

Jatropha is believed to be a suitable crop for wastelands with low inputs. However, an industry cannot be established depending on unreliable feedstock supply based on low-input agriculture. Thus, fertilizers, irrigation and pesticide use will be unavoidable in commercial Jatropha production. Furthermore, transesterification, the major conversion technology, is the major contributor to GHG emissions and energy consumption. To be a suitable alternative of petro-diesel in terms of mitigation of climate change, Jatropha biodiesel needs to be supported by lifecycle data. The environmental benefits of Jatropha biodiesel in comparison to petro-diesel is in doubt [169, 170]. Following measures can improve lifecycle performance of Jatropha biodiesel:

a. Use of bio-fertilizers for Jatropha cultivation and optimization of inputs

b. Irrigation by consuming minimization and/or low-energy

c. Transesterification processes optimization

d. Water footprint consideration

10. Contribution and expectation from Jatropha

With enormous potentials on social, agricultural, environment, sustainable energy production and industrial fronts, Jatropha is attracting interest from researchers and policy makers. Research showed that Jatropha seed has 40–60% oil content with a productivity of 0.1–12 tons per ha [32, 63]. Yields of both fruits and oil depend on species, accession, soil, climate and agronomy [66, 171]. A seed yield of 4–5 mg/ha/yr is expected for the commercial viability of the Jatropha-based biofuel program. With 30–35% oil content and an average seed yield of 3.75 mg ha/yr Jatropha is economically more beneficial to the average yield profile of soybeans and rapeseed [19].

However, the actual seed production of Jatropha in field conditions was poor than expected [20–22]. The reported yields of Jatropha under field conditions in India, Belgium, South Africa and Tanzania are 0.5–1.4 mg/ha/yr, 0.5 mg/ha/yr, 0.35 mg/ha/yr and 2 mg/ha/yr, respectively [23]. The less productivity is because of unavailability of suitable high yielding varieties, large-scale plantation without evaluating the genetic potential of planted materials, consideration of Jatropha as no/low input crop and lack of knowledge on agronomy.

Jatropha seed cake is an excellent source of protein. To add commercial value it is expected to utilize the press cake as an animal feed protein supplement. The presence of toxic compounds hinders its utilization for this purpose. The discovery of non-toxic Jatropha varieties and the detoxification process of toxins are an advantage. Many biological active chemical compounds are extracted from bark, leaves and roots that are expected to be used in the pharmaceutical industry. However, toxicity must be studied before the use of Jatropha products as therapeutic agents or medicines.

Oil from Jatropha has been used as cooking fuel and in soap and cosmetic manufacturing in ancient times. It is reported that Jatropha oil can be utilized as a fuel with diesel engine directly or with slight modification of the engine. Jatropha oil can be converted to biodiesel by chemical reaction called transesterification. This biodiesel can be used by blending with petro-diesel. High viscosity is the major problem of using Jatropha seed oil as fuel.

11. Technological intervention for Jatropha improvement

Jatropha bears multi-dimensional potentials [172]. But it is still behind to compete in commercialization. High yielding Jatropha varieties are not found yet. Lack of agronomic knowledge, unawareness or knowledge gap of farmers and the common belief that no impute crop renders Jatropha's productivity [2, 173]. It requires extensive searching for natural germplasms and systematic breeding programs for genetic improvement. The success of breeding depends on the availability of diverse germplasms [62]. However, some study showed narrow genetic diversity among the worldwide population [174–179]. It limits the success of conventional breeding. Inter-species and inter-generic hybridization, haploid breeding (anther, pollen, ovary culture), somaclonal and germaclonal variation and mutation breeding are biological techniques that have proven records for variation creation and breeding of many important crops. For Jatropha there are some reports on organogenesis only. Other technologies are still unexplored. Genetic resources technology, i.e. marker-assisted selection (MAS), molecular breeding, genomic selection, genome-wide association studies and genetic engineering have been used with confident for many crop breeding programs. Jatropha genomic technology research is still far behind in comparison to other important crops though some reports are available on that [180–186]. Now it is time to explore full potentials of Jatropha by using modern technologies.

12. Conclusions

It is very crucial to find reliable renewable energy sources for healthy economy and environment. Jatropha is a promising candidate of renewable energy as it has interesting characteristics and non-competition with food. A number of big projects has been launched and completed. However, the result is disappointing. Unavailability of good commercial variety, considering low impute and disease resistance crop, knowledge gap, lack of basic research and theoretical assumption mostly without scientific and technological backup are the major reasons of the failure. Thus, Jatropha biofuel industry is confronting a number of challenges.

Efficient oil extraction methods form Jatropha seeds need to be explored. Mechanical pressing is commonly used but it is poor yielding and also affects the oil quality. Solvent extraction needs to use many hazardous solvents. Enzymatic process is good but has a slow reaction rate. It needs to find/develop suitable enzymes and to increase the reaction rate. Conversion of crude Jatropha oil to biofuel (biodiesel or jet fuel) is another challenge. Transesterification and thermal cracking are commonly used. It needs to develop environmentally friendly catalyst with high conversion efficiency. By-product utilization can reduce waste management cost and also add economic value. Seed oil cake and glycerine are the main by-products of the Jatropha industry. It needs appropriate method of glycerine recovery and detoxification method of seed cake for safe use.

Plant breeding in application of biotechnology is the gateway of crop improvement (yield and quality). Diverse germplasm is the basis of a breeding program. Accumulation and utilization of specialised but scattered knowledge is important for Jatropha improvement. Major Jatropha cultivating countries—India, China, Malaysia, Indonesia, Brazil, Mexico and South Africa—can establish an international organization. To design a strategic breeding program for Jatropha improvement, the researchers can share their learning gained by several years of experience.

Biotechnology application in Jatropha breeding is far behind as compared to some other crop. Somaclonal and germaclonal variation are created by *in vitro* mutagenesis, *in vitro* micropropagation, anther and microspore culture, ovary and ovule culture, protoplast culture, nucleolus culture, endosperm culture, and somatic embryogenesis. It needs to explore them because these techniques have proven record for crop improvement. There are a few works on Jatropha genome though; it is still far behind in comparison to other agricultural systems.

Though there are some biotechnology studies, however, Jatropha genome work is far behind than the model and other agricultural systems. Researchers require a high density linkage map for the determination of the association of markers with high oil yield. The *J. curcas* genome is very small (ca.400 Mb). Thus, marker-assisted selection, genome-wide association studies (GWAS) and genomic selection (GS) could be even more attractive. A well-assembled reference genome of Jatropha is indispensable for these applications. Increasing female flowers in inflorescence, reducing toxins and increasing resistance are on priority.

Author details

M. Moniruzzaman[1*], Zahira Yaakob[1], M. Shahinuzzaman[1], Rahima Khatun[1] and A.K.M. Aminul Islam[2]

*Address all correspondence to: monirbge@gmail.com

1 Department of Chemical and Process Engineering, Faculty of Engineering and Built Environment, Universiti Kebangsaan Malaysia, Selangor, Malaysia

2 Department of Genetics and Plant Breeding, Bangabandhu Sheikh Mujibur Rahman Agricultural University, Gazipur, Bangladesh

References

[1] Mabberley, D.J., Mabberley's plant-book: a portable dictionary of plants, their classifications, and uses. 2008. Cambridge University Press, Cambridge CB2 8BS, United Kingdom.

[2] Edrisi, S.A., et al., Jatropha curcas L.: a crucified plant waiting for resurgence. Renewable and Sustainable Energy Reviews, 2015. 41: pp. 855–862.

[3] Carels, N., Towards the domestication of Jatropha: the integration of sciences, in Jatropha, challenges for a new energy crop. 2013. Springer, New York, pp. 263–299.

[4] Sabandar, C.W., et al., Medicinal property, phytochemistry and pharmacology of severalJatrophaspecies (Euphorbiaceae): a review. Phytochemistry, 2013. 85: pp. 7–29.

[5] Dias, L., R. Missio, and D. Dias, Antiquity, botany, origin and domestication of Jatropha curcas (Euphorbiaceae), a plant species with potential for biodiesel production. Genetics and Molecular Research, 2012. 11(3): pp. 2719–2728.

[6] de Sant'Anna, Q., et al., Toward the metabolomics of Jatropha curcas, in Jatropha, challenges for a new energy crop. 2013. Springer, New York, pp. 577–600.

[7] Kumar, A. and S. Sharma, An evaluation of multipurpose oil seed crop for industrial uses (Jatropha curcasL.): a review. Industrial crops and products, 2008. 28(1): pp. 1–10.

[8] Vollmann, J. and M. Laimer, Novel and traditional oil crops and their biorefinery potential. Bioprocessing Technologies in Biorefinery for Sustainable Production of Fuels, Chemicals, and Polymers, 2013: pp. 47–60.

[9] Tikkoo, A., S. Yadav, and N. Kaushik, Effect of irrigation, nitrogen and potassium on seed yield and oil content of Jatropha curcas in coarse textured soils of northwest India. Soil and Tillage Research, 2013. 134: pp. 142–146.

[10] Meher, L., et al., Jatropha curcas as a renewable source for bio-fuels—a review. Renewable and Sustainable Energy Reviews, 2013. 26: pp. 397–407.

[11] Silitonga, A., et al., A global comparative review of biodiesel production from jatropha curcas using different homogeneous acid and alkaline catalysts: study of physical and chemical properties. Renewable and Sustainable Energy Reviews, 2013. 24: pp. 514–533.

[12] Koh, M.Y. and T.I.M. Ghazi, A review of biodiesel production from Jatropha curcas L. oil. Renewable and Sustainable Energy Reviews, 2011. 15(5): pp. 2240–2251.

[13] Mofijur, M., et al., Prospects of biodiesel from Jatropha in Malaysia. Renewable and Sustainable Energy Reviews, 2012. 16(7): pp. 5007–5020.

[14] Singh, B., et al., Agro-technology of Jatropha curcas for diverse environmental conditions in India. Biomass and bioenergy, 2013. 48: pp. 191–202.

[15] Atabani, A., et al., Investigation of physical and chemical properties of potential edible and non-edible feedstocks for biodiesel production, a comparative analysis. Renewable and Sustainable Energy Reviews, 2013. 21: pp. 749–755.

[16] Abhilash, P., et al., Revisited Jatropha curcas as an oil plant of multiple benefits: critical research needs and prospects for the future. Environmental Science and Pollution Research, 2011. 18(1): pp. 127–131.

[17] Karmakar, A., S. Karmakar, and S. Mukherjee, Properties of various plants and animals feedstocks for biodiesel production. Bioresource technology, 2010. 101(19): pp. 7201–7210.

[18] Ong, H., et al., Comparison of palm oil, Jatropha curcas and Calophyllum inophyllum for biodiesel: a review. Renewable and Sustainable Energy Reviews, 2011. 15(8): pp. 3501–3515.

[19] Singh, K., et al., Jatropha curcas: A ten year story from hope to despair. Renewable and Sustainable Energy Reviews, 2014. 35: pp. 356–360.

[20] Singh, B., et al., The field performance of some accessions of Jatropha curcas L.(Biodiesel Plant) on degraded sodic land in North India. International journal of green energy, 2013. 10(10): pp. 1026–040.

[21] Everson, C.S., M.G. Mengistu, and M.B. Gush, A field assessment of the agronomic performance and water use of Jatropha curcas in South Africa. Biomass and Bioenergy, 2013. 59(0): pp. 59–69.

[22] Bailis, R. and G. Kavlak, Environmental implications of Jatropha biofuel from a Silvi-Pastoral production system in Central-West Brazil. Environmental Science & Technology, 2013. 47(14): pp. 8042–8050.

[23] Kant, P. and S. Wu, The extraordinary collapse of Jatropha as a global biofuel. Environmental Science & Technology, 2011. 45(17): pp. 7114–7115.

[24] Del Greco, G.V. and L. Rademakers, The jatropha energy system: an integrated approach to decentralized and sustainable energy production at the village level. Ingegneria Senza Frontiere (Engineers Without Borders–Italy). Accessed on-line, 2008. 22.

[25] Singh, R., et al., SPRERI experience on holistic approach to utilize all parts ofJatropha curcasfruit for energy. Renewable Energy, 2008. 33(8): p. 1868-1873.

[26] Liberalino, A.A.A., et al., Jatropha curcas L. seeds: chemical analysis and toxicity. Arquivos de Biologia e Tecnologia, 1988. 31(4): pp. 539–550.

[27] Gandhi, V., K. Cherian, and M. Mulky, Toxicological studies on ratanjyot oil. Food and Chemical Toxicology, 1995. 33(1): pp. 39–42.

[28] Sharma, G., S. Gupta, and M. Khabiruddin. Cultivation of Jatropha curcas as a future source of hydrocarbon and other industrial products. In Biofuels and industrial products from Jatropha curcas-Proceedings from the Symposium Jatropha. 1997.

[29] Wink, M., et al., 4.1 Phorbol Esters of J. curcas-Biological Activities and Potential Applications. 1997.

[30] Makkar, H. and K. Becker, Potential of Jatropha curcas seed meal as a protein supplement to livestock feed, constraints to its utilization and possible strategies to overcome constraints. Proceeding Jatropha, 1997. 97: p. 37-40.

[31] Openshaw, K., A review ofJatropha curcas: an oil plant of unfulfilled promise. Biomass and Bioenergy, 2000. 19(1): pp. 1–15.

[32] Rivera Lorca, J. and J.K. Vera, Chemical composition of three different varieties of J. curcas from Mexico. Biofuels and industrial products from Jatropha curcas. Dbv-Verlag für die Technische Universität Graz, Graz, Austria, 1997: pp. 47–55.

[33] Akintayo, E., Characteristics and composition of Parkia biglobbossa and Jatropha curcas oils and cakes. Bioresource Technology, 2004. 92(3): pp. 307–310.

[34] Kywe, T.T. and M.M. Oo, Production of biodiesel from Jatropha oil (Jatropha curcas) in pilot plant. Proceeding of World Academy of Science and Technology, 2009. 38: pp. 481–487.

[35] Sotolongo, J.A., et al., Jatropha curcas L. as a source for the production of biodiesel: A Cuban experience. in 15th European biomass conference and exhibition, Berlin, Germany. 2007.

[36] Vyas, D. and R. Singh, Feasibility study of Jatropha seed husk as an open core gasifier feedstock. Renewable Energy, 2007. 32(3): pp. 512–517.

[37] Augustus, G., M. Jayabalan, and G. Seiler, Evaluation and bioinduction of energy components of Jatropha curcas. Biomass and Bioenergy, 2002. 23(3): pp. 161–164.

[38] Kumar, G., et al., Lignocellulose biohydrogen: practical challenges and recent progress. Renewable and Sustainable Energy Reviews, 2015. 44: pp. 728–737.

[39] Kumar, G., et al., Modeling and optimization of biohydrogen production from de-oiled Jatropha using the response surface method. Arabian Journal for Science and Engineering, 2015. 40(1): pp. 15–22.

[40] Kumar, G., et al., Comparative evaluation of hydrogen fermentation of de-oiled Jatropha waste hydrolyzates. International Journal of Hydrogen Energy, 2015. 40(34): pp. 10766–10774.

[41] Lopes, S.L., et al., Bioconversion of Jatropha curcas seed cake to hydrogen by a strain of Enterobacter aerogenes. Fuel, 2015. 139(0): pp. 715–719.

[42] Pratt, J., et al., Malawi Agroforestry Extension Project Marketing & Enterprise Program, Main Report. Malawi Agroforestry, 2002: p. 139.

[43] Henning, R., The Jatropha Manual. A guide to the integrated exploitation of the Jatropha plant in Zambia. Germany: Deutsche Gesellschaft für Technische Zusammenarbeit GTZ, 2000.

[44] Messemaker, L., Assessment of the Jatropha value chain and its potential for pro poor biofuel development in Northern Tanzania. 2008, MSc thesis International development studies at the Faculty of Geosciences, Utrecht University, 2008.

[45] Benge, M., Assessment of the potential of Jatropha curcas,(biodiesel tree,) for energy production and other uses in developing countries. USAID Report, 2006.

[46] Brittaine, R. and N. Lutaladio, Jatropha: a smallholder bioenergy crop: the potential for pro-poor development. Vol. 8. 2010: Food and Agriculture Organization of the United Nations (FAO).

[47] Zhang, X.P., et al., Chemical constituents of the plants from genus Jatropha. Chemistry & Biodiversity, 2009. 6(12): pp. 2166–2183.

[48] Ravindranath, N., et al., Jatrophenone, a novel macrocyclic bioactive diterpene from Jatropha gossypifolia. Chemical and Pharmaceutical Bulletin, 2003. 51(7): pp. 870–871.

[49] Bhagat, R., et al., Anti-inflammatory activity of Jatropha gossypifolia L. leaves in albino mice and Wistar rat. Journal of Scientific and Industrial Research, 2011. 70(4): pp. 289–292.

[50] Apu, A.S., et al., Study of pharmacological activities of methanol extract of Jatropha gossypifolia fruits. Journal of Basic and Clinical Pharmacy, 2012. 4(1): p. 20.

[51] Reena, P., Evaluation of antimicrobial and anti-inflammatory activities of bark of Jatropha gossypifolia. World Journal of Science and Technology, 2011. 1(10).

[52] Oduola, T., et al., Mechanism of action of Jatropha gossypifolia stem latex as a haemostatic agent. European Journal of General Medicine, 2005. 2(4): pp. 140–143.

[53] Singh, D. and A. Singh, The toxicity of four native Indian plants: effect on AChE and acid/alkaline phosphatase level in fishChanna marulius . Chemosphere, 2005. 60(1): pp. 135–140.

[54] Feitosa, C., et al., Acetylcholinesterase inhibition by somes promising Brazilian medicinal plants. Brazilian Journal of Biology, 2011. 71(3): pp. 783–789.

[55] Apu, A.S., et al., Anti-diarrheal Potential of Jatropha gossypifolia (Linn.). Journal of Medical Sciences, 2012. 12(8): pp. 274–279.

[56] Silva, S.d.N., et al., Antispasmodic effect of Jatropha gossypiifolia is mediated through dual blockade of muscarinic receptors and Ca2+ channels. Revista Brasileira de Farmacognosia, 2011. 21(4): pp. 715–720.

[57] Félix-Silva, J., et al., Jatropha gossypiifolia L.(Euphorbiaceae): a review of traditional uses, phytochemistry, pharmacology, and toxicology of this medicinal plant. Evidence-Based Complementary and Alternative Medicine, 2014. 2014 1–32.

[58] Abreu, I.C., et al., Hypotensive and vasorelaxant effects of ethanolic extract from Jatropha gossypiifolia L. in rats. Fitoterapia, 2003. 74(7): pp. 650–657.

[59] Shahwar, D., et al., Antioxidant activities of the selected plants from the family Euphorbiaceae, Lauraceae, Malvaceae and Balsaminaceae. African Journal of Biotechnology, 2010. 9(7): pp. 1086–1096.

[60] Kharat, A., A. Dolui, and S. Das, Free radical scavenging potential of Jatropha gossypifolia. Asian Journal of Chemistry, 2011. 23(2): pp. 799–801.

[61] Gübitz, G.M., M. Mittelbach, and M. Trabi, Exploitation of the tropical oil seed plant Jatropha curcas L. Bioresource Technology, 1999. 67(1): pp. 73–82.

[62] Heller, J., Physic nut. Jatropha curcas L. Promoting the conservation and use of underutilized and neglected crops. 1. Roma: IBPGR, 1996.

[63] Achten, W., et al.,Jatropha bio-diesel production and use. Biomass and Bioenergy, 2008. 32(12): pp. 1063–1084.

[64] Ghosh, Arup, J. S. Patolia, D. R. Chaudhary, Jitendra Chikara, S. N. Rao, Dheerendra Kumar, G. N. Boricha, and A. Zala. "Response of Jatropha curcas under different spacing to Jatropha de-oiled cake." In Expert seminar on Jatropha curcas L. Agronomy and genetics, pp. 26–28. 2007.

[65] Gopinathan, M.C. and R. Sudhakaran, Biofuels: opportunities and challenges in India, in Biofuels. 2011, Springer, New York, pp. 173–209.

[66] Weyerhaeuser, H., et al., Biofuels in China: an analysis of the opportunities and challenges of Jatropha curcas in Southwest China. World Agroforestry Centre, ICRAF Working Paper, 2007. 53.

[67] Zhang, G.-w., Y. Peng, and M. Huang, Existing Problems and Countermeasures for Jatropha curcas Industrialization in China [J]. Journal of Anhui Agricultural Sciences, 2009. 8.

[68] YU, B., et al., The Current Situation and Countermeasures of Jatropha curcas L. in Sichuan Province. Sichuan Forestry Exploration and Design, 2007. 3: p. 005.

[69] Fei, S., X. Chen, and Y. He, Prospects of studies on Jatropha curcas biodiesel in Sichuan. Biomass Chemical Engineering, 2006. 40(12): pp. 193–139.

[70] Terren, M., et al., Attempted cultivation of Jatropha curcas L. in the Lower Senegal River Valley: story of a failure. Tropicultura, 2012. 30(4): pp. 204–208.

[71] Liu, X., et al., Risk management for Jatropha curcas based biodiesel industry of Panzhihua Prefecture in Southwest China. Renewable and Sustainable Energy Reviews, 2012. 16(3): pp. 1721–1734.

[72] Moniruzzaman, M., Z. Yaakob, and R. Khatun, Biotechnology for Jatropha improvement: a worthy exploration. Renewable and Sustainable Energy Reviews, 2016. 54: pp. 1262–1277.

[73] Ye, M., et al., Current situation and prospects of Jatropha curcas as a multipurpose tree in China. Agroforestry systems, 2009. 76(2): pp. 487–497.

[74] WeiGuang, W. and L. NuYun, Forest-based bio-diesel development: target, current status and challenges. Scientia Silvae Sinicae, 2009. 45(11): pp. 141–147.

[75] Ariza-Montobbio, P. and S. Lele, Jatropha plantations for biodiesel in Tamil Nadu, India: viability, livelihood trade-offs, and latent conflict. Ecological Economics, 2010. 70(2): pp. 189–195.

[76] Patolia, J. S., Arup Ghosh, Jitendra Chikara, D. R. Chaudhary, D. R. Parmar, and H. M. Bhuva. "Response of Jatropha curcas grown on wasteland to N and P fertilization." In Expert seminar on Jatropha curcas L. Agronomy and genetics, pp. 26–28. 2007.

[77] Daniel, J., Jatropha Oilseed Production: a realistic approach. BAIF Development Research Foundation.(Available Online: http://www. baif. org. in/aspx_pages/pdf/ Agroforesty/MEDA. pdf), 2008.

[78] Mohapatra, S. and P.K. Panda, Effects of fertilizer application on growth and yield of Jatropha curcas L. in an aeric tropaquept of eastern India. Notulae Scientia Biologicae, 2011. 3(1): p. 95.

[79] Terren, M., et al., Principal disease and insect pests of Jatropha curcas L. in the lower valley of the Senegal River. Tropicultura, 2012. 30(4): pp. 222–229.

[80] Sahai, K., V. Srivastava, and K.K. Rawat, Impact assessment of fruit predation by Scutellera perplexa Westwood on the reproductive allocation of Jatropha. Biomass and Bioenergy, 2011. 35(11): pp. 4684–4689.

[81] Kumar, A. and N. Singh, First report of Maconellicoccus hirsutus Green infestation on Jatropha curcas saplings. Phytoparasitica, 2014. 42(1): pp. 71–73.

[82] Yu L, Z.H., Zhou ZJ, Huang ZQ, Zhou YJ, Wang ZY, Survey of harmful organisms on jatropha curcas L. in Panzhihua City. Modern Agricultural Science and Technology, 2010. 13: pp. 164–169.

[83] Wu, J., et al., Investigation on the diseases and pest insects of jatropha curcas in dry-hot valley. Forest Pest and Disease, 2008. 27(4): pp. 18–21.

[84] Murali Krishna M, V., V. Seshagiri Rao V, and V. Murthy P, Performance evaluation of medium grade low heat rejection diesel engine with carbureted methanol and crude jatropha oil. Renewable and Sustainable Energy Reviews, 2014. 34(C): pp. 122–135.

[85] Subroto, E., et al., Optimization of mechanical oil extraction from Jatropha curcas L. kernel using response surface method. Industrial Crops and Products, 2015. 63(0): pp. 294–302.

[86] Shivani, P., et al., Extraction and analysis of Jatropha curcas L. seed oil. African Journal of Biotechnology, 2013. 10(79): pp. 18210–18213.

[87] Winkler, E., et al., Enzyme-supported oil extraction from Jatropha curcas seeds, in Biotechnology for Fuels and Chemicals. 1997, Springer, New York, pp. 449–456.

[88] Sharma, A., S. Khare, and M. Gupta, Enzyme-assisted aqueous extraction of peanut oil. Journal of the American Oil Chemists' Society, 2002. 79(3): pp. 215–218.

[89] Ahmad, M., et al., Enhancing Jatropha oil extraction yield from the kernels assisted by a xylan-degrading bacterium to preserve protein structure. Applied Microbiology and Biotechnology, 2011. 90(6): pp. 2027–2036.

[90] Abdulla, R. and R. Pogaku, Stability Studies of Immobilized Burkholderia cepacia Lipase and Its Application in Biodiesel Production from Jatropha curcas Oil, in Advances in Biofuels. 2013, Springer, New York, pp. 191–216.

[91] Scott, D., P. Peeters, and S. Gössling, Can tourism deliver its "aspirational" greenhouse gas emission reduction targets? Journal of Sustainable Tourism, 2010. 18(3): pp. 393–408.

[92] Bala, B., Studies on biodiesels from transformation of vegetable oils for diesel engines. Energy Education Science and Technology, 2005. 15(1/2): p. 1.

[93] Schuchardt, U., R. Sercheli, and R.M. Vargas, Transesterification of vegetable oils: a review. Journal of the Brazilian Chemical Society, 1998. 9(3): pp. 199–210.

[94] Knothe, G., J. Krahl, and J. Van Gerpen, The biodiesel handbook. 2015: Elsevier, Urbana Illinois.

[95] Luque, R., et al., Biodiesel as feasible petrol fuel replacement: a multidisciplinary overview. Energy & Environmental Science, 2010. 3(11): pp. 1706–1721.

[96] Kaur, M. and A. Ali, Lithium ion impregnated calcium oxide as nano catalyst for the biodiesel production from karanja and jatropha oils. Renewable Energy, 2011. 36(11): pp. 2866–2871.

[97] Wu, H., et al., Biodiesel production from Jatropha oil using mesoporous molecular sieves supporting K 2 SiO 3 as catalysts for transesterification. Fuel Processing Technology, 2014. 119: pp. 114–120.

[98] Suryaputra, W., et al., Waste capiz (Amusium cristatum) shell as a new heterogeneous catalyst for biodiesel production. Renewable Energy, 2013. 50: pp. 795–799.

[99] Mutreja, V., S. Singh, and A. Ali, Biodiesel from mutton fat using KOH impregnated MgO as heterogeneous catalysts. Renewable Energy, 2011. 36(8): pp. 2253–2258.

[100] Lu, H., et al., Production of biodiesel from Jatropha curcas L. oil. Computers & Chemical Engineering, 2009. 33(5): pp. 1091–1096.

[101] MacLeod, C.S., et al., Evaluation of the activity and stability of alkali-doped metal oxide catalysts for application to an intensified method of biodiesel production. Chemical Engineering Journal, 2008. 135(1): pp. 63–70.

[102] Zhang, Y., et al., Biodiesel production from waste cooking oil: 2. Economic assessment and sensitivity analysis. Bioresource Technology, 2003. 90(3): pp. 229–240.

[103] Marchetti, J., V. Miguel, and A. Errazu, Possible methods for biodiesel production. Renewable and Sustainable Energy Reviews, 2007. 11(6): pp. 1300–1311.

[104] Liu, X., et al., Transesterification of soybean oil to biodiesel using CaO as a solid base catalyst. Fuel, 2008. 87(2): pp. 216–221.

[105] Wen, Z., et al., Biodiesel production from waste cooking oil catalyzed by TiO 2–MgO mixed oxides. Bioresource Technology, 2010. 101(24): pp. 9570–9576.

[106] Lee, H.V., et al., Transesterification of jatropha oil with methanol over Mg–Zn mixed metal oxide catalysts. Energy, 2013. 49: pp. 12–18.

[107] Helwani, Z., et al., Conversion of Jatropha curcas oil into biodiesel using re-crystallized hydrotalcite. Energy Conversion and Management, 2013. 73: pp. 128–134.

[108] Takase, M., et al., Application of zirconia modified with KOH as heterogeneous solid base catalyst to new non-edible oil for biodiesel. Energy Conversion and Management, 2014. 80: pp. 117–125.

[109] Liang, X., et al., Highly efficient procedure for the transesterification of vegetable oil. Renewable Energy, 2009. 34(10): pp. 2215–2217.

[110] Zhang, X. and W. Huang, Biodiesel fuel production through transesterification of Chinese Tallow Kernel Oil using KNO_3/MgO catalyst. Procedia Environmental Sciences, 2011. 11: pp. 757–762.

[111] Manríquez-Ramírez, M., et al., Advances in the transesterification of triglycerides to biodiesel using MgO–NaOH, MgO–KOH and MgO–CeO_2 as solid basic catalysts. Catalysis Today, 2013. 212: pp. 23–30.

[112] Bezergianni, S., A. Kalogianni, and I.A. Vasalos, Hydrocracking of vacuum gas oil-vegetable oil mixtures for biofuels production. Bioresource Technology, 2009. 100(12): pp. 3036–3042.

[113] Christensen, E.D., et al., Analysis of oxygenated compounds in hydrotreated biomass fast pyrolysis oil distillate fractions. Energy & Fuels, 2011. 25(11): pp. 5462–5471.

[114] Shonnard, D.R., L. Williams, and T.N. Kalnes, Camelina-derived jet fuel and diesel: Sustainable advanced biofuels. Environmental Progress & Sustainable Energy, 2010. 29(3): pp. 382–392.

[115] Madras, G., C. Kolluru, and R. Kumar, Synthesis of biodiesel in supercritical fluids. Fuel, 2004. 83(14): pp. 2029–2033.

[116] Liu, J., et al., Hydroprocessing of Jatropha oil over NiMoCe/Al_2O_3 catalyst. International Journal of Hydrogen Energy, 2012. 37(23): pp. 17731–17737.

[117] Zuo, H., et al., Hydrodeoxygenation of methyl palmitate over supported Ni catalysts for diesel-like fuel production. Energy & Fuels, 2012. 26(6): pp. 3747–3755.

[118] Tiwari, R., et al., Hydrotreating and hydrocracking catalysts for processing of waste soya-oil and refinery-oil mixtures. Catalysis Communications, 2011. 12(6): pp. 559–562.

[119] Seames, W.S. and T. Aulich, Method for cold stable biojet fuel. 2015, Google Patents.

[120] Abhari, R., et al., Process for co-producing jet fuel and LPG from renewable sources. 2010, Google Patents.

[121] Krár, M., et al., Fuel purpose hydrotreating of sunflower oil on CoMo/Al_2O_3 catalyst. Bioresource Technology, 2010. 101(23): pp. 9287–9293.

[122] Pérot, G., Hydrotreating catalysts containing zeolites and related materials—mechanistic aspects related to deep desulfurization. Catalysis Today, 2003. 86(1): pp. 111–128.

[123] Wang, Huali, "Biofuels production from hydrotreating of vegetable oil using supported noble metals, and transition metal carbide and nitride" (2012). Wayne State University Dissertations. Paper 485.

[124] Gong, S., et al., Hydrotreating of jatropha oil over alumina based catalysts. Energy & Fuels, 2012. 26(4): pp. 2394–2399.

[125] Shi, N., et al., Hydrodeoxygenation of vegetable oils to liquid alkane fuels over Ni/HZSM-5 catalysts: Methyl hexadecanoate as the model compound. Catalysis Communications, 2012. 20: pp. 80–84.

[126] Ramachandran, K., et al., Recent developments for biodiesel production by ultrasonic assist transesterification using different heterogeneous catalyst: a review. Renewable and Sustainable Energy Reviews, 2013. 22: pp. 410–418.

[127] Borges, M. and L. Díaz, Recent developments on heterogeneous catalysts for biodiesel production by oil esterification and transesterification reactions: a review. Renewable and Sustainable Energy Reviews, 2012. 16(5): pp. 2839–2849.

[128] Liu, Q., et al., One-step hydrodeoxygenation of palm oil to isomerized hydrocarbon fuels over Ni supported on nano-sized SAPO-11 catalysts. Applied Catalysis A: General, 2013. 468: pp. 68–74.

[129] Morgan, P. and P. Roets. The Synthetic Jet Fuel Journey. in 20th World Petroleum Congress. 2011. World Petroleum Congress, 4–8 December, Doha, Qatar.

[130] Bishop, G.J., Aviation Turbine Fuels. Ullmann's Encyclopedia of Industrial Chemistry.

[131] Hancsók, J., et al., Investigation of the production of high cetane number bio gas oil from pre-hydrogenated vegetable oils over Pt/HZSM-22/Al_2O_3. Microporous and Mesoporous Materials, 2007. 101(1): pp. 148–152.

[132] Saifuddin, N., A. Samiuddin, and P. Kumaran, A review on processing technology for biodiesel production. Trends in Applied Sciences Research, 2015. 10(1): p. 1.

[133] Tamunaidu, P. and S. Bhatia, Catalytic cracking of palm oil for the production of biofuels: optimization studies. Bioresource Technology, 2007. 98(18): pp. 3593–3601.

[134] Twaiq, F.A., N.A. Zabidi, and S. Bhatia, Catalytic conversion of palm oil to hydrocarbons: performance of various zeolite catalysts. Industrial & Engineering Chemistry Research, 1999. 38(9): pp. 3230–3237.

[135] Twaiq, F.A., A. Mohamad, and S. Bhatia, Performance of composite catalysts in palm oil cracking for the production of liquid fuels and chemicals. Fuel Processing Technology, 2004. 85(11): pp. 1283–1300.

[136] Leng, T.Y., A.R. Mohamed, and S. Bhatia, Catalytic conversion of palm oil to fuels and chemicals. The Canadian Journal of Chemical Engineering, 1999. 77(1): pp. 156–162.

[137] Ooi, Y.-S., et al., Catalytic cracking of used palm oil and palm oil fatty acids mixture for the production of liquid fuel: kinetic modeling. Energy & Fuels, 2004. 18(5): pp. 1555–1561.

[138] Li, T., et al., Conversion of waste cooking oil to jet biofuel with nickel-based mesoporous zeolite Y catalyst. Bioresource Technology, 2015. 197: pp. 289–294.

[139] Buzetzki, E., et al., The influence of zeolite catalysts on the products of rapeseed oil cracking. Fuel Processing Technology, 2011. 92(8): pp. 1623–1631.

[140] Lang, L., et al., Catalytic activities of K-modified zeolite ZSM-5 supported rhodium catalysts in low-temperature steam reforming of bioethanol. International Journal of Hydrogen Energy, 2015. 40(32): pp. 9924–9934.

[141] Campanella, A. and M.P. Harold, Fast pyrolysis of microalgae in a falling solids reactor: effects of process variables and zeolite catalysts. Biomass and Bioenergy, 2012. 46: pp. 218–232.

[142] Arzamendi, G., et al., Synthesis of biodiesel with heterogeneous NaOH/alumina catalysts: comparison with homogeneous NaOH. Chemical Engineering Journal, 2007. 134(1): pp. 123–130.

[143] Pramanik, K., Properties and use of Jatropha curcas oil and diesel fuel blends in compression ignition engine. Renewable Energy, 2003. 28(2): pp. 239–248.

[144] Agarwal, A.K., Vegetable oils versus diesel fuel: development and use of biodiesel in a compression ignition engine. TIDE, 1998. 8(3): pp. 191–204.

[145] Knothe, G., R.O. Dunn, and M.O. Bagby. Biodiesel: the use of vegetable oils and their derivatives as alternative diesel fuels. in ACS symposium series. 1997. Washington, DC: American Chemical Society, 1974.

[146] Forson, F., E. Oduro, and E. Hammond-Donkoh, Performance of jatropha oil blends in a diesel engine. Renewable Energy, 2004. 29(7): pp. 1135–1145.

[147] Korus, R.A., J. Jo, and C.L. Peterson, A rapid engine test to measure injector fouling in diesel engines using vegetable oil fuels. Journal of the American Oil Chemists' Society, 1985. 62(11): pp. 1563–1564.

[148] Makkar, H.P., G. Francis, and K. Becker, Protein concentrate from Jatropha curcas screw-pressed seed cake and toxic and antinutritional factors in protein concentrate. Journal of the Science of Food and Agriculture, 2008. 88(9): pp. 1542–1548.

[149] Hirota, M., et al., A new tumor promoter from the seed oil of Jatropha curcas L., an intramolecular diester of 12-deoxy-16-hydroxyphorbol. Cancer Research, 1988. 48(20): pp. 5800–5804.

[150] Hajos, G., et al., Biological effects and survival of trypsin inhibitors and the agglutinin from soybean in the small intestine of the rat. Journal of Agricultural and Food Chemistry, 1995. 43(1): pp. 165–170.

[151] Haas, W. and M. Mittelbach, Detoxification experiments with the seed oil from Jatropha curcas L. Industrial Crops and Products, 2000. 12(2): pp. 111–118.

[152] Shahinuzzaman, M., Z. Yaakob, and M. Moniruzzaman, Medicinal and cosmetics soap production from Jatropha oil. Journal of Cosmetic Dermatology, 2016. 15 (2) pp. 185–193.

[153] El Diwani, G., S.A. El Rafei, and S. Hawash, Ozone for phorbol esters removal from Egyptian Jatropha oil seed cake. Advances in Applied Science Research, 2011. 2(4): pp. 221–232.

[154] Najjar, A., et al., Removal of Phorbol Esters Present in Jatropha curcas Kernel by Fungal Isolates. International Journal of Agriculture & Biology, 2014. 16(5).

[155] Gogoi, R., U.K. Niyogi, and A.K. Tyagi, Reduction of phorbol ester content in Jatropha cake using high energy gamma radiation. Journal of Radiation Research and Applied Sciences, 2014. 7(3): pp. 305–309.

[156] Van Gerpen, J., Biodiesel processing and production. Fuel Processing Technology, 2005. 86(10): pp. 1097–1107.

[157] Wall, J., J. Van Gerpen, and J. Thompson, Soap and glycerin removal from biodiesel using waterless processes. Transactions of the ASABE, 2011. 54(2): pp. 535–541.

[158] Berrios, M. and R. Skelton, Comparison of purification methods for biodiesel. Chemical Engineering Journal, 2008. 144(3): pp. 459–465.

[159] Atabani, A.E., et al., A comprehensive review on biodiesel as an alternative energy resource and its characteristics. Renewable and Sustainable Energy Reviews, 2012. 16(4): pp. 2070–2093.

[160] Kaltschmitt, M., G. Reinhardt, and T. Stelzer, Life cycle analysis of biofuels under different environmental aspects. Biomass and Bioenergy, 1997. 12(2): pp. 121–134.

[161] Achten, W.M., et al., Jatropha biodiesel fueling sustainability? Biofuels, Bioproducts and Biorefining, 2007. 1(4): pp. 283–291.

[162] Menichetti, E. and M. Otto, Energy balance and greenhouse gas emissions of biofuels from a life-cycle perspective. Biofuels: Environmental Consequences and Interactions with Changing Land Use. Proceedings of the Scientific Committee on Problems of the Environment (SCOPE) International Biofuels Project Rapid Assessment, Ithaca NY: University of Cornell, 2008: pp. 81–109.

[163] Achten, W.M., et al., Life cycle assessment of Jatropha biodiesel as transportation fuel in rural India. Applied Energy, 2010. 87(12): pp. 3652–3660.

[164] Reinhardt, G., et al., Screening life cycle assessment of Jatropha biodiesel. Heidelberg: Institute for Energy and Environmental Research, 2007.

[165] Prueksakorn, K. and S.H. Gheewala, Full chain energy analysis of biodiesel from Jatropha curcas L. in Thailand. Environmental Science & Technology, 2008. 42(9): pp. 3388–3393.

[166] Ndong, R., et al., Life cycle assessment of biofuels from Jatropha curcas in West Africa: a field study. GCB Bioenergy, 2009. 1(3): pp. 197–210.

[167] Ou, X., et al., Energy consumption and GHG emissions of six biofuel pathways by LCA in (the) People's Republic of China. Applied Energy, 2009. 86: pp. S197–S208.

[168] Kalaivani, K. and N. Balasubramanian, Energy Consumption and Greenhouse Gas Emission Studies of Jatropha Biodiesel Pathway by Life Cycle Assessment in India. Indian Chemical Engineer, 2015: pp. 1–13.

[169] Portugal-Pereira, J., et al., Life cycle assessment of conventional and optimised Jatropha biodiesel fuels. Renewable Energy, 2016. 86: pp. 585–593.

[170] Jingura, R.M. and R. Kamusoko, Evaluation of life-cycle assessment of Jatropha biodiesel. Energy Sources, Part B: Economics, Planning, and Policy, 2016. 11(5): pp. 396–403.

[171] Behera, S.K., et al., Evaluation of plant performance of Jatropha curcas L. under different agro-practices for optimizing biomass–a case study. Biomass and Bioenergy, 2010. 34(1): pp. 30–41.

[172] Islam, A.A., et al., Jatropha curcas L.: A Future Energy Crop with Enormous Potential, in Biomass and bioenergy. 2014, Springer, Switzerland, pp. 31–61.

[173] Divakara, B., et al., Biology and genetic improvement of Jatropha curcas L.: a review. Applied Energy, 2010. 87(3): pp. 732–742.

[174] Ke, C., et al., Genetic relationships among Jatropha curcas L. clones from Panzhihua, China as revealed by RAPD and ISSR. African Journal of Agricultural Research, 2011. 6(11): pp. 2582–2585.

[175] Sun, Q.-B., et al., SSR and AFLP markers reveal low genetic diversity in the biofuel plant in China. Crop Science, 2008. 48(5): pp. 1865–1871.

[176] Satyawan, D. and I.M. Tasma, Genetic diversity analysis of Jatropha curcas provenances using randomly amplified polymorphic DNA markers. Journal AgroBiogen, 2011. 7(1): pp. 47–55.

[177] Tanya, P., et al., Genetic diversity among Jatropha and Jatropha-related species based on ISSR markers. Plant Molecular Biology Reporter, 2011. 29(1): pp. 252–264.

[178] Rosado, T.B., et al., Molecular markers reveal limited genetic diversity in a large germplasm collection of the biofuel crop L. in Brazil. Crop Science, 2010. 50(6): pp. 2372–2382.

[179] Camellia, N.N., A.T. Lee, and N. Abdullah, Genetic relationships and diversity of Jatropha curcas accessions in Malaysia. African Journal of Biotechnology, 2014. 11(13): pp. 3048–3054.

[180] Maravi, D.K., et al., Jatropha (Jatropha curcas L.), in Agrobacterium Protocols. 2015, Springer, New York, pp. 25–35.

[181] Ye, J., et al., Engineering geminivirus resistance in Jatropha curcus. Biotechnology for Biofuels, 2014. 7(1): p. 149.

[182] Ye, J., et al., The Jatropha FT ortholog is a systemic signal regulating growth and flowering time. Biotechnology for Biofuels, 2014. 7(1): p. 91.

[183] Tao, Y.-B., et al., A promoter analysis of MOTHER OF FT AND TFL1 1 (JcMFT1), a seed-preferential gene from the biofuel plant Jatropha curcas. Journal of Plant Research, 2014: pp. 1–12.

[184] Patade, V.Y., et al., RNAi Mediated curcin precursor gene silencing in Jatropha (Jatropha curcas L.). Molecular Biology Reports, 2014. 41(7): pp. 1–8.

[185] Jaganath, B., et al., An efficient in planta transformation of Jatropha curcas (L.) and multiplication of transformed plants through in vivo grafting. Protoplasma, 2014. 251(3): pp. 591–601.

[186] Kim, M.J., et al., Gene silencing of Sugar-dependent 1 (JcSDP1), encoding a patatin-domain triacylglycerol lipase, enhances seed oil accumulation in Jatropha curcas. Biotechnology for Biofuels, 2014. 7(1): 36.

Chapter 13

Review of Continuous Fermentative Hydrogen-Producing Bioreactors from Complex Wastewater

Paula Rúbia Ferreira Rosa and Edson Luiz Silva

Additional information is available at the end of the chapter

http://dx.doi.org/10.5772/110416

Abstract

In recent years, the production of hydrogen through dark fermentation has become increasingly popular because it is a sustainable approach to produce clean energy. Thus, an evaluation of studies reported on hydrogen production from different complex wastewaters will be of immense importance in economizing production technologies. This work presents a review of the advances in the bioreactor and bioprocess design for biohydrogen production from different complex wastewaters. The biohydrogen production is discussed emphasizing the production metabolic pathways, bioreactor configuration and operation, organic loading rate (OLR), pretreatment of wastewater, as well as microbial diversity. Also, in this review, various bioreactor configurations and performance parameters including H_2 yield (HY) and hydrogen production rate (HPR) are evaluated and presented. The work concludes with challenges and prospects of biohydrogen production and claims for more systematic and comprehensive studies on the subject.

Keywords: biogas, global warming, dark fermentation, bioreactor, process parameters

1. Introduction

According to the IPCC [1] (Intergovernmental Panel on Climate Change), global warming of more than 2°C would have serious consequences, such as an increase in the number of extreme climate events. In Copenhagen in 2009, the countries stated their determination to limit global warming to 2°C between 2015 and 2100. To reach this target, climate experts estimate that global greenhouse gas (GHG) emissions need to be reduced by 40–70% by 2050 and that carbon neutrality (zero emissions) needs to be reached by the end of the century at the latest. To reduce global warming, substantial effort is being made at a global scale to explore renewable energy sources that could replace fossil fuels.

Hydrogen gas can be an ideal sustainable energy carrier, which can reduce the over-reliance on fossil fuels. Some of the advantages of hydrogen can be listed as follows: (i) high energy conversion efficiencies, (ii) production from water with no emissions and (iii) abundance [2].

Dark fermentation is a biological approach commonly used to produce H_2 in the absence of light. It is driven by anaerobic bacteria that can produce hydrogen from wastewaters [3]. This technology has attracted attention because it can use a versatile range of substrates, particularly renewable resources that are organically rich such as stillage, sludge, leachate, pomace, stalks and bagasse [4]. Wastewaters generated from various industrial processes are considered to be the ideal substrates because they contain high levels of easily degradable organic material, which results in a net positive energy or economic balance [5]. From the anaerobic digestion process, complex wastewater can be converted into hydrogen, while promoting the treatment of these wastewaters, providing environmental sustainability.

Studies on batch, semi-continuous and continuous hydrogen-producing bioreactors have been conducted. Batch hydrogen fermentation normally brings about lower hydrogen production rates (HPRs) in comparison with its semi-continuous or continuous counterpart. Besides the extensively studied continuous stirred tank reactor (CSTR), numerous biohydrogen bioreactor processes such as anaerobic sequencing batch reactor (ASBR), fixed-bed bioreactor, fluidized-bed bioreactor and upflow anaerobic sludge blanket (UASB) bioreactor have been developed with high production yields and output [6].

Inevitably, performance of hydrogen-producing bioreactor systems and operation are determined by various factors that are associated with environmental conditions, process operating conditions and chemical conditions, such as inoculum, nutrients, hydrogen partial pressure, temperature, hydraulic retention time (HRT) and substrate concentration [6]. Variations in these factors result in different microbial communities, resulting in different hydrogen yields. In this context, this review summarizes the above factors that influence hydrogen production by dark fermentation from different complex wastewaters.

2. Microbiology of hydrogen production: metabolic pathways

Hydrogen can be produced through different metabolic pathways that can be broadly grouped into two distinct categories—light-dependent and light-independent processes. Light-dependent processes include direct or indirect photolysis and photo-fermentation, whereas dark fermentation is a major light-independent process [7]. According to Sinha and Pandey [8], compared to the photosynthetic processes of hydrogen production, fermentation processes have the advantage of a rapid rate of hydrogen production and simplicity of operation.

The anaerobic digestion process generally consists of the four stages, i.e. hydrolysis, fermentation, acetogenesis and methanogenesis (**Figure 1**). In the first two stages, dark fermentation is involved in the production of hydrogen. Various microorganisms are involved in each step and cooperated with each other to achieve carbohydrates that are converted into hydrogen gas, VFAs and alcohols, which are organic pollutants and energy carriers.

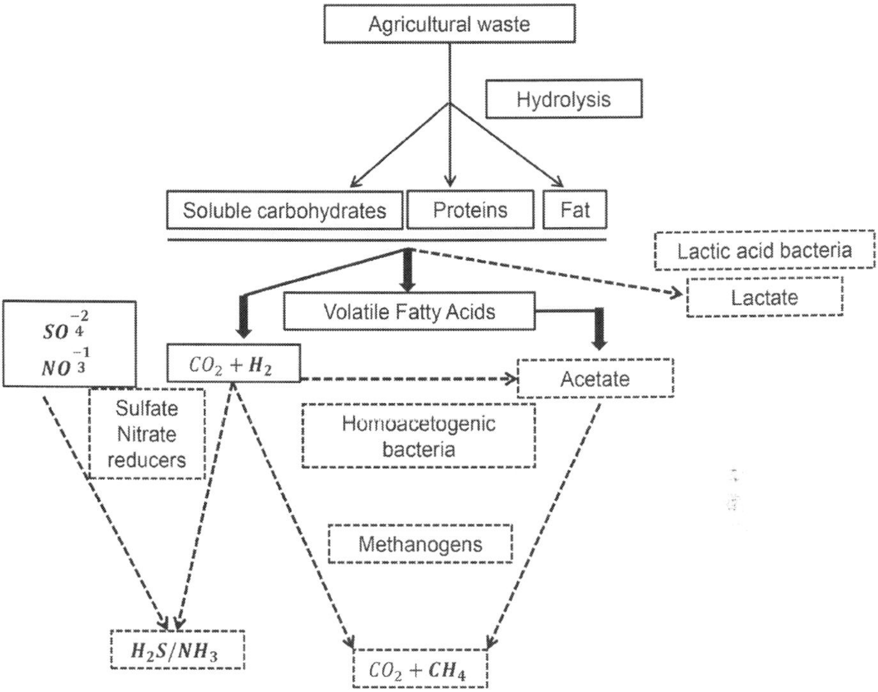

Figure 1. The steps involved in anaerobic digestion [9].

According to Levin et al. [10], carbohydrates are the preferred substrates for the production of hydrogen. Different complex wastewaters have different hydrogen yield per mole of glucose, depending on the metabolic pathway of the final product. When acetic acid is the final product, the maximum theoretical yield is 4 mol/mol glucose (**Table 1**, Eq. (1)). However, when butyrate is the final product, the maximum theoretical yield is 2 mol/mol glucose (**Table 1**, Eq. (2)).

The absence of propionic acid, valeric acid and caproate production ensures higher hydrogen production due to no demand for H_2 formation of this acid (**Table 1**, Eqs. (3) and (11)–(15)). Lactic acid is produced from glucose through three metabolic pathways (**Table 1**, Eqs. (4)–(6)), and in all three metabolic pathways, hydrogen is neither consumed nor produced. The same is true for ethanol production, where the balance of hydrogen is zero (**Table 1**, Eq. (7)).

Hydrogen can be produced simultaneously with ethanol (Eq. (8)) [13, 14]. In addition, there may be a joint production of organic acids (Eqs. (9) and (10)).

Siriwongrungson et al. [15] show that the acetate formed in the acetogenesis may be a consumer of hydrogen. The reducing reaction of hydrogen with carbon dioxide acetate is called homoacetogênese (Eq. (16)). This in turn becomes an important factor in the production of hydrogen, since there is a drop in consumption and performance, through the accumulation of acetate in the medium.

Acidogenesis reactions	Eq. no.	Bacteria
$C_6H_{12}O_6 + 2H_2O \rightarrow 2CH_3COOH + 2CO_2 + 4H_2$	(1)	*Bacteriodes, Clostridium, Butyrivibrie, Eubacterium, Bifidobacterium, Lactobacillus, Acetobacterium, Butyribacterium, Eubacterium, Peptostreptococcus*
$C_6H_{12}O_6 + 2H_2O \rightarrow CH_3CH_2CH_2COOH + 2CO_2 + 2H_2$	(2)	
$C_6H_{12}O_6 + 2H_2 \rightarrow CH_3CH_2COOH + 2H_2O$	(3)	
$C_6H_{12}O_6 \rightarrow 2CH_3CHOHCOOH + 2\ CO_2$	(4)	
$C_6H_{12}O_6 \rightarrow CH_3CHOHCOOH + CH_3CH_2OH + CO_2$	(5)	
$C_6H_{12}O_6 \rightarrow 3CH_3COOH + 2CH_3CHOHCOOH$	(6)	
$C_6H_{12}O_6 \rightarrow 2CH_3CH_2OH + 2CO_2$	(7)	
$C_6H_{12}O_6 + H_2O \rightarrow CH_3CH_2OH + CH_3COOH + 2H_2 + 2CO_2$	(8)	
$C_6H_{12}O_6 \rightarrow 2H_2 + 2CO_2 + (1/2)\ CH_3COOH + (3/4)CH_3(CH_2)2COOH$	(9)	
$C_6H_{12}O_6 \rightarrow (4/3)\ CH_3CH_2COOH + (2/3)CH_3COOH + (2/3)CO_2 + (2/3)H_2O$	(10)	
$CH_3CH_2COOH + 2CO_2 + 6H_2 \rightarrow CH_3(CH_2)_3COOH + 4H_2O$	(11)	
$CH_3CH_2COOH + CH_3(CH_2)2COOH \rightarrow CH_3(CH_2)_3COOH + CH_3COOH$	(12)	
$CH_3CH_2COOH + CH_3COOH + H_2 \rightarrow CH_3(CH_2)_3COOH + 2H_2O$	(13)	
$CH3(CH_2)_2COOH + CH_3COOH + 2H_2 \rightarrow CH_3(CH_2)_4COOH\ 2H_2O$	(14)	
$CH3(CH_2)_2COOH + CH_3COOH + 2H_2 \rightarrow CH_3(CH_2)_4COOH + 2H_2O$	(15)	
Acetogenic reactions		
$CO_2 + 4H_2 \rightarrow CH_3COOH + 2\ H_2O$	(16)	*Desulfovibrio, Syntrophobacter wolinii, Syntrophomonas*
$CH_3CHOHCOOH + H2O \rightarrow CH_3COOH + CO_2 + 2H_2$	(17)	
$CH_3CH_2OH + H2O \rightarrow CH_3COOH + 2H_2$	(18)	
$CH_3CH_2COOH + 2\ H_2O \rightarrow CH_3COOH + CO_2 + 3\ H_2$	(19)	
$CH_3(CH_2)_2COOH + 2\ H_2O \rightarrow 2\ CH_3COOH + 2H_2$	(20)	
Methanogenic reactions		
$4\ H_2 + CO_2 \rightarrow CH_4 + 2\ H_2O$	(21)	*Methanobacterium formicicum, M. bryantii, Methanobrevibacter ruminantium, M. arboriphilus, Methanospirillum hungatei, Methanosarcina barkeri*
$CH_3COOH \rightarrow CH_4 + CO_2$	(22)	
$2CH_3(CH_2)_2COOH + 2H_2O + CO_2 \rightarrow 4CH_3COOH + CH_4$	(23)	
$4HCOOH \rightarrow 3CH_4 + CO_2 + 2H_2O$	(24)	
$4CH_3OH \rightarrow 3CH_4 + CO_2 + 2H_2O$	(25)	
$2CH_3CH_2OH + CO_2 \rightarrow CH_4 + 2CH_3COOH$	(26)	

Adapted from Abbasi et al. [11] and Saady et al. [12].

Table 1. Reactions during acetogenic hydrogen fermentative.

In the step of acetogenesis, the hydrogen could be formatted from lactic acid, ethanol, propionic acid and butyric acid (Eqs. (17)–(20)).

Hydrogen could be consumed for archaea hydrogenotrophic (Eq. (21)). Approximately 70% of all the methane produced in anaerobic digestion process stems from Eq. (22). Furthermore, methane is formed from acetate, butyrate, formate, ethanol and methanol (Eqs. (22)–(26)).

Among the fermentative anaerobes, *Clostridia* have been well known and studied extensively, not for their hydrogen production capability but for their role in the industrial solvent production from various carbohydrates [7]. Hydrogen production by these bacteria is highly dependent on the process conditions such as pH, hydraulic retention time (HRT), and gas partial pressure, which affect metabolic balance. Thus, fermentation end-products produced by microorganism depend on the environmental conditions in which it grows [10].

There are also some bacteria that consumed hydrogen, such as *Lactobacillus* spp. and *Bifidobacterium* spp. [4]. Moreover, the major H_2-consuming microorganisms other than hydrogenotrophic methanogens are homoacetogens, such as *Methanobacterium*.

Depending on the pathway, the theoretical biogas composition is around 67% of H_2 (acetate pathway) or 50% of H_2 (butyrate pathway). The various metabolic pathways that may establish can either be promoted or inhibited, depending on the adopted operating conditions, which govern the production of specific volatile fatty acids (VFAs) and alcohols including acetate, propionate, butyrate, lactate and ethanol [16].

3. Dark fermentation from complex wastewaters

Dark fermentation is a biological approach commonly used to produce H_2 in the absence of light and hence the configuration of the bioreactor is simpler and cheaper. Hydrogen production by dark fermentation has several other advantages such as the ability to produce hydrogen from organic waste and therefore control and stabilize biological waste which has a potential danger of contamination. For instance, dark fermentation can be integrated into wastewater treatment systems to produce H_2 from wastewater. Producing hydrogen from organic waste has a potential to reduce hydrogen production costs since organic waste (including wastewater) is cheap and easily available [2]. Moreover, the regulatory need of treatment of wastewater prior to disposal is making them an ideal commodity to produce biohydrogen from the anaerobic treatment [17].

The main source for the fermentative H_2 production is complex wastewater containing carbohydrate substances. Crucial points for improving the efficiency of hydrogen production that are frequently emphasized throughout literature are associated with facilitated access to cheap wastewater, such as vinasse, cassava wastewater, cheese whey, glycerol, sago wastewater and textile wastewater. There are several articles in the literature that demonstrate the hydrogen production from these wastewaters, indicated in **Table 2**, including the process parameters such as substrate concentration, pH, temperature, HRT, reactor type and seed sludge.

Numerous works have been focused on vinasse. This wastewater, one of the major by-products of the ethanol production process with nearly 14 L of vinasse produced per litre of ethanol, can cause extensive pollution due to its high organic load (up to 40 g COD/L) [18]. Cassava wastewater and cheese whey, main components of agro industrial processes, are considered as highly polluting due to their high organic load and the volume generated, representing a significant environmental impact for the agro-industry [19, 20]. They were also used for the successful H_2 production [21–25].

Substrate	Reactor	Inoculum	Range of pH	*T*	Substrate conc. (mg COD/L)	HRT (h)	OLR (kg COD/m^3 d)	Maximum HY (mmol/g COD)	HPR (L/dL)	Methane production	Reference
Mixture of glucose and cheese whey	AFBR	Sludge from poultry slaughterhouse	4.0–4.5	30°C	5000	6 h	20	1.6	2.4		Ferreira Rosa et al. [21]
Cheese whey powder	UASB	Full-scale methanogenic UASB reactor treating wastewater from a confectionery factory	4.5–5.63	–		13–3 h	20–48	–	1.62	0.2–0.6 **L/dL**	Carrillo-Reyes et al. [22]
Cassava wastewater	UASB	Sludge collected from the bottom of a first anaerobic pond treating a cassava wastewater	5.5	37°C	–		10–30	39.83 l H_2/kg COD removed	0.39		Intanoo et al. [23]
Cassava flour wastewater	Continuous multiple tube reactor	Natural fermentation	6.5	25°C	4	4	24	2.07 mol/mol substrate	1.94		Gomes et al. [24]
Cassava processing wastewater	AFBR	Anaerobic sludge from a UASB reactor was used for the treatment of swine wastewater	4.5–5.0	30°C	2000–15,000	12–10	4–30	2.0 (12 h)	1.66 (12 h)	14–27% 0.9 L CH_4 d_1 L_1 - 1.5 L CH_4 -d_1 L_1	Rosa e al. [25]
Molasses wastewater	Continuous mixed immobilized sludge reactor	Anaerobic sludge obtained from a local municipal wastewater treatment plant	4–5	35°C	2–6	6 h	8–32	130.57 mmol/mol	12.51 mmol/h L		Han et al. [26]
Tapioca wastewater	ABR	Anaerobic mixed cultures	pH initial—9 effluent = 5.2–5.8	32.3°C	–	24, 18, 12, 6 and 3 h	16–130 kg m^{-3} d^{-1}	0.745 mmol/g DQO (OLR = 31)	0.883 (6 h HRT)	0.63–3.7%	Thanwised et al. [27]

Substrate	Reactor	Inoculum	Range of pH	*T*	Substrate conc. (mg COD/L)	HRT (h)	OLR (kg COD/m³ d)	Maximum HY (mmol/g COD)	HPR (L/dL)	Methane production	Reference
Crude glycerol	UASB	Sludge from UASB reactor of a seafood wastewater treatment system	Initial = 8	40°C	10–30	12–2	–	44.27 mmol H_2/g glycerol	242.15 mmol H_2/L/d		Chookaew et al. [28]
Beverage industry wastewater	CSTR	Anaerobic sludge	5.6–6.3 without any pH control	37°C	–	8–1.5	60–320	1.7 mol/mol hexose utilized	55 L/L-d		Sivagurunathan et al. [29]
Sugarcane vinasse	AFBR	Sludge from an upflow anaerobic sludge blanket (UASB) reactor used for the treatment of swine	4–5	30°C	5000–10,000	6–1 h	20–240	3.07	13.68	0–40%	Reis et al. [30]
Washing wastewater of beverage production process	Continuously stirred anaerobic bioreactor (CSABR)	Seed sludge from municipal wastewater treatment plant	5.5	37°C	5	1 h		0.36 mol/mol	11.39		Liu et al. [31]
Alcohol wastewater	UASB	Sludge from the UASB reactor treating an alcohol wastewater	5.5	37°C	–	1.93–0.72	23–62	125.1 ml /g COD removed	18 L/d		Poontaweegeratigarn et al. [32]
Tequila vinasses	ASBR	Sludge from an UASB treating the wastewater from a brewery plant	5.5	25 and 35°C	0.5–5	24–12 h		–	1.2	35–44%	Buitrón and Carvajal [33]
Tofu-processing wastewater	CSTR	Sludge from wastewater treatment plants	5.5	35°C	20 g COD/L	24–6 h	–	–	1.73		Lay et al. [34]

Substrate	Reactor	Inoculum	Range of pH	*T*	Substrate conc. (mg COD/L)	HRT (h)	OLR (kg COD/m³ d)	Maximum HY (mmol/g COD)	HPR (L/dL)	Methane production	Reference
Brewery wastewater	Batch	Sludge from a full-scale upflow anaerobic sludge blanket reactor treating citrate-producing wastewater	4–8	25–45°C	2–12	–	–	158 mL/g COD	59 mL/h		Shi et al. [35]
Alcohol distillery wastewater	ASBR	Sludge from the anaerobic tank of Red Bull Distillery	5.5	37°C	20–60	32–13	15–112.5	172 ml H_2/g COD removed,	3.3	6.5–35%	Searmsirimongkolet al. [36]
Palm oil mill effluent	UASB	Seed sludge	5.5	37°C	10–40	32–8 h	–	0.38 L H_2/g COD added	8.76		Singh et al. [37]
Glycerol	UASB	UASB granules obtained from a UASB reactor	5.5	37°C	–	–	25–75	410 (mmol H_2/mol glycerol)	9 mmol H_2/L h		Reungsang et al. [38]
Sugarcane vinasse	Up flow anaerobic packed bed reactors (APBR)	Natural wastewater fermentation process	5.4–5.7	25°C	–	24 h	36.3	Maximum 1.8 average 0.3	0.509		Ferraz Júnior et al. [39]
Textile wastewater	Batch	Sludge anaerobic from the treatment plant	Initial 7.0	37°C	20			1.37 mol H_2/ molreducing sugar	0.312 L d/l		Li et al. [40]
Cheese whey	Batch	Anaerobic sludge	Initial 8	36°C				3.3 mol/mol lactose	16.2 mL/h	–	Seo et al. [41]
Rice mill wastewater	Batch	*E. aerogenes* RM 08	Initial 7.0 final 5.1	33°C				0.97 mol H_2/mol of suga	134.6 mL/h	–	Ramprakash and Muthukumar [42]
Mixture of sugar cane stillage and glucose	AFBR	Sludge from a granular sludge of a thermophilic upflow anaerobic sludge blanket reactor	4.1–4.3	55°C	5000–5300 mg COD/L	8, 6, 4, 2 and 1 h	26–216	5.73 mmol g COD added (HRT 4 h)	18.72		Santos et al. [43]

http://dx.doi.org/10.5772/110416

Substrate	Reactor	Inoculum	Range of pH	*T*	Substrate conc. (mg COD/L)	HRT (h)	OLR (kg COD/m³ d)	Maximum HY (mmol/g COD)	HPR (L/dL)	Methane production	Reference
Corn starch wastewater	Bach	*Bacillus cereus* and *Brevundimonas naejangsanensis* isolated from sludge anaerobic	Initial 6.5	35°C	10–20	–		1.88 mol/mol glucose	3.96	–	Wang et al. [44]
Coffee drink manufacturing wastewater	CSTR	Anaerobic sludge	5.5	35°C	20	12–6	–	0.2 mol/mol	0.34		Jung et al. [45]
Sago wastewater	Batch	Fresh cattle dung	7.0	30°C	0.5–5 (% w/v)	–		323.4 mL g^{-1} starch	3.48		Sen et al. [46]
Soft-drink wastewater	Upflow anaerobic packed bed reactor	Natural fermentation	Initial 6.5	25°C	2.3	0.5	–	3.5 mol H_2 mol of sucrose	9.6		Peixoto et al. [47]
Condensed molasses	Continuously stirred anaerobic bioreactor (CSABR)	Municipal sewage treatment plant	5.5	37°C	40–60	8–0.5 h	–	5.3	14.04 (HRT 0.5 h)		Chu et al. [48]
Glycerol	Batch	Anaerobic sludge	6.5	–		–	–	2.2 mmol/L	–		Trchounian et al. [49]

Table 2. Studies of anaerobic biohydrogen production processes using complex wastewater.

The performance parameters were hydrogen production yield (HY) and hydrogen production rate (HPR). The process parameters including pH [22, 26], hydraulic retention time [22, 27–32], temperature [33–35], substrate concentration [30, 31, 36–38], different sludge [21], support materials [39], pretreatment of wastewater [24, 25, 40–42], use of co-substrate [21, 25, 30, 43, 44], inoculum pretreatment [22, 45, 46], addition of nutrients [44, 47], reactor configuration [45, 48] and effects of some heavy metal ions [49] have already been evaluated. Several types of wastewaters listed in this review could produce hydrogen with a HY range of 0.74–5.3 mmol-H_2/g-COD and a HPR range of 0.03–14.04 L/L/d. Among the wastewaters studied, one of highest HY of 5.3 mmol/g COD was obtained using continuously stirred anaerobic bioreactor (CSABR) from condensed molasses; it was successfully operated for 300 days [48].

4. Simultaneous hydrogen and methane production in a single-stage biosystem

The key point in the fermentative production of hydrogen is the inhibition of the methanogenic step so that the formed hydrogen is not consumed for the formation of methane. Two routes can lead to the formation of methane: the route acetoclastic from acetic acid and methanol (methanogenic acetoclastic or acetotrophic microorganism (**Table 1**, Eqs. (22) and (25)) and the hydrogenotrophic route from H_2 (hydrogen consumers microorganisms or hydrogenotrophic (**Table 1**, Eq. (21)).

Among the forms of control, methanogenic activity can be used to maintain the acidic pH medium for cultivation and processing of the inoculum in order to inactivate methanogens [50]. Furthermore, for continuous reactors, the reduction of HRT and consequently higher organic loading rate (OLR), they can avoid the use of H_2 as a substrate for methanogenesis.

Recently produced hythane (H_2+ methane) in a single-stage biosystem, using complex wastewaters with low pH, shorts HRT and high concentration of organic matter suggested that some archaea can survive at conditions that do not favour methanogens [22, 25, 27, 30, 33, 36]. The hythane production has also received much commercial attention in the transportation sector [51], and a production in a single stage has the advantage of being economically more viable due to economic financial, energy and manpower, than the hythane production by two-stage fermentation [52].

Kim et al. [53] conducted a study on the influence of pH on the activity of users consuming hydrogen methanogens. According to the authors, the formed methane left in consumers of hydrogen archaeas, which are commonly inhibited at pH below 5.0, proved to be more tolerant of acidic conditions than other methane-producing microorganisms.

Carrillo-Reyes et al. [22] evaluated the reduction of pH (5.63–4.5) in UASB reactors fed cheese whey, with a HRT of 6 h and OLR of 20 kg COD/m^3.d (**Table 2**). The authors reported that the strategy of reducing the pH to 4.5 to avoid methane production was not efficient. This fact did not favour the hydrogen production and even caused a sharp drop in the total gas production. Similar results were found by Taconi et al. [54], who found a 30% increase in methanogenic activity when the pH was decreased from 7 to 4.5.

As maintaining the pH in acidic conditions does not guarantee the inactivation of methanogens, the heat treatment of the inoculum is not conclusive. For instance, the acetoclásticas microorganisms can survive thermal shock, leading to the consumption of hydrogen to acetic acid formation [50]. The formed acetic acid is then converted into methane (**Table 1**, Eq. (23)).

This fact is reported by Luo et al. [55] who used cassava stillage as the substrate for hydrogen production and found that thermal pretreatment of inoculum does not improve the yield of hydrogen in continuous reactors under mesophilic temperatures. The study analysed the effect of different pretreatments of the inoculum such as acid treatment, heat treatment and shock load in repeated batch tests, demonstrated that inoculum pretreatment could not permanently inhibit methanogenesis either. According to the authors, the methane inhibition only occurs by proper control of fermentation, pH and temperature.

Given the resistance of methanogenic archaea of pretreatment of inoculum, Carrillo-Reyes et al. [22] showed that repeated heat treatment of the granular sludge was the only strategy that completely inhibited methane production, leading to high volumetric hydrogen production rates (1.67 L H_2/L-d). In the same study, the authors use a strategy to decrease methane production: the shock loads (from 20 to 30 g COD/L-d) was a more effective strategy to decrease the methane production rate (75%) and to increase the hydrogen production rate (172%), without stopping reactor operation.

Methanogens were detected in different hydrogen-producing reactors operated at low pH (values between 4.0 and 5.63) and with high organic loading rate (**Table 2**) revealing that they can survive under these extreme conditions.

In ASBRs, Buitrón and Carvajal [33] reported the production of methane (35–44%) concomitant with the production of H_2 when employed with HRT of 24 h. They found that the higher the concentration of vinasse, the greater the percentage of methane achieved. According to the authors, methanogens could be already present in vinasse and before the source of organic acids, and H_2 produced by the reactor found an environment conducive to development on the other hand, using similar wastewater, and even reactor. Searmsirimongkol et al. [36] evaluated the effect of concentration (20, 30, 40 g/L) on the hydrogen production. The highest methane yields (approximately 40% methane content of the biogas) were found at lower concentration (20 g/L); however, concentration higher than 60 g/L did not verify the presence of methane. Serious methanogeneses were reported in high rate reactors, such as UASB [22] and AFBR [25, 30].

The hydraulic retention time (HRT) is also an important parameter in the fermentation processes. Higher rates of volumetric hydrogen production and increased percentages of hydrogen in biogas can be obtained by decreasing the HRT and thus increasing the organic loading rate (OLR) [56]. In addition, low HRT could suppress methane producers and inhibit methanogenesis. However, in many complex wastewaters, this behaviour is not checked. Exemplifying, Rosa et al. [57] evaluated the effects of different hydraulic retention times (HRTs) of 4, 2 and 1 h and varying sources of inoculum (sludge from swine and sludge from poultry) on the hydrogen production in two AFBRs from cheese whey. When the HRT was reduced, methane was produced concurrently with hydrogen in both reactors, with maximum methane production of 0.68 L CH_4/h/L with an applied HRT of 1 h. Carrillo-Reyes et al. [56] found that the application of a OLR of 20 g COD/L/d and a gradual decrease of HRT from 24 to 6 h led to a decrease in H_2 production from 0.03 to 0.015 LH_2/L/h, due to the presence of methane. According to the authors, the delay in the production of methane from this reactor, when compared to other reactors in their study, was due to the application of high substrate concentrations. The maximum methane yields of 0.02 L/h/L were obtained in reactors with the application of HRT of 6 h, and OLR from 5 to 20 g COD/L. Other studies have also found the simultaneous hydrogen and methane production in short HRTs, from different wastewaters, such as stillage [30, 33], rich in starch wastewater [25, 27].

These results indicated that the low HRT in different configurations of reactors might reduce microbial richness through the washout of microbes and increase microbial diversity through accelerating the proliferation of non-hydrogen-producing microorganism. So, methanogens could adapt to the conditions imposed in hydrogen-producing reactors (low pH, high OLR and low HRT). In spite of the negative effect of these organisms in hydrogen production, they may have an important application in the production of hythane (H_2 and CH_4) using wastewaters with low pH and high concentration of organic matter.

5. Bioreactor configuration

The reactor configuration and the improvement of operating parameters is essential to obtain best hydrogen production rates, indicating that the system performance is largely influenced by the retention of biomass reactor [58]. The batch modes of operation and continuity have been reported in the literature for producing hydrogen. Most batch studies have the advantage of being easily operated, flexible, generating a series of studies with different wastes to produce hydrogen [9]. However, these reactors provide lower H_2 production rates as compared to continuous systems.

Continuous stirred tank reactors (CSTR) are the most common continuous system used for hydrogen production by dark fermentation from olive milk wastewater, cheese whey and condensed molasses (**Table 2**). Reactors upflow anaerobic sludge blanket (UASB), anaerobic fluidized bed (AFBR) and anaerobic packed bed reactor (APBR) also are used for the production of hydrogen in different complex wastewater. The advantages and disadvantages of different types of bioreactors for H_2 production are listed in **Table 3** [59, 60].

Glycerol [49], sago wastewater [46] and brewery wastewater [35] were proved to be feasible substrates by batch tests showing the maximum HY of 2.2 mmol/L, 323.4 mL/g starch and 158 mL/g COD, respectively. In continuous H_2 production, the main reactor used was AFBR [25, 30, 43], UASB [22, 23, 28, 32, 37, 38] and CSTR [29, 34, 45].

In fermentative hydrogen production, the HRT, and in turn the OLR, affect the substrate conversion efficiency, the type of active microbial population as well as the metabolic pathways established in the system [16]. In the following sections, a discussion of literature findings about the influence of these parameters is presented.

Reactor type	Advantages	Drawbacks
CSTR	Simple construction, easy to operate and control	Low biomass retention
UASB	Good retention of biomass in all reactor areas (bedand sludge blanket)	Slow development of granules
AFBR	Good retention of biomass Good mass transfer due to efficient mixing	Excessive shear stress can detach biomass Energy required for fluidization bed
APBR	Good retention of biomass	Clogging Lower mass transfer than FBR

Table 3. Bioreactors for H_2 production: advantages and drawbacks.

5.1. Influence of OLR

The parameters that constitute the OLR are the concentration of organic matter and HRT. For it is a design variable which determines the capacity and the reactor operating conditions. Changes in OLR have a considerable influence on the diversity of the microbial population and on the metabolism pathways of bacteria that may favour hydrogen production [61].

According to De Gioannis et al. [16], there is a discrepancy in the literature regarding the effect of OLR and HY. According to these authors, the OLR is affected by the accumulation of acid, pH changes and variations in the composition that subsequently change the metabolic pathways.

5.1.1. Substrate concentration

The substrate concentration should be selected in order to meet the needs of microbial growth and hydrogen production and its increase can ensure a stable production of hydrogen in high yield [43]. However, concentrations of organic matter in excess decrease substrate conversion and the yield of hydrogen due to the accumulation of inhibitory compounds in the medium, reducing the competitiveness of hydrogen producers for other microorganisms [3, 43].

In the batch tests, optimal substrate concentration varied and was deeply influenced by other operational parameters such as the pH. When the pH was not controlled, HY usually decreased with increasing substrate concentration due to low pH condition. In contrast, finding the optimal substrate concentration in continuous operation mode is more meaningful and practical, since the batch mode does not take into consideration the hydrodynamic effect, steady state of the substrate concentration and pH condition for bacterial growth [62].

Higher feeding concentrations of the substrate could increase H_2 production [22, 26]; however, excessive substrate concentrations may decrease this capacity [25, 28, 31, 34, 36, 37]. Chu et al. [48] in a suspended sludge bioreactor producing H_2 fed with condensed molasses fermentation soluble, increased the H_2 production rate by 2.3 times by elevating the substrate concentration from 40 to 60 g COD/L at a HRT of 2 h. Already, in continuous mixed immobilized sludge reactor from molasses wastewater, Han et al. [26] increased the HPR 3.36 times by elevating the substrate concentration from 2 to 6 g COD/L at a HRT of 6 h. In contrast, when varying the tofu-processing wastewater concentration from 10 to 40 g COD/L in a batch reactor, Lay et al. [34] found that 20 g COD/L was the optimum concentration for H_2 production.

Most studies reported that hydrogen production from complex wastewaters had substrate concentrations lower than 40 g COD/L (**Table 2**). Often it is noted that higher concentration of any substrate leads to a drop in HY [34]. Moreover, it has been reported that in some cases hydrogen production can be inhibited by the toxicity of the complex wastewaters. This fact is noted by Searmsirimongkol et al. [36] who then diluted alcohol distillery wastewater to obtain various feed COD values of 20,000, 40,000 and 60,000 mg/L. The highest concentrations of hydrogen production resulted in inhibition due to the presence of high potassium concentration. Already Liu et al. [31] showed that SO_3^{2-} affected the hydrogen production at the substrate concentration of 10 g total sugar/L process, when the performance of hydrogen production decreases, HPR was reduced from 34.59 to 6.50 L/L/d, yield was reduced from 0.92 hexose to 0.08 mol H_2/mol hexose, when SO_3^{2-} iincreased from 0 to 80 mg/L. Sulphate-reducing bacteria (SRBs) causes hydrogen gas converting hydrogen sulphide become less efficient in hydrogen production.

A maximum hydrogen production of 11.39 L/d/L was obtained at HRT 1.0 h (a concentration of 10 g) from washing wastewater of beverage production process with continuously stirred anaerobic bioreactor [31]. The authors suggested that the hydrogen-producing bacteria (HPBs) were adaptive to the system.

High substrate concentration allows more energy-efficient operation but product inhibition is likely to set the upper limit. Certain level of metabolic products in the dark fermentation may inhibit H_2 producing pathway as well as microbial activity [58].

5.1.2. HRT

HRT indicates the time that the organic matter remains in the reactor. This time depends on the metabolism rate of organic matter by microbial community and may vary according to the process. HRT can be used to select a producer of hydrogen community depending on the substrate used.

HRT is also an important parameter in the fermentation process. Higher rates of volumetric hydrogen production and increased percentages of hydrogen in biogas can be obtained by decreasing the HRT and thus increasing the organic loading rate (OLR) [56]. Shortening hydraulic retention times (HRTs) is a well-used and effective operation strategy to enhance hydrogen production from organic wastewater and solid wastes because of its ability to exclude methanogens which have longer generation time.

In most studies on continuously dark fermentative hydrogen production, continuous systems are expected to operate at a low HRT 36–12 h [27, 33, 34, 37, 39] and very low of HRT 12–2 h [21, 22, 25, 26, 28–30, 43, 45, 48] for obtaining a high biohydrogen production that can be operated at extremely low HRT 2–0.5 h [30, 47, 48] with immobilized cell in the biohydrogen.

As shown in **Table 2**, the range of organic loading rate (OLR) was 16–320 kg/m^3/d equivalent by a gradual decrease in HRT from 32 to 0.5 h. When considering the variation in hydrogen production with respect to the HRT, it can be seen that the HRT greatly affected microbial activity and metabolic products, leading to variations in gas production rate, gas composition and hydrogen production rate [36].

With regard to the microbial community, short HRT is also preferred from beverage wastewater [29, 31], sugarcane stillage and glucose [43] and crude glycerol [28]. In contrast, most studies had a drop in hydrogen production because: of too low mixing and poor contact of glycerol with the microorganisms [28]; of the occurrence of OLR shock from tapioca wastewater [27]; of longer reaction time, which allowed for more time to metabolize the Tequila vinasses [33]; microbial cells were washed out from the system as a result from the toxicity of VFA accumulation from alcohol wastewater [32] and of lactate accumulation from tofu-processing wastewater [34].

A maximum hydrogen production of 55 L/d/L was obtained at HRT 1.5 h (an OLR of 320 g/L-d hexose equivalent) from beverage industry wastewater (20 g/L hexose equivalent) with CSTR [29]. This HPR value is much higher than those of other complex wastewaters employed in fermentative hydrogen production.

Therefore, it is essential to define a range of OLR, which will enable to achieve constant efficiency in the biological reactor, or an optimum OLR value for maximum H_2 yield. As a result, the fermentative routes and final metabolites products may be modified due to the OLR applied, as well as the conversion efficiency of the substrate and the microbial community established in the system [16].

6. Strategies for improved hydrogen production

Fermentative hydrogen production is a very complex process and is influenced by many factors such as inoculum, substrate, reactor type, temperature and pH. The effects of these factors on hydrogen production have been reported by a great number of studies throughout the world in the last few years.

Wang and Wan [63] showed that there usually existed some disagreements on the optimal condition of a given factor for fermentative hydrogen production, thus more researches in this respect are recommended.

6.1. Pretreatment of complex wastewater

To enhance the fermentations of some complex wastewaters, such as cassava wastewater, tofu-processing wastewater, corn starch wastewater and textile wastewater, pretreatment must be done to make the process feasible and sustainable. These processes include various combinations of biological, physical and chemical treatment processes [24, 25, 34, 40–42]. Each of these pretreatment methods has a unique purpose and will depend on the wastewater used.

Starch can be hydrolyzed into glucose and maltose by acid or enzymatic hydrolysis followed by biological conversion of the carbohydrates into organic acids and then into hydrogen gas [64]. Moreover, mixed culture could produce more various hydrolases which could utilize complex substrates present in wastewater than pure culture [65]. Rosa et al. [25] used the technique of acid hydrolysis with sulphuric acid and heated the cassava wastewater at 120°C for 30 min before being used as a substrate. A maximum hydrogen yield of 2.0 mmol/g COD was achieved with OLRs of 10 kg COD/m^3/d.

The heat treatment of complex wastewater rich in starch (corn starch wastewaters, rice mill wastewater, cassava wastewater) is common, with the purpose to remove the mixed population of microorganism in the wastewater, which could either compete with biohydrogen producers or inhibit their growth [24, 42, 44]. In these studies, the heat treatment was made in 120°C with times between 15 and 25 min. This pretreatment was also used for tofu-processing wastewater, but at temperature 70°C for 30 min to inhibit the hydrogen-consuming bacteria [34].

Lactic acid bacteria (LABs) are members of the autochthonous microbiota of cassava and are responsible for the fermentation of the root; furthermore, LAB reduces cassava toxicity and prevents post-harvest deterioration [66]. However, in hydrogen-producing reactors, a few LAB strains may have an inhibitory effect due to their bacteriocins, which are antimicrobial peptides that have a deleterious effect on H_2-producing bacteria [67]. Gomes et al. [24] conducted pre-treatment heat (121°C; 15 min) in order to eliminate probable negative effects of the presence of LAB in the cassava wastewater used. The bacteriocins as well as their degradation products were detected in both the raw and heat-treated cassava wastewater samples. Their presence suggests that the poor results of hydrogen production observed in all assays could be attributed to these compounds, and demonstrated that the heat treatment of wastewaters may not completely deactivate bacteriocins. In contrast, Seo et al. [41] evaluated the effect of different pretreatment of cheese whey for hydrogen production (heat pretreatment; sonication pretreatment; and hydrodynamic cavitation). All the treated samples exhibited H_2 production activity, suggesting the fact that LABs, which exist predominantly in the raw cheese whey and produce lactic acid, were effectively suppressed. The maximum H_2 yield of 1.89 mol H_2/mol lactose was obtained from the cheese whey pretreated with hydrodynamic cavitation for 15 min.

The production of bio-H_2, particularly from more complex wastewater, such as textile wastewater, has been treated with activated carbon. This technique is available for wastewater industries, solvent recovery, chemical catalyst, gold extraction, gas separation and liquid adsorption. Li et al. [40] used the textile wastewater, hydrolyzed by α-amylase with a concentration of 0.2 mL/L for 20 min. After α-amylase hydrolysis, the hydrolysate was pretreated with activated carbon and cation exchange resin with a concentration of 1% w/w for 30 min. The removal efficiency of ion concentration was 95.85%. After that, the hydrolysate was fed into the batch reactor and the best hydrogen yield was 1.37 mol H_2/mol, reducing sugar.

The application of pretreatment to the complex wastewater was tested in an attempt to overcome eventual limitations these wastewater. The selection of suitable hydrolysis method and/or control of inhibitors production will improve the fermentation, resulting in a positive effect and improving the degradability of the complex wastewater during the biological process [24, 25, 42].

6.2. Nutrients

Excluding the main substrate, carbohydrate materials, dark fermentative hydrogen production requires nutrients for bacterial activity like all biological treatment processes. The nutrients include nitrogen (N), phosphorous (P), ferrous (Fe) and some trace metals. Among the many kinds of nutrients, N is the most essential one for bacterial growth. Optimal C/N ratio is 47 according to Lin and Lay [68]. P and Fe concentrations affect the metabolic pathway of *Clostridium* sp., and hydrogen production potential decreases when their concentrations are limited.

Appropriate ratios of carbon and nitrogen, carbon and phosphorus, and between carbon and sulphate increase bioproduction of hydrogen by modifying metabolic pathways associated with the nutritional requirement of microorganisms [69]. Argun et al. [70] observed that an adequate nitrogen concentration depends on the phosphorus concentration in the medium. That is, systems with a low phosphorus concentration require a low nitrogen concentration and vice versa. However, in their research, the best hydrogen yield of 281 mL H_2/g starch was obtained at a C/N ratio of 200 and a C/P ratio of 1000, namely, for lower concentrations of nutrients. High nitrogen and phosphorous concentrations could inhibit hydrogen formation by dark fermentation, which likely alters the metabolic pathway [47]. In contrast, low at C/N and the pH values below 3.5 suggests that surplus carbon source could cause rapid acidification and influence the metabolism and growth of microorganism [44].

Some complex wastewater, such as cheese whey [21, 22], Tequila vinasses [33], cassava wastewater [24, 25], soft-drink wastewater [47] and corn starch wastewaters [44], added nutrients to ensure that all the required components were present.

Peixoto et al. [47] showed a similar example when added urea (COD:N of 100:0.7) was used as the nitrogen source in one of their upflow fixed-bed reactors. Under that condition, the hydrogen production ceased completely after 8 days of operation. In contrast, the reactor with a COD:N ratio of 100:0.3 produced hydrogen continuously for 70 days with an average hydrogen yield of 3.5 mol H_2/mol substrate. These authors suggested that the excessive cell growth caused by the addition of nutrients affected the reactor's hydrodynamic pattern, hindering the liquid-gas transfer mass of hydrogen. In addition, the decrease of the HRT increased the production of non-reduced compounds.

Searmsirimongkol et al. [36] evaluated hydrogen production using as source substrate wastewater from ethanol processing produced from sugarcane in anaerobic sequencing batch reactor (ASBR). Through concentration of 40 g/L, OLR of 60 kg COD/m^3/d, HRY of 16 h and pH 5.5, at 37°, reached 3320 mL H_2/L/d and 172 mL H_2/g COD removed. The high concentrations of potassium and sulphate observed in raw stillage (with COD of 150 g/L), 8.8 and 7.0 g/L, respectively, show the need to dilute the affluent to avoid toxic effect to the hydrogen-producing bacteria. At concentrations above 40 g COD/L, system performance decreased in terms of hydrogen production due to higher concentrations of PO4-3 and SO4-2. Gomes et al. [24] showed that hydrogen production from cassava wastewater quickly decreased and terminated even in the presence of the heat-treated wastewater with or without nutrient supplementation. The authors suggested that the problems were not due to lack of nutrients, but due to the presence of lactic bacteria.

6.3. Temperature

Temperature affects the growth rate and the metabolic pathways of microorganisms, and is considered one of the most important operating parameters which affect fermentative production of hydrogen. Microorganisms are capable of producing hydrogen at a temperature ranging from 15 to 85°C. Fermentative reactions for hydrogen production are mainly conducted at mesophilic (25–40°C) and thermophilic (40–65°C) temperatures, while few studies have been conducted in hyperthermophilic temperatures (above 80°C) [8].

As shown in **Table 1**, most of the studies were conducted under mesophilic conditions (25–40°C). Only a few studies [35, 43] were conducted under thermophilic conditions (45–55°C). High temperature can promote hydrolysis and simplify microbial diversity in a manner favourable to H_2 production, but it can also bring about monotonous microbial diversity, resulting in incomplete substrate degradation, especially in the treatment of actual waste. Also, operation at high temperature places an economic burden, as it requires a tight and closed structure and immense energy to heat and maintain the temperature of the reactor. Therefore, the temperature effect must be thoroughly investigated considering not only the H_2 fermentation performance but also substrate degradation and economic factors [62].

Few studies have evaluated the effect of temperature, and the substrates during the investigation of the effect of temperature on fermentative hydrogen production were tofu-processing wastewater [34], brewery wastewater [35] and Tequila vinasses [33].

It should be noted that fermentative processes operating under thermophilic conditions have some advantages over mesophilic processes. This is due to: (i) higher temperature has lower solubility gas (Henry's law); (ii) the hydrogen synthesis pathways are less affected by the partial pressure of hydrogen (pH_2) [10] and (iii) the rates of chemical and enzymatic reactions are higher [71]. However, according to Mohan et al. [72], the optimal temperature for the production of hydrogen depends on the nature of the biocatalyst and the type of wastewater to be used as a substrate.

The effect of the temperature (25 and 3))5°C) on hydrogen production from Tequila vinasse was studied using a sequencing batch reactor, with HRT of 24 h [33]. A maximum HPR of 50.5 mL H_2/h/L and an average hydrogen content in the biogas of 29.2 ± 8.8% were obtained when the reactor was fed with 3 g COD/L, at 35 °C and 12-h HRT. It is 6.2 times greater than the temperature of 25°C under the same conditions.

Lay et al. [34] used two different temperatures (35 and 55°C) and two different seed sludges to evaluate the hydrogen production performance and obtain the best criteria for maximum production from tofu-processing wastewater. The temperature variation did not affect the HY significantly. The maximum HY of 61.2 mL/g COD was obtained at 35°C. Similar values were obtained with the 55°C) under the same conditions (HY of 58.8 mL/g COD).

6.4. Use of co-substrates

The use of co-substrates is motivated by other objectives being pursued concomitantly, including (a) combined treatment of different waste streams, (b) ability to treat residues otherwise difficult to manage individually, (c) dilution of potentially toxic/inhibitory compounds, (d) optimization of the conditions for hydrogen production and (e) optimization of the carbohydrate/protein ratio [16].

The literature also reports that simple substrates, such as glucose, have been used in mixtures with other complex substrates in the search for optimal conditions for hydrogen production: mixture of sugarcane stillage and glucose [30, 43]; glucose and cheese whey [21]; glucose and cassava wastewater [25]; and corn starch wastewaters [44]. The strategy of using mixed substrates demonstrates the high interest among researchers in evaluating the feasibility of hydrogen production through waste fermentation in the presence of glucose.

Wang et al. [44] using corn starch wastewaters exhibited an efficient H_2 yield which was found to be 76.0 and 31.7% higher than that of using corn starch and cassava starch, respectively. Moreover, in the study of Ferreira Rosa et al. [21] showed that the use of mixed substrates also favoured the production of hydrogen, when compared using glucose as an individual substrate. The co-fermentation of the cheese whey and glucose mixture was favourable for the concomitant production of hydrogen and ethanol, with yields of up to 1.7 mmol H_2/g COD and 3.45 mol EtOH/g COD in AFBR.

Most studies of co-fermentation focused on the performance of hydrogen production in AFBRs. It is interesting to note that there was a variation in biogas composition when the carbon source was changed from a mixture of glucose/wastewater [21, 30, 43]. Even with different operating conditions and wastewater, the same pattern of behaviour was observed, indicating that the substrate mixtures are a preferable carbon source compared with glucose.

Chen et al. [73] reported the inhibition of anaerobic processes, suggesting that to effect a better adaptation of microbial community, prior to use more complex substrate is placed on a simpler carbon source until their total consumption. Acclimation of anaerobic microorganisms both increases their tolerance to the toxicants shock and enhances toxicant biodegradability.

Co-fermentation from wastewaters with glucose and adaptation of microorganisms to inhibitory substances can significantly improve the wastewater treatment efficiency and hydrogen production. Possibly a favourable environment for the development of microorganisms has been created, with the presence of simple substrates and nutrients. However, the costs for pure carbohydrate sources are high for practical-scale hydrogen production, which can only be viable when based on renewable and low cost sources [6]. Studies to analyse the nutrients of different wastewaters, in order to get a better rate C:N and C:P, could make viable the combination of two complex wastewaters. This would make it feasible to process hydrogen production, due to the lower cost of substrates.

7. Microbial diversity

Table 4 shows that a limited number of reports co-exist on microbial communities producing hydrogen from complex wastewaters. These studies analysed the composition of microbial communities by cloning and sequencing the 16s rRNA from sugarcane vinasse, 454-pyrosequencing data analysis from sugarcane vinasse, fluorescent *in situ* hybridization (FISH) from glycerol and affiliation of band sequence from denaturing gradient gel electrophoresis (DGGE) from beverage wastewater.

Substrate	Microorganisms	GenBankaccess	Relative abundance	Microbiological analysis	Reference
Beverage wastewater	*Clostridium* sp.	NR_042144.1 NR_044718.2 NR_042144.1 NR_026100.1 NR_104822.1NR_074511.1JF499889NR_074482.1		Affiliation of band sequence (retrieved from DGGE gel)	Sivagurunathan et al. [29]
	Klebsiellaoxytoca	NR_102982.1			
	Selenomonas lacticifex	AF373024.1			
Sugarcane vinasse	Uncultured *Prevotella* sp.	JX575984.1	7	Cloning and sequencing the 16s rRNA	Reis et al. [30]
	Uncultured Prevotellaceae bacterium	JF806757.1	55		
	Megasphaera sp.	HM990965.2	28		
	Uncultured bacterium	JQ072158.1	7		
	Uncultured *Clostridia* bacterium	EU887973.1	13		
Glycerol	*Enterobacter* sp.		27.1	Fluorescent in situ hybridization (FISH)	Reungsang et al. [38]
	Firmicutes bacteria		18.88		
Sugarcane vinasse	*Pectinatus*		54.1	454-pyrosequencing data analysis	Ferraz Junior et al. [39]
	Clostridium		12.8		
	Megasphaera		3.3		
	Propionispora		3.2		
	Order Burkholderiales		3.6		
	Family Comamonadaceae		18		
Textile wastewater	*Clostridium butyricum*			PCR-DGGE) with partial 16S rRNA genes followed by their sequencing	Li et al. [40]
	Klebsiella oxytoca				
	Clostridium sp.				
Sugar cane stillage	*Clostridium cellulosi*	NR044624.1	7	Cloning and sequencing the 16s rRNA	Santos et al. [43]
	Thermoanaerobacterium	JX442957.1JX984979.1JX984974.1HM585225.1AF247003.1	62		
	Uncultured bacterium clone D8-50C-C4-3	HQ266872.1	20		
	Lactobacillus sp.	AB016864.1DQ523489.2	4		
	Moorella sp.	AB086398.1	2		
Soft-drink wastewater	*Clostridium* sp.	DQ196630		Amplified and sequenced from the DGGE samples	Peixoto et al. [47]
	Klebsiella sp.	EU196756			
	Enterobacter sp.	EU430750			

http://dx.doi.org/10.5772/110416

Substrate	Microorganisms	GenBankaccess	Relative abundance	Microbiological analysis	Reference
Condensed molasses fermentation solubles	*Clostridium butyricum*			Affiliation of band sequence (retrieved from DGGE gel)	Chu et al. [48]
	Megaspharea sp.				
	Corynebacterium glutamicum				

Table 4. Microbial diversity from complex wastewaters.

Hydrogen can be efficiently and economically obtained from dark fermentation by hydrogen-producing bacteria (HPB) [74]. *Clostridium* and *Enterobacter* were the most widely used microorganisms for fermentative hydrogen production in mesophilic conditions, and *Thermoanaerobacterium* genus under thermophilic conditions [75]. The members of genus *Clostridium* are Gram-positive, and contain endospore-forming rods that produce hydrogen. Already, *Enterobacter* are Gram-negative, rod-shaped and facultative anaerobes [63].

Among the fermentative anaerobes, *Clostridia* have been well known and studied extensively, not for their hydrogen production capability but for their role in the production of industrial solvent from various carbohydrates [7]. This is a common sense, and numerous studies have already been conducted considering the investigation and identification of *Clostridium* sp. with hydrogen yield productive capacity. However, in the production of hydrogen from complex wastewater it has been shown that it is possible to produce hydrogen from other bacteria beyond the genus *Clostridium*. Ferraz Júnior et al. [39] showed by 454-pyrosequence analysis, organisms affiliated with the *Clostridium* and *Pectinatus* genera were dominant in the sample associated with hydrogen production from sugarcane vinasse. In contrast, from the same wastewater, Reis et al. [30] showed by cloning and sequencing the 16s rRNA that 55% belonged to the phylum Bacteroidetes and uncultured Prevotella, and 28% belonged to the phylum Firmicutes genus Megasphaera. Also, the presence of 3% of uncultured *Clostridia* also belonged to the phylum Firmicutes. Under thermophilic conditions, both *Thermoanaerobacterium* sp. and *Clostridium* sp. were efficient hydrogen producers [43].

Many studies reported in the literature that evaluated the microbial community did not report a direct association between microorganisms found and hydrogen production. The likely cause for this is due to the diversity of other organisms found, different *Clostridium*. In contrast, Sivagurunathan et al. [29] reported that the *Clostridium* species dynamics were not significantly affected, but total microbial community structure changed with respect to HRT variation as evident from PCR-DGGE analyses. Moreover, the appearance of *Selenomonas* spp. in a CSTR at low OLR improved the HY, whereas the disappearance of *Selenomonas* spp. at high OLR improved the HPR, but gave a drop in HY from beverage industry wastewaters.

Other organisms have also been found in complex wastewater, such as *Klebsiella oxytoca* and *Enterobacter* sp., indicating the presence of predominant hydrogen-producing bacteria from textile wastewater and glycerol [38, 40].

8. Conclusions and perspectives

The analysis of over 35 literature references on fermentative hydrogen production from complex wastewater has shown that numerous process parameters have the potential of affecting the evolution of the metabolic pathways involved, in turn affecting the process kinetics and the conversion yield.

The production of hydrogen from wastewaters should contribute technologically to the fate of some wastewater, opening the possibility for them to be used as raw material to produce bioenergy. Thus, the discovery of new raw materials for the production of a sustainable fuel contributes to the consolidation of the sector. However, this review showed that there usually existed some disagreements on the optimal condition of a given factor for fermentative hydrogen production from complex wastewaters, thus more researches in this respect are recommended.

Author details

Paula Rúbia Ferreira Rosa and Edson Luiz Silva*

*Address all correspondence to: edsilva@ufscar.br

Department of Chemical Engineering, Federal University of São Carlos, São Carlos, São Paulo, Brazil

References

[1] IPCC- Intergovernmental panel on Climate Change. 2015. Available at: http:// https://www.ipcc.ch/ [accessed 05.02.15]

[2] Dincer I, Acar C. Review and evaluation of hydrogen production methods for better sustainability. International Journal of Hydrogen Energy. 2015;40:11094-11111. DOI: 10.1016/j.ijhydene.2014.12.035

[3] Barca C, Soric A, Ranava D, Giudici-Orticoni M-T, Ferrasse J-H. Anaerobic biofilm reactors for dark fermentative hydrogen production from wastewater: A review. Bioresource Technology. 2015;**185**:386–398. DOI: 10.1016/j.biortech.2015.02.063

[4] Wong YM, Wu TY, Juan JC. A review of sustainable hydrogen production using seed sludge via dark fermentation. Renewable and Sustainable Energy Reviews. 2014;**34**:471–482. DOI: 10.1016/j.rser.2014.03.008

[5] Mohan SV. Harnessing of biohydrogen from wastewater treatment using mixed fermentative consortia: Process evaluation towards optimization. International Journal of Hydrogen Energy. 2009;**34**:7460–7474. DOI: 10.1016/j.ijhydene.2009.05.062

[6] Show KY, Lee DJ, Chang JS. Bioreactor and process design for biohydrogen production. Bioresource Technology. 2011;**102**:8524–8533. DOI: 10.1016/j.biortech.2011.04.055

[7] Show KY, Lee DJ, Tay JH, Lin CY, Chang JS. Biohydrogen production: Current perspectives and the way forward. International Journal of Hydrogen Energy. 2012;**37**:15616–15631. DOI: 10.1016/j.ijhydene.2012.04.109

[8] Sinha P, Pandey A. An evaluative report and challenges for fermentative biohydrogen production. International Journal of Hydrogen Energy. 2011;**36**:7460–7478. DOI: 10.1016/j.ijhydene.2011.03.077

[9] Guo XM, Trably E, Latrille E, Carrere E, Steyer JP. Hydrogen production from agricultural waste by dark fermentation: A review. International Journal of Hydrogen Energy. 2010;**35**:10660–10673. DOI: 10.1016/j.ijhydene.2010.03.008

[10] Levin DB, Pitt L, Love M. Biohydrogen production: Prospects and limitations to practical application. International Journal of Hydrogen Energy. 2004;**29**:173–185. DOI: 10.1016/S0360-3199(03)00094-6

[11] Abbasi T, Tauseef SM, Abbasi SA. Anaerobic digestion for global warming control and energy generation—An overview. Renewable and Sustainable Energy Reviews. 2012;**16**:3228–3242. DOI: 0.1016/j.rser.2012.02.046

[12] Noori M, Saady C. Homoacetogenesis during hydrogen production by mixed cultures dark fermentation: Unresolved challenge. International Journal of Hydrogen Energy. 2013;**38**:13172– 13191. DOI: 10.1016/j.ijhydene.2013.07.122

[13] Hwang MH, Jang NJ, Hyun SH, Kim IS. Anaerobic bio-hydrogen production from ethanol fermentation: The role of pH. Journal of Biotechnology. 2004;**111**:297–309. DOI: 10.1016/j.jbiotec.2004.04.024

[14] Barros AR, Silva EL. Hydrogen and ethanol production in anaerobic fluidized bed reactors: Performance evaluation for three support materials under different operating conditions. Biochemical Engineering Journal. 2012;**61**:59–65. DOI: 10.1016/j.bej.2011.12.002

[15] Siriwongrungson V, Zeng RJ, Angelidaki I. Homoacetogenesis as the alternative pathway for H_2 sink during thermophilic anaerobic degradation of butyrate under suppressed methanogenesis. Water Research. 2007;**41**:4204–4210.

[16] De Gioannis G, Muntoni A, Polettinib A, Pomi R. A review of dark fermentative hydrogen production from biodegradable municipal waste fractions. Waste Management. 2013;**33**:1345–1361. DOI: 10.1016/j.wasman.2013.02.019

[17] Mohan SR. Structure and growth of research on biohydrogen generation using wastewater. International Journal of Hydrogen Energy. 2015;**40**:16056–16069. DOI: 10.1016/j.ijhydene.2015.08.072

[18] Siqueira LM, Damiano ESG, Silva EL. Influence of organic loading rate on the anaerobic treatment of sugarcane vinasse and biogas production in fluidized bed reactor. Journal of Environmental Science and Health A. 2013;**48**:1707–1716.

[19] Oliveira MA, Reis EM, Nozaki J. Biological treatment of wastewater from the cassava meal industry. Environmental Research. 2001;**85**:177–183. DOI: 10.1006/enrs.2000.4118

[20] Carvalho F, Prazeres AR, Rivas J. Cheese whey wastewater: Characterization and treatment. Science of the Total Environment. 2013;**445–446**:385–396. DOI: 10.1016/j.scitotenv.2012.12.038

[21] Ferreira Rosa PR, Santos SC, Silva EL. Different ratios of carbon sources in the fermentation of cheese whey and glucose as substrates for hydrogen and ethanol production in continuous reactors. International Journal of Hydrogen Energy. 2014;**39**:1288–1296. DOI: 10.1016/j.ijhydene.2013.11.011

[22] Carrillo-Reyes J, Celis LB, Alatriste-Mondrago F, Razo-Flores E. Decreasing methane production in hydrogenogenic. Biomass and Bioenergy. 2014;**63**:101–108. DOI: 10.1016/j.biombioe.2014.01.050

[23] Intanoo P, Chaimongkol P, Chavadej S. Hydrogen and methane production from cassava wastewater using two-stage upflow anaerobic sludge blanket reactors (UASB) with an emphasis on maximum hydrogen production. International Journal of Hydrogen Energy. 2016;**41**:6107–6114. DOI: 10.1016/j.ijhydene.2016.10.125

[24] Gomes SD, Fuess LT, Manunga T, Gomes PCFL, Zaiat M. Bacteriocins of lactic acid bacteria as a hindering factor for biohydrogen production from cassava flour wastewater in a continuous multiple tube reactor. International Journal of Hydrogen Energy. 2016;**41**:8120–8131. DOI: 10.1016/j.ijhydene.2015.11.186

[25] Rosa PRF, Gomes BC, Varesche MBA, Silva EL. Characterization and antimicrobial activity of lactic acid bacteria from fermentative bioreactors during hydrogen production using cassava processing wastewater. Chemical Engineering Journal. 2016;**284**:1–9. DOI: 10.1016/j.cej.2015.08.088

[26] Han W, Wang B, Zhou Y, Wang DS, Wang Y, Yue L-R, Li H-F, Ren NQ. Fermentative hydrogen production from molasses wastewater in a continuous mixed immobilized sludge reactor. Bioresource Technology. 2012;**110**:219–223. DOI: 10.1016/j.biortech.2012.01.057

[27] Thanwised P, Wirojanagud W, Reungsang A. Effect of hydraulic retention time on hydrogen production and chemical oxygen demand removal from tapioca wastewater using anaerobic mixed cultures in anaerobic baffled reactor (ABR). International Journal of Hydrogen Energy. 2012;**37**:15503–15510. DOI: 10.1016/j.ijhydene.2012.02.068

[28] Chookaew T, O-Thong S, Prasertsan P. Biohydrogen production from crude glycerol by immobilized *Klebsiella* sp. TR17 in a UASB reactor and bacterial quantification under non-sterile conditions. International Journal of Hydrogen Energy. 2014;**39**:9580–9587. DOI: 10.1016/j.ijhydene.2014.04.083

[29] Sivagurunathan P, Sen B, Lin C-Y. High-rate fermentative hydrogen production from beverage wastewater. Applied Energy. 2015;**147**:1–9. DOI: 10.1016/j.apenergy.2015.01.136

[30] Reis CM, Carosia MF, Sakamoto IK, Varesche MBA, Silva EL. Evaluation of hydrogen and methane production from sugarcane vinasse in an anaerobic fluidized bed

reactor. International Journal of Hydrogen Energy. 2015;**40**:8498–8509. DOI: 10.1016/j.ijhydene.2015.04.136

[31] Liu C-M, Zheng J-L, Wu S-Y, Chu C-Y. Fermentative hydrogen production potential from washing wastewater of beverage production process. International Journal of Hydrogen Energy. 2016;**41**:4466–4473. DOI: 10.1016/j.ijhydene.2015.08.079

[32] Poontaweegeratigarn T, Chavadej S, Rangsunvigit P. Hydrogen production from alcohol wastewater by upflow anaerobic sludge blanket reactors under mesophilic temperature. International Scholarly and Scientific Research & Innovation. 2012;**6**:293–296.

[33] Buitrón G, Carvajal C. Biohydrogen production from Tequila vinasses in an anaerobic sequencing batch reactor: Effect of initial substrate concentration, temperature and hydraulic retention time. Bioresource Technology. 2010;**101**:9071–9077. DOI: 10.1016/j.biortech.2010.06.127

[34] Lay C-H, Sen B, Huang S-C, Chen C-C, Lin C-Y. Sustainable bioenergy production from tofu-processing wastewater by anaerobic hydrogen fermentation for onsite energy recovery. Renewable Energy. 2013;**58**:60–67. DOI: 10.1016/j.renene.2013.03.011

[35] Shi X-Y, Jin D-W, Sun Q-Y, Li W-W. Optimization of conditions for hydrogen production from brewery wastewater by anaerobic sludge using desirability function approach. Renewable Energy. 2010;**35**:1493–1498. DOI: 10.1016/j.renene.2010.01.003

[36] Searmsirimongkol P, Rangsunvigit P, Leethochawalit M, Chavadej S. Hydrogen production from alcohol distillery wastewater containing high potassium and sulfate using an anaerobic sequencing batch reactor. International Journal of Hydrogen Energy. 2011;**36**:12810–12821. DOI: 10.1016/j.ijhydene.2011.07.080

[37] Singh L, Wahid ZA, Siddiqui MF, Ahmad A, Rahim MHA, Sakinah M. Application of immobilized upflow anaerobic sludge blanket reactor using Clostridium LS2 for enhanced biohydrogen production and treatment efficiency of palm oil mill effluent. International Journal of Hydrogen energy. 2013;**38**:2221–2229. DOI: 10.1016/j.ijhydene.2012.12.004

[38] Reungsang A, Sittijunda S, O-thong S. Bio-hydrogen production from glycerol by immobilized Enterobacter aerogenes ATCC 13048 on heat-treated UASB granules as affected by organic loading rate. International Journal of Hydrogen Energy. 2013;**38**:6970–6979. DOI: 10.1016/j.ijhydene.2013.03.082

[39] Ferraz Júnior ADN, Etchebehere C, Zaiat M. Mesophilic hydrogen production in acidogenic packed-bed reactors (APBR) using raw sugarcane vinasse as substrate: Influence of support materials. Anaerobe. 2015;**34**:94–105. DOI: 10.1016/j.anaerobe.2015.04.008

[40] Li Y-C, Chu C-Y, Wu S-Y, Tsai C-Y, Wang C-C . Feasible pretreatment of textile wastewater for dark fermentative hydrogen production. International Journal of Hydrogen energy. 2012;**37**:15511–15517. DOI: 10.1016/j.ijhydene.2012.03.131

[41] Seo YW, Yun W-M, Lee H, Han JI. Pretreatment of cheese whey for hydrogen production using a simple hydrodynamic cavitation system under alkaline condition. Fuel. 2015;**150**:202–207. DOI: 10.1016/j.fuel.2015.01.100

[42] Ramprakash B, Muthukumar S. Comparative study on the performance of various pretreatment and hydrolysis methods for the production of biohydrogen using Enterobacteraerogenes RM 08 from rice mill wastewater. International journal of Hydrogen energy. 2015;**40**:9106–9112. DOI: 10.1016/j.ijhydene.2015.05.027

[43] Santos SC, Rosa PRF, Sakamoto IK, Varesche MBA, Silva EL. Continuous thermophilic hydrogen production and microbial community analysis from anaerobic digestion of diluted sugar cane stillage. International Journal of Hydrogen Energy. 2014;**39**:9000–9011. DOI: 10.1016/j.ijhydene.2014.03.241

[44] Wang S, Zhang T, Su H. Enhanced hydrogen production from corn starch wastewater as nitrogen source by mixed cultures. Renewable Energy. 2016;**96**:1135–1141. DOI: 10.1016/j.renene.2015.11.072

[45] Jung K-W, Kim D-H, Shin H-S. Continuous fermentative hydrogen production from coffee drink manufacturing wastewater by applying UASB reactor. International journal of Hydrogen Energy. 2010;**35**:13370–13378. DOI: 10.1016/j.ijhydene.2009.11.120

[46] Sen B, Suttar RR. Mesophilic fermentative hydrogen production from sago starch-processing wastewater using enriched mixed cultures. International Journal of Hydrogen Energy. 2012;**37**:15588–15597. DOI: 10.1016/j.ijhydene.2012.04.027

[47] Peixoto G, Saavedra NK, Varesche MBA, ZaiatM. Hydrogen production from softdrink wastewater in an upflow anaerobic packed-bed reactor. International Journal of Hydrogen Energy. 2011;**36**:8953–8966. DOI: 10.1016/j.ijhydene.2011.05.014

[48] Chu CY, Wu SY, Hsieh PC, Lin CY. Biohydrogen production from immobilized cells and suspended sludge systems with condensed molasses fermentation soluble. International Journal of Hydrogen Energy. 2011;**36**:14078–14085. DOI: 10.1016/j.ijhydene.2011.04.101

[49] Trchounian K, Poladyan A, Trchounian A. Optimizing strategy for Escherichia coli growth and hydrogen production during glycerol fermentation in batch culture: Effects of some heavy metal ions and their mixtures. Applied Energy. 2016;**177**:335–340. DOI: 10.1016/j.apenergy.2016.05.129

[50] Oh SE, Van Ginkel S, Logan BE. The Relative effectiveness of pH control and heat treatment for enhancing biohydrogen gas production. Environmental Science and Technology. 2003;**37**:5186–5190. DOI: 10.1021/es034291y

[51] Liu Z, Zhang C, Lu Y, Wu X, Wang L, Wang L, Han B, Xing X-H. States and challenges for high-value biohythane production from waste biomass by dark fermentation technology. Bioresource Technology. 2013;135:292-303. Doi: 10.1016/j.biortech.2012.10.027.

[52] Pasupuleti SB, Mohan SK . Single-stage fermentation process for high-value biohythane production with the treatment of distillery spent-wash. Bioresource Technology. 2015;**189**:177–185. DOI: 10.1016/j.biortech.2015.03.128

[53] Kim IS, Hwang MH, Jang N, Hyun SH, Lee ST. Effect of low pH on the activity of hydrogen utilizing methanogen in bio-hydrogen process. International Journal of Hydrogen Energyy. 2004;**29**:1133–1140. DOI: 10.1016/j.ijhydene.2003.08.017

[54] Taconi KA, Zappi ME, Todd French WT, Brown LR. Feasibility of methanogenic digestion applied to a low pH acetic acid solution. Bioresource Technology. 2007;**98**:1579–1585. DOI: 10.1016/j.biortech.2006.06.019

[55] Luo G, Karakashev Di, Xie L, Zhou Q, Angelidaki I. Long-term effect of inoculum pretreatment on fermentative hydrogen production by repeated batch cultivations: Homoacetogenesis and methanogenesis as competitors to hydrogen production. Biotechnology and Bioengineering. 2011;**108**:1816–1827.

[56] Carrillo-Reyes J, Celis LB, Alatriste-Mondragona F, Razo-Flores E. Different start-up strategies to enhance biohydrogen production from cheese whey in UASB reactors. International Journal of Hydrogen Energy. 2012;**37**:5591–5601. DOI: 10.1016/j.ijhydene.2012.01.004

[57] Rosa PRF, Santos SC, Sakamoto IK, Varesh MBA, Silva EL. Hydrogen production from cheese whey with ethanol-type fermentation: Effect of hydraulic retention time on the microbial community composition. Bioresource Technology. 2014;**161**:10–19. DOI: 10.1016/j.biortech.2014.03.020

[58] Ren N-R, Tang J, Liu B-F, Guo W-Q. Biological hydrogen production in continuous stirred tank reactor systems with suspended and attached microbial growth. International Journal of Hydrogen Energy. 2010;**35**:2807–2813. DOI: 10.1016/j.ijhydene.2009.05.010

[59] Chernicharo CAL. Anaerobic Reactors—Biological Wastewater Treatment Series—Volume Four. 1st ed. London, New York: IWA Publishing; 2007. 175 p.

[60] Campos JR. Sewage treatment by anaerobic process and disposition controlled soil. 1st ed. Rio de Janeiro: ABES; 1999. 435 p.

[61] Hafez H, Nakhla G, Naggar MHE, Elbeshbishy E, Baghchehsaraee B. Effect of organic loading on a novel hydrogen bioreactor. International Journal of Hydrogen Energy. 2010;**35**:81–92. DOI: 10.1016/j.ijhydene.2009.10.051

[62] Jung K-W, Kim D-H, Kim S-H, Shin H-S. Bioreactor design for continuous dark fermentative hydrogen production. Bioresource Technology. 2011;**102**:8612–8620. DOI: 10.1016/j.biortech.2011.03.056

[63] Wang J, Wan W. Factors influencing fermentative hydrogen production: A review. International Journal of Hydrogen Energy. 2009;**34**:799–811. DOI: 10.1016/j.ijhydene.2008.11.015

[64] Kapdan IK, Kargi F. Biohydrogen production from waste materials. Enzyme and Microbial Technology. 2006;**38**:569–582.

[65] Mohan SV, Mohanakrishna G, Ramanaiah SV, Sarma PN. Simultaneous biohydrogen production and wastewater treatment in biofilm configured anaerobic periodic discontinuous batch reactor using distillery. International Journal of Hydrogen Energy. 2008; 33:550-558.

[66] Lacerda IC, Miranda RL, Borell BM, Nunes AC, Nardi RM, Lachance MA, Rosa CA. Lactic acid bacteria and yeasts associated with spontaneous fermentations during the production of sour cassava starch in Brazil. International Journal of Food Microbiology. 2005;**105**:213–219.

[67] Noike T, Takabatake H, Mizuno O, Ohba M. Inhibition of hydrogen fermentation of organic wastes by lactic acid bacteria. International Journal of Hydrogen Energy. 2002;**27**:1367–1371.

[68] Lin CY, Lay CH. Carbon/nitrogen-ratio effect on fermentative hydrogen production by mixed microflora. International Journal of Hydrogen Energy. 2004;**29**:41–45. DOI: 10.1016/S0360-3199(03)83-1

[69] Intanoo P, Rangsunvigit P, Namprohm W, Thamprajamchit B, Chavadej J, Chavadej S. Hydrogen production from alcohol wastewater by an anaerobic sequencing batch reactor under thermophilic operation: Nitrogen and phosphorous uptakes and transformation. International Journal of Hydrogen Energy. 2012;**37**:11104–11112. DOI: 10.1016/j.ijhydene.2012.04.129

[70] Argun H, Kargi F, Kapdan I, Oztekin R. Biohydrogen production by dark fermentation of wheat powder solution: Effects of C/N and C/P ratio on hydrogen yield and formation rate. International Journal of Hydrogen Energy. 2008;**33**:1813–1819. DOI: 10.1016/j.ijhydene.2008.01.038

[71] Hallenbeck PC. Fundamentals of the fermentative production of hydrogen technology. Water Science and Technology. 2005;**52**:21–19

[72] Mohan SV, Chiranjeevi P, Mohanakrishna G. A rapid and simple protocol for evaluating biohydrogen production potential (BHP) of wastewater with simultaneous process optimization. International Journal of Hydrogen Energy. 2012;**37**:3130–3141. DOI: 10.1016/j.ijhydene.2011.10.109

[73] Chen Y, Cheng JJ, Creamer KS. Inhibition of anaerobic digestion process: A review. Bioresource Technology. 2008;**99**:4044–4064. DOI: 10.1016/j.biortech.2007.01.057

[74] Luo G, Xie L, Zou Z, Wang W, Zhou Q. Evaluation of pretreatment methods on mixed inoculum for both batch and continuous thermophilic biohydrogen production from cassava stillage. Bioresource Technology. 2010;**101**:959–964. DOI: 10.1016/j.biortech.2009.08.090

[75] Li RY, Zhang T, Fang HHP. Application of molecular techniques on heterotrophic hydrogen production research. Bioresource Technology. 2011;**102**:8445–8456. DOI: 10.1016/j.biortech.2011.02.072

Chapter 14

Bifunctional Heterogeneous Catalysts for Biodiesel Production using Low Cost Feedstocks: A Future Perspective

Anita Ramli, Muhammad Farooq, Abdul Naeem,
Saleem Khan, Muhammad Hummayun, Azhar Iqbal,
Sohail Ahmed and Liaqat Ali Shah

Additional information is available at the end of the chapter

http://dx.doi.org/10.5772/65553

Abstract

Currently, the fossil fuel sources are the major contributors to the world's energy mix. However, these conventional energy sources are depleting very fast due to their finite nature and extensive uses. An addition to their finite nature, environmental problems related to their uses are getting progressively worse and worse, initiating challenging debates for scientific communities. Biodiesel, a renewable fuel, has shown promising prospects due to its strong socioeconomic benefits and motivations in most of the countries of the world. Bifunctional heterogeneous catalysts are strongly recommended for biodiesel production from different feedstocks to simplify the process. This review highlights the challenges and opportunities associated with the heterogeneous catalysts and some recommendations to design an efficient bifunctional heterogeneous catalyst for economical biodiesel production from waste cooking oil.

Keywords: biodiesel, bifunctional heterogeneous catalyst, recommendation

1. Introduction

Deep concerns regarding fast depletion of conventional energy resources and their associated environmental issues are hot debates for both developed and developing countries. Moreover, these conventional energy sources are located in politically unstable regions, creating issues about scarcity of supplies and instabilities in the international oil prices [1]. It is, therefore essential to explore some alternative, potential, renewable, and sustainable source of energy that would be secure, economical, and environment friendly [2–4]. Among explored biofuels, biodiesel has received a great deal of attention as compared to other biofuels due to its domestic and renewable origin, nontoxic nature, biodegradability, environmental benefits, and excellent lubricity [5, 6].

A variety of oleo-chemical feedstocks such as vegetable oils (edible and non-edible oils), waste cooking oils (WCOs), animal fats, and greases have been identified as potential raw oils for biodiesel production in different countries (**Figure 1**) [7]. At present, biodiesel is mainly produced from edible oils all over the world (more than 95%) [8, 9]. However, the extensive use of edible oils for biodiesel production might lead to chronic food crisis in poor and developing countries. Therefore, to avoid the food problems, World Food Organization (WFO) has also legislated strict regulations over the use of edible oils for fuel purpose. The use of non-edible oils or waste cooking oils is therefore considered to be the possible solution to overcome the food issues as well as lower the biodiesel production cost-effectively [10–14]. In addition, the use of waste cooking will further solve the environmental problems that arise from its disposal [15, 16].

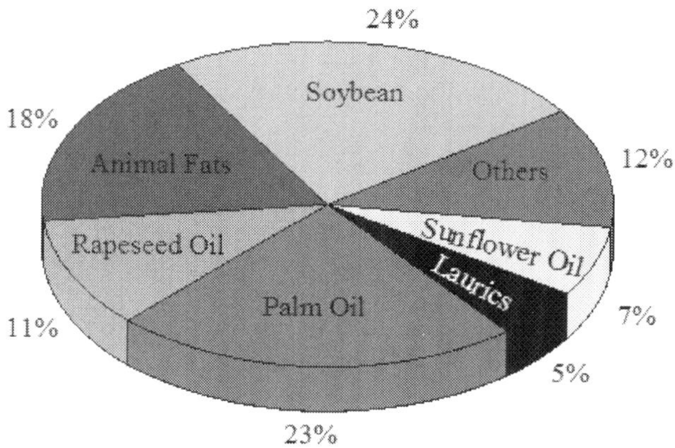

Figure 1. Production of biodiesel from different feed stocks [7].

Commercially, biodiesel production is carried out from vegetable oils or animal fats using homogeneous catalysts all over the world [17, 18]. Homogeneous base catalysts possess high catalytic activity under mild reaction conditions (from 40 to 65°C in normal atmospheric pressure) [19]. Similarly, homogeneous acid catalysts are also used in the biodiesel production from high free fatty acids feedstocks. However, homogeneous catalysts bear some technical issues such as soap formation, reactor corrosion, difficult to recover, and the production of large amount of polluted water, which in turn increase the overall biodiesel production cost and hazards [20, 21]. Therefore, special attention is being paid to the involvement of the heterogeneous catalyst for biodiesel production due to their green and recyclable catalytic activities. Heterogeneous catalysis has the ability to mitigate the various challenges encountered with the use of homogeneous catalysts for biodiesel production from low cost feedstocks. Heterogeneous catalysts bear several technical advantages such as easy separation and purification of the reaction products, reduced reactor corrosion, low sensitivity towards free fatty acids and moisture contents [22, 23]. In general, solid base catalysts show high activity, relatively shorter reaction times, and require lower reaction temperatures as compared to solid acid catalysts [24]. However, solid acid/ base catalysts have several limitations such as feedstock specification, deactivation due to leaching, and reusability for sustainable biodiesel production.

Recently, the concept of bifunctional heterogeneous catalysts has been introduced in the biodiesel technology for efficient biodiesel production from different feedstocks. Several studies have been carried out on the use of the bifunctional heterogeneous catalyst for biodiesel production from low cost feedstocks [25–28]. Bifunctional heterogeneous catalysts exhibit both acid and base character, therefore can simultaneously carry out esterification of free fatty acids and transesterification of triglycerides to develop cleaner and economical processes for biodiesel production. More importantly, a bifunctional heterogeneous catalyst can easily be modified to introduce the desired physicochemical properties so that the presence of free fatty acids or water does not adversely affect the reaction steps during the transesterification process.

This study aims to review the role of bifunctional heterogeneous catalysts in biodiesel production from different feedstocks for a sustainable energy process. In addition, the possible recommendations will also be discussed in the light of existing problems associated with the current biodiesel technology to design a robust catalyst for a sustainable biodiesel technology.

2. Biodiesel and its production scenario

According to American Society for Testing and Materials (ASTM), biodiesel is a mixture of mono-alkyl esters of long-chain fatty acids derived from vegetable oils (edible and non-edible origin) or animal fats. Biodiesel is regarded as a clean fuel, emitting negligible amount of pollutants into the environment and sufficiently reduces the emission of pollutants when blended with diesel [29–31]. A number of methods are currently used for biodiesel production from different feedstocks to overcome the high viscosity of the vegetable oils as a fuel. However, there are four main processes employed for biodiesel production (**Figure 2**): direct use and blending of raw oils [32, 33], micro-emulsions [34, 35], thermal cracking [36, 37], and transesterification [38–40]. Among them, the most commonly used method for converting oils into biodiesel is transesterification because the fuel produced by this method shows good compatibility with the existing engine [41, 42]. The vegetable oils and fats are extremely viscous (10–17 times viscous than petroleum diesel fuel), therefore posing several problems as alternative engine fuels [43]. The main objective of the transesterification process is to lower the viscosity of the oil close to that of petro-diesel. In transesterification, vegetable oils or animal fats are chemically converted into their corresponding fatty acid esters in the presence of alcohol using a suitable catalyst [44, 45].

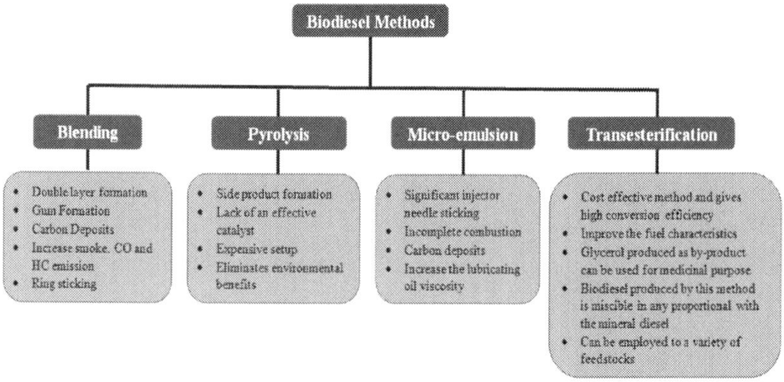

Figure 2. Methods used for biodiesel production.

The cost of biodiesel is widely regarded as a principal barrier in the development of the sustainable energy process. It has been reported that the feedstock cost is the major factor contributing to the final biodiesel cost and this cost represents approximately 70–95% of the total cost of biodiesel production [46–48]. Therefore, several efforts have been devoted to explore and select an ideal feedstock for economically viable biodiesel production. As a result, some countries of the world have focused on the use non-edible oils as a feedstock for biodiesel production to avoid food problems. However, the use of non-edible oils for biodiesel production cannot be the ultimate solution for a sustainable energy process as large plantation land areas would be required for large-scale non-edible oils production. This may disturb the entire animal and plant ecosystems [45, 47]. Therefore, it is important to search for cost effective biodiesel feedstock such as waste cooking oil, which is cheap and easily available all over the world. The utilization of waste cooking oil as a fuel effectively resolves the environmental problems associated with its direct discharge into the drainage system [49].

Similarly, another important factor contributing to the final cost of biodiesel is the technological challenge, involving exploration of highly effective catalysts and their corresponding processes for a sustainable biodiesel technology. A wide range of catalysts (homogeneous/heterogeneous) can be employed for biodiesel production from different feedstocks as shown in **Figure 3**. Heterogeneous catalysis has been considered to be the best choice for the biodiesel technology in the near future. Heterogeneous catalysts can easily be recovered, recycled, and have environmental friendly behavior. Recently, bifunctional heterogeneous catalysts have gained worldwide interest for biodiesel production due to their excellent performance in biodiesel production from low cost feedstocks. A bifunctional heterogeneous catalyst has the ability to carry out simultaneous esterification of free fatty acids and transesterification of triglycerides present in the waste cooking oil efficiently.

http://dx.doi.org/10.5772/65553

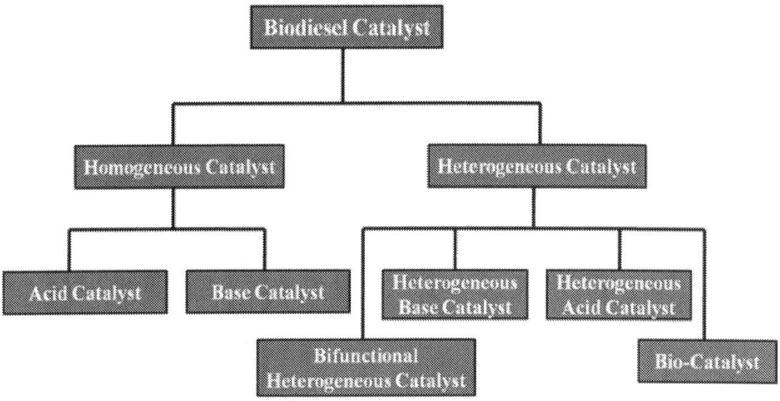

Figure 3. Catalysts used for biodiesel production.

3. Feedstocks for biodiesel production

Biodiesel is mainly produced by tranesterifying vegetable oils or animal fats commercially. However, compared to petroleum-derived diesel, the high cost of biodiesel is a major obstacle to its commercialization, which is 1.5–3 times higher than petroleum derived diesel [14]. The feedstocks mainly contribute to a major portion of the overall biodiesel production cost. Therefore, it is essential to select a feedstock for biodiesel synthesis, which is cheap, domestically available, and not compete with the food materials.

The choice of feedstocks for biodiesel production depends on two main factors, its availability and cost. Biodiesel is produced from different biological raw materials such as vegetable oils (edible and non-edible oils), animal fats, algal lipids, waste cooking oil, etc. [50–52]. Edible vegetable oils such as canola oil and soybean oil in USA, palm oil in Malaysia, rapeseed oil in Europe, and corn oil have been used as feedstocks for biodiesel production and found to be good diesel substitutes. Non-edible vegetable oils, such as oil from *Pongamia pinnata* (Karanja or Honge), *Jatropha curcas* (Jatropha or Ratanjyote) and *Madhuca indica* (Mahua) have also been found to be suitable feedstocks for biodiesel production [53, 54].

Similarly, algal lipids as feedstocks for biodiesel production are also gaining interest all over the world [55]. Algae convert carbon dioxide into sugar and proteins in the presence of sunlight. However, in the absence of nitrogen they mainly produce oil. A microalgae *Chlorella protothecoides*, has been grown under autotrophic and heterotrophic conditions to obtain lipids as a raw material for biodiesel industries. The lipid content in the heterotrophic cells reached 55.20% as compared to 14.57% in autotrophic cells [56].

Vegetable oils (edible and non-edible oils) are the predominant raw materials for the production of biodiesel, because they are renewable, potentially an inexhaustible source of energy, possess environmentally friendly characters, and can be produced on a large scale [16]. More than 95% of the biodiesel production is made from edible oils in different countries. It has been reported that approximately 70–95% of the total biodiesel production cost is related to the cost of the raw materials (vegetable oil or animal fats) [11, 21, 22]. According to reports of Food and Agriculture Organization (FAO), esculent plants containing oil are used for the production of biodiesel among which about 84% of the biodiesel is from rapeseed oil (RSO) and the remaining is from sunflower (13%), palm oil (1%), soybean, and others (2%) [2].

Moreover, the use of edible oils for fuel production is puzzling as more and more of the global food demand rises. There are also issues of deforestation and ecological imbalance while diverting the virgin forests and arable lands to large-scale biofuel production feedstocks [57]. In other words, the sustainability of edible oils as a biodiesel feed is under threat. Thus, the biodiesel production technology faces several challenges that must be overcome to make it sustainable.

The use of waste cooking oils for biodiesel production instead of edible oils can be a promising choice to enhance the economic viability of biodiesel production on a large scale. It has been reported that the biodiesel production cost can be reduced effectively from 60 to 70% by using waste cooking oil [24]. Since waste oil is easily available at a relatively low price, therefore can be a workable feedstock for biodiesel production to make the biodiesel competitive in price with petroleum-based diesel.

Huge amounts of waste cooking oils are produced all over the world every day, especially in the developed countries. Such a large amount of waste cooking oil production can lead to several disposal problems and contamination to water and land resources and ultimately to environmental pollution. Therefore, we need to search a green utilization of waste cooking oil to avoid the disposal problems. The utilization of waste cooking oil, as feedstock for biodiesel production is not only economical and guarantees food safety but will also minimize the environmental problems associated with their disposal [58]. The estimated waste cooking oil production in the selected countries is depicted in **Table 1**. Recently, several studies have been made to investigate the biodiesel production from waste cooking oil using heterogeneous catalysts for sustainable energy production. The details of these studies are given in **Table 2**.

Country	Waste cooking oil produced (million tons/year)
United States	10.00
China	4.50
Europe	0.70–1.00
Japan	0.45–0.57
Malaysia	0.50
Canada	0.12
Taiwan	0.07
Ireland	0.153

Table 1. Annual production of waste cooking oil in the selected countries [59].

Catalyst	Time (min)	Temperature (°C)	Catalyst loading (wt%)	Methanol:oil ratio	Yield (%)	Reference
CaO	120	65	0.78	12:1	≥ 99 (soap formation occurred)	[60]
Oil Palm Ash (K_2O major component)	30	60	5.35	18:1	71.74	[61]
Quick Lime Bit (solid base catalyst)	120	60	1.3	150:300 (v/v)	96.5	[62]
CaO	180	60	2	7:1	≥ 80.0	[63]
CaO-La_2O_3	180	58	5	20:1	~ 96.0	[64]
Mg-Al hydrotalcite	360	120	6	24:1	≥ 90	[65]
Mg-Al hydrotalcite or Mg-MCM-41	300	60	0.5	65 mL methanol and 5 g oil	97 or 87	[66]
ZnO-La_2O_3	180	200	3	180:126 (wt/wt)	96	[67]
K_3PO_4 (tri-potassium phosphate)	120	60	4	6:1	97.3	[68]
Zn_3La_1(Lanthanum modified Zno	180	200	2.4	26:1	92.3	[69]
CaO	120	80	3	6:1	≥ 84	[70]
Heterogeneous base catalyst						
Na-Mg-Al hydrotalcite (HT-Na)	480	60	7	9:1	67	[71]
$Mg_{1-x}Zn_{1+x}O_2$	255	188	2.5	9:1	≥ 80	[72]
SrO	0.67	MW oven of 1100 output with a magnetic stirring	0.276	6:1	99.8	[73]
Mg-Al hydrotalcite (2:1)	240	65	2.5	9:1	77.2	[74]
Hydrated lime	120	60	3.6	0.17 (vol)	100	[75]
CaO-ZrO_2	120	65	10	30:1	92	[76]
CaO (Clamshell)	180	60	3	6.03:1	> 89	[77]
KOH/Al_2O_3	120	70	5	9:1	96.8	[78]
Calcined snail shell (CaO is major component)	480	65	2.0	6.03:1	87.28	[79]
$CaAl_2$-CLDH (hydrocalumite)	300	65	5	≤6:1	≥ 95	[80]

Catalyst	Time (min)	Temperature (°C)	Catalyst loading (wt%)	Methanol:oil ratio	Yield (%)	Reference
Calcined eggshell waste (99.2 wt% CaO)	4	900 W Microwave power	15	18:1	96.7	[81]
Waste mud crab shells + cockleshells + boiler ash	30	65	3	15:1	99	[82]
Heterogeneous acid catalyst						
$K_4Zn_4[Fe(CN)_6]_3{\cdot}6H_2O{\cdot}$ 2(t-BuOH) [Fe–Zn double-metal cyanide (DMC) complexes]	480	170	3	15:1	98.0	[83]
D-glucose-derived solid acid	720	80	0.5	5.54:5 (wt/wt)	≥ 90	[84]
$SO_4^{2-}/TiO_2–SiO_2$	120	200	3	9:1	92	[48]
Zinc stearate/SiO_2	600	200	3	18:1	98	[85]
Starch-derived solid acid	600	80	10	30:1	92	[86]
$H_3PW_{12}O_{40}.6H_2O$ (heteropolyacid)	840	65	0.0375	70:1	97	[87]
$Al_{0.9}H_{0.3}PW_{12}O_{40}$ (aluminum dodecatungstophosphate)	840	65	3	34:1	96.1	[88]
$Zn_{1.2}H_{0.6}PW_{12}O_{40}$	720	65	2.5	28:1	97.2	[89]
$Zr_{0.7}H_{0.2}PW_{12}O_{40}$ (heteropolyacid, ZrHPW)	320	65	2.1	20:1	96.7	[90]
SO_4^{2-}/SnO_2-SiO_2	180	150	3	15:1	92.3	[91]
WO_x/Al_2O_3	120	110	1	0.3 (wt/wt)	97.5	[92]
SO_4^{2-}/ZrO_2	240	120	3	9:1	93.6	[93]
$Fe_2(SO_4)_3/C$	180	95.15	3.5	18:1	98	[94]
12-Tungstophosphoric acid (TPA)/Nb_2O_5 (25 wt % TPA)	1200	200	3	18:1	92	[95]
SO_4^{2-}/SnO_2-SiO_2	90	150	6	15:1	88.2	[96]
Vegetable oil asphalt-based solid acid	270	220	0.2	16.8:1	94.8	[97]
$Al(HSO_4)_3$	50	220	0.5	16:1	81	[98]
Arenesulfonic acid-functionalized mesoporous silica SBA-15	120	160	8	30:1	~ 80	[99]
SO_4^{2-}/SnO_2-SiO_2	60	150	6	Methanol:ethanol:oil: 9:6:1	81.4	[100]

Catalyst	Time (min)	Temperature (°C)	Catalyst loading (wt%)	Methanol:oil ratio	Yield (%)	Reference
Silica-sulfuric acid	30	60	1.5	6:1	90	[101]
$Al_2O_7Si_2.2H_2O$ (kaolinite)	120	160	3	31:1	92.4	[102]
SO_4^{2-}/ZrO_2-Al_2O_3	240	200	0.3	67.9:1	94.8?	[103]
$H_3PW_{12}O_{40}{\cdot}6H_2O$	60	30 (1.3 kW reboiler duty)		67.9:1	93.98	[104]
$Cs_{2.5}PW_{12}O_{40}$ [Cs-doped heteropoly acid]	40	260 (pressure: 20 MPa)	3	40:1	92	[105]
$Cs_{2.5}H_{0.5}PW_{12}O_{40}$/ionic liquids	220	70	1	6:1	97.3	[106]

Table 2. Summary of various heterogeneous catalysts used in the transesterification reaction of waste cooking oil.

The catalysts mentioned in **Table 2**, show good catalytic activity in biodiesel production from low cost waste cooking oil feedstocks. However, several problems such as separation, recycling, soap formation, leaching, reactor corrosion, etc., are associated with these catalysts while using for biodiesel production from low cost feedstocks containing high free fatty acids and water contents. In order to overcome these problems, bifunctional heterogeneous catalysts have been formulated to produce biodiesel from low cost feedstocks for a sustainable energy process. Bifunctional heterogeneous catalysts show very good efficiency in low cost feedstock conversion into biodiesel by carrying transesterification of triglycerides and esterification of free fatty acids present in the feedstock without any substantial loss in their activity due to the water content present.

4. Bifunctional heterogeneous catalysts for biodiesel production

Several studies have been carried out to address the technical challenges associated with the biodiesel technology for large-scale and profitable biodiesel production. First, the high cost of edible vegetable oil as the source of triglycerides plays a key role in process profitability. Therefore, to reduce the biodiesel production costs and make it competitive with petroleum based diesel, low cost feedstocks, such as non-edible oils, waste frying oils, and animal fats, may be utilized as raw materials. However, the major problem associated with such feedstock is higher amounts of free fatty acids and water, resulting in soap formation in the presence of an alkali catalyst. Thus, additional steps are required in order to remove any water and either the free fatty acids or soap from the reaction mixture.

Therefore, synthesis of novel heterogeneous catalysts with desired physical and chemical properties for biodiesel production form low-grade feedstocks is one of the focuses of the catalytic scientist. In this context, the bifunctional heterogeneous catalyst with acid-base character has attained a great focus for various organic reactions over the past decade. Bifunctional heterogeneous catalysts carry out simultaneous esterification of free fatty acids and transesterification of TG present in the oil effectively without being affected by the water content present or produced during the biodiesel formation due to the presence of both active acids and bases sites on the surface of catalyst, thereby effectively reducing the biodiesel production cost. Moreover, bifunctional heterogeneous catalysts can easily be tuned to include desired catalyst properties so that the presence of free fatty acids or water does not adversely affect the reaction steps and the biodiesel yield during the transesterification of triglycerides. The general mechanism for bifunctional heterogeneous catalysts is presented in **Figure 4**.

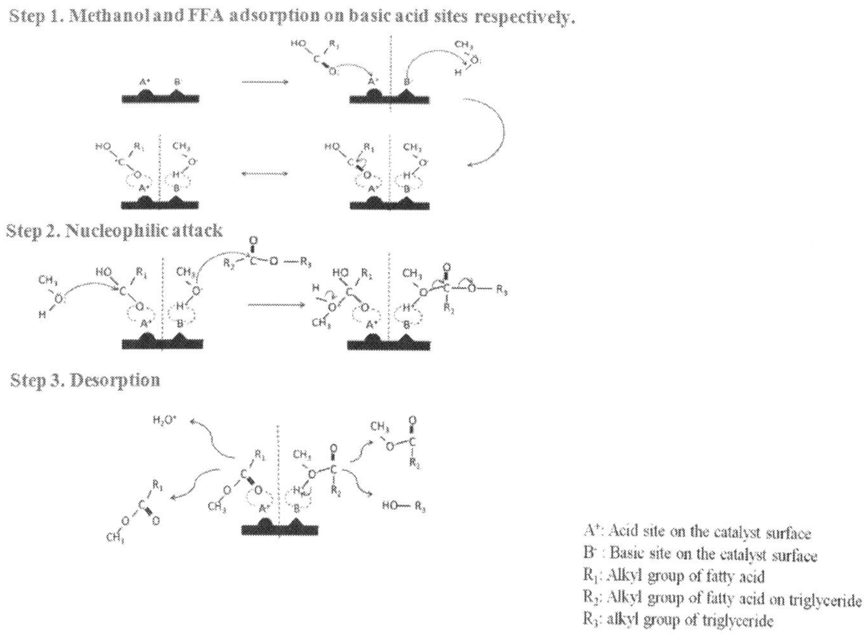

Figure 4. General mechanism for simultaneous esterification and transesterification reactions on bifunctional heterogeneous catalyst [107].

Several attempts have been made to investigate the catalytic activity of this type of catalysts in the transesterification reaction of low-grade feedstocks for biodiesel production. These catalysts show good catalytic performance during the biodiesel production from different feedstocks. The serious problem associated with the heterogeneous catalyst is its deactivation, which in turn reduce the catalyst reusability and chemical stability. These two factors play a key role in the biodiesel production cost, which is the main hurdle for its commercialization. In the view of the literature presented, it shows that the biodiesel technology is still not so mature to contribute more for overcoming the energy needs. Therefore, efforts are still in progress all over the world to develop an efficient bifunctional catalyst for biodiesel production

http://dx.doi.org/10.5772/65553

to overcome the serious disadvantages associated with the present biodiesel production technology. The bifunctional heterogeneous catalysts developed in the last few years are studied in detail below.

Yan et al. [64] synthesized novel zinc and lanthanum mixed oxides as heterogeneous catalysts for biodiesel production from unrefined or waste oils. They also studied the physicochemical properties of the synthesized catalysts, effects of the metal oxide molar ratio, free fatty acids and water contents in feedstock, molar ratio of methanol and oil, and reaction temperature, on the yield of biodiesel. The authors found that a strong interaction between the Zn and La species existed with enhanced catalyst activities. Moreover, lanthanum enhanced the zinc oxide distribution, thereby, increased the surface acid and base sites which in turn enhanced the catalyst ability for simultaneous transesterification and esterification reactions. They reported that the catalyst with a 3:1 ratio of zinc to lanthanum could effectively catalyze both transesterification/esterification reactions, and with negligible hydrolysis activity in oil as well as in biodiesel. The biodiesel yield of 96% of fatty acid methyl esters (FAME) was recorded within 3 h using waste oils in reaction temperature range of 170–220°C.

Macario et al. [108] produced biodiesel from waste oilseed fruits with methanol in the presence of heterogeneous/homogeneous systems using acid and basic catalysts. The authors synthesized catalysts with strong acid sites (USY, BEA, FAU-X) and catalysts with weak acid sites (MCM-41 and ITQ-6) by the hydrothermal process. Potassium was introduced into different materials by ionic exchange such as K-MCM-41 and K-ITQ-6 to prepare bifunctional catalysts (acid–base catalysts). They found that highest triglyceride conversion and biodiesel yield values could be achieved using K-ITQ-6 catalysts in reaction time of 24 h at 180°C temperature. They further demonstrated that deactivation of the catalyst occurred due to potassium leaching. The catalyst exhibited good regeneration viability and could be reused for biodiesel production from low-quality oil for cheaper biodiesel production.

Cannilla et al. [109] investigated the catalytic performance of the $MnCeO_x$ type bifunctional catalyst using refined sunflower oil for biodiesel production with methanol. They synthesized a series of manganese-ceria catalysts with the Mn/Ce atomic ratio ranging between 0.4 and 3.4 by the redox-precipitation method. NH_3-TPD and CO_2-TPD were utilized for acid-base active sites measurements on the surface of $MnCeO_x$ catalysts. It was found that $MnCeO_x$ catalytic systems exhibited high surface area, high chemical and thermal stability and high Mn dispersion, thereby resulted in high biodiesel production by the transesterification reaction of sunflower oil with methanol at a temperature of 140°C in 5 h of reaction time at low catalyst/oil ratio (1 wt.%). Moreover, the catalyst showed excellent catalytic stability and the deactivation phenomenon was found to be lower than that employed for acid catalytic reactions. They concluded that catalyst performance was the result of a synergic role played by both the surface acid/base character and textural porosity.

Wen et al. [110] prepared a TiO_2–MgO bifunctional mixed oxide catalyst by the sol–gel method to carry out simultaneous transesterification and esterification reactions in waste cooking oil into biodiesel. They found that the catalyst withone molar ratio of Ti and calcined at 923 K exhibited high activity and stability in the biodiesel reaction. They suggested that titanium improved the stability of the catalyst as a result of defects induced by the substitution of Ti ions for Mg ions in the magnesia lattice. Furthermore, they found that the MT-1-923 catalyst provided maximum biodiesel yield of 92.3% in the first use and increased after reuse of fourth time. The increase in the catalytic activity was attributed to increase in the specific surface area and average pore diameter after regeneration. The TiO_2-MgO mixed oxide catalyst showed good potential in large-scale biodiesel production from waste cooking oil.

Borges et al. [111] studied biodiesel production from sunflower oil and frying oil with a bifunctional heterogeneous catalyst derived from natural porous silica, pumice. The bifunctional activity of the catalyst was enhanced by introduction of K into the catalyst. They further investigated the dependence of the reaction variables such as temperature, reaction time, catalyst loading, and methanol/oil molar ratio on biodiesel yield using sunflower oil and waste oil as feedstocks. The natural material, pumice exchanged with a KOH aqueous solution demonstrated to be an efficient bifunctional heterogeneous particulate catalyst for simultaneous transesterification of triglycerides and esterification of free fatty acids present in sunflower oil and waste frying oil at low temperature (55°C). Moreover, the reusability studies showed that there was no considerable change in the activity of the catalyst even after five regenerations, demonstrating it to be a stable material.

Misra et al. [112] synthesized bifunctional heterogeneous catalysts by the conventional sol gel method for biodiesel production from different feedstocks such as soy, canola, coffee and waste vegetable oils containing variable amounts of free fatty acids (0–30 wt%). The authors reported that the synthesized bifunctional Quntinite-3T catalyst could effectively converted the free fatty acids and triglycerides (TGs) simultaneously into biodiesel in a single step of batch reactor. Similarly, Quntinite-3T also showed substantial reusability up to five times with more than 95% catalytic activity.

Omar et al. [107] studied the catalytic activity of alkaline modified zirconia (Mg/ZrO_2, Ca/ZrO_2, Sr/ZrO_2, and Ba/ZrO_2) to identify an efficient bifunctional heterogeneous catalyst for biodiesel production from waste cooking oil (WCO). Physicochemical properties of the synthesized catalysts were analyzed by BET surface area, XRD, FESEM and CO_2-NH_3-TPD. The authors reported that among different synthesized catalysts, Sr/ZrO_2 exhibited higher catalytic activity due to the presence of optimal active basic/acidic sites, facilitating simultaneous transesterification and esterification reactions in WCO. The results demonstrated that the methyl ester (ME) yield of 79.7% could be obtained by using 2.7 wt% catalyst loading (Sr/ZrO_2), 29:1 methanol to oil molar ratio, 169 min of reaction time and 115.5°C temperature.

Jiménez-López et al. [113] reported that incorporation of the WOx species into a mesoporous zirconium-doped silica could lead to an effective catalyst formulation that could catalyze both transesterification and esterification reactions in the used oil. During their investigations, they found that the catalyst calcined at 700°C with 15 wt% WO_3 loading provided the biodiesel yield higher than 80% after 2.5 h reaction time at reaction temperature of 200°C in the presence of methanol. Moreover, the catalytic activity could be maintained even in the presence of 5 wt% of water and after three cycles of re-utilization, without any further treatment of the catalyst.

Salinas et al. [114] investigated the catalytic performance of potassium supported on titania as a catalyst for the production of biodiesel from canola oil. The authors reported that low loadings of potassium lead to the formation of weak basic sites on the acid support (titania), therefore exhibited bifunctional behavior to produce biodiesel from low cost feedstocks. Furthermore, they added that the synthesized catalyst presented interesting activities with robust character since there was no need for in situ pre-treatment or inert reaction environment.

Farooq et al. [115] employed mixed oxide supported bifunctional heterogeneous catalysts in biodiesel production from waste cooking oil. The author reported that the synthesized catalysts show improved transesterification activities and provided the maximum biodiesel yield of 91.4% in reaction time of 4 h at a reaction temperature of 100°C, methanol to oil molar ratio of 27:1 and an agitation speed of 500 rpm. Moreover, the synthesized catalyst showed substantial chemical stability and could be reused for at least eight times without major loss in its catalytic activity. The catalyst deactivation in the higher run was attributed to strongly adsorbed organic molecules onto the active sites and leaching of the various active metals during biodiesel production from waste cooking oil. They further concluded that the physicochemical properties of the biodiesel produced from waste cooking oil comply with the international standard specifications.

Similarly, Taufiq and his research team [116] modified La_2O_3 with Bi_2O_3 (1–7 wt%) to develop an efficient catalyst for simultaneous esterification and transesterification reactions at atmospheric pressure. The catalysts were characterized by X-ray diffraction (XRD), BET surface area, desorption of CO_2 (TPD-CO_2) and NH_3 (TPD-NH_3). The authors found that the bismuth concentration significantly affected the performance of the catalyst and La_2O_3-Bi_2O_3 mixed with 5 wt% bismuth exhibited excellent transesterification activity for biodiesel production from jatropha oil. The bifunctional catalyst, under optimal reaction condition of methanol/oil molar ratio of 15:1, 2 wt% of the catalyst amount, reaction temperature of 150°C and reaction time of 4 h, provided the highest conversion of 93% from jatropha oil. This catalyst maintained 87% of FAME conversion after three successive recycling experiments. The decrease in the catalytic activity of the La_2O_3-Bi_2O_3 catalyst was related to the decrease in the concentration of Bi and Li metals in the catalyst when used in transesterification of jatropha oil.

Taufiq [28] and his research group also reported a CaO-La_2O_3 mixed oxide based bifunctional heterogeneous catalyst for biodiesel production from jatropha oil. They studied the stability of the CaO-La_2O_3 binary system in detail for the sustainable biodiesel process. The authors mentioned that the metal–metal oxide network between Ca and La resulted in well dispersion of CaO on the composite surface and thereby, increased the number of active acidic and basic sites as compared to that of bulk CaO and La_2O_3 metal oxide which in turns enhanced the catalytic activity. Furthermore, the biodiesel conversion increased with the increase in the Ca/La atomic ratio up to 8.0, where the stability of the CaO-La_2O_3 binary system decreased at the Ca/La atomic ratio of 10.0 due to highly saturation of CaO on the catalyst surface. The authors reported highest biodiesel yield of 98.76% at 160°C, 3 h, 25 methanol/oil molar ratio and 3 wt

%. In addition, the $CaO-La_2O_3$ binary system was stable even after four cycles with negligible leaching of Ca^{2+} ion into the reaction medium.

Alhassan et al. [117] reported $Fe_2O_3-MnO-SO_4^{2-}/ZrO_2$ nano sized bifunctional heterogeneous catalysts for biodiesel production from low-grade waste cooking oil. The physicochemical properties of the synthesized catalysts were studied by using X-ray diffraction (XRD), Temperature Programmed Desorption of NH_3 (TPD-NH_3/CO_2), Thermal gravimetric analysis (TGA), Fourier Transform Infrared Spectroscopy (FT-IR), Brunner–Emmett–Teller (BET) surface area measurement, Energy Dispersive X-ray Spectroscopy (EDS), Transmission Electron Microscopy (TEM) and X-ray Photoelectron Spectroscopy (XPS). The authors reported that the catalyst could be reused six times without any substantial loss in its catalytic activity with the maximum yield of 96.5 ± 0.02% at the optimized conditions of the reaction temperature of 180°C; stirring speed of 600 rpm, 1:20 M ratio of oil to alcohol and 3 wt/wt% catalyst loading.

Recently, zinc oxide (ZnO) nanostar, synthesized by the microwave-assisted surfactant free hydrolysis method was applied as the catalyst for biodiesel synthesis through one-step simultaneous esterification and transesterification from high free fatty acid by Kwong et al. [118]. The author reported that the ZnO nanostar catalyst showed high stability and robustness at the end of the reaction, providing a biodiesel yield of 97.3% via simultaneous esterification and transesterification in low grade feedstock like waste cooking oil and crude plant oils. They pointed out that the high efficiency of the synthesized catalyst could be attributed to the in-situ formation of the ZnOl intermediate and the ZnGly deposited on the catalyst surface to form a new co-catalyst.

In the view of above discussion, bifunctional heterogeneous catalysts seem to be a promising technology for biodiesel production from low-grade feedstocks, but after several time usage their catalytic activity topple down due to deactivation by strong adsorption of organics and leaching of the various active metals during transesterification of low-cost feedstocks. Hence, still there is enough gap for improvement to develop a robust bifunctional heterogeneous catalyst and incorporate desired physicochemical properties with enhanced catalytic activity, selectivity, and stability for economical biodiesel production from low cost feedstocks.

5. Recommendation

Environmentally friendly and efficient heterogeneous catalysts for a sustainable biodiesel technology are getting more and more attention among the research communities. The invention of the novel heterogeneous catalysts that have desirable physical and chemical properties, more stable, durable, and efficient under ambient conditions is one of the primary goals in focus of many research communities. The following steps may be recommended to design an efficient catalyst for sustainable biodiesel production from different biomass.

1. An appropriate catalytic configuration and preparation method may be adopted to design an efficient bifunctional catalyst with desired physical and chemical properties, which in terms help to achieve good thermal and mechanical stability.

2. The catalyst structure-activity correlation should be thoroughly defined by evaluating the physicochemical properties of the synthesized catalyst using different analytical techniques to develop a robust bifunctional heterogeneous catalyst with desired physicochemical properties to carry simultaneous esterification and transesterification reactions in the low-grade feedstocks. The physicochemical properties of the bifunctional heterogeneous catalyst may be tuned to get optimum number of active sites, which in terms would improve the catalytic activity, stability during biodiesel reaction from low grade feedstocks for sustainable energy process.

3. The use of waste materials such as animal bones, naturally available clays, etc., should be utilized as raw materials for catalyst preparation. The use of raw materials in the biodiesel technology will further contribute to low-cost biodiesel production. These materials may be modified to develop an effective catalyst which may be used several times for biodiesel production from low cost feedstocks.

Currently, our research group is also working on the modification of naturally available clay for biodiesel production from waste cooking oil. We have introduced different metals and groups into selected clays by the modified impregnation method to develop bifunctional activity of the catalyst with enhanced catalyst stability and activity for sustainable biodiesel technology.

6. Conclusion

The use of bifunctional heterogeneous catalysts for biodiesel production seems to be a promising technology for the efficient biodiesel production from low cost feedstocks, such as waste cooking oil for sustainable energy production. Therefore, it is essential to design a bifunctional heterogeneous catalyst with desired physical and chemical properties, which show potential ability to carry simultaneous esterification and transesterification reactions in low grade feedstocks for sustainable biodiesel production. The catalyst structure-activity correlation should be properly studied to modify the physicochemical properties of the synthesized catalyst to design an effective heterogeneous catalyst for biodiesel production. Moreover, the use of waste cooking oils as feedstocks and waste materials for catalyst preparation is very interesting, and may further make the biodiesel production process economically feasible and will greatly decrease the production cost of biodiesel.

Acknowledgements

The authors would like to express their sincere gratitude to Universiti Teknologi PETRONAS, Malaysia.

Author details

Anita Ramli[1*], Muhammad Farooq[1,2*], Abdul Naeem[2], Saleem Khan[2], Muhammad Hummayun[2], Azhar Iqbal[2], Sohail Ahmed[1] and Liaqat Ali Shah[2]

*Address all correspondence to: anita_ramli@petronas.com.my and farooq_khann@yahoo.com

1 Department of Fundamental and Applied Sciences, Universiti Teknologi PETRONAS (UTP), Tronoh, Perak, Malaysia

2 National Centre of Excellence in Physical Chemistry, University of Peshawar, Peshawar, Pakistan

References

[1] Karmee SK, Lin CSK. Lipids from food waste as feedstock for biodiesel production: Case Hong Kong. Lipid Technol. 2014;26:206–209.

[2] Wang R, Sun L, Xie X, Ma L, Liu L, Liu X, Ji N, Xie G. Biodiesel production from *Stauntonia chinensis* seed oil (waste from food processing): Heterogeneous catalysis by modified calcite, biodiesel purification, and fuel properties. Ind. Crop. Prod. 2014;62:8–13.

[3] Fatih, Demirbas M, Balat M, Balat H. Biowastes-to-biofuels. Energy Convers. Manage. 2011;52:1815–1828.

[4] Ashnani MHM, Johari A, Hashim H, Hasani E. A source of renewable energy in Malaysia, why biodiesel? Renew. Sust. Energ. Rev. 2014;35(0):244–257.

[5] MacLeod CS, Harvey AP, Lee AF, Wilson K. Evaluation of the activity and stability of alkali-doped metal oxide catalysts for application to an intensified method of biodiesel production. Chem. Eng. J. 2008;135(1–2):63–70.

[6] Luque R, Lovett JC, Datta B, Clancy J, Campelo JM, Romero AA. Biodiesel as feasible petrol fuel replacement: A multidisciplinary overview. Energy Environ. Sci. 2010;3:1706–1721.

[7] Gnanaprakasam A, Sivakumar VM, Surendhar A, Thirumarimurugan M, Kannadasan T. Recent strategy of biodiesel production from waste cooking oil and process influencing parameters: A review. J. Energy. 2013;2013:10.

[8] Bankovic-Ilic, IB, Stamenkovic OS, Veljkovic VB. Biodiesel production from non-edible plant oils. Renew. Sust. Energ. Rev. 2012;16:3621–3647.

[9] Chhetri AB, Watts KC, Islam MR. Waste cooking oil as an alternate feedstock for biodiesel production. Energ, 2008;1:3–18.

[10] Karmee SK, Linardi D, Lee J, Lin CSK. Conversion of lipid from food waste to biodiesel. Waste Manage. 2015;41:169–173.

[11] Gui MM, Lee KT, Bhatia S. Feasibility of edible oil vs non-edible oil vs waste edible oil as biodiesel feedstock. Energy, 2008;33(11):1646–1653.

[12] Felizardo P, Correia MJN, Raposo I. Production of biodiesel from waste frying oils. Waste Manage. 2006;26:487–494.

[13] Öner C, Altun S. Biodiesel production from inedible animal tallow and an experimental investigation of its use as alternative fuel in a direct injection diesel engine. Appl. Energy, 2009;86:2114–2120.

[14] Viola E, Blasi A, Valerio V, Guidi I, Zimbardi F, Braccio G, Giordano G. Biodiesel from fried vegetable oils via transesterification by heterogeneous catalysis. Catal. Today, 2012;179(1):185–190.

[15] Zhang Y, DubÃ© MA, McLean DD, Kates M. Biodiesel production from waste cooking oil: 2. Economic assessment and sensitivity analysis. Bioresour. Technol. 2003;90(3):229–240.

[16] Balat, M, Balat H. Progress in biodiesel processing, Appl. Energy, 2010;87:1815–1835.

[17] Shay EG. Diesel fuel from vegetable oils: Status and opportunities. Biomass Bioenerg. 1993;4(4):227–242.

[18] Lee AF, Wilson K. Recent developments in heterogeneous catalysis for the sustainable production of biodiesel. Catal. Today, 2015;242, Part A:3–18.

[19] Borges ME, DÃaz L. Recent developments on heterogeneous catalysts for biodiesel production by oil esterification and transesterification reactions: A review, Renew. Sust. Energ. Rev. 2012;16:2839–2849.

[20] Istadi I, Anggoro DD, Buchori L, Rahmawati DA, Intaningrum D. Active acid catalyst of sulphated zinc oxide for transesterification of soybean oil with methanol to biodiesel. Procedia. Environ. Sci. 2015;23(0):385–393.

[21] Sharma M, Khan AA, Puri SK, Tuli DK. Wood ash as a potential heterogeneous catalyst for biodiesel synthesis. Biomass Bioenerg. 2012;41(0):94–106.

[22] Farooq M, Ramli A, Naeem A. Biodiesel production from low FFA waste cooking oil using heterogeneous catalyst derived from chicken bones. Renew. Energ. 2015;76(0): 362–368.

[23] Islam A, Taufiq-Yap YH, Chu C-M, Chan E-S, Ravindra P. Studies on design of heterogeneous catalysts for biodiesel production. Process Saf. Environ. Prot. 2013;91(1–2): 131–144.

[24] Jiang ST, Zhang FJ, Pan LJ. Sodium phosphate as a solid catalyst for biodiesel preparation. Braz. J. Chem. Eng. 2010;27(1):137–144.

[25] Farooq M, Ramli A, Naeem A, Saleem Khan M. Effect of different metal oxides on the catalytic activity of γ-Al_2O_3-MgO supported bifunctional heterogeneous catalyst in biodiesel production from WCO. RSC Adv. 2016;6(2):872–881.

[26] Ali RM, El Latif MMA, Farag HA. Preparation and characterization of $CaSO_4$–SiO_2–CaO/SO_4^{2-} composite for biodiesel production. Am. J. Appl. Chem. 2015;3(3–1):38–45.

[27] Pereira CO, Portilho MF, Henriques CA, Zotin FMZ. SnSO4 as catalyst for simultaneous transesterification and esterification of acid soybean oil. J. Braz. Chem. Soc. 2014;25:2409–2416.

[28] Lee HV, Juan JC, Taufiq-Yap YH. Preparation and application of binary acid-base CaO-La_2O_3 catalyst for biodiesel production. Renew. Energ. 2015;74(0):124–132.

[29] Hoekman SK, Robbins C. Review of the effects of biodiesel on NO_x emissions. Fuel Process. Technol. 2012;96:237–249.

[30] Qing Shu HY, Bao Liu, Lihua Zhu, Caixia Zhang, Jinfu Wang. Synthesis of biodiesel from model acidic oil catalyzed by a novel solid acid catalyst SO_4^{2-}/C/Ce^{4+}. Fuel. 2015;143:547–554.

[31] Sousa LL, Lucena IL, Fernandes FAN. Transesterification of castor oil: effect of the acid value and neutralization of the oil with glycerol. Fuel Process. Technol. 2010; 91:194–196.

[32] de Almeida VF, GarcÃa-Moreno PJ, Guadix A, Guadix EM. Biodiesel production from mixtures of waste fish oil, palm oil and waste frying oil: Optimization of fuel properties. Fuel Process. Technol. 2015;133(0):152–160.

[33] Boehman AL. Biodiesel production and processing. Fuel Process. Technol. 2005;86(10): 1057–1058.

[34] Keskin A, Guru M, Altiparmak D, Aydin K. Using of cotton oil soapstock biodiesel-diesel fuel blends as an alternative diesel fuel. Renew. Energ. 2008;33(4):553–557.

[35] Ramadhas AS, Jayaraj S, Muraleedharan C. Use of vegetable oils as I.C. engine fuels:A review. Renew. Energ. 2004;29(5):727–742.

[36] Amais RS, Garcia EE, Monteiro MR, Nogueira ARA, Nobrega JA. Direct analysis of biodiesel microemulsions using an inductively coupled plasma mass spectrometry. Microchem. J. 2010;96(1):146–150.

[37] Ma F, Hanna MA. Biodiesel production: A review. Bioresour. Technol. 1999;70(1):1–15.

[38] Brennan L, Owende P. Biofuels from microalga:eâ€" a review of technologies for production, processing, and extractions of biofuels and co-products. Renew. Sust. Energ. Rev. 2010;14(2):557–577.

[39] Naik SN, Goud VV, Rout PK, Dalai AK. Production of first and second generation biofuels: A comprehensive review. Renew. Sust. Energ. Rev. 2010;14(2):578–597.

[40] Bharathiraja B, Chakravarthy M, Kumar RR, Yuvaraj D, Jayamuthunagai J, Kumar R , Palani S. Biodiesel production using chemical and biological methods- A review of process, catalyst, acyl acceptor, source and process variables. Renew. Sust. Energ. Rev. 2014; 38:368–382.

[41] Kouzu M, Hidaka J-s. Transesterification of vegetable oil into biodiesel catalyzed by CaO: A review. Fuel. 2012;93(0):1–12.

[42] Moussavou Mounguengui RdoW, Brunschwig C, Barea B, Villeneuve P, Blin J. Are plant lipases a promising alternative to catalyze transesterification for biodiesel production? Prog. Energy Combust. Sci. 2013;39(5):441–456.

[43] Demirbas A. Comparison of transesterification methods for production of biodiesel from vegetable oils and fats. Energy Convers. Manage, 2008;49(1):125–130.

[44] Takase M, Zhao T, Zhang M, Chen Y, Liu H, Yang L, Wu X. An expatiate review of neem, jatropha, rubber and karanja as multipurpose non-edible biodiesel resources and comparison of their fuel, engine and emission properties. Renew. Sust. Energ. Rev. 2015;43(0):495–520.

[45] Reed TB, Graboski MS, Gaur S. Development and commercialization of oxygenated diesel fuels from waste vegetable oils. Biomass Bioenerg. 1992 3(2):111–115.

[46] Karmee SK. Liquid biofuels from food waste: Current trends, prospect and limitation. Renew. Sust. Energ. Rev. 2016;53:945–953.

[47] Meira M , Quintella CM, Ribeiro EMO, Silva HRG, Guimaraes AK, Overview of the challenges in the production of biodiesel, Biomass Conversion and Biorefinery, 5 (2015) 321–329.

[48] Fadhil AB, Dheyab MM, Abdul-Qader A-QY. Purification of biodiesel using activated carbons produced from spent tea waste, Journal of the Association of Arab Universities for Basic and Applied Sciences 2012, 11; 45–49.

[49] Hoque ME, Singh A, Chuan YL. Biodiesel from low cost feedstocks: The effects of process parameters on the biodiesel yield. Biomass Bioenerg. 2011;35(4):1582--1587.

[50] Chen K-S, Lin Y-C, Hsu K-H, Wang H-K. Improving biodiesel yields from waste cooking oil by using sodium methoxide and a microwave heating system, Energy 2012, 38:151–156.

[51] Balat M. Potential alternatives to edible oils for biodiesel production: A review of current work. Energy Convers. Manage. 2011;52(2):1479–1492.

[52] Borugadda VB, Goud VV. Biodiesel production from renewable feedstocks: Status and opportunities. Renew. Sust. Energ. Rev. 2012;16(7):4763–4784.

[53] Talebian-Kiakalaieh A, Amin NAS, Mazaheri H. A review on novel processes of biodiesel production from waste cooking oil. Appl. Energy, 2013;104(0):683–710.

[54] Zhang H, Wang Q, Mortimer SR. Waste cooking oil as an energy resource: Review of Chinese policies. Renew. Sust. Energ. Rev. 2012;16(7):5225–5231.

[55] Demirbas A. Progress and recent trends in biodiesel fuels. Energy Convers. Manage. 2009;50(1):14–34.

[56] Sharma Y, Singh B. Development of biodiesel: Current scenario. Renew. Sust. Energ. Rev. 2009;13(6):1646–1651

[57] Yusuf NNAN, Kamarudin SK, Yaakub Z. Overview on the current trends in biodiesel production. Energy Convers. Manage. 2011;52(7):2741–2751.

[58] Xiang Y, Wang L, Jiao Y. Ultrasound strengthened biodiesel production from waste cooking oil using modified coal fly ash as catalyst. J. Environ. Chem. Eng. 2016;4(1): 818–824.

[59] Diya uddeen BH, Abdul Aziz AR, Daud WMAW, Chakrabarti MH. Performance evaluation of biodiesel from used domestic waste oils: A review. Process Saf. Environ. Prot. 2013;90(3):164–179.

[60] Kouzu M, Kasuno T, Tajika M, Sugimoto Y, Yamanaka S, Hidaka J. Calcium oxide as a solid base catalyst for transesterification of soybean oil and its application to biodiesel production. Fuel, 2008;87(12):2798–2806.

[61] Chin LH, Hameed BH, Ahmad AL. Process optimization for biodiesel production from waste cooking palm oil (Elaeis guineensis) using response surface methodology. Energ. Fuel, 2009;23(2):1040–1044.

[62] Kouzu M, Hidaka J-s, Komichi Y, Nakano H, Yamamoto M. A process to transesterify vegetable oil with methanol in the presence of quick lime bit functioning as solid base catalyst. Fuel, 2009;88(10):1983–1990.

[63] Omar WNNW, Nordin N, Mohamed M, Amin NAS. A two-step biodiesel production from waste cooking oil: Optimization of pre-treatment step. J. Appl. Sci. 2009;9(17): 3098–3103.

[64] Yan S, Kim M, Salley SO, Ng KYS. Oil transesterification over calcium oxides modified with lanthanum. Appl. Catal. A: Gen. 2009;360(2):163–170.

[65] Brito A, Borges ME, Garín M, Hernández A. Biodiesel production from waste oil using Mg-Al layered double hydroxide catalysts. Energ. Fuel, 2009;23(6):2952–2958.

[66] Georgogianni KG, Katsoulidis AP, Pomonis PJ, Kontominas MG. Transesterification of soybean frying oil to biodiesel using heterogeneous catalysts. Fuel Process. Technol. 2009;90(5):671–676.

[67] Yan S, Salley SO, Simon Ng KY. Simultaneous transesterification and esterification of unrefined or waste oils over ZnO-La_2O_3 catalysts. Appl. Catal. A: Gen. 2009;353(2):203–212.

[68] Guan G, Kusakabe K, Yamasaki S. Tri-potassium phosphate as a solid catalyst for biodiesel production from waste cooking oil. Fuel Process. Technol. 2009; 90(4):520–524.

[69] Yan S, Mohan S, DiMaggio C, Kim M, Ng KYS, Salley SO. Long term activity of modified ZnO nanoparticles for transesterification. Fuel 2010;89(10):2844–2852.

[70] Son SM, Kusakabe K, Guan G. Biodiesel synthesis and properties from sunflower and waste cooking oils using CaO catalyst under reflux conditions. J.Appl. Sci. 2010; 10(24): 3191–3198.

[71] Del Remedio Hernández M, Reyes-Labarta JA, Valdés FJ. New heterogeneous catalytic transesterification of vegetable and used frying oil. Ind. Eng. Chem. Res. 2010;49(19): 9068–9076.

[72] Olutoye MA, Hameed BH. Synthesis of fatty acid methyl ester from used vegetable cooking oil by solid reusable $Mg_{1-x}Zn_{1+x}O_2$ catalyst. Bioresour. Technol. 2011b;102(4): 3819–3826.

[73] Koberg M, Abu-Much R, Gedanken A. Optimization of bio-diesel production from soybean and wastes of cooked oil: Combining dielectric microwave irradiation and a SrO catalyst. Bioresour. Technol. 2011;102(2):1073–1078.

[74] Gomes JFP, Puna JFB, Gonçalves LM, Bordado JCM. Study on the use of MgAl hydrotalcites as solid heterogeneous catalysts for biodiesel production. Energy 2011;36(12): 6770–6778.

[75] Sánchez-Cantú M, Pérez-Díaz LM, Rosales R, Ramírez E, Apreza-Sies A, Pala-Rosas I, Rubio-Rosas En, Aguilar-Franco M, Valente JS. Commercial hydrated lime as a cost-effective solid base for the transesterification of wasted soybean oil with methanol for biodiesel production. Energ. Fuel, 2011;25(7):3275–3282.

[76] Molaei Dehkordi A, Ghasemi M. Transesterification of waste cooking oil to biodiesel using Ca and Zr mixed oxides as heterogeneous base catalysts. Fuel Process. Technol. 2012;97(0):45–51.

[77] Nair P, Singh B, Upadhyay SN, Sharma YC. Synthesis of biodiesel from low FFA waste frying oil using calcium oxide derived from mereterix as a heterogeneous catalyst. J. Clean. Prod. 2012;29–30(0):82–90.

[78] Agarwal M, Chauhan G, Chaurasia SP, Singh K. Study of catalytic behavior of KOH as homogeneous and heterogeneous catalyst for biodiesel production. J. Taiwan. Inst. Chem. 2012;43(1):89–94.

[79] Birla A, Singh B, Upadhyay SN, Sharma YC. Kinetics studies of synthesis of biodiesel from waste frying oil using a heterogeneous catalyst derived from snail shell. Bioresour. Technol. 2012;106(0):95–100.

[80] Sankaranarayanan S, Antonyraj CA, Kannan S. Transesterification of edible, non-edible and used cooking oils for biodiesel production using calcined layered double hydroxides as reusable base catalysts. Bioresour. Technol. 2012;109(0):57–62.

[81] Khemthong P, Luadthong C, Nualpaeng W, Changsuwan P, Tongprem P, Viriya-empikul N, Faungnawakij K. Industrial eggshell wastes as the heterogeneous catalysts for microwave-assisted biodiesel production. Catal. Today, 2012;190(1):112–116.

[82] Boey P-L, Ganesan S, Maniam GP, Khairuddean M. Catalysts derived from waste sources in the production of biodiesel using waste cooking oil. Catal. Today, 2012;190(1): 117–121.

[83] Sreeprasanth PS, Srivastava R, Srinivas D, Ratnasamy P. Hydrophobic, solid acid catalysts for production of biofuels and lubricants. Appl. Catal. A 2006;314(2):148–159.

[84] Zong M-H, Duan Z-Q, Lou W-Y, Smith TJ, Wu H. Preparation of a sugar catalyst and its use for highly efficient production of biodiesel. Green Chem. 2007;9(5):434–437.

[85] Jacobson K, Gopinath R, Meher LC, Dalai AK. Solid acid catalyzed biodiesel production from waste cooking oil. Appl. Catal. B 2008;85(1–2):86–91.

[86] Lou W-Y, Zong M-H, Duan Z-Q. Efficient production of biodiesel from high free fatty acid-containing waste oils using various carbohydrate-derived solid acid catalysts. Bioresour. Technol. 2008;99(18):8752–8758.

[87] Cao F, Chen Y, Zhai F, Li J, Wang J, Wang X, Wang S, Zhu W. Biodiesel production from high acid value waste frying oil catalyzed by superacid heteropolyacid. Biotechnol. Bioeng. 2008;101(1):93–100.

[88] Wang J, Chen Y, Wang X, Cao F. Aluminum dodecatungstophosphate nanotube as a solid acid catalyst one-pot production of biodiesel from waste cooking oil. BioRes. 2009;4(4):1477–1486.

[89] Li J, Wang X, Zhu W, Cao F. $Zn_{1.2}H_{0.6}PW_{12}O_{40}$ nanotubes with double acid sites as heterogeneous catalysts for the production of biodiesel from waste cooking oil. ChemSusChem. 2009;2(2):177–183.

[90] Zhang X, Li J, Chen Y, Wang J, Feng L, Wang X, Cao F. Heteropolyacid nanoreactor with double acid sites as a highly efficient and reusable catalyst for the transesterification of waste cooking oil. Energ. Fuels, 2009;23(9):4640–4646.

[91] Lam MK, Lee KT, Mohamed AR. Sulfated tin oxide as solid superacid catalyst for transesterification of waste cooking oil: An optimization study. Appl. Catal. B 2009;93(1–2):134–139.

[92] Komintarachat C, Chuepeng S. Solid acid catalyst for biodiesel production from waste used cooking oils. Ind. Eng. Chem. 2009;48(20):9350–9353.

[93] Fu B, Gao L, Niu L, Wei R, Xiao G. Biodiesel from waste cooking oil via heterogeneous superacid catalyst SO_4^{2-}/ZrO_2. Energ. Fuel, 2008;23(1):569–572.

[94] Gan M, Pan D, Ma L, Yue E, Hong J. The kinetics of the esterification of free fatty acids in waste cooking oil using $Fe_2(SO_4)_3$/C catalyst. Chin. J. Chem. Eng. 2009;17(1):83–87.

[95] Srilatha K, Issariyakul T, Lingaiah N, Sai Prasad PS, Kozinski J, Dalai AK. Efficient esterification and transesterification of used cooking oil using 12-Tungstophosphoric acid (TPA)/Nb_2O_5 catalyst. Energ. Fuel, 2010;24(9):4748–4755.

[96] Lam MK, Lee KT. Accelerating transesterification reaction with biodiesel as co-solvent: A case study for solid acid sulfated tin oxide catalyst. Fuel, 2010;89(12):3866–3870.

[97] Shu Q, Gao J, Nawaz Z, Liao Y, Wang D, Wang J. Synthesis of biodiesel from waste vegetable oil with large amounts of free fatty acids using a carbon-based solid acid catalyst. Appl. Energ. 2010;87(8):2589–2596.

[98] Ramachandran K, Sivakumar P, Suganya T, Renganathan S. Production of biodiesel from mixed waste vegetable oil using an aluminium hydrogen sulphate as a heterogeneous acid catalyst. Bioresour. Technol. 2011;102(15):7289–7293.

[99] Morales G, Bautista LF, Melero JA, Iglesias J, and Sánchez-Vázquez R. Low-grade oils and fats: Effect of several impurities on biodiesel production over sulfonic acid heterogeneous catalysts. Bioresour. Technol. 2011;102(20):9571–9578.

[100] Lam MK, Lee KT. Mixed methanol–ethanol technology to produce greener biodiesel from waste cooking oil: A breakthrough for $SO_4^{2-}/SnO_2–SiO_2$ catalyst. Fuel Process. Technol. 2011;92(8):1639–1645.

[101] Otadi M, Shahraki A, Goharrokhi M, Bandarchian F. Reduction of free fatty acids of waste oil by acid-catalyzed esterification. Procedia Eng. 2011;18:168–174.

[102] Ramírez-Ortiz J M-VJ, Rosales MM. Used frying oil for biodiesel production kaolinite as catalyst. World Acad. Sci. Eng. Technol. 2011;80:977–980.

[103] Wang Y, Ma S, Wang L, Tang S, Riley WW, Reaney MJ. Solid superacid catalyzed glycerol esterification of free fatty acids in waste cooking oil for biodiesel production. Eur. J. Lipid Sci. Technol. 2012;114(3):315–324.

[104] Noshadi I, Amin NAS, Parnas RS. Continuous production of biodiesel from waste cooking oil in a reactive distillation column catalyzed by solid heteropolyacid: Optimization using response surface methodology (RSM). Fuel, 2012;94(0):156–164.

[105] Sin H-Y, An S-H, Sheikh R, Park YH, Bae S-Y. Transesterification of used vegetable oils with a Cs-doped heteropolyacid catalyst in supercritical methanol. Fuel, 2012;96(0): 572–578.

[106] Wu J, Gao Y, Zhang W, Tan Y, Tang A, Men Y, Tang B. Esterification of cooking oil for biodiesel production using composites $Cs_{2.5}H_{0.5}PW_{12}O_{40}$/ionic liquids catalysts. Appl. Petrochem. Res. 2014;4(3):305–312.

[107] Wan Omar WNN, Amin NAS. Biodiesel production from waste cooking oil over alkaline modified zirconia catalyst. Fuel Process. Technol. 2011;92(12):2397–2405.

[108] Macario A, Giordano G, Onida B, Cocina D, Tagarelli A, Giuffre AM. Biodiesel production process by homogeneous/heterogeneous catalytic system using an acid/base catalyst. Appl. Catal. A 2010;378(2):160–168.

[109] Cannilla C, Bonura G, Rombi E, Arena F, Frusteri F. Highly effective MnCeOx catalysts for biodiesel production by transesterification of vegetable oils with methanol. Appl. Catal. A 2010;382(2):158–166.

[110] Wen Z, Yu X, Tu S-T, Yan J, Dahlquist E. Biodiesel production from waste cooking oil catalyzed by TiO_2-MgO mixed oxides. Bioresour. Technol. 2010;101(24):9570–9576.

[111] Borges ME, Díaz L, Alvarez-Galván MC, Brito A. High performance heterogeneous catalyst for biodiesel production from vegetal and waste oil at low temperature. Appl. Catal. B 2011;102(1–2):310–315.

[112] Kondamudi N, Mohapatra SK, Misra M. Quintinite as a bifunctional heterogeneous catalyst for biodiesel synthesis. Appl. Catal. A 2011;393(1–2):36–43.

[113] Jiménez-López A, Jiménez-Morales I, Santamaría-González J, Maireles-Torres P. Biodiesel production from sunflower oil by tungsten oxide supported on zirconium doped MCM-41 silica. J. Mol. Catal. A: Chem. 2011;335(1–2):205–209.

[114] Salinas D, Araya P, Guerrero S. Study of potassium-supported TiO_2 catalysts for the production of biodiesel. Appl. Catal. B 2012;117–118(0):260–267.

[115] Farooq M, Ramli A, Subbarao D. Biodiesel production from waste cooking oil using bifunctional heterogeneous solid catalysts. J. Clean. Prod. 2013;59:131–140.

[116] Rabiah Nizah MF, Taufiq-Yap YH, Rashid U, Teo SH, Shajaratun Nur ZA, Islam A. Production of biodiesel from non-edible Jatropha curcas oil via transesterification using Bi_2O_3-La_2O_3 catalyst. Energy Convers. Manage. 2014;88(0):1257–1262.

[117] Alhassan FH, Rashid U, Taufiq-Yap YH. Synthesis of waste cooking oil-based biodiesel via effectual recyclable bi-functional Fe_2O_3-MnO-SO_4^{2-}/ZrO_2 nanoparticle solid catalyst. Fuel, 2015;142(0):38–45.

[118] Kwong T-L, Yung K-F. One-step production of biodiesel through simultaneous esterification and transesterification from highly acidic unrefined feedstock over efficient and recyclable ZnO nanostar catalyst. Renew. Energy, 2016;90:450–457.

Chapter 15

Role of Mass-Transfer Interfacial Area in the Biodiesel Production Performance of Acid-Catalyzed Esterification

Devjyoti Nath, Adisorn Aroonwilas and
Amornvadee Veawab

Additional information is available at the end of the chapter

http://dx.doi.org/10.5772/65657

Abstract

This work investigated the role of mass-transfer interfacial area in the biodiesel production using the acid-catalyzed esterification process. The interfacial area between alcohol and oil feedstock was determined by conducting acid-catalyzed esterification experiments using methanol and oleic acid (as free fatty acid) under ranges of five process parameters: reaction temperature (45–65°C), agitation speed (200–400 rpm), methanol-to-oil ratio (3:1–9:1 mol/mol), catalyst concentration (0.5–2.0%), and concentration of free fatty acid (5–30%). Effects of these parameters on the biodiesel conversion rate and the interfacial area were quantified. An empirical correlation for the interfacial area was developed as a function of process parameters. Results show that the enhancement of biodiesel production rate is attributed to reaction kinetics and/or interfacial area. The interfacial area is the sole contributor to the increase in biodiesel production rate due to the increase in methanol-to-oil ratio and agitation speed. Both kinetics and interfacial area contribute to the increase in biodiesel production rate due to the reaction temperature and catalyst concentration. The interfacial area plays negligible role in the change in biodiesel production rate due to the free fatty acid content.

Keywords: biodiesel, esterification, mass-transfer interfacial area, reactor design, hydrodynamics, parametric effects

1. Introduction

Energy use is considered to be the most fundamental requirement for various human activities, especially in the industrial, transportation, and agricultural sectors. Among different kinds of fuels, petroleum constitutes the majority of the world's energy supply. However, petroleum is a finite and nonrenewable energy source, which has already caused serious environmental pollution. Therefore, a sustainable, affordable, and environmentally friendly alternative to petroleum is urgently needed.

Biodiesel is considered to be an important alternative to conventional petroleum-based diesel [1]. It is nontoxic, and biodegradable, offers lower emissions of a number of air pollutants, can be used in typical diesel engines without any major modifications, and has greater lubricity than conventional diesel, thus reducing corrosion in diesel engines [2]. Biodiesel can be used in its pure form called B100 (100% biodiesel) or in a blend with different proportions of conventional diesel fuels. Common blends include B20 (20% biodiesel and 80% conventional diesel), which are much closer to diesel fuel properties than B100 and B5 (5% biodiesel and 95% conventional diesel).

In the typical biodiesel production process, the catalyzed transesterification or esterification reaction is carried out in a multiphase reactor where the two immiscible chemical reactants (oil feedstock and alcohol-containing catalyst) are brought into contact using different agitation or mixing mechanisms. Because the cost of biodiesel production depends heavily on the cost of raw materials, using low-quality feedstocks such as waste cooking oils or nonedible oils instead of high-quality feedstocks will significantly reduce the biodiesel production cost. However, low-quality feedstocks have high content of free fatty acids (FFAs), which can react with the alkaline catalyst and produce soaps. This side reaction in the alkali-catalyzed transesterification process will reduce the catalyst efficiency and the biodiesel conversion rate. Additionally, the formation of soaps will make the later purification process difficult. As a result, the undesired side reactions caused by FFAs will increase the cost of biodiesel production. Therefore, when using low-quality feedstocks for biodiesel production, the content of FFAs must be reduced to an acceptable level (typically below 1% according to references [3–5]) before the alkali-catalyzed transesterification process. One efficient method for removing the FFAs from feedstocks is esterification, where the FFAs react with alcohol to form ester and water as products.

Due to the nature of the multiphase reaction, the efficiency or rate of biodiesel production relies heavily on two primary factors: (i) the kinetics of catalyzed transesterification or esterification reactions and (ii) the hydrodynamics of liquid-liquid mixing promoted by reactor design and operation. In order to arrive at a high-efficiency and optimized biodiesel reactor, these two fundamental features must be understood. To date, a large quantity of biodiesel research works has been carried out in many different aspects, such as production rate and the quality of biodiesel products derived from different feedstocks, kinetic studies to find optimal reaction conditions for achieving higher yields, and use of enzyme and heterogeneous catalysts as an alternative to the conventional homogeneous catalysts [6–11]. Most kinetic works reported biodiesel conversion profiles as a function of reaction time under specific reaction conditions and for specific types of reactor design and operation. As such, the reported kinetic data essentially reflect the combined performance of both reaction kinetics and hydrodynamics of liquid-liquid reaction systems.

Despite its importance to the development of high-performance reactors, the knowledge of hydrodynamics or mass-transfer interfacial area (a_e) between the two immiscible reactants during biodiesel reaction is very limited. Only one study by Stamenkovic et al. [12] relates to the interfacial area in the biodiesel production process. In their work, the effect of agitation intensity during the base-catalyzed transesterification of sunflower oil was investigated under a specific reaction condition, that is, 20°C and alcohol-to-oil ratio of 6:1. There are no other studies reporting the interfacial area for the acid-catalyzed reaction system.

Therefore, the objectives of this work are: (i) to extend knowledge of interfacial area formed between immiscible reactants during the acid-catalyzed esterification reaction which can be used for the design of a high-efficiency reactor, (ii) to investigate the role of process parameters on interfacial area in the esterification process, and (iii) to develop an empirical correlation for interfacial area estimation as a function of process parameters. To achieve these objectives, a series of esterification experiments were performed using a stirred reactor operated under variable ranges of reaction conditions (**Table 1**). The experimental results were obtained in forms of free fatty acid (FFA) conversion profiles which were subsequently used for determining the interfacial area values.

Process parameter	Range
Reaction temperature (°C)	45–65
Agitation speed (rpm)	200–400
Methanol-to-oil ratio (mol:mol)	3:1–9:1
Catalyst concentration (wt%)	0.5–2.0
Free fatty acid concentration (%)	5–30
Type of free fatty acid	Oleic acid
Type of catalyst	Sulfuric acid

Table 1. Summary of test conditions for esterification experiments.

2. Methods

2.1. Materials

Two sets of chemicals were used in the experiments: (i) reactants and an acid catalyst for the esterification reaction and (ii) supporting chemicals for liquid sample analysis. For esterification experiments, canola oil was used as the base ingredient of oil feedstock. Oleic acid (90%) from Sigma-Aldrich (Oakville, Ontario) was used as the representative of free fatty acids (FFAs) commonly found in the feedstock. A predetermined amount of oleic acid was added to the base canola oil in order to simulate low-quality feedstock. Sulfuric acid (98%) was used as the acid catalyst, and methanol (99.98%) was chosen to represent the alcohol reactant. Both sulfuric acid and methanol were purchased from Fisher Scientific (Ottawa, Ontario). For liquid sample analysis, toluene (99.9%), isopropyl alcohol (99.9%), and potassium hydroxide (0.1 N) were used for titrations to determine the acid number or FFA content of the oil phase.

2.2. Experimental setup

The mass-transfer interfacial area (a_e) and reaction kinetics between oil feedstock and methanol were determined by carrying out esterification experiments in a bench-scale reaction system. As shown in **Figure 1**, the reaction system consists of a 500-mL glass reactor that is jacketed for heating/cooling (Ace Glass Inc., USA), a mechanical agitator powered by a variable-speed drive (Cole-Parmer, Canada), and a water bath with a temperature controller/circulator (Cole-Parmer, Canada). The reactor was designed for operating pressures and temperatures of up to 35 psig and 100°C, respectively. The reactor head has three connecting ports: one for the mechanical agitator, one for sampling collection, and one for temperature measurement. A glass bearing with PTFE coupling was connected to the reactor head to accommodate the agitator. The sampling port was equipped with a silicone rubber septum, thus making possible the collection of liquid samples without interrupting the reaction progress. A K-type thermocouple connected to a handheld meter was used for monitoring reaction temperature. During the experiments, a heating medium (i.e., water from the temperature-controlled water bath) was circulated through the reactor jacket in order to keep reaction temperature constant.

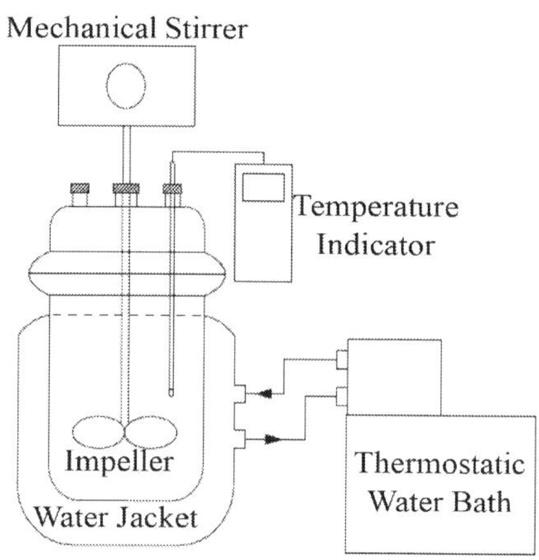

Figure 1. A schematic diagram of the bench-scale reaction system for esterification experiments.

2.3. Experimental procedures

The experiments were conducted in two different modes: (i) esterification tests with a well-defined interfacial area between oil feedstock and methanol, and (ii) esterification tests with the complete mixing between the two reactants. The first mode of experiments provided the true kinetic features of the esterification reaction, while the second gave the reaction performance that integrates both kinetic and hydrodynamic effects of the reaction system.

For the experiments with a fixed interfacial area (first mode), the canola oil was mixed with oleic acid to simulate a low-quality feedstock containing different levels of FFA. A 250 mL of the prepared feedstock was then transferred into the 500-mL glass reactor and maintained at a desired reaction temperature. An impeller or agitator was placed in the middle of this oil phase and set at a particular mixing speed in order to keep the oil phase homogenized but yet the oil-surface undisturbed. Meanwhile, a predetermined amount of H_2SO_4 (catalyst) was mixed with methanol to form a catalyst/methanol mixture with a desired catalyst concentration. For each experimental run, a 93 mL of catalyst/methanol mixture was used to ensure an excessive amount of methanol (more than 40 mol/mol ratio) available for reacting with FFA in the oil phase. Prior to the reaction, the catalyst/methanol mixture was heated to the desired reaction temperature in a water bath. Once the reaction temperature was reached, the methanol mixture was transferred into the glass reactor to start the esterification reaction. In order to keep the interface between the oil phase and the methanol phase undisturbed, a separating funnel was used to smoothly transfer the preheated catalyst/methanol mixture into the reactor. For each experiment, the reaction temperature was controlled by the water bath. The reaction was timed until it reached its equilibrium. During the experiment, a series of samples were collected from the oil phase at different time intervals. Each sample was transferred into a test tube and then immersed in cold water at 4°C to quench the reaction immediately. For better separation of the final mixture, the samples were centrifuged for 5 min at 3000 rpm, and then, the top layer sample was collected and sent for analysis.

For the experiments with the complete mixing (second mode), each esterification experiment also began with the preparation of low-quality feedstock by mixing canola oil and oleic acid at a specific ratio. The FFA content of the prepared feedstock was analyzed in terms of acid number in accordance with the ASTM D974-04 standard, the details of which are provided in the next subsection. Following the preparation, a known amount of feedstock was charged to the reactor and heated to the desired reaction temperature with an accuracy of ±1°C. The feedstock was also stirred by the agitator at a fixed speed. Once the reaction temperature was reached, a predetermined amount of methanol/sulfuric acid mixture (with a given catalyst concentration) was rapidly injected into the reactor to start the esterification reaction. Prior to injection, this alcohol/catalyst mixture was preheated to the reaction temperature in order to avoid unwanted fluctuation in reaction temperature, especially at the beginning of the test. Each experimental run was carried out for at least 70 min at the desired temperature and agitation speed. A series of liquid samples (3 mL) were collected from the reactor at a regular time interval during the experiment. These liquid samples were then analyzed for their acid number so as to determine the depletion of FFA as a function of time.

2.4. Sample analysis

A 3-mL liquid sample collected from the reactor was transferred to a test tube where 6 mL of de-ionized water was added. The tube was then capped and shaken vigorously to promote complete contact between water and the sample. This allowed the methanol and catalyst to combine with water, thus separating them from the sample. After being shaken, the test tube was placed in a centrifuge operating at 4000 rpm for 10 min. The centrifugal force helped develop two liquid layers, that is, the top layer for oil and the bottom layer for a mixture of water, methanol, and catalyst. The top layer was then withdrawn from the test tube for FFA content analysis by ASTM D974-04. A 2-mL sample was taken from the oil phase, weighed for its mass, and then dissolved in a 100-mL titration solvent (a mixture of toluene, water, and isopropyl alcohol with a volumetric mixing ratio of 100:1:99). Then, p-naphtholbenzein (the titration indicator) was added into the sample which was eventually titrated with 0.1 N potassium hydroxide (KOH) solution. Results from titration were then used for calculating the acid number (in mg KOH/g oil) based on the following equation:

$$\text{Acid number} = \left(\frac{(A - B) \times M \times 56.1}{W} \right) \tag{1}$$

where A is the volume of KOH solution required for the titration of the sample in mL, B is the volume of KOH solution required for the titration of 100 mL of titration solvent in mL, M is the molarity of the KOH solution, and W is the weight of the sample in grams. The acid number was then converted to a FFA content value.

2.5. Data analysis

Data obtained from each esterification experiment were composed of a set of FFA content values (or acid numbers) taken at different reaction times. These data were subsequently used for determining mass-transfer interfacial area (a_e) formed during esterification reaction. The following demonstrates how kinetic and mass-flux equations were used for the analysis of a_e.

The rate of esterification reaction is essentially the rate of FFA conversion into fatty acid methyl ester (FAME). With the stoichiometric ratio of 1:1, the conversion rate can be expressed as a function of reactant concentrations (i.e., C_{FFA} for free fatty acid and C_{Alc} for alcohol):

$$rate = -\frac{dC_{FFA}}{dt} = kC_{FFA}C_{Alc} \tag{2}$$

where k is the reaction rate constant varying with reaction temperature. Because an excess amount of alcohol for reaction was used in this experimental study, the conversion rate can be rewritten in the pseudo–first-order form:

$$rate = -\frac{dC_{FFA}}{dt} = k'C_{FFA} \tag{3}$$

http://dx.doi.org/10.5772/65657

where k' is the pseudo–first-order constant (kC_{Alc}). It should be noted that, in an immiscible reaction system (i.e., oil and alcohol), the reaction rate also depends upon the measure of dispersion or interfacial contact between two immiscible reactants. Due to the involvement of the interface between oil and alcohol, the rate of FFA conversion can also be expressed in terms of the mass-transfer flux of FFA (N_{FFA}):

$$-\frac{dC_{FFA}}{dt} = a_e N_{FFA} \tag{4}$$

where a_e is the interfacial area per unit volume of the reaction system. By combining Eqs. (3) and (4), the mass-transfer flux can be written as a function of FFA concentration:

$$N_{FFA} = \left(\frac{k'}{a_e}\right) C_{FFA} \tag{5}$$

Because the magnitude of constant k' is proportional to the degree of contact between oil and alcohol, the $\left(\frac{k'}{a_e}\right)$ ratio in Eq. (5) can be considered to be a constant value, suggesting that mass-transfer flux, N_{FFA}, at a given C_{FFA} concentration should have a fixed value. Then, Eq. (5) can be rewritten as:

$$N_{FFA} = \left(\frac{k'}{a_e}\right) C_{FFA} = \left(\frac{k'}{a_e}\right)_{Ref} C_{FFA} = \text{constant} \tag{6}$$

where $\left(\frac{k'}{a_e}\right)_{Ref}$ is the ratio derived from the reference esterification experiments with the well-defined interfacial area (the first mode experiments). With a known N_{FFA} flux, Eq. (4) can be rewritten as:

$$-\frac{dC_{FFA}}{dt} = a_e \left(\frac{k'}{a_e}\right)_{Ref} C_{FFA} \tag{7}$$

Integrating the above equation results in the following equation:

$$ln\left(\frac{C_{FFA,0}}{C_{FFA}}\right) = a_e \left(\frac{k'}{a_e}\right)_{Ref} t \tag{8}$$

where $C_{FFA,0}$ is the initial FFA concentration. To determine a_e under a given reaction condition, a plot between $ln\left(\frac{C_{FFA,0}}{C_{FFA}}\right)$ and reaction time (t) was developed using the experimental data. The values of the $\left(\frac{k'}{a_e}\right)_{Ref}$ ratio were obtained as a function of reaction temperature and catalyst concentration and reported in a separate work. [13]

3. Results and discussion

3.1. Parametric effects on FFA conversion rate and mass-transfer interfacial area

3.1.1. Effect of reaction temperature

The effect of reaction temperature was observed from the experiments carried out at three different temperatures: 45°C, 55°C, and 65°C and for oil feedstock containing 5%, 15%, and 30% FFA. Other experimental conditions were fixed at 0.5 wt% H_2SO_4 catalyst, 6:1 methanol-to-oil ratio, and 300 rpm agitation speed. Results in **Figure 2a**, **b** show that the conversion of FFA proceeded rapidly at the beginning of the reaction period. As much as 80% conversion (based on initial FFA concentration) was observed within the first 20 min. Then, the conversion rate diminished significantly when FFA conversion approached the plateau. Both figures also show that the FFA conversion rate (or slope of FFA conversion profiles at the first reaction period) increased with reaction temperature regardless of the initial FFA concentration. The increasing conversion rate was quantified and presented in terms of percent improvement compared to the conversion rate at 45°C, as shown in **Figure 3**. It appears that the conversion rate could be enhanced as much as 160% when the reaction temperature was raised from 45 to 65°C. Both kinetic and hydrodynamic factors (a_e) contribute to the rate improvement. Between the two factors, the kinetics plays the major role in controlling the conversion of FFA.

As for the role of temperature on a_e, results in **Figure 2c** show that a_e increases with temperature. The a_e could increase approximately 30 - 60% when the temperature increases from 45°C to 65°C. This is due to the decrease in liquid density and viscosity with increasing temperature. The dependence of density and viscosity of oil on temperature was previously reported by [14, 15]. According to [16], the rate of any reactions in an immiscible liquid-liquid system is controlled by the mass transfer of chemical species across the interface between the two liquids. For the FFA esterification, mass-transfer interfacial area is dependent upon the dispersion level of methanol in the oil feedstock, which is usually controlled by mixing characteristics (e.g., flow, shear, and turbulence). Such mixing characteristics are ultimately dependent upon physical properties, especially the density and viscosity of liquids. This is supported by the fact that Reynolds number (*Re*) is a function of density and viscosity [17]. Therefore, an increase in reaction temperature causes the density and viscosity of liquids to drop, thus allowing methanol to easily disperse in oil.

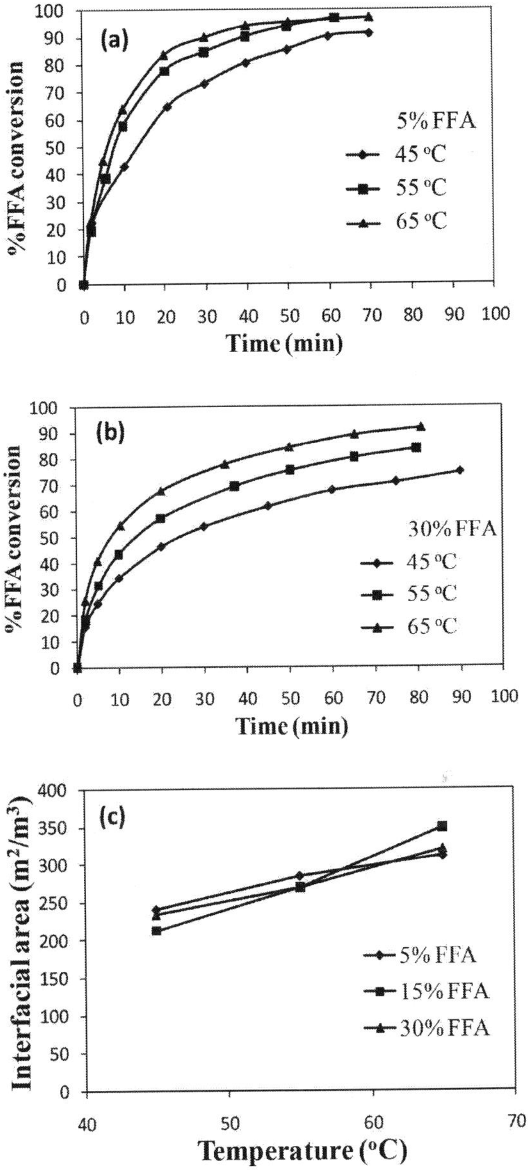

Figure 2. Effect of temperature on esterification performance: (a) FFA conversion profiles for initial FFA concentration of 5%; (b) FFA conversion profiles for initial FFA concentration of 30%; and (c) interfacial area at different temperatures (300 rpm agitation speed, 0.5 wt% of catalyst, 6:1 methanol-to-oil ratio).

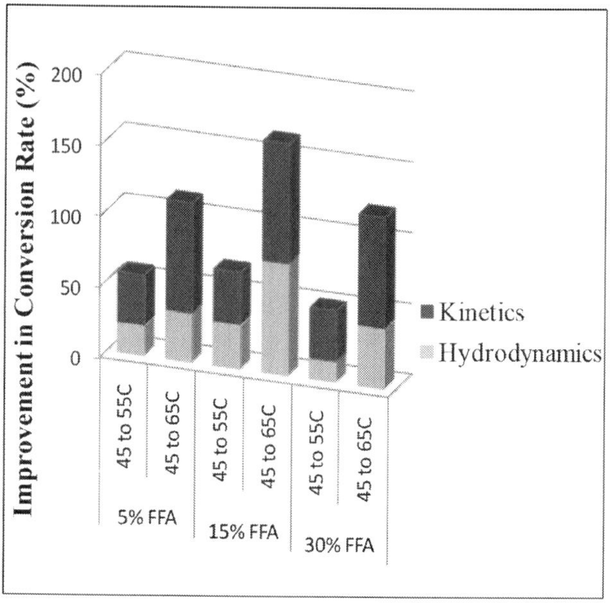

Figure 3. Hydrodynamic and kinetic contributions for effect of reaction temperature on FFA conversion rate (300 rpm agitation speed, 0.5 wt% of H_2SO_4, and 6:1 mol/mol methanol-to-oil ratio).

3.1.2. Effect of methanol-to-oil ratio

The effect of methanol-to-oil ratio was investigated under 0.5 wt% H_2SO_4, 300 rpm agitation speed, 45°C and 65°C reaction temperature, for three different FFA concentrations (5%, 15%, and 30%). It was found that methanol-to-oil ratio has a significant impact on FFA conversion performance. An increase in methanol-to-oil ratio enhances the conversion rate for all test conditions. From **Figure 4a**, **b** FFA conversion rate could be improved by as much as 30 - 35% when methanol-to-oil ratio increases from 3:1 to 9:1. The increasing conversion rate is due to a significant increase in interfacial area a_e. As shown in **Figure 4c**, **d**, the area a_e increases by 2.1 5.3 times when methanol-to-oil ratio increases from 3:1 to 9:1. This is due to the greater amount of methanol available for dispersion in the oil phase.

Based on the analysis shown in **Figure 5**, the improvement in FFA conversion rate due to increasing methanol-to-oil ratio is primarily caused by a_e, not reaction kinetics. This is because the increasing methanol-to-oil ratio leads to more dispersion of methanol, which in turn provides a greater interfacial area for esterification reaction. On the contrary, increasing the amount of methanol in oil does not result in any changes in concentration of methanol at the reaction interface; thus, the reaction kinetics is unaffected.

http://dx.doi.org/10.5772/65657

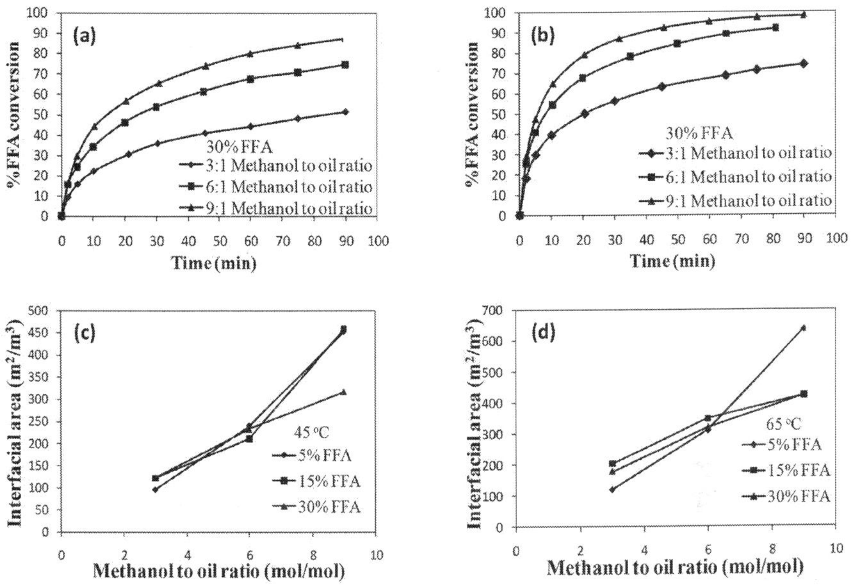

Figure 4. Effect of methanol-to-oil ratio on esterification performance: (a) FFA conversion profiles at 45°C for initial FFA concentration of 30%; (b) FFA conversion profiles at 65°C for initial FFA concentration of 30%; (c) interfacial area plotted against methanol-to-oil ratio at 45°C; and (d) interfacial area plotted against methanol-to-oil ratio at 65°C (300 rpm agitation speed, 0.5 wt% of catalyst).

3.1.3. *Effect of agitation speed*

The effect of agitation speed on FFA conversion was investigated by varying the agitation speed from 200 to 300 rpm and further to 400 rpm. The investigation was done for three different FFA concentrations (5%, 15%, and 30%) at 0.5 wt% H_2SO_4, 6:1 methanol-to-oil ratio, and 45°C and 65°C. Results show that agitation speed has an impact on FFA conversion performance. As shown in **Figure 6a**, **b**, increasing agitation speed from 200 to 300 rpm leads to a significant increase in the conversion rate. For instance, the rate could be improved by 150% at the reaction temperature of 45°C for oil feedstock containing 5% FFA (**Figure 7**). However, it should be noted that raising agitation speed further from 300 to 400 rpm leads to only a small increase in the rate of FFA conversion. It is apparent that the improvement under fixed reaction conditions (excluding agitation speed) was solely caused by an increase in a_e, not reaction kinetics. Raising agitation speed induces more turbulence, thereby creating smaller size methanol droplets in oil and in turn providing a greater a_e for esterification reaction. The increase in a_e is evidenced in **Figure 6c**, **d**.

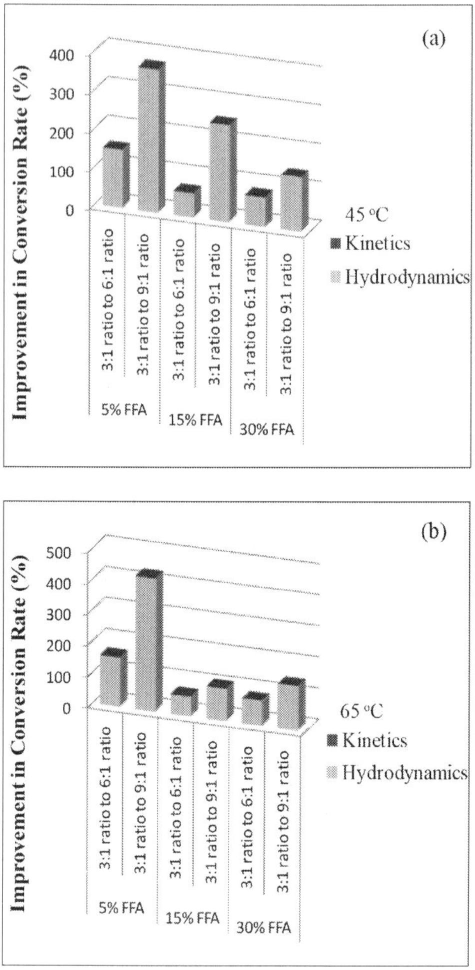

Figure 5. Hydrodynamic and kinetic contributions for the effect of methanol-to-oil ratio on FFA conversion rate: (a) reaction temperature of 45°C; (b) reaction temperature of 65°C (test conditions = 300 rpm agitation speed and 0.5 wt% of H_2SO_4).

It should be noted that the degree of rate improvement also depends on reaction temperature. This exhibits an interaction effect between agitation speed and temperature. The effect of agitation speed at a lower reaction temperature (45°C) is much greater than the effect at the higher temperature (65°C). This behavior can be explained by comparing the magnitude of interfacial area formed at these two temperatures. From **Figure 6c, d** it can be seen that the higher temperature (65°C) tends to offer a greater area, a_e, than the lower temperature (45°C) does. This is due to the reduction in density and viscosity of liquid mixtures with an increase in temperature. Therefore, increasing agitation speed at 65°C, where the higher a_e is already established, does not yield a much greater improvement in conversion rate.

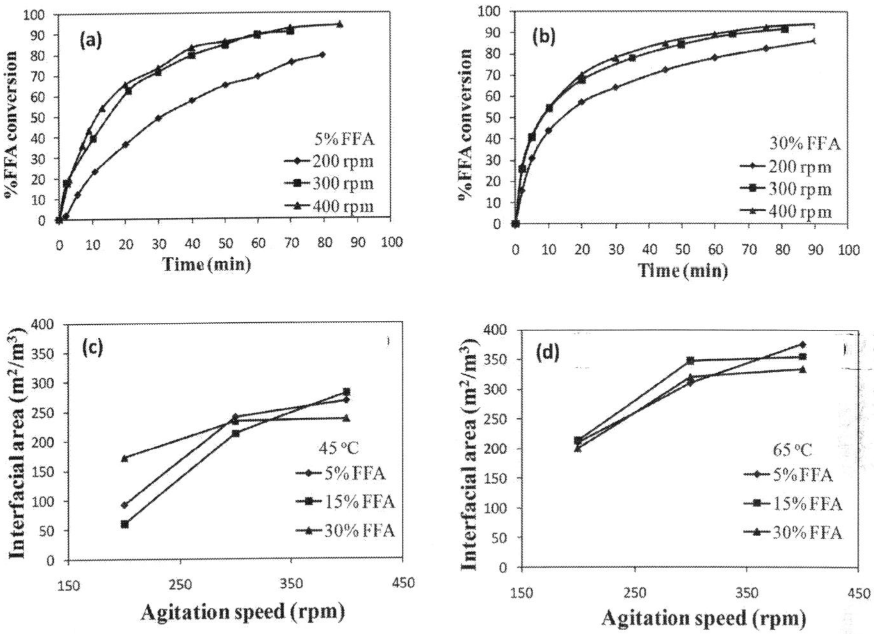

Figure 6. Effect of agitation speed on esterification performance: (a) FFA conversion profiles at 45°C for initial FFA concentration of 5%; (b) FFA conversion profiles at 65°C for initial FFA concentration of 30%; (c) interfacial area plotted against agitation speed at 45°C; and (d) interfacial area plotted against agitation speed at 65°C (6:1 methanol-to-oil ratio, 0.5 wt% of catalyst).

As mentioned previously, raising agitation speed beyond 300 rpm does not have much impact on the conversion rate of FFA. This can be explained by considering the conventional power correlation for agitated reaction. According to McCabe et al. [18], the power number, N_P, for the typical stirred reactor (i.e., an index that reflects friction preventing the impeller rotation) tends to decrease with the Reynolds number (*Re*), especially at low and moderate turbulence regions, while it remains virtually unaffected by the Reynolds number under highly turbulent conditions. This suggests that the effect of agitation speed should be gradually diminished with the increasing level of system turbulence. This behavior was observed in this work. The a_e increases considerably due to the significant reduction in friction on the impeller when agitation speed increases from 200 to 300 rpm. However, when agitation speed increases from 300 to 400 rpm, despite the increase in turbulence, the friction on the impeller does not diminish much further. This indicates that the friction may reach its minimum for a given system geometry that accounts for the design and dimensions of the reaction system as well as the type of fluid in the reactor. As such, the degree of mixing does not improve, causing the interfacial area, a_e, to remain unchanged. This in turn results in the stabilization of the FFA conversion rate.

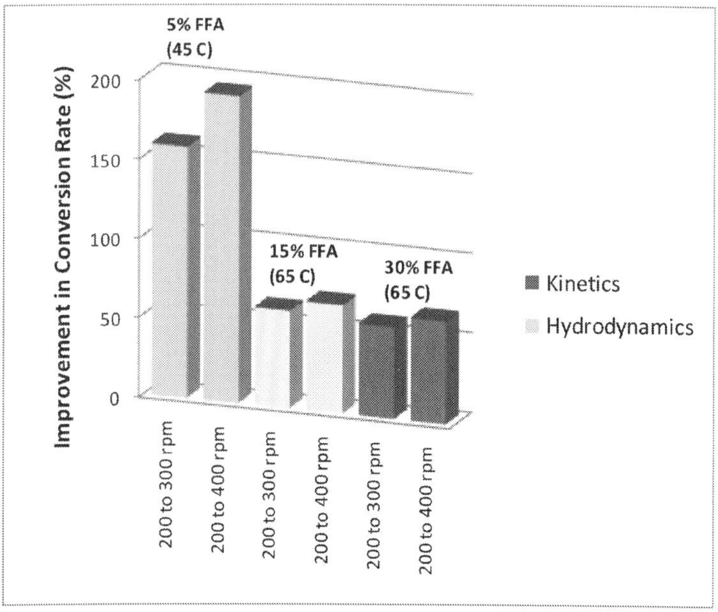

Figure 7. Hydrodynamic and kinetic contributions for the effect of agitation speed on FFA conversion rate (0.5 wt% of H_2SO_4 and 6:1 methanol-to-oil ratio).

3.1.4. *Effect of catalyst concentration*

The effect of catalyst concentration was studied by varying H_2SO_4 concentration from 0.5 to 2.0 wt%. The effect was examined for three FFA concentrations (5%, 15%, and 30%) and two reaction temperatures (45°C and 65°C) at 6:1 methanol-to-oil ratio and 300 rpm agitation speed. Results in **Figure 8a, b** show that an increase in H_2SO_4 concentration leads to an enhancement of FFA conversion performance for all test conditions. For instance, the conversion rate can be improved by 70% when H_2SO_4 concentration increases from 0.5 to 2.0 wt% at 45°C. Both hydrodynamics and kinetics were found to contribute to such improvement as shown in **Figure 9**. The hydrodynamic contribution (or an increase in a_e) results from the reduction in liquid viscosity. Note that the hydrodynamic contribution is not as significant as the kinetic contribution at a higher temperature (i.e., 65°C). This is supported by the results in **Figure 8c, d** which show that the change in a_e with H_2SO_4 concentration is relatively small at the higher temperature.

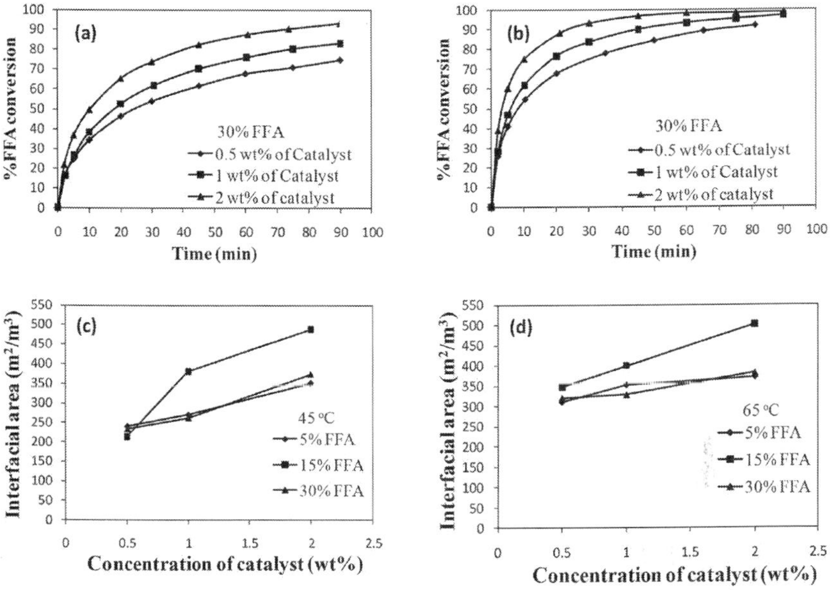

Figure 8. Effect of catalyst concentration on esterification performance: (a) FFA conversion profiles at 45°C for initial FFA concentration of 30%; (b) FFA conversion profiles at 65°C for initial FFA concentration of 30%; (c) interfacial area plotted against catalyst concentration at 45°C; and (d) interfacial area plotted against catalyst concentration at 65°C (300 rpm agitation speed, 6:1 methanol-to-oil ratio).

3.1.5. Effect of FFA concentration in oil feedstock

The effect of FFA concentration was examined over ranges of operating conditions, that is, 45 - 65°C reaction temperature, 200 - 400 rpm agitation speed, 3:1 - 6:1 methanol-to-oil ratio, and 0.5 - 2.0 wt% catalyst concentration. The results in **Figure 10** show that FFA concentration plays an important role in the FFA conversion performance. An increase in FFA concentration causes the conversion rate to decrease. However, it should be noted that the hydrodynamics of the reaction system in this case does not contribute to the changes in FFA conversion rate since the interfacial area, a_e, does not vary with FFA concentration in oil (**Figure 11**). It seems that the unaffected a_e is a result of the invariable physical properties of oil feedstock. According to Kulkarni et al. [19] and Zhou et al. [20], the viscosity and density of canola oil (base ingredient of oil feedstock) and oleic acid (FFA) are in similar ranges. The density of canola oil and oleic acid is 0.912 and 0.90 g/mL, while the viscosity of canola oil and oleic acid is 33.4 and 34.8 cP, respectively. Due to the similar properties of the two ingredients, increasing FFA concentration from 5 to 30% does not considerably alter the viscosity and density of the oil mixture. The unchanged oil properties help establish the stable turbulence level within the reaction system, thus keeping the interfacial area, a_e, relatively unchanged.

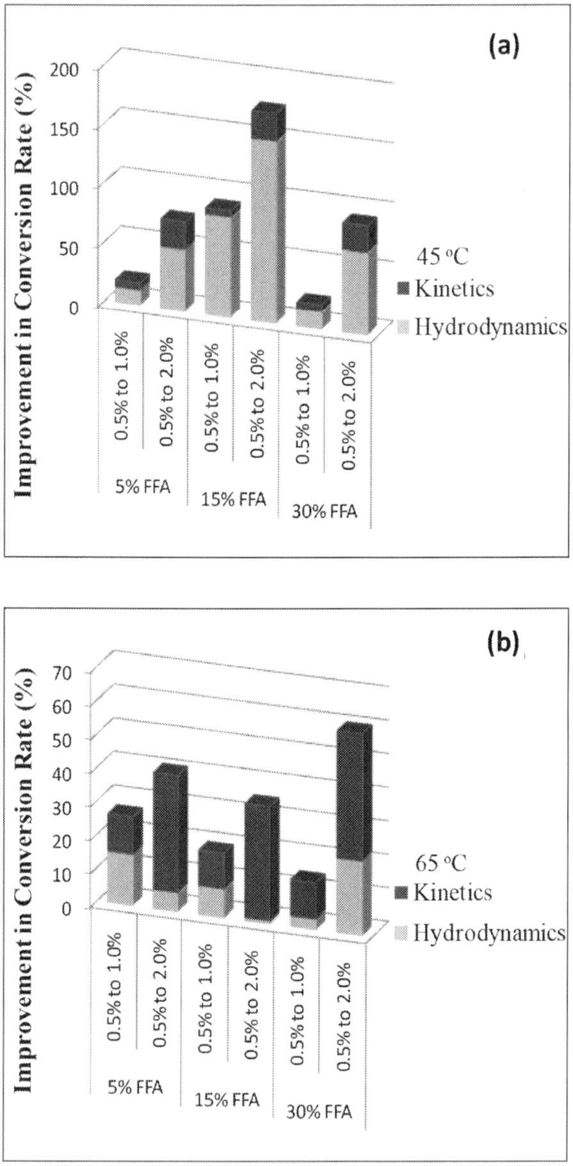

Figure 9. Hydrodynamic and kinetic contributions for the effect of catalyst concentration on FFA conversion rate: (a) reaction temperature of 45°C; (b) reaction temperature of 65°C (test conditions = 300 rpm agitation speed and 6:1 methanol-to-oil ratio).

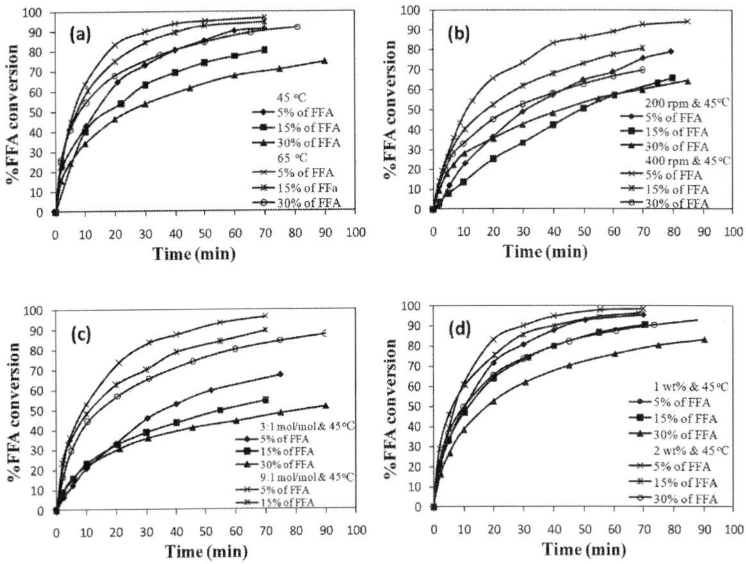

Figure 10. Effect of FFA concentration on esterification performance: (a) based on temperature data series; (b) based on agitation speed data series; (c) based on catalyst concentration data series; and (d) based on methanol-to-oil ratio data series.

3.2. Empirical correlation for mass-transfer interfacial area

The effects of process parameters on the interfacial area reported earlier were correlated in the form of an empirical equation that would facilitate the design of a biodiesel reactor. Development of the correlation was focused primarily on four important parameters controlling the interfacial area between methanol and oil feedstock, that is, reaction temperature, agitation speed, methanol-to-oil ratio, and catalyst concentration. Firstly, the effect of each process parameter was regressed individually to arrive at the best mathematical expression offering simplicity and the lowest data deviation. Four types of mathematical expressions were considered in this screening step: linear, exponential, logarithmic, and power forms. It was found that most parametric effects can be described by linear expressions, except for the effect of agitation speed, the nonlinear behavior of which can be expressed well by the logarithmic equation. Values of average absolute deviation (%AAD) and R^2 derived from individual regressions are summarized in **Table 2**.

Based on the selected equations in the screening step, an overall empirical correlation that combines all four parametric effects was formulated and expressed in the following form:

$$a_e = k_1 T + k_2 \ln(n - k_3) + k_4 R + k_5 c + k_6 \tag{9}$$

where k_1 to k_6 are correlation constants, T is reaction temperature in K, n is agitation speed in rpm, R is methanol-to-oil ratio in mol/mol, and c is catalyst concentration in wt%. The calculated interfacial area (a_e) is presented in m^2/m^3 units. Based on all experimental data obtained in this study, a computer-software package called "*NLREG*" was used for regression to arrive at values of correlation constants (k_1 to k_6) as listed in **Table 3**. It should be noted that this empirical correlation is capable of predicting methanol-oil interfacial area with an average absolute deviation (AAD) of 12%. A good agreement between the calculated a_e values and experimental data can be observed from a parity plot in **Figure 12**, which shows a R^2 value of 0.88.

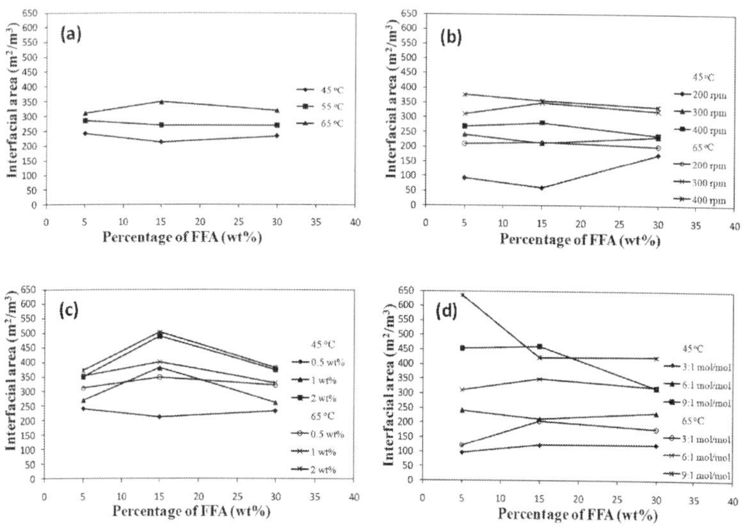

Figure 11. Effect of FFA concentration on mass-transfer interfacial area: (a) based on temperature data series; (b) based on agitation speed data series; (c) based on catalyst concentration data series; and (d) based on methanol-to-oil ratio data series.

Process parameter	Mathematical expression	%AAD	R^2
Temperature (T)	$a_e = k_1T + k_2$	4.06	0.91
Methanol-to-oil ratio (R)	$a_e = k_1R + k_2$	13.34 (45°C)	0.89 (45°C)
		14.68 (65°C)	0.82 (65°C)
Agitation speed (n)	$a_e = k_1\ln(n - k_2) + k_3$	16.81 (45°C)	0.83 (45°C)
		3.64 (65°C)	0.96 (65°C)
Catalyst concentration (c)	$a_e = k_1c + k_2$	11.09 (45°C)	0.69 (45°C)
		11.09 (65°C)	0.48 (65°C)

Table 2. Results of individual regressions for parametric effects.

http://dx.doi.org/10.5772/65657

Correlation constant	Value
k_1	3.85
k_2	32.50
k_3	198.04
k_4	51.98
k_5	53.01
k_6	−1465.72

Table 3. Correlation constants for Eq. (9).

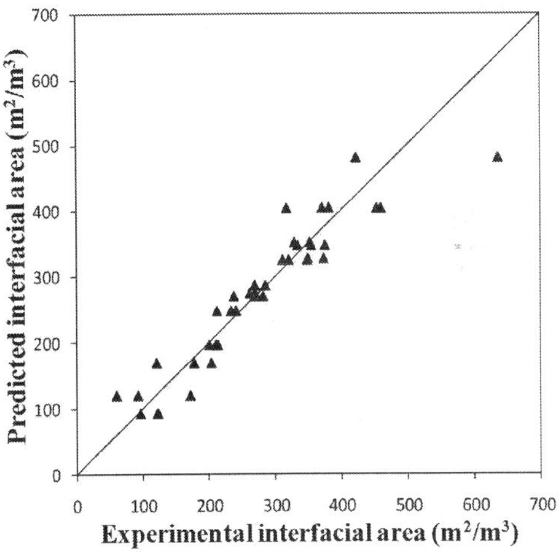

Figure 12. Parity plot between experimental data and calculated interfacial area.

4. Conclusions

Mass-transfer interfacial area plays an important role in the performance of acid-catalyzed esterification-based biodiesel production. Increasing the interfacial area enhances rate of biodiesel production (or rate of free fatty acid conversion). The magnitude of the interfacial area varies with process parameters, except free fatty acid content in oil feedstock. The interfacial area increases with increasing reaction temperature, agitation speed, methanol-to-oil ratio, and catalyst concentration, thus resulting in the increase in biodiesel production rate.

The increase in the biodiesel production rate may or may not be solely attributed to the available interfacial area. It can be attributed to both reaction kinetics and interfacial area. The interfacial area is the exclusive contributor to the increase in the biodiesel production rate when the agitation speed or the methanol-to-oil ratio increases. Both interfacial area and kinetics contribute to the enhancement of biodiesel production rate when the reaction temperature or the catalyst concentration increases.

Acknowledgements

The authors would like to thank the Natural Sciences and Engineering Research Council of Canada (NSERC) and the City of Regina for their financial support and collaboration.

Author details

Devjyoti Nath, Adisorn Aroonwilas and Amornvadee Veawab*

*Address all correspondence to: veawab@uregina.ca

Energy Technology Laboratory, Faculty of Engineering and Applied Science, University of Regina, Regina, Saskatchewan, Canada

References

[1] Santacesaria E, Tesser R, Di Serio M, Guida M, Gaetano D, Garcia Agreda A. Kinetics and mass transfer of free fatty acids esterification with methanol in a tubular packed bed reactor: A key pretreatment in biodiesel production. Industrial and Engineering Chemistry Research. 2007; 46(15): 5113–5121.

[2] Bhatnagar A K, Kaul S, Chhibber V K, Gupta A K. HFRR studies on methyl esters of nonedible vegetable oils. Energy Fuels. 2006; 20: 1341–1344.

[3] Freedman B E H P, Pryde E H, Mounts T L. Variables affecting the yields of fatty esters from transesterified vegetable oils. Journal of the American Oil Chemists' Society. 1984; 61(10): 1638–1643.

[4] Liu K S. Preparation of fatty acid methyl esters for gas-chromatographic analysis of lipids in biological materials. Journal of the American Oil Chemists' Society. 1994; 71(11): 1179–1187.

[5] Ma F, Clements L D, Hanna M A.Biodiesel fuel from animal fat: ancillary studies on transesterification of beef tallow. Industrial and Engineering Chemistry Research. 1998; 37(9): 3768–3771.

[6] Sendzikiene E, Makareviciene V, Janulis P, Kitrys S. Kinetics of free fatty acids esterification with methanol in the production of biodiesel fuel. European Journal of Lipid Science and Technology. 2004; 106: 831–836.

[7] Berrios M, Siles J, Martin M A, Martin A. A kinetic study of the esterification of free fatty acids (FFA) in sunflower oil. Fuel. 2007; 86: 2383–2388.

[8] Aranda D A G, Santos R T P, Tapanes N C O, Ramos A L D, Antunes O A C. Acid-catalyzed homogeneous esterification reaction for biodiesel production from palm fatty acids. Catalysis Letters. 2008; 122: 20–25.

[9] Thiruvengadaravi K V, Nandagopal J, Bala V S S, Kirupha S D, Vijayalakshmi P, Sivanesan S. Kinetic study of the esterification of free fatty acids in non-edible Pongamia pinnata oil using acid catalyst. Indian Journal of Science and Technology. 2009; 2(12): 20–24.

[10] Berrios M, Martin M A, Chica A F, Martin A. Study of esterification and transesterification in biodiesel production from used frying oils in a closed system. Chemical Engineering Journal. 2010; 160: 473–479.

[11] Yadav P K S, Singh O, Singh R P. Palm fatty acid biodiesel: process optimization and study of reaction kinetics. Journal of Oleo Science. 2010; 59(11): 575–580.

[12] Stamenkovic O S, Lazic M L, Todorovic Z B, Veljkovic V B, Skala D U. The effect of agitation intensity on alkali-catalyzed methanolysis of sunflower oil. Bioresource Technology. 2007; 98: 2688–2699.

[13] Zhou L. Reaction kinetics of biodiesel production by using low quality feedstock [thesis]. University of Regina: Canada; 2013.

[14] Ceriani R, Goncalves C B, Rabelo J, Caruso M, Cunha A C C, Cavaleri F W, Batista E A C, Meirelles A J A. Group contribution model for predicting viscosity of fatty compounds. Journal of Chemical Engineering Data. 2007; 52: 965–972.

[15] Ceriani R, Paiva F R, Goncalves C B, Batista E A C, Meirelles A J A. Densities and viscosities of vegetable oils of nutritional value. Journal of Chemical Engineering Data. 2008; 53: 1846–1853.

[16] Paul E L, Atiemo-Obeng V A, Kresta S M. Handbook of Industrial Mixing: Science and Practice. New Jersey: John Wiley & Sons Inc; 2004.

[17] McCabe W L, Smith J C, Harriott P. Unit Operations of Chemical Engineering. 5th ed. New York: McGraw-Hill Inc; 1993.

[18] McCabe W L, Smith J C, Harriott P. Unit Operations of Chemical Engineering. 7th ed. New York: McGraw-Hill Inc; 2005.

[19] Kulkarni M G, Dalai A K, Bakhshi N N. Transesterification of canola oil in mixed methanol/ethanol system and use of esters as lubricity additive. Bioresource Technology. 2007; 98: 2027–2033.

[20] Zhou Z, Li M, Yan Y. Viscosity of oil and surfactant binary solution. Colloids and Surfaces A: Physicochemical and Engineering Aspects. 2000; 175: 263–266.

Chapter 16

Biodiesel Compatibility with Elastomers and Steel

Salete Martins Alves, Valdicleide Silva e Mello and
Franklin Kaic Dutra-Pereira

Additional information is available at the end of the chapter

http://dx.doi.org/10.5772/65551

Abstract

This chapter describes the compatibility of biodiesel with automotive components, such as metallic and polymeric materials. It consists of a survey of literature as well as research results obtained by the authors. Aspects as wear, corrosion, and degradation materials are discussed.

Keywords: biodiesel, biofuel, elastomers, steel, corrosion, lubricity

1. Introduction

The great attention is attributed to biodiesel in recent years, due to its renewable character and sustainable use that can minimize damage to the environment. They can reduce emissions of pollutants gas and particulate materials compared to diesel from petroleum. Also, they have biodegradable and nontoxic character. In order to meet market requirements, in recent years a significant advance in installed capacity of the biodiesel industry was observed. The use of this fuel and its increase in the energetic matrix implies in constant research and development of technology to ensure its safety use. In the Brazilian energy matrix, the biodiesel use is regulated in 7% biodiesel blended with diesel. However, in this year, the biodiesel proportion will increase to 10%. Associated with growth in biodiesel demand, the fuel quality control is a greater concern, because of its natural process of degradation, corrosion or tampering, and consequently of their blends with diesel.

On the other hand, due to the unsaturated molecules present in your chemical composition, some adverse effects were reported by various authors [1–6]. Most of them are focused on the corrosive character because it is more oxidative and causes enhanced corrosion and material degradation. However, at low concentrations, no serial problems were reported to parts of the engine.

The correct material selection minimizes the corrosion problem presented by biodiesel. For example, the material used to biodiesel transportation and storage is commonly stainless steel because it has a good corrosion resistance, as well as benefits cost relation. It had been cited that they have an excellent compatibility with corrosive fluids. Some metallic substrates are used in automotive systems like tanks and carbon steel plates (covered or not by zinc), iron-zinc alloys, aluminum-zinc or nickel-zinc, lead, and tin [7–9].

Studies about the biodiesel compatibility with the other kinds of materials are some important, especially because of their injection process in the automotive application. In this step, it gets in contact with different materials such as metallic, ferrous, and even elastomeric.

The chapter describes studies of the biodiesel compatibility with some components of fuel injection system and materials used to storage and transporting of this fuel, focusing on elastomers and metals degradation after biodiesel and diesel contact.

2. Fuel and biofuel compatibility with seals of injection systems of diesel engine

Some studies had shown that when biodiesel was used as fuel in diesel engines, the injection system has been suffering some damages, such as swelling in the elastomeric seals in injection distribution, which may result in leakage of fuel [10].

Swelling results showed an incompatibility between the elastomer and the fuel, which cause a substantial loss of elastomeric properties and thus loss of sealing ability. Faced with this issue, some questions arise associated with the use of blends of biodiesel vehicle: How to evaluate the deterioration of elastomers applied to the injection system of a diesel engine caused by the use of biofuel? How the addition of the biodiesel in diesel influences the mechanical properties loss and swelling of elastomers? How are biodiesel compatibility and their blends with elastomers?

Searching answers for these questions, many authors have exposed the elastomeric materials, which are used in sealing automotive, in contact with fuel. The compatibility methods are described in the following sections of this chapter, as well as the results of degradation and compatibility of some fuels with these stamps.

2.1. Conventional compatibility test

The compatibility evaluation is commonly performed by immersion test, which consists of immersed sample in a suitable solvent, analyzing the damage by swelling by gravimetric analysis. Hasseb et al. studied the degradation of different elastomers in contact with palm biodiesel [9]. After immersion test, they found that some properties such as tensile strength, elongation and toughness were significantly reduced for both the nitrile rubber (NBR) and chloroprene rubber (CR), while minor changes were found to fluoro-VITON. Bessee and Fey assessed the influence of methylic soybean biodiesel blends in the mechanical properties of the elastomer, such as hardness, tensile strength, elongation, and swell [11]. They observed changes for nitrile, nylon 6/6, and high-density polypropylene exhibited in the mechanical properties listed above. On the other hand, this behavior is not observed for VITON. Trakarn-pruk and Porntangjitlikit investigated the impact of biodiesel on the properties of six types of elastomers commonly found in fuel systems (NBR, HNBR, NBR/PVC, rubber, acrylic copoly-mer FKM, and FKM terpolymer). Biodiesel is mixed with diesel to prepare B10 (10% mixed with diesel) [12]. The study showed little impact on properties for the polymer FKM and FKM terpolymer, ensuring consumer confidence in the use of B10 for contact with these polymers.

Michal Dubovsky reviewed the changes in the physical and mechanical properties of rubber mixtures (produced from NBR) immersed in blends of biodiesel-diesel, B10, B50, B75, and B100 at room temperature (23°C) for 3000 h and at 100°C for 500 h [13]. After immersion tests, the greatest degradation of samples exposed to a higher concentration of biodiesel (B100) at 100°C was observed. So rate used the SAE J1748 standard for assessing the compatibility of natural rubber, nylon, and EPDM (monomer ethylene-propylene diene) immersed in high oil biodiesel FFA (high content of free fatty acids) for 500 h at 55 ± 2°C. EPDM and nylon change significantly after immersion tests, but no changes were observed to natural rubber [14].

Lei Zhu studied the NBR compatibility with nine different fuels: diesel, PME, WCOME, PME, (C12:0), (C16:0), (C18:0), (C18:1 M), and (C18:1 E). After 168 h of immersion test at ambient temperature (25°C), the changes in mass, volume, and mechanical properties of NBR samples showed that biodiesel has a higher solvent power than diesel fuel. Mei Sze Loo investigated the effect of fatigue on the nitrile rubber [15]. The elastomers were immersed in the conven-tional diesel engine for 3 months and palm biodiesel for 10 days, and the same degree of swelling was obtained before the application of uniaxial fatigue loading. The authors through the stretch-N curves found that the swollen rubbers B100 had a shorter life when compared with swollen rubbers diesel [16].

2.2. Compressive immersion and pressurized tests

In order to analyze the changes in shaft seals suffered by fuel contact, some authors conducted tests and simulated that the real conditions of contact in these shaft seals are submitted.

In their investigations, Chai et al. [17] submitted two kinds of rubbers (NBR and CR) to a test tensioned by a set of plates that pressed the seal. The device consisted of four rectangular stainless steel plates with spacer bars between them. The spacer bars are designed to introduce pre-compression on the rubber specimens while they are immersed into biodiesel. Then, the device with rubbers was immersed in different palm biodiesel blends (B0, B25, B75, and B100) for 30 and 90 days, respectively. After the tests, it was found that for the types of rubber used, there was an increase in mass as well as a change in volume when the exposure time is increased from 30 to 90 days, especially when using the higher percentage of biodiesel content (B100). However, it is also noted that the pre-compression stress applied in the test reduced the amount of swelling, as compared to rubber without application of tension. That is, the swelling of NBR and CR increases with increasing biodiesel content and decreases with increasing pre-compressive stress [15].

Chai et al. and Mello [17–19] were performed immersion tests in innovative pressurized fluid device fluid, to simulate and investigate the degradation of a rubber seal in a fuel injection pump. She studied two elastomers NBR and VITON, as a sealing material. Pressurized fluids used were: diesel, biodiesel soybean, sunflower and palm oil, and its mixture with diesel (B5 and B20). The system is available in BR 10 2014 028966 6 patent.

In this study, the elastomeric seals degradation because fuel contact was investigated by testing for static and pressurized immersion. Thus it is possible to identify changes in the seal degradation under compression conditions. Below, schematic drawing device has been described.

The compression was employed using m steel cylinder by applying a preload 2500N, calculated by the required torque to maintain the closed cylinder during the expansion (**Figure 1**). O-rings (NBR and VITON) were compressed within the cylinder. All equipment was subjected to the pressure of fluid 200 bar for 5 h. The pressure amounted to 80% of the less-resistant elastomer pressurized NBR. The fluids used were diesel and biodiesel of soybean, sunflower and palm oil, and its mixture with diesel (B5 and B20).

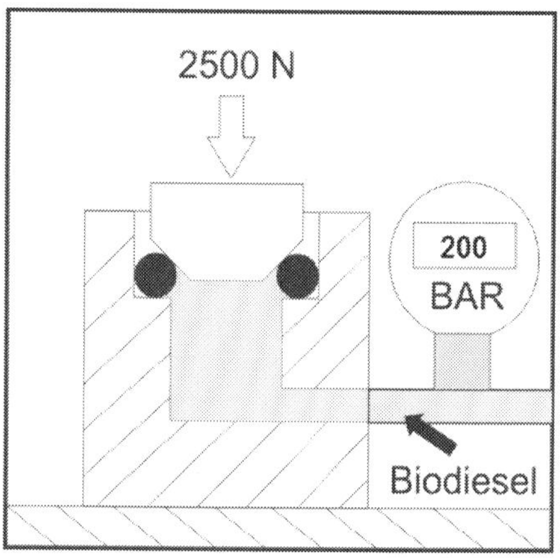

Figure 1. Simulation of the fuel injection system compression.

2.3. Comparative analysis of the compatibility of seals with fuel by the static immersion and pressurized method

Samples of elastomeric seals, NBR (nitrile rubber) and VITON (fluorine-carbon rubber), were studied by two exposure methods in order to verify the influence of the two approaches in the compatibility of the seals with fuel and degradation mechanisms suffered by seals.

In comparison purposes with the norm, the seals were also exposed to the solvent provided in the standard (toluene) with the same temperature ASTM D3616-95, but for 100 h. After the samples were weighed in the precision balance, the samples were weighed before and after the immersion time to determinate the weight loss and were dried in an oven at 108°C for 24 h, and weighed again. The evaluation of the compatibility of the elastomer with the fuel occurs based on the analysis of mechanical properties changes, swelling degree and morphological analysis of the seals after exposure to fluids.

2.3.1. Swelling rate

The **Figure 2** concerns the comparison of elastomers tested, regarding their changes in weight after static immersion (ASTM D3616-95) in soybean, sunflower, and palm biodiesel, and its blends.

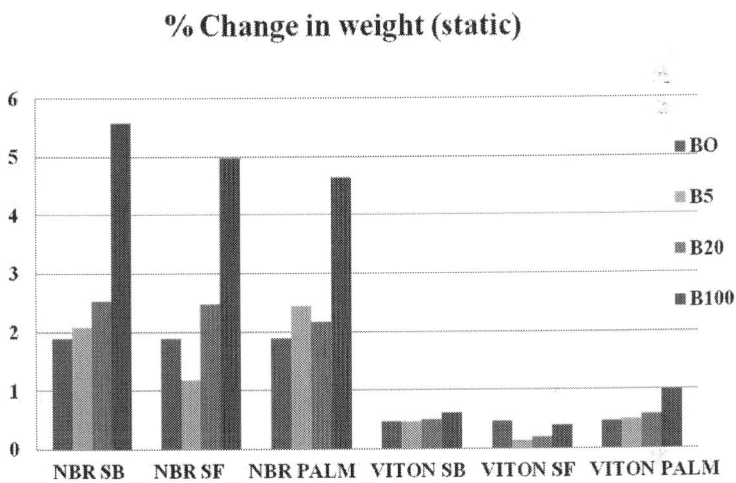

Figure 2. Change in weight of elastomers after static display of biodiesel.

The increase in weight for NBR shows an exponential trend to increased biodiesel concentration for all tested biodiesels. For VITON, moderate increase is observed for all fuels. These results are similar to those found in studies [10].

The nitrile elastomer (NBR) showed significant weight changes in contact with pure biodiesel (B100) due to the swelling ability that increased the absorption of fluid compared to soluble elastomer components. Different biofuels (soybean and palm biodiesel) did not show a big difference in the swelling degree of elastomers. Thus, it can be concluded that differences in the biodiesel composition do not affect in elastomer damage. However, the degree of swelling evaluated for sunflower shows a significant increase due to the moisture content in the biodiesel. **Figure 3** gives the change in the weight of elastomers in contact with pressurized fluids.

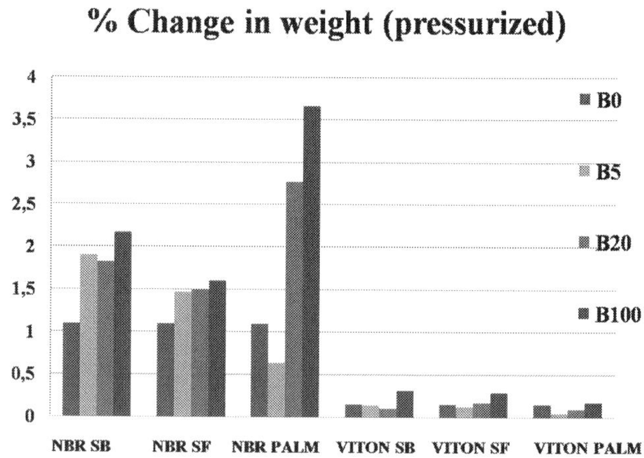

Figure 3. Change in weight of the elastomer after exposure of pressurized biodiesel.

Analogous to static swelling test, the volume change shows the same behavior for the elastomers in the pressure test, even with a time lesser than the static test exposure. Similarly, the VITON elastomer showed substantially constant, this being more compatible with the fuel investigated. This indicates that the pressure contact role should be considered when analyzing systems in which this parameter is present. Based on these results, we found the significant influence of the pressurized contact in the elastomer properties, despite the short-time exposure.

2.3.2. *Mechanical properties*

The results in the hardness changes for the elastomers are shown in **Figure 4**. For the NBR, the hardness in the condition B0 and B100 (only soybean and sunflower biodiesel) decreases compared to the hardness of the untested material. To contact the palm biodiesel, a reduction in this property for all blends (B0, B5, B20, and B100) occurs. For VITON, significant changes in this property with the use of all biodiesels and their blends are not observed.

The elastomers based on carbon and silica with materials used as fillers may serve to improve the hardness properties, abrasion resistance, tear resistance, and tensile strength. Also, the physical properties of elastomeric materials are determinate by addition of curing agents and accelerators, because they promote cross-linking between polymer chains or backbone.

Trakarnpruk and Porntangjitlikit [12] explain that biodiesel can be absorbed by the polymer that swells and therefore reduces entanglement of the polymer chain. Haseeb et al. [10] suggest an interaction between biodiesel with fillers and curing systems used in the production of elastomers.

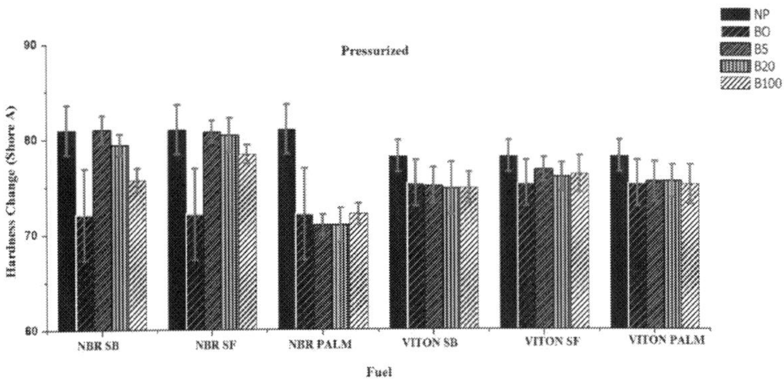

Figure 4. Change in hardness of the elastomers.

According to Haseeb et al. [10], after exposure of different elastomers biodiesel, these cross-linking agents and/or fill appear to react with various components of biodiesel and thus, deteriorate the mechanical properties.

The results of tensile strength tests are shown in **Figure 5**. After the pressure test, there was a greater loss in tensile strength compared to NBR nonpressurized material. Similarly, they observed different behaviors for various fuels, such as palm oil biodiesel, and it, in its pure form, showed a little loss in tensile strength compared to soybean biodiesel, in particular, for the conditions with B20 and B100. This loss was even more pronounced for biodiesel sunflower and all their blends.

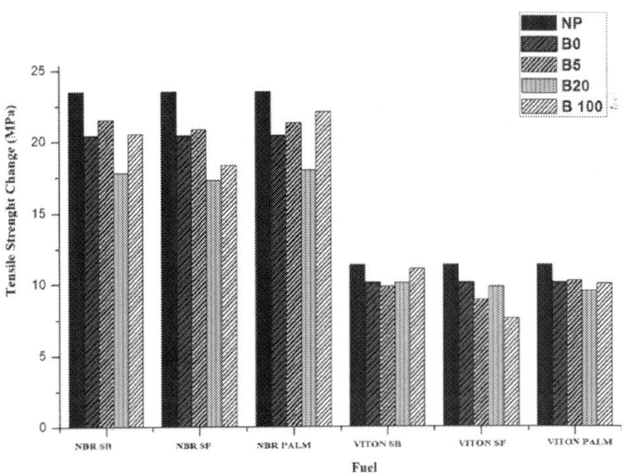

Figure 5. Tensile strength change.

The study concludes that the compatibility of the elastomer with biodiesels showed greater losses in mechanical properties, for investigated elastomers (NBR and VITON), even this displaying lower mechanical properties in conditions not tested.

It also concludes that testing of pressurized biodiesel contact with elastomers showed significant influences on changes in materials, despite the short-time exposure. This demonstrates the importance of pressure in studies of degradation of elastomers.

3. Corrosion caused by biofuel

The damage resulting from corrosive processes includes not only the need to replace metal parts in industries but also environmental problems with improper waste disposal of waste from these processes. Few studies report about corrosion of metallic materials in organic media, although of broad applications.

Biodiesel is a hygroscopic fuel, with greater capacity of water absorption than petroleum diesel. Studies reported by Kovács et al. [4] demonstrated that biodiesel is 30 times more hygroscopic than regular diesel. This feature depends on the feedstock, which favors the oxidation reactions. According to Fazal et al. [20], these reactions change the fuel properties and increase its destructive potential. The water absorbed can act directly on the corrosion of materials; it can cause biodiesel hydrolysis reactions, increasing, therefore, the metal corrosion, and promote microbial growth and, consequently, microbial corrosion.

According to Kovács et al. [4] biodiesel must avoid to its aging and/or oxidation during storage and should not deteriorate of the storage tank materials. Commercial automotive fuels contain additives carefully formulated to meet the stability of the product requirements

Ferrous and nonferrous metal parts after contact with biodiesel corrode through the chemical and electrochemical attack. According to Singh et al. [3], corrosion and wear are caused by contact of metallic materials with biodiesel. However, different materials present different corrosion behaviors in biodiesel. Ferrous alloys have better compatibility with biodiesel than nonferrous alloys. On the other hand, the copper alloys are more susceptible to corrosion than ferrous alloys. The fluorocarbons, a new group of compounds, have a high corrosion resistance.

The use of biodiesel plays considerable economic importance to national and global energy matrix. The corrosion knowledge related to fuel is a relevant issue for investigation, due the damage and cost caused by corrosive processes. Thus, the challenge is to find reliable methods to quantify the biodiesel corrosiveness, materials compatibility, and way to prevent corrosion.

The Brazilian National Agency of Petroleum (ANP) recommends the use of the ASTM D130-04 test standard to evaluate the fuel corrosiveness. This standard consists in the immersion of copper strip (clean and polished) in fuel for 1 h at 37.8°C. After this time, the piece is removed and compared against a color chart. In Refs. [4, 21–23] assess the corrosion of metals by gravimetric techniques such as ASTM G1 rules (2003) and ASTM G31 (2004) and by electrochemical techniques such as potentiodynamic polarization, linear polarization resistance, and electrochemical impedance spectroscopy (EIS).

3.1. Evaluation of corrosion by gravimetric techniques

The immersion tests or gravimetric technique is an almost used method to determinate corrosion rate, corrosion speed, and thickness loss when it is important to investigate the influence of organic or inorganic fluid on metallic materials. ASTM G3172 describes this test (2004) and consists of some steps, such as sample preparation (sanding and polishing), immersion time (hours), pickling (metal immersion in HCl solution 0.2 mol/L for 120 s), and weighing (before and after immersion test).

The degradation of different automotive materials such as copper, brass, aluminum, and cast iron was evaluated by Fazal et al. [23]. These materials were immersed in palm biodiesel and diesel at room temperature for 2880 h. The results showed that biodiesel is more corrosive than diesel, once the corrosion rate of copper (Cu), brass (BS), aluminum (Al), and cast iron (CI) increased when immersed in biodiesel. Also, corrosion rate in biodiesel for the studied materials was: copper (0.39278 mpy), brass (0.209898 mpy), aluminum (0.173055 mpy), and cast iron (0.112232 mpy). Also, the surface damage was analyzed by scanning electron microscopy. They concluded that the corrosion attacks on biodiesel-exposed metal surfaces are greater than diesel. The surface damage of aluminum was less than copper, brass, and cast iron. The corrosion was uniform on the metal surface.

The temperature influence on biodiesel corrosiveness was studied by [1]. Mild steel coupons were immersed in diesel (B0), blend palm biodiesel and diesel (B50) and palm biodiesel (B100) at 27, 50, and 80°C. They observed that corrosion rate increases with an increase in temperature for all analyzed fuel. This fact can be attributed to fast dissolution of corrosion products. EDS analysis showed the presence of oxygen on coupons surfaces indicating the formation of iron oxides or iron carbide. The metal surface was degraded by subsequent formation of other oxides and their dissolution in fuel. The biodiesel attacks more the metal surface than diesel fuel.

Dutra-Pereira [6], his master thesis, investigated the stainless AISI 316 corrosion when in contact with different biodiesel. These biodiesels were synthesized from vegetable oils (soybean, sunflower, and castor) and methanol/ethanol. The influence of alcohol used in biodiesel formation on corrosion was studied. The experimental procedure followed ASTM G3172 (2004) and time immersion of 2160 h. **Figure 6** shows the mass loss of stainless samples by corrosion related to biodiesel. The nomenclature of biodiesel was adopted considering the precursor oil and alcohol used in transesterification, p.e. soybean methylic (the biodiesel was synthesized from soybean oil and methanol), and B7 is a commercial fuel in Brazil (a mix of 7% of biodiesel and 93% of diesel).

From **Figure 6**, it is possible to verify that the loss mass is a little bigger when ethanol was used in biodiesel synthesis, and it can be justified for two carbons in the carbon chain. Considering the precursor oil, it is noted that soybean biodiesel promotes less mass loss than other biodiesel. The B7 has more mass loss indicating that the mix of diesel and biodiesel increases the corrosiveness of fuel, once diesel doesn't promote corrosion in stainless [22].

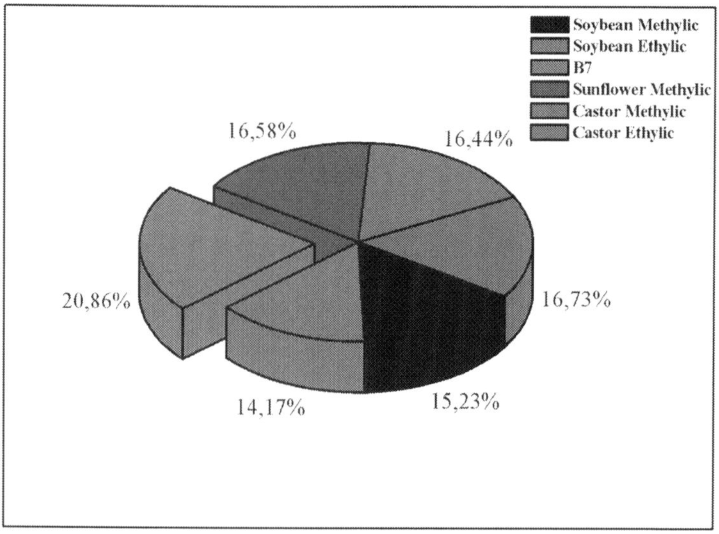

Figure 6. Loss mass due to the corrosion process [6].

The corrosion attack on stainless steel surface was analyzed by Ref. [6] by scanning electronic micrographs (**Figure 7**). Comparing with as-received steel and post-corrosion samples, it is possible to see little damage in surfaces. Basically, the corrosion occurs in pitting form.

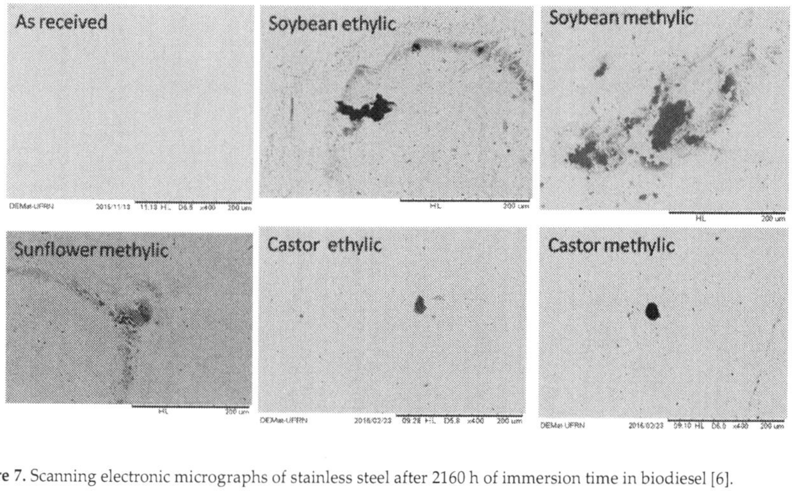

Figure 7. Scanning electronic micrographs of stainless steel after 2160 h of immersion time in biodiesel [6].

In **Figure 8**, it is clear that all biofuels corrode the metal surfaces, and the corrosion rate increases with time immersion. Although soy biodiesel and sunflower have similar molecular structure, the corrosion rate is different, being higher for sunflower biodiesel.

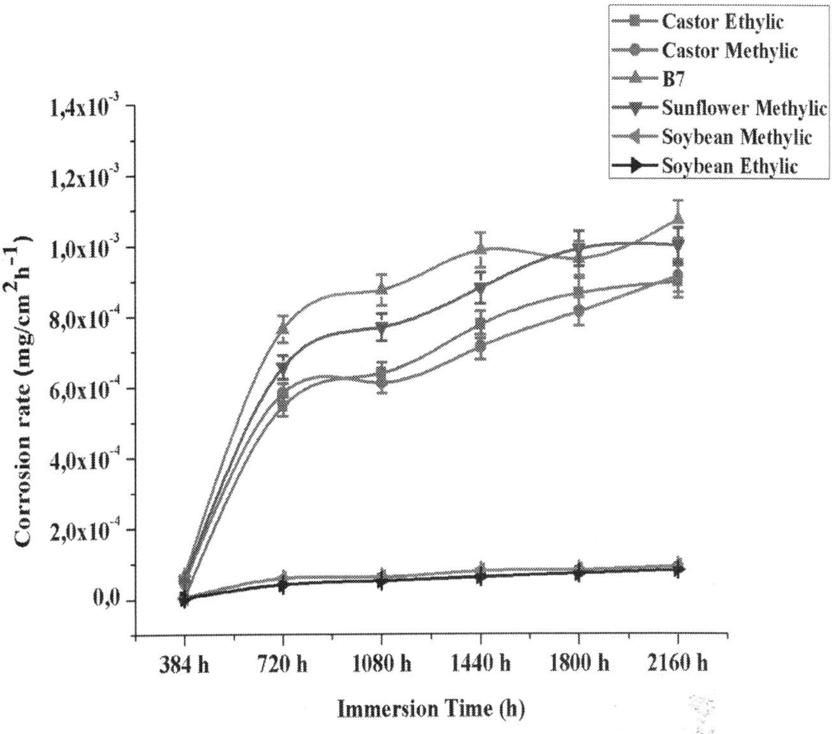

Figure 8. Corrosion rate in function of immersion time [6].

According to Fazal et al. [24], some factors influence the biodiesel corrosiveness: (i) biodiesel composition (ester and its polarity, unsaturated components, presence of oxygen, and contaminants), (ii) environment (temperature, air, light, moisture, and exposure to different metals), and nature of components (auto-oxidation, hygroscopic nature, microbial growth, and affinity of exposed metal).

The corrosion involves chemical and electrochemical reactions, which can accelerate the metal wear and promote corrosion. However, there are few studies in the literature about corrosion of automotive materials in contact with biodiesel [20, 22, 25].

As mentioned before, there are no specific techniques to analyze corrosive aspects of contact between biodiesel and metallic materials. Thus, it is important to study new methods that can evaluate the biodiesel corrosiveness, such as electrochemical techniques.

3.2. Evaluation of corrosion by electrochemical methods

As mentioned before, the most common method to assess the biodiesel corrosiveness in literature is gravimetric by immersion tests. However, this approach is only qualitative and did not characterize the trend and mechanism of corrosion of metals in contact with fluids. Also, it is important to consider that, usually, the metals corrosion is described as an electrochemical mechanism. Based on this, some electrochemical techniques and results are presented in this topic.

According to Aquino [2], the evaluation of nonaqueous fluid, as biodiesel, through electrochemical techniques is a challenge because of the high resistivity (or low conductivity) of media making difficult the determination of quantitative parameters of corrosion. The most usual electrochemical techniques are potentiodynamic polarization and electrochemical impedance spectroscopy (EIS).

Electrochemical impedance spectroscopy (EIS) is a nondestructive technique and consists "in a transient method where an excitation is applied to the system and the response (as a function of frequency) is observed" [26]. In corrosion studies, the diagram more used to analyze the information from EIS test is the Nyquist diagram (**Figure 9**), where the real part of impedance is plotted on the x-axis and the imaginary part in the y-axis of a chart. In this plot, the y-axis is negative, and each point on the Nyquist plot is the impedance Z at one frequency. The Nyquist diagram is useful to recognize the process type.

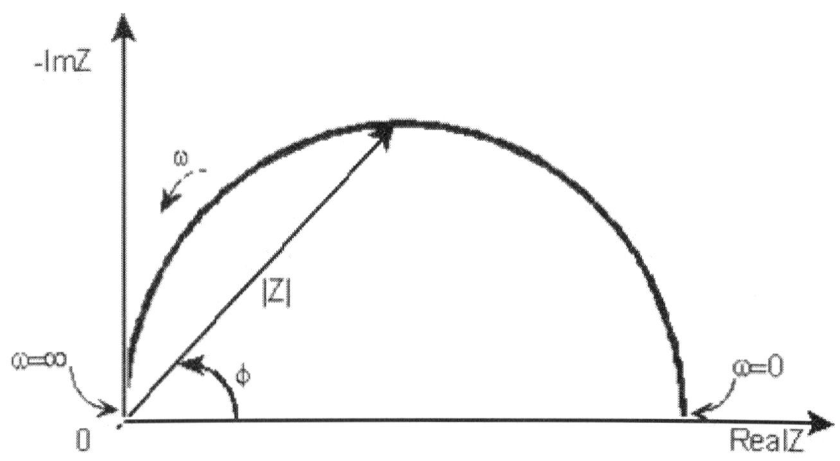

Figure 9. An example of Nyquist plot.

The potentiodynamic polarization is a technique that obtains polarization curve and scanned the electrode potential continuously. This curve (**Figure 10**) gives some important characteristics of the electrochemical behavior of metal in contact with fluids, like biodiesel.

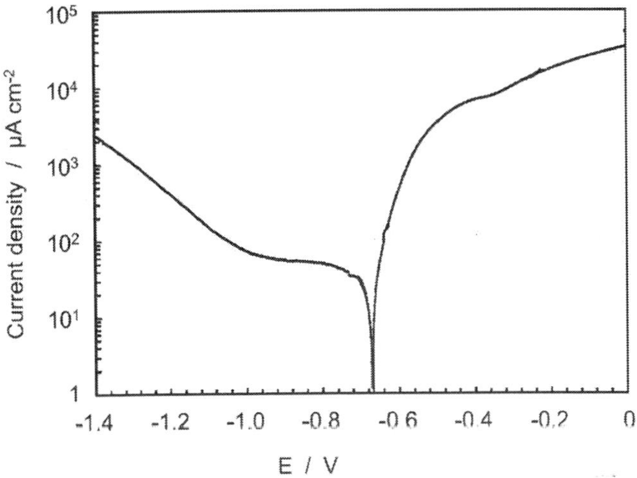

Figure 10. Typical polarization curve.

Few works are found in literature about the use of the electrochemical technique to evaluate corrosiveness of biofuel. In Diaz-Ballote [27] studied the corrosive biodiesel effect on aluminum by conventional electrochemical techniques. They verified that the corrosion current density and free corrosion potential depend on the purity of the biodiesel. Also, the biodiesel quality was determined by the provided electrochemical data.

As corrosion of steel equipment in biodiesel factory is a major problem, Torres et al. [5] analyzed AISI 316L steel corrosion in a biodiesel plant using electrical resistance probes at strategic points in the process. This procedure was developed to monitor the corrosion.

The biodiesel corrosiveness was studied by Aquino [2] and also gravimetric and electrochemical techniques. The electrochemical characterization was performed by electrochemical impedance spectroscopy (EIS) in order to evaluate the corrosion behavior of metal in biodiesel, without the addition of supporting electrolyte. The results of electrochemical characterization by EIS indicated that it could be used as an interesting tool to evaluate the biodiesel quality as well as corrosion behavior of metal in biodiesel, but the EIS for this purpose must be investigated more.

Dutra-Pereira [6] studied the corrosion of stainless steel AISI 304 in biodiesel using two electrochemical techniques: potentiodynamic polarization and electrochemical impedance spectroscopy. The experimental setup is shown in **Figure 11**. The potentiodynamic polarization curves were obtained with a scanning velocity of 1 mV s^{-1}, with the curves was possible to know the corrosion potential. EIS diagrams were achieved in frequency interval from 100 kHz to 0.004 Hz with 0.01 V of amplitude.

According to Dutra-Pereira [6], the results of electrochemical tests demonstrated that the stainless steels are incompatible with the biofuels because they oxidize in the presence of the organic medium. The diagrams of Nyquist allowed observing well-defined capacitive arcs, presenting behavior second order. The values of Rp (polarization resistance) prove that the immersion test changes the metal properties. The Rp value is a good number to compare fuel corrosiveness, less Rp means high corrosion rate. Also, polarization curves showed that occurred the passivation breaking, allowing the metals dissolution, such as chromium. This fact is evidenced by high potential values.

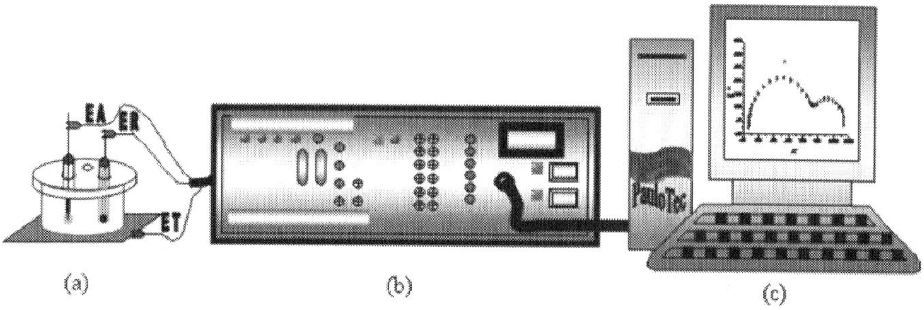

Figure 11. Experimental setup to analyze corrosion in biodiesel (a) electrochemical cell, (b) potentiostat/galvanostat, (c) microcomputer with AUTOLAB software (GPES-4 and FRA) [6].

4. Biodiesel lubricity

The diesel engines require the fuel to have lubricating properties, avoiding the direct contact between pieces in movement. Biodiesel presents superior lubricity than diesel, becoming an alternative to replace the diesel.

In their studies, Mello [18] evaluated the effect of the low-sulfur diesel (LSD) and high-sulfur diesel (HSD) on diesel lubricity. Also, they studied the biodiesel addition in diesel. A lower lubricity was detected for diesel with low-sulfur than with high-sulfur content. For blends from soybean and sunflower biodiesel, the wear scar diameters (WSD) were lower showing greater lubricity.

Another parameter that affects the lubricity is the temperature. Wadumesthrige et al. [28] observed that the lubricity decreases with increasing temperature between 20 and 70°C, for blends of 2% of biodiesel in LSD. However, for high temperature (80–90°C), lubricity increases. The positive effect on lubricity at high temperatures is due to the increased molecular motion of polar components, allowing their better distribution on the metal sometimes positively or sometimes negatively.

The influence of temperature, concentration, and oil precursor type (using in biodiesel synthesis) was investigated by Mello [18]. The researcher used Box-Behnken statistical tool as a method to evaluate these variables and their combination on biodiesel lubricity. In her experimental design were assessed the input parameters: the type of the biodiesel (soybean,

http://dx.doi.org/10.5772/65551

sunflower and palm), the concentration (5, 20, and 100%), and the temperature of contact (25, 40, and 60°C). The analyzed output parameters were the percentage of film formation, the coefficient of friction, and wear scar diameter (WSD) of the ball, and these output parameters were obtained by tribometer HFRR. Levels and factors used as test parameters are shown in **Figure 12**, with their real and coded values.

Variable	Symbol	Coded levels		
		Low	Central	High
		-1	0	1
Biodiesel	x1	Soybean	Sunflow er	Palm
Concentration (%)	x2	B5	B20	B100
Temperature (°C)	x3	25	40	60

Figure 12. Provision levels and actual and coded factors [18].

The trend to the coefficient of friction, in function of variables, is shown in **Figure 13**. In the contour surface generated for the fuel, the lower results of coefficient of friction for higher concentration levels were observed, as proposed in the literature. This fact occurs due to the oxygen present in the ester molecule and the presence of carboxylic acids which improve the lubricity. In fuels, higher coefficients of friction were observed for sunflower biodiesel, probably due to the moisture present in this fuel. According to Fazal et al. [29], the high moisture absorption seems to act as a factor that potentiates the corrosiveness of biodiesel.

In addition, it is possible to observe the decrease in coefficient of friction at high-temperature level; at the higher temperature, molecular motion for polar components increases enough, enabling these to be more evenly distributed on the metal surfaces and, therefore, enhancing lubricity. Also, the chemical adsorption of polar compounds on the metal surface is greater at high temperatures.

The response surface generated for WSD (**Figure 14**) confirms the influence of concentration and moisture of biodiesel in the lubricity. Small WSD values were found for all fuels in the upper level of concentration, as observed in friction coefficient and higher WSD values to sunflower biofuel due to the most moisture level.

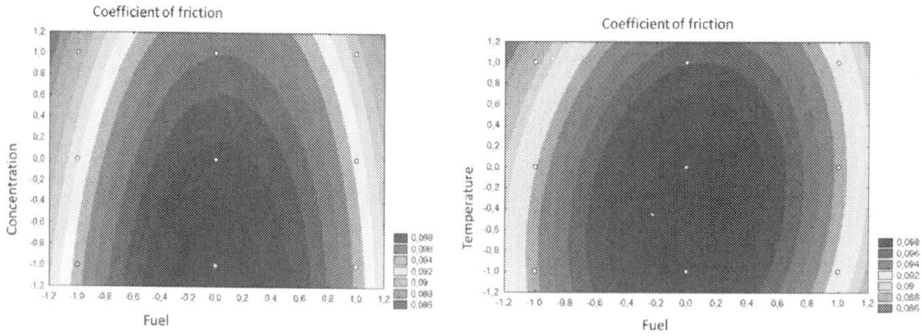

Figure 13. The coefficient of friction response [18].

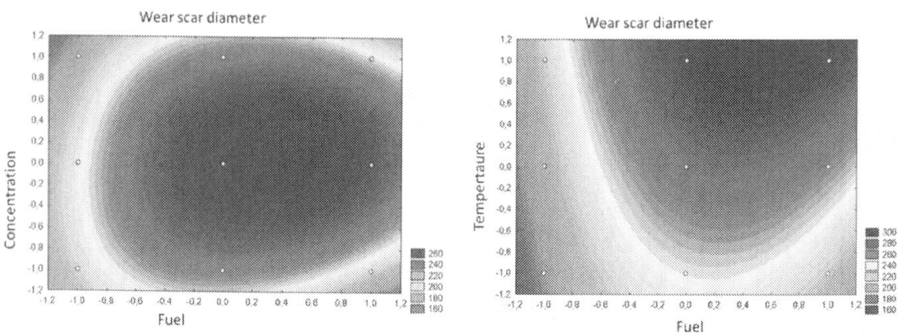

Figure 14. Wear scar diameter response [18].

The effect of fuel temperature observed in WSD confirms the results obtained by Wadumesthrige et al. [28], and better lubricity ability was noted for fuel at high temperatures. **Figure 14** shows that lubricity increases at the upper-level fuel temperature.

Figure 15 presents the response surface generated for a percentage of film formation. It is possible to verify that high rate of film formation is reached for the concentration levels above of center and high point. The biodiesel lubricity is due to the presence of a polarity-imparting heteroatom, the oxygen, and the presence of a carbonyl moiety. The influence of fuel temperature on percentage of film formation presents a more uniform response surface, showing that the percentage of the film has its optimum performance for analyzed temperature. For these variables, there is not a significant influence on the concentration in the lubricity of the higher to an intermediate point. However, there is a negative impact of the synergism by lowest levels of temperature and concentration. This synergism may compromise the lubrication system. It is a definite point because the working temperature of the engine is high than studied in this work.

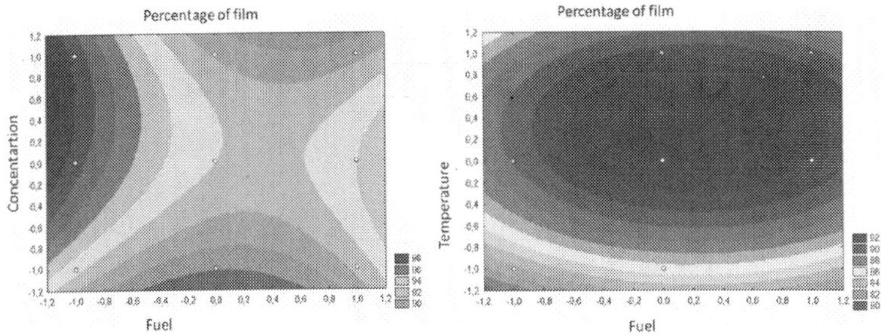

Figure 15. Percentage of film response [18].

5. Final considerations

The biodiesel use is consolidated in some countries like Brazil. Thus, it is essential to research in technologies to produce this fuel and to minimize the damage during its use in a diesel engine. In this issue, investigation about compatibility between biodiesel and materials engine is crucial to avoid premature wear and maintenance in this engine.

As addressed in this chapter, development of methodologies that described better or simulated the real contacts between biofuel and different materials is mentioned, as per example device that represents the biodiesel in contact with the elastomer of the injection system.

The corrosion has demonstrated a big challenge to biodiesel use because of its nature that attacks some metals of diesel automobiles. In this context to know, the corrosion behavior of metal in biodiesel is necessary, and thus, electrochemical techniques appear a promissory method to evaluate corrosion mechanism and quantify biodiesel corrosiveness.

Another important aspect of biodiesel is its lubricity that decreases the wear in the injection system, besides it can restore lubricity of low-sulfur diesel. Thus, it is possible to conclude that with some research to adequate the diesel engine to biodiesel and to understand better its characteristics, the biofuel is suitable to replace diesel fuel.

Author details

Salete Martins Alves[1*], Valdicleide Silva e Mello[2] and Franklin Kaic Dutra-Pereira[2]

*Address all correspondence to: saletealves@ect.ufrn.br

1 School of Sciences and Technology, Federal University of Rio Grande do Norte, Natal, Brazil

2 Federal University of Rio Grande do Norte-UFRN, Natal, RN, Brazil

References

[1] Fazal, M. A.; Haseeb, A. S. M. A.; Masjuki, H. H. Effect of temperature on the corrosion behaviour of mild steel upon exposure to palm biodiesel. Energy. 2011;36:3328–3334.

[2] Aquino, I. P. Evaluation of biodiesel corrosiveness by gravimetric and electrochemical techniques. Doctor thesis—Escola Politécnica da Universidade de São Paulo. São Paulo SP. 243 p. 2012 (in portuguese)

[3] Singh, B.; Sharma, Y. C. A.; Korstad, J. A critical review on corrosion of compression ignition (CI) engine parts by biodiesel and biodiesel blends and its inhibition. Renewable and Sustainable Energy Reviews. 2012;16:3401–3408.

[4] Kovács, A.; Tóth, J.; Isaák, G.; Keresztényi, I. Aspects of storage and corrosion characteristics of biodiesel. Fuel Processing Technology. 2015;134;59–64.

[5] Torres, C. E. D. A. S.; Costa, C. G. F.; Pereira, A. P.; De Castro, M. D. M. R.; Lins, V. D. F. C. Corrosion failure analysis in a biodiesel plant using electrical resistance probes. Engineering Failure Analysis. 2016;66:365–372.

[6] Dutra-Pereira, F. K. Evaluation of corrosive action of different biodiesels on the stainless steel AISI 316 using electrochemical methods and planning statistical. [Dissertation] [Masters degree in Engineering Mechanics]. Federal University of Rio Grande do Norte, UFRN. Natal, RN. 160 p. 2016 (in Portuguese).

[7] Ambrozin, A. R. P.; Monteiro, M. R.; Kuri, S. E. Metallic corrosion related to mineral fuels and biofuels utilization. Quimica Nova. 2009;32(7):1910–1916.(in portuguese)

[8] Susuki, C. Comparative Study of alternatives to development, design and manufacturing of fuel tank for passagers car into General Motors of Brazil. Master These (Master in Automotive Engineering)—Escola Politécnica da Universidade Estadual de São Paulo, São Paulo, SP. 136 p. 2007. (in portuguese)

[9] Haseeb, A. S. M. A.; Jum, T. S.; Fazal, M. A. Degradation of physical properties of different elastomers upon exposure to palm biodiesel. Energy. 2011;36:1814–1819.

[10] Haseeb, A. S. M. A.; Fazal, M. A. Jahihul, M. F. Masjuki, H.H. Compatiblity of automotive materials in biodiesel: a review. Fuel. 2010;90:922–931.

[11] Bessee, G. B.; Fey, J. P. Society of automotive engineering, Technical Paper No. 971690. 1997.

[12] Trakarnpruk, W.; Porntangjitlikit, S. Palm oil biodiesel synthesized with potassium loaded calcined hydrotalcite and effect of biodiesel blend on elastomer properties. Renewable Energy. 2008;33:1558–1563.

[13] Dubovsky, M.; Bozek, M.; Olsovsky, M. Degradation of aviation sealing materials in rapeseed biodiesel. Journal of Applied Polymer Science. 2015; 132 (28);42254

[14] Sorate, K. A.; Bhale, P. V.; Dhaolakiya, B. Z. A material compatibility study of automotive elastomers with high FFA based Biodiesel. Energy Procedia. 2015;75:105–110.

[15] Loo, M. S.; Le Camb, J. B.; Andriyana, A.; Robin, E.; Afifi, A. M. Fatigue of swollen elastomers. International Journal of Fatigue. 2015;74:132–141.

[16] Zhu, L.; Cheung, C. S.; Zhang, W. G.; Huang, Z. Compatibility of different biodiesel composition with acrylonitrilebutadiene rubber (NBR). Fuel. 2015;158:288–292.

[17] Chai, A. B.; Andriyana, A.; Verron, E.; Johan, M. R.; Haseeb, A. S. M. A. Development of a compression test device for investigating interaction between diffusion of biodiesel and large deformation in rubber. Polymer Testing. 2011;30:867–875.

[18] Mello, V. S. Study of physical-chemical biodiesel properties and its influence in deterioration of elastomers of injection system of diesel engine. [Dissertation] [Masters degree in Engineering Mechanics]. Federal University of Rio Grande do Norte, UFRN. Natal, RN. 102 p. 2013 (in Portuguese).

[19] Alves, S. M.; Mello, V. S.; Medeiros, J.S. Palm and soybean biodiesel compatibility with fuel system elastomers. Tribology International. 2013;65:74–80.

[20] Fazal, M. A.; Haseeb, A. S. M. A.; Masjuki, H. H. Comparative corrosion characteristics of petroleum diesel and palm biodiesel for automotive materials. Fuel Processing Technology. 2010;91:1308–1315.

[21] Wang, W., Jenkisn, P. E.; Ren, Z. Electrochemical corrosion of carbon stell exposed to biodiesel/simulated seawater mixture. Corrosion Science. 2015;57:215–219.

[22] Gallina, A. L.; Stroparo, E. C.; Cunha, M. T.; Rodrigues, P. R. P. A. The corrosion of austenitic 304 stainless steel in biodiesel. Revista Escola de Minas. 2010;63(1):71–75 (in portugese).

[23] Fazal, M. A.; Haseeb, A. S. M. A.; Masjuki, H. H. Degradation of automotive materials in palm biodiesel. Energy. 2012;40:76–83.

[24] Fazal, M. A.; Haseeb, A. S. M. A.; Masjuki, H. H. A critical review on the tribological compatibility of automotive materials in palm biodiesel. Energy Conversion and Management. 2014;79:180–186.

[25] Chew, K. V.; Haseeb, A. S. M. A.; Masiuki, H. H.; Fazal, M. A.; Gupta, M. Corrosion of magnesium and aluminum in palm biodiesel: a comparative evaluation. Energy. 2013: 57; 478–483.

[26] Fayyad, E. M.; Almaadeed, M. A.; Jones, A.; Abdullah, A. M. Evaluation techniques for the corrosion resistance of self-healing coatings. International Journal of Electrochemical Science. 2014;9:4989–5011.

[27] Diaz-Ballote, L.; López-Sansores, J. F.; López-Maldonado, L.; Mesias-Garfias, L. F. Corrosion behavior of aluminum exposed to a biodiesel. Electrochemistry Communications. 2009;11:41–44.

[28] Wadumesthrige, K.; Ara, M.; Salley, O.S.; Simon, K.Y. Investigation of lubricity characteristics of biodiesel in petroleum and synthetic fuel. Energy & Fuels. 2009;23:2229–22344.

[29] Fazal, M. A; Hasseb, A. S. M. A; Masjuki, H. H. Biodiesel feasibility study: an evaluation of material compatibility; performance; emission and engine durability. Renewable and Sustainable Energy Reviews. 2011;15:1314–1324.

Chapter 17

Biofuel Additives: Conversion of Glycerol with Benzyl Alcohol over SBA-15 with Sulfonic Acid Groups

Pedro Canhão and Jose E. Castanheiro

Additional information is available at the end of the chapter

http://dx.doi.org/10.5772/65800

Abstract

The etherification of glycerol with benzyl alcohol was carried out over mesostructured silica, SBA-15, with sulfonic acid groups. The products of glycerol etherification are ethers (glycerol mono-ether, glycerol di-ether and tri-glycerol ether). It was prepared with different catalysts, consisting of SBA-15 with different amounts of sulfonic groups (SBA-15, $[SO_3H]1$-SBA-15, $[SO_3H]2$-SBA-15, and $[SO_3H]3$-SBA-15). It was observed that the activity increased with the amount of sulfonic acid groups on SBA-15 until a maximum ($[SO_3H]2$-SBA-15). However, with high amount of acid groups, a decrease in catalytic activity was observed. The effect of different parameters, such as catalysts loading, temperature, and initial concentration of glycerol, was studied in order to optimize the reaction conditions. Catalyst $[SO_3H]2$-SBA-15 showed good activity after four uses.

Keywords: biodiesel, glycerol, benzyl alcohol, SBA-15-SO_3H, etherification

1. Introduction

Biodiesel is defined as mono-alkyl esters of fatty acids, which can be obtained from different feedstocks (animal fats and vegetable oils) [1–7]. Biodiesel is a renewable, biodegradable fuel with lower sulfur content, environmentally less toxicity, and better lubrication.

Biodiesel production can be carried out by transesterification of triglycerides or by esterification of free fatty acids with methanol or ethanol in the presence of base or acid catalysts [1–7]. **Figure 1** represents the transesterification of triglycerides with different alcohols.

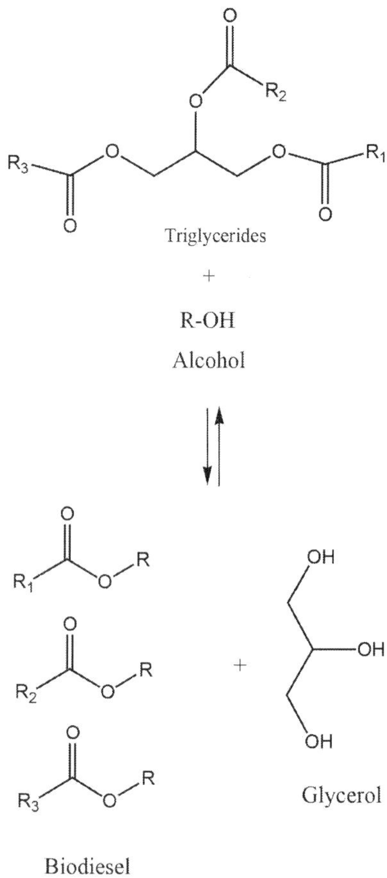

Figure 1. Representation of transesterification of triglycerides with alcohols into biodiesel and glycerol.

In the previous years, an increase in biodiesel production has been observed, and consequently, a large amount of glycerol has been produced. So, it is imperative to develop different processes to produce glycerol as a product with high commercial value. The main use of glycerol is in personal care and cosmetics, but its use as a valuable feedstock for new products and processes is growing in importance [8–13]. Glycerol ethers have many potential uses, such as fuel additives, solvents, and cryogenics. Reactions of glycerol with isobutene or *tert*-butanol under acid catalysis conditions afford *tert*-butyl-glycerol ethers, which have potential for blending with diesel [12].

Traditionally, strong homogeneous acid catalysts have been used. In order to become the process a "green process," the homogenous catalysts have been replaced by heterogeneous ones [13].

Gu et al. [14] reported the etherification of glycerol with different alcohols catalyzed by acid-functionalized silica. They reported yields varying from 61 to 96% of the mono and di-glycerol ethers, using batch reaction conditions. These results prompted us to report some preliminary data of glycerol benzylation with benzyl alcohol, using different types of heterogeneous acid catalysts, aiming to produce mono, di, and tribenzyl glycerol ethers (**Figure 2**).

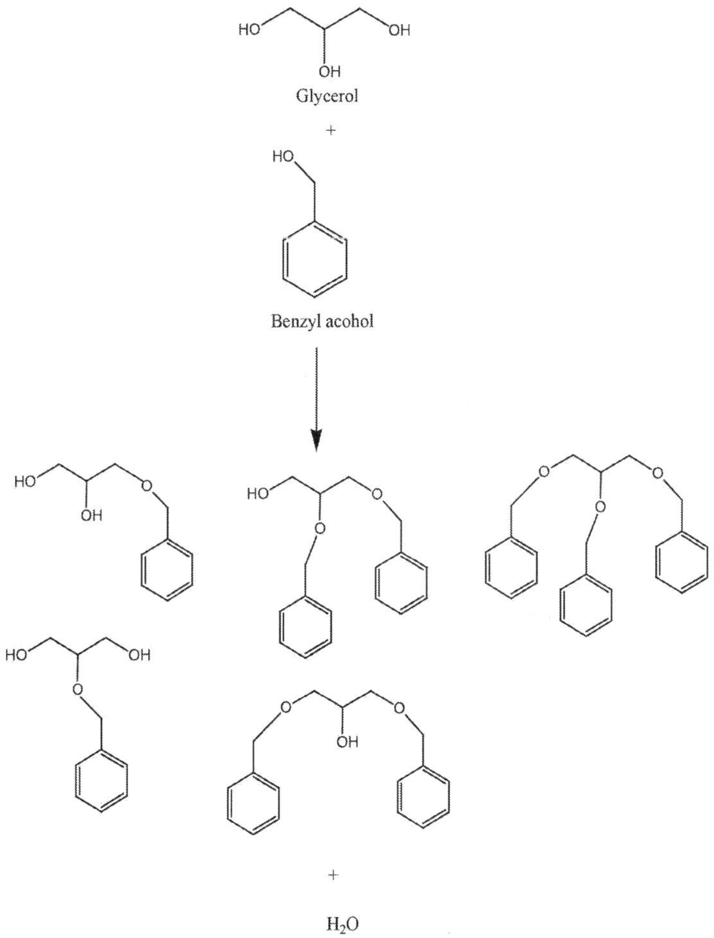

Figure 2. Etherification of glycerol with benzyl alcohol.

Etherification of glycerol with benzyl alcohol was carried out in the presence of zirconia modified with sulfuric acid (1 and 2 mol dm^{-3}). The products were mono- and dibenzyl glycerol ethers. The catalytic tests were carried out at different temperatures and initial reactant mass ratios. The catalyst prepared with the highest sulfuric acid concentration showed the highest activity [15].

Due to high surface areas and controlled pore sizes, mesoporous materials, such as SBA-15, PMO, and MCM-41, have been used in heterogeneous catalysis as catalyst supports. This kind of materials can be functionalized with different organic groups on the surface. There are different techniques to change the materials: by grafting or co-condensation. These modified materials can be used as catalysts in different chemical reactions [16–18].

In the present work, we studied the etherification of glycerol with benzyl alcohol over SBA-15 with sulfonic acid groups.

2. Experimental

2.1. Preparation of catalysts

The catalyst samples were prepared according to Grieken et al. [19].

2.2. Catalysts characterization

Micromeritics ASAP 2010 apparatus was used to determine the nitrogen adsorption isotherm at 77 K.

A CHNS Elemental Analyser 1112 series Thermo Finnigan instrument was used to determine the amount of sulfur present in SBA-15.

Cation-exchange capacities of catalysts were determined by potentiometrical titration. An aqueous solution of sodium chloride (NaCl, 2M) was used as a cationic-exchange agent.

X-ray diffraction (XRD) patterns of the catalysts were obtained by using a Rigaku powder diffractometer.

2.3. Catalytic experiment

The catalytic experiments were carried out in a stirred batch reactor at 80°C. In a typical experiment, the reactor was loaded with 10 mL of benzyl alcohol, 4 g of glycerol, and 0.2 g of catalyst.

The catalytic stability of $[SO_3H]2$-SBA-15 was carried out in the same conditions with the same sample. The catalyst was separated from reaction mixture by centrifugation. After this operation, the catalyst was washed with acetone, and it was dried at 80°C overnight.

Undecano was used as the internal standard. Samples were taken periodically and analyzed by GC, using a Hewlett Packard instrument equipped with a 30 m × 0.25 mm HP-5 column.

3. Results and discussion

3.1. Catalyst characterization

SBA-15 and SBA-15 with sulfonic acid groups show a typical IV adsorption isotherm, according to the IUPAC classification. **Table 1** reports the physicochemical characterization of materials. It was observed that the surface area (S_{BET}) and the pore volume decreased with the amount of sulfonic acid groups immobilized on SBA-15. It can be also observed that the amount of sulfur, determined by elemental analysis, is similar to the amount of sulfonic acid groups, determined by acid-base titration (**Table 1**).

Material	Amount of S (mmol/g)	H^+ (meq/g)	$S_{BET}{}^{a}$ (m^2/g)	Vp^{b} (cm^3/g)
SBA-15	–	–	950 ± 3	0.99 ± 0.02
$[SO_3H]$1-SBA-15	0.06 ± 0.01	0.05 ± 0.01	935 ± 5	0.95 ± 0.03
$[SO_3H]$2-SBA-15	0.14 ± 0.02	0.13 ± 0.01	875 ± 2	0.88 ± 0.01
$[SO_3H]$3-SBA-15	0.20 ± 0.01	0.19 ± 0.02	820 ± 4	0.83 ± 0.02

[a] BET.

[b] $(p/p^{o}) = 0.98$.

Table 1. Physicochemical characterization of materials.

Figure 3 shows the powder X-ray diffraction patterns of SBA-15 and SBA-15 with sulfonic acid. It can be observed that the materials with sulfonic acid exhibit a hexagonal pore structure, characteristic of SBA-15 materials:

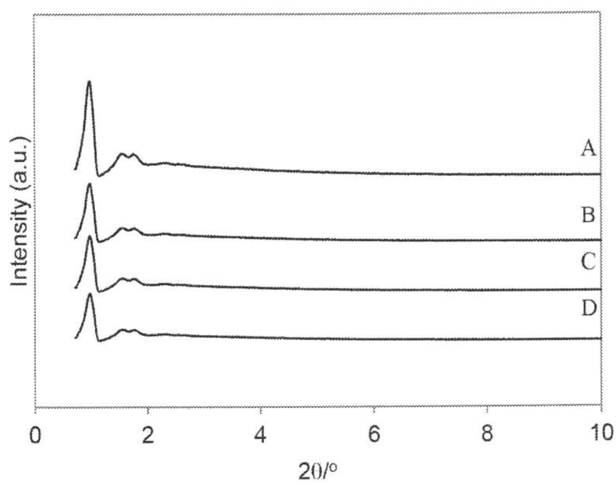

Figure 3. X-ray diffraction of materials. (A) SBA-15; (B) $[SO_3H]$1-SBA-15; (C) $[SO_3H]$2-SBA-15; and (D) $[SO_3H]$3-SBA-15.

3.2. Catalytic experiments

Figure 2 shows the products obtained in the glycerol etherification with benzyl alcohol. The products resulting from glycerol etherification reaction are mono-glycerol ether, di-glycerol ether, and tri-glycerol ether.

Figure 4 shows the initial activity of the catalysts in the glycerol etherification reaction with benzyl alcohol. It is observed that the catalytic activity increases from the material SBA-15 to the catalyst sample [SO_3H]2-SBA-15, which can be explained by the amount of sulfonic acid groups present on the SBA-15 surface (**Table 1**). However, when the amount of sulfonic acid groups increases (from catalyst sample [SO_3H]2-SBA-15 to [SO_3H]3-SBA-15) the catalytic activity decreases, which can be explained by the decrease of accessibility to the active centers. In fact, a decrease in S_{BET} and total pore volume with the amount of sulfonic acid groups were observed (**Table 1**).

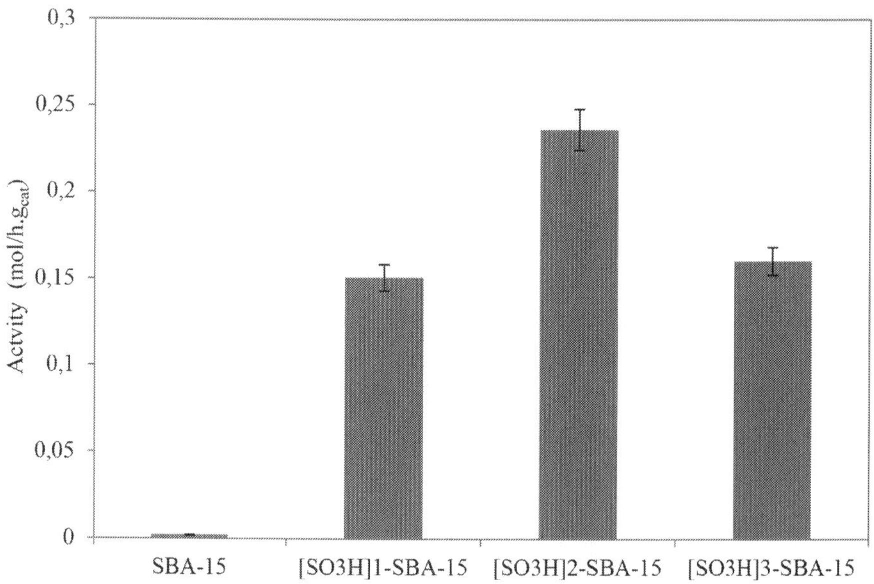

Figure 4. Etherification of glycerol with benzyl alcohol over [SO_3H]-SBA-15 catalysts.

Table 2 shows the values of glycerol conversion and selectivity of catalysts for the different products obtained from glycerol etherification with benzyl alcohol, after 7 h of reaction. It is observed that the sample of [SO_3H]2-SBA-15 got the highest conversion.

The selectivity mono-ether ranges from 72%, in the presence of SBA-15, to 78%, in the presence of the catalyst [SO_3H]1-SBA-15. It is observed that the selectivity for tri-ether is reduced. This behavior of this compound may be explained by obtaining consecutive reactions.

Catalysts	Conversion (%)[a]	Selectivity		
		Mono-ether	Di-ether	Tri-ether
SBA-15	5 ± 0.1	72 ± 1	20 ± 1	8 ± 1
$[SO_3H]$1-SBA-15	73 ± 2	78 ± 2	16 ± 2	6 ± 1
$[SO_3H]$2-SBA-15	89 ± 1	70 ± 1	20 ± 1	10 ± 1
$[SO_3H]$3-SBA-15	86 ± 0.5	71 ± 1	18 ± 1	11 ± 1

[a] Glycerol conversion after 7 h of reaction.

Table 2. Conversion of glycerol and selectivity to the glycerol etherification products over sulfonic acid groups presents on SBA-15 surface.

The effect of different parameters (catalyst loading, initial glycerol concentration, and temperature) in etherification reaction with $[SO_3H]$2-SBA-15 catalyst was also studied in order to optimize the reaction.

3.2.1. *Effect catalyst loading*

In order to study the effect of catalyst loading $[SO_3H]$2-SBA-15 in glycerol conversion, at 80°C, different experiments were carried out. The amount of catalyst ranges from 0.05 to 0.20 g. The initial concentration of glycerol (2.9 mol dm^{-3}) was kept constant. **Figure 5** shows the conversion of glycerol versus time. It was observed that when the catalyst loading increases, equilibrium conversion can be obtained faster. This behavior could be explained by the total number of active sites, with the increase in the amount of catalyst used in the reaction. However, when the catalyst amount increases from 0.1 to 0.2 g, only a slight increase in the glycerol conversion was observed.

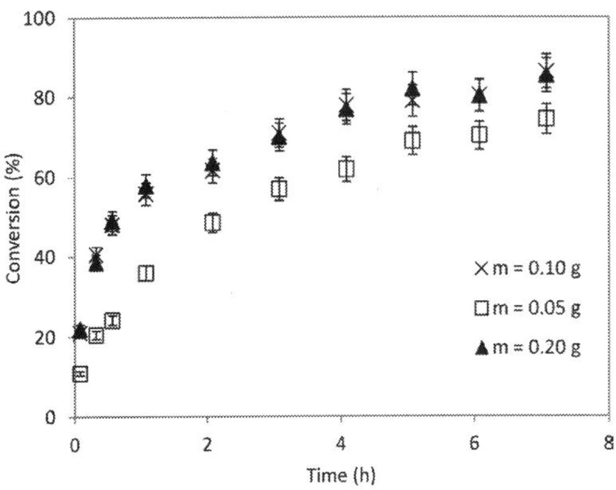

Figure 5. Etherification of glycerol with benzyl alcohol in the presence of SBA-15 with sulfonic acid groups (catalyst $[SO_3H]$2-SBA-15). Effect of catalyst amount. Conversion (%) versus time (h).

It was also observed that the increase in catalyst loading has no effect on the equilibrium conversion (**Figure 5**).

The amount of effect of catalyst $[SO_3H]2$-SBA-15 on selectivity to the different compounds was also studied. In all catalyst tests with the sample $[SO_3H]2$-SBA-15, similar values of selectivity to mono-ether (70%, at 70% of glycerol conversion) were observed.

3.2.2. *Effect of the initial glycerol concentration*

The initial concentration of glycerol ranged between 1.7 and 2.9 mol dm^{-3}, while the reaction temperature ($T = 80°C$) and the catalyst amount ($m = 0.10$ g) were kept constant. The etherification reaction of glycerol was performed with the catalyst $[SO_3H]2$-SBA-15. The results are shown in **Figure 6**. It was found that the conversion of glycerol increases with the increase in the initial glycerol concentration under the same reaction conditions. This behavior may be explained by the increased reaction rate with the concentration of glycerol.

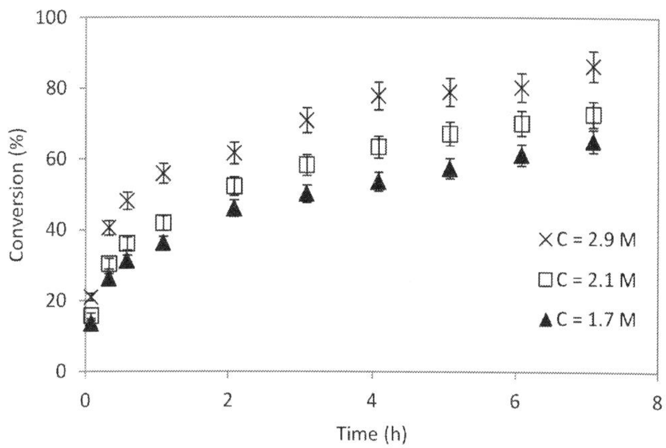

Figure 6. Etherification of glycerol with benzyl alcohol in the presence of SBA-15 with sulfonic acid groups (catalyst $[SO_3H]2$-SBA-15). Effect of catalyst amount. Conversion (%) versus time (h).

The effect of glycerol initial concentration for the different compounds was also studied. Similar values of selectivity to mono-ether (about 72%, 70% glycerol conversion) were observed.

3.2.3. *Effect of temperature*

In this work, we also studied the effect of temperature on the glycerol etherification. Catalytic tests with the catalyst $[SO_3H]2$-SBA-15 were performed at different temperatures, whereas the initial concentration of glycerol (2.9 mol dm^{-3}) and the catalyst amount ($m = 0.10$ g) were kept constant. **Figure 7** shows the effect of temperature on the glycerol conversion. An increase in glycerol conversion with temperature was observed.

The effect of temperature on catalyst selectivity $[SO_3H]2$-SBA-15 for the different compounds was also studied. Increasing the temperature resulted in a decrease in the selectivity to monoether (about 81%, 70% glycerol conversion (T = 55°C) to 60%, the glycerol conversion (T = 110°C)). This behavior can be explained by the increase in the reaction rate.

The catalytic stability of $[SO_3H]2$-SBA-15 catalyst was also studied. Different catalytic experiments were also carried out. A slight decrease in catalytic activity was observed after the second run (**Figure 8**).

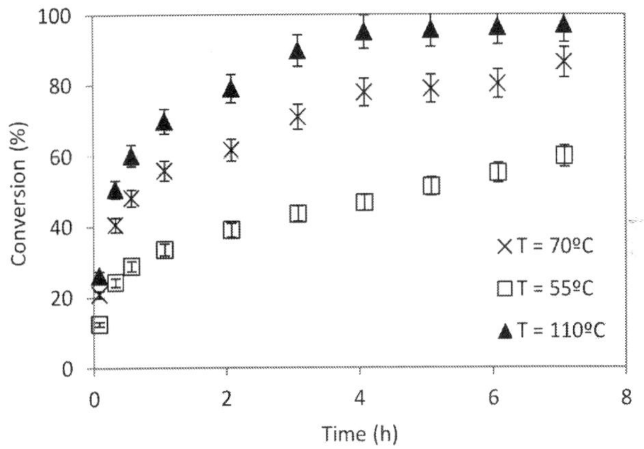

Figure 7. Etherification of glycerol with benzyl alcohol in the presence of SBA-15 with sulfonic acid groups (catalyst $[SO_3H]2$-SBA-15). Effect of temperature. Conversion (%) versus time (h).

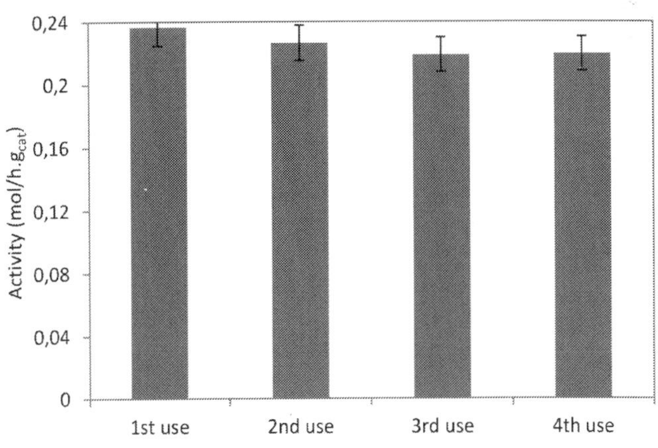

Figure 8. Catalytic stability of $[SO_3H]2$-SBA-15 catalyst in etherification of glycerol with benzyl alcohol.

4. Conclusions

The etherification of glycerol with benzyl alcohol can be achieved with the use of heterogeneous acid catalysts, consisting in SBA-15-SO_3H. The products of the glycerol etherification reaction are mono-glycerol ether, di-glycerol ether, and tri-glycerol ether.

The catalytic activity increased with the amount of sulfonic acid groups on SBA-15 surface, until a maximum. After this value, the catalytic activity decreases.

Different reaction parameters were optimized. It was observed that increasing the catalyst loading allows faster equilibrium conversion. However, the increases in catalyst loading do not affect the equilibrium conversion. Another conclusion is that the glycerol conversion increased with increase in the initial glycerol concentration. An increase in the glycerol conversion with temperature was observed.

After the second use ([SO_3H]2-SBA-15), the catalyst tends to stabilize.

Acknowledgements

The study was supported by Project PTDC/CTM-POL/114579/2009 and is gratefully acknowledged.

Author details

Pedro Canhão and Jose E. Castanheiro*

*Address all correspondence to: jefc@uevora.pt

Evora Chemistry Centre, Chemistry Department, Evora University, Évora, Portugal

References

[1] Ma, F., Hanna, M.A. Biodiesel production: A review. Bioresource Technology 1999; 70, 1–15.

[2] Lotero, J.E., Liu, Y., Lopez, D.E., Suwannakarn, K., Bruce, D.A., Goodwin, J.G. Synthesis of biodiesel via acid catalysis. Industrial and Engineering Chemical Research 2005; 44, 5353–5363.

[3] Corma, A., Iborra, S., Velty, A. Chemical routes for the transformation of biomass into chemicals. Chemical Reviews 2007; 107, 2411–2502.

[4] Knothe G. Improving biodiesel fuel properties by modifying fatty ester composition. Energy Environmental Science 2009; 2, 759–766.

[5] Yusuf, N.N.A.N., Kamarudin, S.K., Yaakub, Z. Over view on the current trends in biodiesel production. Energy Conversion Management 2011; 52, 2741–2751.

[6] Avhad, M.R., Marchetti, J.M. A review on recent advancement in catalytic materials for biodiesel production. Renewable and Sustainable Energy Reviews 2015; 50, 696–718.

[7] Singh, S P., Singh, D. Biodiesel production through the use of different sources and characterization of oils and their esters as the substitute of diesel: A review. Renewable and Sustainable Energy Reviews 2010; 14, 200–216.

[8] Anitha, M., Kamarudin, S.K., Kofli, N.T. The potential of glycerol as a value-added commodity. Chemical Engineering Journal 2016; 295, 119–130.

[9] Calero, J., Luna, D., Sancho, E.D., Luna, C., Bautista, F.M., Romero, A.A., Posadillo, A., Berbel, J., Verdugo-Escamilla, C. An overview on glycerol-free processes for the production of renewable liquid biofuels, applicable in diesel engines. Renewable and Sustainable Energy Reviews 2015; 42, 1437–1452.

[10] Trifoi, A.R., Agachi, P.S., Pap, T. Glycerol acetals and ketals as possible diesel additives: a review of their synthesis protocols. Renewable and Sustainable Energy Reviews 2016; 62, 804–814.

[11] Kong, P.S., Aroua, M.K., Daud, W.M.A.W. Conversion of crude and pure glycerol into derivatives: a feasibility evaluation. Renewable and Sustainable Energy Reviews 2016; 63, 533–555.

[12] Zheng, Y., Chen, X., Shen, Y. Commodity chemicals derived from glycerol, an important biorefinery feedstock. Chemical Reviews 2008; 108, 5253–5277.

[13] Rahmat, N., Abdullah, A.Z., Mohamed, A.R. Recent progress on innovative and potential technologies for glycerol transformation into fuel additives: a critical review. Renewable and Sustainable Energy Reviews 2010; 14, 987–1000.

[14] Gu, Y., Azzouzi, A., Pouilloux, Y., Jérôme, F., Barrault, J. Heterogeneously catalyzed etherification of glycerol: new pathways for transformation of glycerol to more valuable chemicals. Green Chemistry 2008; 10, 164–167.

[15] Jaworskia, M.A., Vega, S.R., Siria, G.J., Casella, M.L., Salvador, A.R., López, A.S. Glycerol etherification with benzyl alcohol over sulfated zirconia catalysts. Applied Catalysis A: General 2015; 505, 36–43.

[16] Melero, J.A., Grieken, R.V., Morales, G. Advances in the synthesis and catalytic applications of organosulfonic-functionalized mesostructured materials. Chemical Reviews 2006; 106, 3790–3812.

[17] Ciesla, U., Schuth, F. Ordered mesoporous materials. Microporous and Mesoporous Materials 1999; 27, 131–149.

[18] Taguchi, A., Schuth, F. Ordered mesoporous materials in catalysis. Microporous and Mesoporous Materials 2005; 77, 1–45.

[19] Grieken, R.V., Melero, J.A.,Morales, G. Etherification of benzyl alcohols with 1-hexanol over organosulfonic acid mesostructured materials. Journal of Molecular Catalysis A: Chemical 2006; 256, 29–36.

Chapter 18

Thermodynamic Properties of Propanol and Butanol as Oxygenate Additives to Biofuels

Eduardo A. Montero, Fernando Aguilar,
Natalia Muñoz-Rujas and Fatima E. M. Alaoui

Additional information is available at the end of the chapter

http://dx.doi.org/10.5772/66297

Abstract

Alternative and renewable energy technologies are being sought throughout the world to reduce pollutant emissions and increase the efficiency of energy use. Oxygenate second-generation biofuels fuels lead to a reduction in pollutant emissions and their thermodynamic and transport properties allow that the facilities for transport, storage and distribution of fuels could be used without modification. Higher alcohols, like propanol and butanol, enhance the octane number, boosting the anti-knock effect in gasoline. Then the compression ratio of the engines can be increased without risk of knocking, leading to higher delivery of power. From the combustion point of view, the production of carbon monoxide and volatile hydrocarbons from the combustion of alcohols is less than the one of gasoline. This chapter covers mixtures of butanol and propanol with hydrocarbons. The properties reviewed are excess volume or density (VE), vapour-liquid equilibrium (VLE), and heat capacity (C_p).

Keywords: butanol, propanol, biofuel, density, enthalpy, phase equilibrium, heat capacity

1. Introduction

Biofuels, as environmental friendly fluids, have been paid much attention over the last decades. They contribute to diminish the greenhouse gas emissions due to its neutral carbon dioxide balance. Moreover, some oxygenated compounds are used as biofuel additives as they lead to a reduction in pollutant emissions and to an increase in the energy efficiency of vehicle engines [1, 2].

Some alcohols and ethers, as oxygenated compounds additives, are added to present gasoline with the aim of reducing the emission of gases that produce environmental impact. The advantages of these oxygenates can be classified into several categories. First, they can be obtained from renewable, agricultural and raw materials, reducing the dependence of fossil sources [3]. Second, they enhance the octane number, boosting the anti-knock effect in gasoline. Then, the compression ratio of the engines can be increased without risk of knocking, leading to higher delivery of power. From the combustion point of view, the production of carbon monoxide and volatile hydrocarbons from the combustion of alcohols is less than the one of gasoline. Amongst the thermodynamic properties, the heat of vaporization of alcohols is high and leads to a reduction in the peak temperature of combustion, which means lower emissions of nitrogen oxides.

Alternative and renewable energy technologies are being sought to reduce pollutant emissions and increase the efficiency of energy use. Propanol and butanol have been proposed as an alternative to conventional gasoline and diesel fuels [4, 5]. They are higher member of the series of alcohols with each molecule containing three or four carbon atoms rather than two as in ethanol. The EN standards of the European Union (EU) and the World-Wide Fuel Charter (WWFC) for gasoline include, for example, 2-propanol, 2-methyl-2 propanol (also known as tert-butyl alcohol, TBA), and 2-methyl-1 propanol [6, 7] as gasoline components.

The traditional production and consumption of bioethanol have found an alternative with the second-generation biofuels, such as biobutanol. For example, 85% ethanol, E85, needs some modification of the internal combustion engines specifications, unlike butanols, which can work directly in present engines. The energy content per volume unit of butanol is similar to the one of gasoline, and higher than the same for ethanol. Concerning the contribution to the anti-knocking effect, butanol behaves almost the same as other alcohols like methanol or ethanol. And in the presence of water, the mixture butanol/gasoline shows lesser tendency to separation of phases than the mixture ethanol/gasoline. Then, all the facilities for transport, storage and distribution of fuels can be used without modification. Butanol, which can be synthesized chemically or biologically, is an alternative transportation fuel since it has properties that would allow its use in existing engines with minor hardware modifications [5]. For practical purposes, ASTM D7862 [8] gives specifications for blends of butanol with gasoline at 1–12.5% in volume for automotive spark ignition engines. Three butanol isomers are covered by the specification, 1-butanol, 2-butanol, and 2-methyl-1-propanol, while specifically excludes 2-methyl-2-propanol (TBA).

Besides its use as fuel component, its industrial uses covers a broad range of applications as solvent or as reactive for the production of other chemicals. Applications, chemicals and products that use butanol include solvents, plasticizers, coatings, chemical intermediate or raw material, textiles, cleaners, cosmetics, drugs and antibiotics, hormones, and vitamins.

Since the 1950s, most butanol is obtained from fossil sources [6]. 1-butanol and/or 2-butanol could be obtained from reduction of butyraldehyde with hydrogen, which is previously

http://dx.doi.org/10.5772/66297

obtained by hydroformulation reaction of propene (propyelene). Meanwhile, propylene oxide production leads to isobutene, from which TBA could be derived. Butanol from biomass is called biobutanol [9], and it can be used in unmodified gasoline engines. Biobutanol can be produced by fermentation of biomass by the ABE process [9, 10]. The process uses the bacterium *Clostridium acetobutylicum,* the bacterium for the production of acetone from starch. The butanol was a by-product of this fermentation. Other by-products as acetic, lactic and propionic acids, isopropanol and ethanol, as well as a certain amount of H_2, are generated by the process. *Ralstonia eutropha* can also be used to produce biobutanol, by means of an electro-bioreactor and the input of carbon dioxide and electricity.

According to DuPont [11], existing bioethanol plants can be converted to biobutanol production with low economic cost. The main modification could affect to the fermentation process, with minor changes in distillation, as both alcohols use the same stocks: food energy crops (sugar beets, sugar cane, corn grain, wheat, etc.), non-food energy crops (switchgrass, cellulose, etc.) and agricultural by-products (straw, corn stalks, etc.).

Biopropanol is a rarely discussed biofuel. Tough propanol is included as regular component of gasolines [6], its frequent use as chemical solvent makes it rare to consider it as a fuel. Biopropanol could be produced from microbial fermentation of biomass (cellulose), but the process is extremely inefficient [12]. The issues with microbial production of biopropanol are analogous to the issues with microbial production of biobutanol, so if biobutanol becomes a more practical biofuel to produce, then biopropanol will also become more feasible.

This paper concerns thermodynamic properties of 1-propanol, 2-propanol, 1-butanol, 2-butanol and TBA. Accurate experimental data on thermodynamic properties should be available to check and develop predictive empirical equations, models and simulation programs. Industrial processes as storage, transport, separation and mixing processes also need reliable data for its design. As a result, the experimental literature reviews on properties of pure compounds and its mixtures with characteristic hydrocarbons can provide valuable information about the fluid behaviour under various temperature and pressure conditions.

The paper presents the literature review of available data on thermodynamic properties (density, vapour-liquid equilibrium, specific heat,) of the mixtures of 1-propanol, 1-butanol, TBA and its mixtures with hydrocarbons representatives of gasoline. Density has to do with the volumetric behaviour of the mixtures under pressure and temperature conditions and is the primary data to check equations of state. The vapour-liquid equilibria, which allows the calculation of the Gibbs function, deal with the equilibrium between the liquid and vapour phase under fixed pressure and temperature conditions. And the heat capacity gives information related to the sensible energy storage of the liquids. The review includes only the interval of temperature and pressure of every property reported. The wider is the range of pressure and temperature of the measured properties, so it would be the reliability of the applications of predictive and equations and models. Discussion of further data (uncertainties, experimental apparatus, etc.) would require more space than available. Interested readers should access the literature references to check these issues.

2. The literature review

Thermodynamic properties of liquid propanol and butanol and its liquid mixtures with some hydrocarbon have been obtained from the literature search using online library databases (Web of Science©, Scopus©, NIST© Standard Reference Database) and high impact electronic journals.

Special attention is given to alcohol + hydrocarbon mixtures. As stated, 1-propanol, 1-butanol and TBA have been selected as alcohols. As representative of hydrocarbons, n-heptane, 2,2,4 trimethylpentane (iso-octane), cyclohexane, methyl-cyclohexane, benzene, toluene and 1-hexene have been chosen. They represent linear, branched and cyclic alkanes, aromatics, as well as olefins, which are regular components of gasoline. **Table 1** presents the list of selected compounds.

Compound	CAS number	Chemical formula
Alcohols		
1-Propanol	71-23-8	C_3H_8O
1-Butanol	71-36-3	$C_4H_{10}O$
Tert-butyl alcohol (TBA)	75-65-0	$C_4H_{10}O$
Hydrocarbons		
Heptane	142-82-5	C_7H_{16}
2,2,4 trimethylpentane (TMP)	540-84-1	C_8H_{18}
Cyclohexane	110-82-7	C_6H_{12}
Methyl cyclohexane	108-87-2	C_7H_{14}
Benzene	71-43-2	C_6H_6
Toluene	108-88-3	C_7H_8
1-Hexene	592-41-6	C_6H_{12}

Table 1. Selected alcohols and hydrocarbons.

Concerning properties, there is a huge amount of available thermodynamic data for pure compounds. With respect to the mixtures, density data are shown in **Table 2** for binary mixtures alcohol (1) + hydrocarbon (2). **Tables 3** and **4** present the vapour-liquid equilibria selected for mixtures alcohol (1) + hydrocarbon (2) and alcohol (1) + hydrocarbon (2) + hydrocarbon (3). Finally, heat capacity data for binary mixtures alcohol (1) + hydrocarbon (2) are provided in **Table 5**.

Substance 1	Substance 2	References	Year	T_{min}/K	T_{max}/K	P_{min}/kPa	P_{max}/kPa
1-Propanol	Heptane	[13]	1967	298.15	298.15	101	101
	Heptane	[14]	1967	348.15	348.15	101	101
	Heptane	[15]	1977	298.15	298.15	101	101
	Heptane	[16]	1982	423.11	523.11	422	5495
	Heptane	[17]	1983	298.15	298.15	101	101

Substance 1	Substance 2	References	Year	T_{min}/K	T_{max}/K	P_{min}/kPa	P_{max}/kPa
	Heptane	[18]	1993	313.15	313.15	101	101
	Heptane	[19]	1994	278.15	308.15	101	101
	Heptane	[20]	1995	298.15	298.15	101	101
	Heptane	[21]	1996	298.15	308.15	101	101
	Heptane	[22]	1997	298.15	298.15	101	101
	Heptane	[23]	1998	278.15	308.15	101	101
	Heptane	[24]	2003	308.15	308.15	101	101
	Heptane	[25]	2004	293.15	318.21	101	101
	Heptane	[26]	2005	298.15	298.15	101	101
	Heptane	[27]	2005	298.15	298.15	101	101
	2,2,4 trimethylpentane	[28]	1981	298.15	298.15	101	101
	2,2,4 trimethylpentane	[29]	2007	298.15	298.15	101	101
	2,2,4 trimethylpentane	[30]	2007	303.15	303.15	101	101
	2,2,4 trimethylpentane	[31]	2012	298.15	298.15	101	101
	2,2,4 trimethylpentane	[32]	2015	298.15	323.15	101	101
	Cyclohexane	[33]	1979	298.15	298.15	101	101
	Cyclohexane	[34]	1980	298.15	298.15	101	101
	Cyclohexane	[35]	1991	298.15	298.15	101	101
	Cyclohexane	[36]	1996	298.15	308.15	101	101
	Cyclohexane	[37]	1997	298.15	303.15	101	101
	Cyclohexane	[38]	1998	303.15	303.15	101	101
	Cyclohexane	[24]	2003	308.15	308.15	101	101
	Cyclohexane	[39]	2004	298.15	298.15	101	101
	Cyclohexane	[26]	2005	298.15	298.15	101	101
	Cyclohexane	[40]	2007	293.15	303.15	101	101
	Cyclohexane	[41]	2008	303.15	303.15	101	101
	Cyclohexane	[42]	2016	303.15	313.15	101	101
	Methylcyclohexane	[43]	1977	303.15	303.15	101	101
	Methylcyclohexane	[44]	1996	298.15	298.15	101	101
	Methylcyclohexane	[45]	2006	298.15	308.15	101	101
	Benzene	[46]	1969	298.15	298.15	101	101
	Benzene	[33]	1979	298.15	298.15	101	101
	Benzene	[34]	1980	298.15	298.15	101	101
	Benzene	[47]	1993	308.15	308.15	101	101
	Benzene	[58]	1994	298.15	298.15	101	101
	Benzene	[59]	2001	303.15	303.15	101	101
	Benzene	[24]	2003	308.15	308.15	101	101
	Benzene	[39]	2004	298.15	298.15	101	101
	Benzene	[50]	2007	288.15	313.15	101	101
	Benzene	[51]	2008	298.15	298.15	101	101
	Benzene	[52]	2009	298.15	298.15	101	101

Substance 1	Substance 2	References	Year	T_{min}/K	T_{max}/K	P_{min}/kPa	P_{max}/kPa
	Benzene	[53]	2015	303.15	303.15	101	101
	Toluene	[54]	1980	298.15	298.15	101	101
	Toluene	[47]	1993	308.15	308.15	101	101
	Toluene	[48]	1994	298.15	298.15	101	101
	Toluene	[55]	2000	303.15	313.15	101	101
	Toluene	[24]	2003	308.15	308.15	101	101
	Toluene	[56]	2005	303.15	333.15	100	30000
	Toluene	[57]	2006	298.15	298.15	101	101
	Toluene	[58]	2006	303.15	333.15	101	101
	Toluene	[59]	2008	298.15	298.15	101	101
	Toluene	[53]	2015	303.15	303.15	101	101
	1-Hexene	[60]	1993	298.15	298.15	101	101
	1-Hexene	[61]	2010	298.15	298.15	101	101
1-Butanol	n-Heptane	[15]	1977	298.15	298.15	101	101
	n-Heptane	[62]	1979	298.15	298.15	101	101
	n-Heptane	[17]	1983	298.15	298.15	101	101
	n-Heptane	[63]	1984	298.15	298.15	101	101
	n-Heptane	[64]	1994	313.15	313.15	101	101
	n-Heptane	[21]	1996	298.15	308.15	101	101
	n-Heptane	[65]	1997	293.15	293.15	101	101
	n-Heptane	[66]	1997	288.15	298.15	101	101
	n-Heptane	[24]	2003	308.15	308.15	101	101
	n-Heptane	[67]	2003	316.85	458.15	4930	4930
	n-Heptane	[26]	2005	298.15	298.15	101	101
	n-Heptane	[68]	2006	288.15	308.15	101	101
	n-Heptane	[69]	2009	288.15	308.15	101	101
	2,2,4 trimethylpentane	[65]	1997	293.15	293.15	101	101
	2,2,4 trimethylpentane	[66]	1997	288.15	298.15	101	101
	2,2,4 trimethylpentane	[31]	2012	298.15	298.15	101	101
	2,2,4 trimethylpentane	[70]	2013	298.15	328.15	101	101
	Cyclohexane	[33]	1979	298.15	298.15	101	101
	Cyclohexane	[34]	1980	298.15	298.15	101	101
	Cyclohexane	[71]	1983	298.15	318.15	101	101
	Cyclohexane	[72]	1995	293.15	313.15	101	101
	Cyclohexane	[38]	1998	303.15	303.15	101	101
	Cyclohexane	[73]	2001	298.15	298.15	101	101
	Cyclohexane	[24]	2003	308.15	308.15	101	101
	Cyclohexane	[74]	2005	298.15	313.15	101	101
	Cyclohexane	[40]	2007	293.15	303.15	101	101
	Cyclohexane	[75]	2010	293.15	293.15	101	101
	Cyclohexane	[76]	2014	293.15	333.15	100	100000

Substance 1	Substance 2	References	Year	T_{min}/K	T_{max}/K	P_{min}/kPa	P_{max}/kPa
	Cyclohexane	[42]	2016	303.15	313.15	101	101
	Methylcyclohexane	[43]	1977	303.15	303.15	101	101
	Methylcyclohexane	[77]	1989	298.15	298.15	101	101
	Methylcyclohexane	[78]	2004	303.15	303.15	101	101
	Methylcyclohexane	[45]	2006	298.15	308.15	101	101
	Benzene	[46]	1969	298.15	298.15	101	101
	Benzene	[33]	1979	298.15	298.15	101	101
	Benzene	[34]	1980	298.15	298.15	101	101
	Benzene	[79]	1993	298.15	298.15	101	101
	Benzene	[80]	1994	298.15	308.15	101	101
	Benzene	[81]	1996	308.15	308.15	101	101
	Benzene	[49]	2001	303.15	303.15	101	101
	Benzene	[21]	2003	308.15	308.15	101	101
	Benzene	[82]	2004	303.15	303.15	101	101
	Benzene	[83]	2008	288.15	313.15	101	101
	Toluene	[84]	1940	298.15	298.15	101	101
	Toluene	[54]	1980	298.15	298.15	101	101
	Toluene	[81]	1996	308.15	308.15	101	101
	Toluene	[55]	2000	303.15	313.15	101	101
	Toluene	[24]	2003	308.15	308.15	101	101
	Toluene	[70]	2013	298.15	328.15	101	101
	Toluene	[85]	2015	298.15	328.15	101	101
	1-Hexene	[86]	2013	273.15	333.15	101	101
TBA	n-Heptane	[62]	1979	299.15	299.15	101	101
	n-Heptane	[64]	1994	313.15	313.15	101	101
	n-Heptane	[87]	2011	303.15	323.15	101	101
	2,2,4 trimethylpentane	[88]	1999	298.15	298.15	101	101
	2,2,4 trimethylpentane	[89]	2001	298.15	298.15	101	101
	2,2,4 trimethylpentane	[90]	2005	298.15	318.15	101	101
	Cyclohexane	[71]	1983	298.15	318.15	101	101
	Cyclohexane	[72]	1995	293.15	313.15	101	101
	Methylcyclohexane	[88]	1999	298.15	298.15	101	101
	Benzene	[79]	1993	298.15	298.15	101	101
	Benzene	[91]	1995	313.15	313.15	101	101
	Benzene	[81]	1996	308.15	308.15	101	101
	Benzene	[82]	2004	303.15	303.15	101	101
	Toluene	[81]	1996	308.15	308.15	101	101
	Toluene	[88]	1999	298.15	298.15	101	101
	Toluene	[55]	2000	303.15	313.15	101	101

Table 2. Reported density ($g \cdot cm^{-3}$) for binary mixtures alcohol (1) + hydrocarbon (2).

Substance 1	Substance 2	References	Year	T_{min}/K	T_{max}/K	P_{min}/kPa	P_{max}/kPa
1-propanol	Heptane	[92]	1966	357.72	371.52	101.32	101.32
	Heptane	[14]	1967	347.97	347.97	39.72	73.63
	Heptane	[13]	1967	303.13	333.12	3.92	39.81
	Heptane	[93]	1980	278.16	303.14	1.67	10.17
	Heptane	[16]	1982	423.15	573.15	200	5066
	Heptane	[94]	1991	313.15	313.15	10.95	16.52
	Heptane	[95]	1992	313.15	313.15	9.638	16.428
	Heptane	[96]	1993	303.15	303.15	5.42	10.24
	Heptane	[97]	1995	379.38	475.45	204.5	1032.8
	Heptane	[98]	1995	316.78	357.58	19.60	101.33
	Heptane	[99]	2000	298.15	298.15	–	–
	Heptane	[100]	2004	303.15	343.15	–	–
	2,2,4, trimethylpentane	[28]	1981	328.36	348.50	15.98	72.75
	2,2,4, trimethylpentane	[101]	1994	357.88	365.46	101.3	101.3
	2,2,4, trimethylpentane	[102]	1994	343.15	343.15	42.04	60.07
	2,2,4, trimethylpentane	[29]	2007	303.15	303.15	4.88	10.45
	2,2,4, trimethylpentane	[103]	2011	318.15	318.15	9.00	21.13
	Cyclohexane	[104]	1977	298.15	298.15	2.79	14.29
	Cyclohexane	[105]	1986	347.66	369.17	101.33	101.33
	Cyclohexane	[98]	1995	347.58	347.58	101.33	101.33
	Cyclohexane	[106]	1996	298.15	308.15	2.63	22.1
	Cyclohexane	[107]	1997	323.15	333.15	27.92	61.17
	Cyclohexane	[99]	2000	313.15	343.15	–	–
	Cyclohexane	[108]	2000	298.15	298.15	101.32	101.32
	Methylcyclohexane	[109]	1969	360.13	366.83	101	101
	Methylcyclohexane	[110]	1989	332.98	332.98	29.12	38.60
	Methylcyclohexane	[111]	1997	358.75	373.60	101.3	101.3
	Benzene	[112]	1947	298.94	363.52	13.33	99.99
	Benzene	[113]	1963	349.12	365.92	101	101
	Benzene	[114]	1964	493.16	558.18	2419.4	4904.2
	Benzene	[104]	1977	298.15	298.15	2.79	13.04
	Benzene	[105]	1986	350.03	361.85	101.33	101.33
	Benzene	[115]	1987	313.15	313.15	7.01	25.98
	Benzene	[107]	1997	323.15	333.15	22.02	56.70
	Benzene	[116]	2001	313.15	313.15	7.047	26.069
	Benzene	[117]	2006	313.15	313.15	7.00	25.91
	Benzene	[52]	2008	323.15	323.15	17.98	39.89
	Benzene	[51]	2008	329.45	368.35	50	94
	Toluene	[115]	1987	313.15	313.15	7.01	11.35
	Toluene	[118]	1996	298.15	298.15	2.63	5.39
	Toluene	[119]	2003	293.15	370.15	–	–

http://dx.doi.org/10.5772/66297

Substance 1	Substance 2	References	Year	T_{min}/K	T_{max}/K	P_{min}/kPa	P_{max}/kPa
	Toluene	[59]	2008	333.15	333.15	20.27	27.41
	Toluene	[120]	2009	323.15	323.15	13.42	17.85
1-butanol	Heptane	[121]	1966	361.92	376.92	91.2	91.2
	Heptane	[122]	1967	387.93	434.34	192.65	496.63
	Heptane	[63]	1984	333.15	363.15	8.01	89.49
	Heptane	[123]	1990	353.15	373.15	101.32	101.32
	Heptane	[124]	1994	313.15	313.15	4.39	13.22
	Heptane	[98]	1995	312.34	357.58	12.93	74.47
	Heptane	[125]	1996	328.45	366.55	25.63	101.38
	Heptane	[126]	1997	303.15	303.15	1.35	8.26
	Heptane	[99]	2000	298.15	298.15	–	–
	Heptane	[127]	2001	365.05	389.05	95	95
	Heptane	[100]	2004	303.15	343.15	–	–
	Heptane	[138]	2010	349.00	387.75	53.3	91.3
	Heptane	[129]	2012	313.15	313.15	2.51	13.22
	2,2,4-trimethylpentane	[130]	2006	308.15	318.15	2	17
	2,2,4-trimethylpentane	[103]	2011	318.15	318.15	7.2	16.4
	2,2,4-trimethylpentane	[129]	2012	313.15	313.15	2.55	13.71
	2,2,4-trimethylpentane	[131]	2013	313.15	313.15	11.55	13.50
	Cyclohexane	[132]	1968	353.15	383.12	21.23	229.21
	Cyclohexane	[133]	1982	293.15	293.15	–	–
	Cyclohexane	[71]	1983	318.15	318.15	3.41	30.59
	Cyclohexane	[134]	1990	312.8	389.9	–	–
	Cyclohexane	[98]	1995	352.7	352.7	101.33	101.33
	Cyclohexane	[108]	2000	298.15	298.15	101.32	101.32
	Cyclohexane	[99]	2000	313.15	343.15	–	–
	Cyclohexane	[135]	2001	350.95	389.05	95	95
	Cyclohexane	[136]	2002	325.6	386.12	40.0	101.3
	Cyclohexane	[129]	2012	313.15	313.15	2.51	24.83
	Methylcyclohexane	[109]	1969	369.75	385.65	101	101
	Methylcyclohexane	[110]	1989	332.98	332.98	11.07	29.77
	Methylcyclohexane	[111]	1997	368.45	390.50	101.3	101.3
	Benzene	[137]	1939	298.15	298.15	0.85	12.59
	Benzene	[128]	1963	353.21	390.83	101.32	101.32
	Benzene	[114]	1964	513.17	558.18	2032.6	4751.2
	Benzene	[115]	1987	313.15	313.15	2.52	24.37
	Benzene	[79]	1993	298.15	298.15	0.82	12.83
	Benzene	[139]	1995	354.03	425.26	105	303
	Benzene	[140]	2004	308.15	308.15	4.03	20.28
	Benzene	[141]	2006	313.15	313.15	2.49	24.37
	Toluene	[84]	1940	376.12	390.83	101	101

Substance 1	Substance 2	References	Year	T_{min}/K	T_{max}/K	P_{min}/kPa	P_{max}/kPa
	Toluene	[142]	1963	378.63	390.83	101.33	101.33
	Toluene	[115]	1987	313.15	313.15	2.52	8.48
	Toluene	[134]	1990	349.5	389.9	–	–
	Toluene	[143]	1997	360.9	389.1	56.4	94.0
	Toluene	[119]	2003	323.15	390.15	–	–
	Toluene	[140]	2004	308.15	308.15	2.49	6.39
	Toluene	[129]	2012	313.15	313.15	2.49	8.39
	1-hexene	[131]	2013	313.15	313.15	2.48	44.99
TBA	Heptane	[145]	1982	313.15	313.15	12.33	19.23
	Heptane	[146]	1983	352.47	371.42	101	101
	Heptane	[124]	1994	313.15	313.15	14.81	19.20
	Heptane	[147]	1995	351.40	368.23	101.3	101.3
	Heptane	[127]	2001	352.25	369.45	95	95
	2,2,4-trimethylpentane	[148]	1999	352.4	372.5	101.3	101.3
	2,2,4-trimethylpentane	[89]	2001	318.13	339.28	15.85	59.49
	2,2,4-trimethylpentane	[149]	2006	353.35	370.55	95.8	95.8
	Cyclohexane	[150]	1976	344.43	354.33	101	101
	Cyclohexane	[71]	1983	318.15	318.15	18.11	36.23
	Cyclohexane	[151]	1985	328.19	343.28	30.43	95.1
	Cyclohexane	[98]	1995	295.35	344.28	13.46	101.40
	Methylcyclohexane	[110]	1989	332.98	332.98	30.44	46.33
	Methylcyclohexane	[148]	1999	353.1	374.0	101.3	101.3
	Benzene	[152]	1902	347.10	347.10	100.66	101.32
	Benzene	[153]	1933	347.10	347.10	–	–
	Benzene	[137]	1939	298.15	298.15	5.60	13.96
	Benzene	[154]	1969	318.15	318.15	18.12	34.32
	Benzene	[155]	1977	346.58	353.98	101	101
	Benzene	[79]	1993	298.15	298.15	5.59	14.83
	Benzene	[98]	1995	296.66	347.19	13.59	101.51
	Benzene	[156]	1998	308.15	308.15	10.18	23.48
	Toluene	[156]	1998	308.15	308.15	6.40	12.97
	Toluene	[148]	1999	355.4	383.8	101.3	101.3

Table 3. Reported vapour-liquid equilibria for binary mixtures alcohol (1) + hydrocarbon (2).

Substance 1	Substance 2	Substance 3	References	Year	T_{min}/K	T_{max}/K	P_{min}/kPa	P_{max}/kPa
1-butanol	2,2,4-trimethylpentane	1-hexene	[131]	2013	313.15	313.15	2.48	44.99
1-butanol	toluene	1-hexene	[144]	2015	313.15	313.15	2.51	44.99
TBA	Cyclohexane	Benzene	[98]	1995	294.91	344.25	13.61	101.36

Table 4. Reported vapour-liquid equilibria for ternary mixtures alcohol (1) + hydrocarbon (2) + hydrocarbon (3).

Substance 1	Substance 2	References	Year	T_{min}/K	T_{max}/K	P_{min}/kPa	P_{max}/kPa
1-propanol	Heptane	[157]	1976	298.15	298.15	101	101
	Heptane	[158]	1981	184.97	300.00	101	101
	Heptane	[159]	1993	298.15	298.15	101	101
1-butanol	Heptane	[159]	1993	298.15	298.15	101	101
	2,2,4 Trimethylpentane	[160]	2012	293.15	313.15	101	25,000
	Cyclohexane	[76]	2014	293.15	313.15	101	25,000
	Toluene	[161]	1991	298.15	368.15	101	101
	1-Hexene	[86]	2013	293.15	313.15	101	25,000

Table 5. Reported heat capacity for binary mixtures alcohol (1) + hydrocarbon (2).

3. Discussion

3.1. Density of mixtures 1-propanol, or 1-butanol, + hydrocarbon

Table 2 presents density data for the selected mixtures alcohol (1) + hydrocarbon (2). Fifty-nine references correspond to mixtures 1-propanol (1) + hydrocarbon (2) and 51 to the one 1-butanol (1) + hydrocarbon (2), while only 16 references have been found for TBA (1) + hydrocarbon (2).

For 1-propanol (1) + hydrocarbon (2), only atmospheric pressure density data have been found for the binary mixtures, except Refs. [16, 56] that are above 5 MPa. The highest pressure, 30 MPa, is reported by Zeberg-Mikkelsen and Andersen [56]. Temperatures above 350 K are only measured by Zawisza and Vejrosta [16]. Concerning 1-butanol (1) + hydrocarbon (2), Refs. [67, 76] report pressure above the atmospheric pressure. Hundred Megapascal is the maximum pressure measured in Ref. [76]. Reference [67] also reports temperature above 350 K. Finally, mixtures TBA (1) + hydrocarbon are reported only at atmospheric pressure and moderate temperatures, being 323.15 K the highest measured temperature [87]. No data were found for the mixture TBA (1) + 1-hexene (2).

3.2. Vapour-liquid equilibrium of mixtures 1-propanol, or 1-butanol, + hydrocarbon

With respect to the binary mixtures, **Table 3** shows 43 references for VLE data on 1-propanol (1) + hydrocarbon (2), 47 for 1-butanol (1) + hydrocarbon (2) and 24 for TBA (1) + hydrocarbon (2). No references for the mixtures 1-propanol (1) + 1-hexene (2) and TBA (1) + 1-hexene (2) were found, while [131] was the only one for 1-butanol (1) + 1-hexene (2). Most references were found for pressures lower or equal to atmospheric pressure. Studies done in Refs. [97, 122, 132] were measured at moderate pressures, below 1.0 MPa, and only Ref. [114] reports pressure close to 5 MPa for both mixtures 1-propanol (1), or 1-butanol (1), + benzene (2).

Concerning temperature, most measurements were performed at low and moderate temperatures. Within the interval 350–400 K, we found a limited number of 27 set of data [51, 63, 84, 98, 105, 109, 111–113, 119, 121, 123, 125, 127, 128, 132, 134–136, 142, 143, 146–150, 155]. Only Refs. [16, 97, 114, 122, 139] report temperatures between 400 and 573 K.

Only three references were found reporting VLE data of ternary mixtures, as shown in **Table 4**, at atmospheric or lower pressures. Temperatures were moderate, with maximum at 344 K measured in Ref. [98]. No ternary mixture with 1-propanol was found.

3.3. Heat capacity of mixtures 1-propanol, or 1-butanol, + hydrocarbon

Only eight references reporting heat capacity of binary mixtures alcohol (1) + hydrocarbon (2) are cited. Three of them correspond to the binary mixture 1-propanol (1) + heptane (2) at atmospheric pressure and at moderate temperatures (up to 300 K). No other mixture of 1-propanol with the any of selected hydrocarbons was found.

While the heat capacity of 1-butanol with heptane, 2,2,4 trimethylpentane, cyclohexane, toluene and 1-hexane was measured by several authors. It must be pointed out that some measurements [86, 160, 161] have been performed at pressures up to 25 MPa and temperature of 313 K.

4. Conclusion

The literature review on thermodynamic properties of liquid mixtures of 1-propanol, 1-butanol and TBA with representative hydrocarbons has been reported. Seven hydrocarbons (linear, branched and cyclic alkanes, aromatics, and olefins) have been selected as representative of present and future unleaded gasoline. The review covers density, vapour-liquid equilibrium and heat capacity of mixtures.

The review of density data shows a big amount of data at low pressure and moderate temperatures. Only two references report data above 30 MPa at a maximum temperature of 333 K. And at temperatures above 450 K, the maximum pressure is 5.5 MPa. With respect to the vapour-liquid equilibrium, only one reference shows measurements over 555 K at 5 MPa. Heat capacity data of mixtures are very scarce, tough some high pressure and high temperature data can be found for some alcohol + hydrocarbon mixtures.

The performance of fuels and biofuels in engines and other devices shows a trend of increasing pressure and temperature, which leads to the need of more reliable predictive models for complex mixtures at such conditions. Availability of high pressure and high temperature thermodynamic properties is then a requisite for the implementation of these equation and models. The review shows a lack of reliable data at high pressure and high temperature thermodynamic data, which serve as a basis for the development of predictive equations and models.

Author details

Eduardo A. Montero[1*], Fernando Aguilar[1], Natalia Muñoz-Rujas[1] and Fatima E. M. Alaoui[2]

*Address all correspondence to: emontero@ubu.es

1 University of Burgos, Spain

2 University Chouaib Doukkali-El Jadida, Morocco

References

[1] Directive 2009/28/EC of the European Parliament and of the Council on the promotion of the use of energy from renewable sources.

[2] Directive 2009/30/EC of the European Parliament and of the Council as regards the specification of petrol, diesel and gas-oil and introducing a mechanism to monitor and reduce greenhouse gas emissions.

[3] Ezeji TC, Blaschek HP. Butanol Production from Lignocellulosic Biomass. In: Blaschek HP, Ezeji TC, Sheffran J, editors. Biofuels from Agricultural Wastes and Byproducts. Ames: Wiley-Blackwell; 2010. pp. 19–37.

[4] Agarwal AK. Biofuels (alcohols and biodiesel) applications as fuels for internal combustion engines. Progress in Energy and Combustion Science. 2007; 33: 233–271. doi:10.1016/j.pecs.2006.08.003

[5] Surisetty VR, Dalai AK, Kozinski J. Alcohols as alternative fuels: An overview. Applied Catalysis A: General. 2011; 404: 1–11. doi:10.1016/j.apcata.2011.07.021

[6] Elvers B, editor. Handbook of Fuels: Energy for Transportation. Weinheim: Wiley-VCH; 2008. 356 p.

[7] De Klerk A. Fischer-Tropsch Refining. Weinheim: Wiley-VCH; 2011. 620 p.

[8] ASTM D7862 – 13. Standard Specification for Butanol for Blending with Gasoline for Use as Automotive Spark-Ignition Engine Fuel.

[9] Köpke M, Dürre P. Biochemical Production of Biobutanol. In: Luque R, Campelo J, Clark J, editors. Handbook of Biofuels Production. Cambridge, UK: Woodhead; 2011. pp. 221–257.

[10] Mousdale D. Introduction to Biofuels. Boca Raton: CRC Press; 2010. 429 p.

[11] Butamax: The Biofuel Company [Internet]. 2016. Available from: http://www.butamax.com/ [Accessed: 2016-09-12]

[12] Minteer SD. Biochemical Production of Other Bioalcohols: Biomethanol, Biopropanol, Bioglycerol and Bioethylene Glycol. In: Luque R, Campelo J, Clark J, editors. Handbook of Biofuels Production. Cambridge, UK: Woodhead; 2011. pp. 258–265.

[13] Van Ness HC, Soczek CA, Peloquin GL, Machado RL. Thermodynamic excess properties of three alcohol-hydrocarbon systems. Journal of Chemical and Engineering Data. 1967; 12: 217–224. doi:10.1021/je60033a017

[14] Lee LL, Scheller WA. Isothermal vapor-liquid equilibrium data for the system heptane + 1-propanol at 75°C and decane + 1-butanol at 100°C. Journal of Chemical and Engineering Data. 1967; 12: 497. doi:10.1021/je60035a009

[15] Treszczanowicz AJ, Benson GC. Excess volumes for n-alkanols I. Binary mixtures of methanol, ethanol, n-propanol, and n-butanol + n-heptane. Journal of Chemical Thermodynamics. 1977; 9: 1189–1197. doi:10.1016/0021-9614(77)90119-7

[16] Zawisza A, Vejrosta J. High-pressure liquid-vapor equilibria, critical stat, and p(V, T, x) up to 573.15 K and 5.066 MPa for (heptane + propan-1-ol). Journal of Chemical Thermodynamics. 1982; 14: 239–249. doi:10.1016/0021-9614(82)90014-3

[17] Kumaran MK, Benson GC. Limiting partial molar volumes of ethanol, propan-1-ol, butan-1-ol, pentan- 1-ol, and hexan-1-ol in n-heptane at 298.15 K. Journal of Chemical Thermodynamics. 1983; 15: 245–248. doi:10.1016/0021-9614(83)90115-5

[18] Zielkiewicz J. Excess molar volumes of (heptane + ethanol + propan-1-ol) at the temperature 313.15 K. Journal of Chemical Thermodynamics. 1993; 25: 1243–1248. doi: 10.1006/jcht.1993.1122

[19] Romani L, Peleteiro J, Iglesias TP, Carballo E, Escudero R, Legido JL. Temperature dependence of the volumetric properties of binary mixtures containing alcohols (1-propanol, 1-pentanol, 1-heptanol) + heptane. Journal of Chemical and Engineering Data. 1994; 39: 19–22. doi:10.1021/je00013a006

[20] Keller M, Schnabel S, Heintz A. Thermodynamics of the ternary mixture propan-1-ol + tetrahydrofuran + n- heptane at 298.15 K. Experimental results and ERAS model calculations of GE, HE and VE. Fluid Phase Equilibria. 1995; 110: 231–265. doi: 10.1016/0378-3812(95)02757-6

[21] Sastry NV, Valand MK. Densities, speeds of sound, viscosities, and relative permittivities for 1-propanol + and 1-butanol + heptane at 298.15 K and 308.15 K. Journal of Chemical and Engineering Data. 1996; 41: 1421–1425. doi:10.1021/je960135d

[22] Orge B, Iglesias M, Rodriguez A, Canosa JM, Tojo J. Mixing properties of (methanol, ethanol, or 1-propanol) with (n-pentane, n-hexane, n-heptane and n-octane) at 298.15 K. Fluid Phase Equilibria. 1997; 133: 213–227. doi:10.1016/S0378-3812(97)00031-9

[23] Jiménez E, Franjo C, Segade L, Legido JL, Paz Andrade MI. Viscosities and densities for the 1-propanol + n-heptane system at several temperatures. Journal of Solution Chemistry. 1998; 27: 569–579. doi:10.1023/A:1022686707250

[24] Bahadur PNV, Sastry NV. Densities, sound speeds, excess volumes, and excess isentropic compressibilities of methyl acrylate + 1-propanol (or 1-butanol) + hydrocarbons (n-hexane, n-heptane, cyclohexane, benzene, and toluene) at 308.15 K. International Journal of Thermophysics. 2003; 24: 447–462. doi:10.1023/A:1022920005042

[25] Dzida M. The effect of pressure on the thermodynamic properties of propan-1-ol + n-heptane mixtures. Journal of Solution Chemistry. 2004; 33: 529–548. doi:10.1023/B:JOSL.0000037774.94690.d1

[26] El-Hefnawy M, Sameshima K, Matsushita T, Tanaka R. Apparent dipole moments of 1-alkanols in cyclohexane and n-heptane, and excess molar volumes of (1-alkanol + cyclohexane or n-heptane) at 298.15 K. Journal of Solution Chemistry. 2005; 34: 43–69. doi:10.1007/s10953-005-2072-1

[27] Mato MM, Cebreiro SM, Verdes PV, Pallas AV, Legido JL, Paz Andrade MI. Experimental and predicted excess molar volumes of the ternary system tertbutylmethylether +1-propanol+heptane at 298.15 K. Journal of Thermal Analysis and Calorimetry. 2005; 80: 345–349. doi:10.1007/s10973-005-0658-x

[28] Berro C, Neau E, Rogalski M. Vapor-liquid equilibrium of the systems 1-propanol + 2,2,4-trimethylpentane and 2-propanol + n-hexane. Fluid Phase Equilibria. 1981; 7: 41–54. doi:10.1016/0378-3812(81)87004-5

[29] Hwang IC, Lim HM, Park SJ, Han KJ, Park IH. Isothermal vapor liquid equilibrium at 303.15 K and excess molar volumes at 298.15 K for the ternary system of propyl vinyl ether + 1-propanol + 2,2,4-trimethyl-pentane and its binary sub-systems. Fluid Phase Equilibria. 2007; 259: 146–152. doi:10.1016/0378-3812(81)87004-5

[30] Hwang IC, Jo MY, Kwak HY, Park SJ, Han KJ. Isothermal VLE and VE at 303.15 K for the binary and ternary mixtures of Di-isopropyl ether (DIPE) + 1-propanol + 2,2,4-trimethylpentane. Journal of Chemical and Engineering Data. 2007; 52: 2503–2508. doi:10.1021/je700460w

[31] Faneite AM, Garces SI, Aular JA, Urdaneta MR, Soto D. Excess molar volumes, excess molar enthalpies and refractive index deviations for binary mixtures of propan-1-ol, butan-1-ol and pentan-1-ol with 2,2,4-trimethylpentane at 298.15 K. Fluid Phase Equilibria. 2012; 334: 117–127. doi:10.1016/j.fluid.2012.06.031

[32] Wang X, Wang X, Song B. Densities and viscosities of binary mixtures of 2,2,4-trimethylpentane + 1-propanol, + 1-pentanol, + 1-hexanol, and + 1-heptanol from (298.15 to 323.15) K. Journal of Chemical and Engineering Data. 2015; 60: 1664–1673. doi:10.1021/je501041r

[33] Edward JT, Farrell PG, Shahidi F. Effect of solvent (benzene, ethanol, cyclohexane) on the partial molar volumes of organic compounds. Canadian Journal of Chemistry. 1979; 57: 2887–2891. doi:10.1139/v79-469

[34] Arce A, Blanco A, Antorrena G, Quintela MD. Excess physical properties of ternary mixtures. Benzene-cyclohexane-alcohol systems at 25°C. Anales de Química. 1980; 76: 405–413.

[35] Letcher TM, Mercer-Chalmers J. Excess molar volumes of (a cycloalkane + an alkanol) at 298.15 K. Canadian Journal of Chemistry. 1991; 69: 1259–1260. doi:10.1139/v91-188

[36] Singh KC, Kalra KC, Maken S, Gupta V. Excess enthalpies and volumes of mixing of 1-propanol or 2-propanol + cyclohexane at 298.15 and 308.15 K. Fluid Phase Equilibria. 1996; 123: 271–281. doi:10.1016/S0378-3812(96)90037-0

[37] Lee LS, Chuang ML. Excess volumes of cyclohexane with 2-propanone, 2-butanone, 3-pentanone, 4-methyl-2-pentanone, 1-propanol, and 2-propanol and ethanoic acid + 1-propanol systems. Journal of Chemical and Engineering Data. 1997; 42: 850–853. doi: 10.1021/je9603335

[38] Oswal SL, Prajapati KD. Speeds of sound, isentropic compressibilities, and excess molar volumes of an alkanol + cycloalkane at 303.15 K. 1. Results for alkan-1-ols + cyclohexane. Journal of Chemical and Engineering Data. 1998; 43: 367–372. doi:10.1021/je970235z

[39] Tanaka R, Yokoyama T. Apparent dipole moments of 1-butanol, 1-propanol, and chlorobenzene in cyclohexane or benzene, and excess molar volumes of (1-propanol or chlorobenzene + cyclohexane or benzene) at T = 298.15 K. Journal of Solution Chemistry. 2004; 33: 1061–1072. doi:10.1023/B:JOSL.0000048056.49122.6d

[40] González B, Calvar N, Domínguez A, Tojo J. Dynamic viscosities of binary mixtures of cycloalkanes with primary alcohols at T = (293.15, 298.15, and 303.15) K: New UNIFAC-VISCO interaction parameters. Journal of Chemical Thermodynamics. 2007; 39: 322–334. doi:10.1016/j.jct.2006.06.008

[41] Gardas RL, Oswal S. Volumetric and transport properties of ternary mixtures containing 1-propanol, triethylamine or tri-n-butylamine and cyclohexane at 303.15 K: Experimental data, correlation and prediction by ERAS model. Thermochimica Acta. 2008; 479: 17–27. doi:10.1016/j.tca.2008.09.006

[42] Umadevi M, Kesavasamy R, Rathina K, Mahalakshmi R. Studies on liquid-liquid interactions of some ternary mixtures by density, viscosity and ultrasonic speed measurements. Journal of Molecular Liquids. 2016; 219: 820–828. DOI: 10.1016/j.molliq.2016.03.085.

[43] Rao MVP, Naidu PR. Viscosities of binary liquid mixtures of 1-propanol, 1-butanol, 1-pentanol and 1-hexanol with methylcyclohexane, cyclopentanone and cyclohexanone. Indian Journal of Chemistry - Section A: Inorganic, Bio-inorganic, Physical, Theoretical & Analytical Chemistry. 1977; 15: 239.

[44] Letcher TM, Nevine JA. Excess molar volumes of chloro- or methylcyclohexane + an alkanol at 298.15 K. Thermochimica Acta. 1996; 287: 287–292. doi:10.1016/S0040-6031(96)03001-8

[45] Aralaguppi MI, Baragi JG. Physico-chemical and excess properties of the binary mixtures of methylcyclohexane + ethanol, + propan-1-ol, + propan-2-ol, + butan-1-ol, + 2-methyl-1-propanol, or 3-methyl-1-butanol at T = (298.15, 303.15, and 308.15) K. Journal of Chemical Thermodynamics. 2006; 38: 434–442. doi:10.1016/j.jct.2005.06.011

[46] Coca Prados J. Physical-chemical properties of liquid binary mixtures. Viscosity and heat of mixture. Acta Salmanticensia. Ciencias. 1969; 33: 7–39.

[47] Yadav BL, Maken S, Kalra KC, Singh KC. Excess volumes of (an alkanol + an aromatic hydrocarbon) at the temperature 308.15 K. Journal of Chemical Thermodynamics. 1993; 25: 1345–1350. doi:10.1006/jcht.1993.1135

[48] Singh KC, Kalra KC, Maken S, Yadav BL. Excess volumes of 1-propanol and 2-propanol with aromatic hydrocarbons at 298.15 K. Journal of Chemical and Engineering Data. 1994; 39: 241–244. doi:10.1021/je00014a011

[49] Kumar KS, Reddy NV. Excess volumes of binary mixtures of benzene 1-alkanols at 303.15 K. Physics and Chemistry of Liquids. 2001; 39: 117–123. doi: 10.1080/00319100108030331

[50] Kijevcanin ML, Puric IM, Radovic IR, Djordjevic BD, Serbanovic SP. Densities and excess molar volumes of the binary 1-propanol + chloroform and 1-propanol + benzene and ternary 1-propanol + chloroform + benzene mixtures at (288.15, 293.15, 298.15, 303.15, 308.15, and 313.15) K. Journal of Chemical and Engineering Data. 2007; 52: 2067–2071. doi:10.1021/je700254t

[51] Mejía A, Segura H, Cartes M. Vapor-liquid equilibrium, densities, and interfacial tensions for the system benzene + propan-1-ol. Physics and Chemistry of Liquids. 2008; 46: 185–200. doi:10.1080/00319100701459350

[52] Kim HD, Hwang IC, Park SJ. Isothermal vapor liquid equilibrium at 323.15 K and excess molar volumes and refractive indices at 298.15 K for the ternary system propyl vinyl ether + 1-propanol + benzene and its binary sub-systems. Fluid Phase Equilibria. 2008; 274: 73–79. doi:10.1016/j.fluid.2008.09.005

[53] Gahlyan S, Rani M, Lee I, Moon I, Maken SK. Measurement and correlation of excess molar volumes for mixtures of 1-propanol and aromatic hydrocarbons. Korean . Journal of Chemical Engineering. 2015; 32: 168–177. DOI : 10.1007/s11814-014-0200-6

[54] Swamy GN, Dharmaraju G, Raman GK. Excess volumes of toluene mixtures with some alcohols at 303.15 K. Canadian Journal of Chemistry. 1980; 58: 229–230. doi:10.1139/v80-037

[55] Nikam PS, Jagdale BS, Sawant AB, Hasan M. Densities and viscosities of binary mixtures of toluene with methanol, ethanol, propan-1-ol, butan-1-ol, pentan-1-ol, and 2-methylpropan-2-ol at (303.15, 308.15, 313.15) K. Journal of Chemical and Engineering Data. 2000; 45: 559–563. doi:10.1021/je990317i

[56] Zeberg-Mikkelsen CK, Andersen SI. Density measurements under pressure for the binary system 1-propanol + toluene. Journal of Chemical and Engineering Data. 2005; 50: 524–528. doi:10.1021/je049685z

[57] Atik Z. Experimental and predicted volumetric and refractive index properties of ternary mixtures of iodoethane with toluene and alcohols at temperature 298.15 K and pressure 101 kPa. Journal of Chemical Thermodynamics. 2006; 38: 201–208. doi:10.1016/j.jct.2005.05.004

[58] Zéberg-Mikkelsen CK, Lugo L, Fernandez J. Corrigendum to "Density measurements under pressure for the binary system (ethanol + methylcyclohexane)". Journal of Chemical Thermodynamics. 2006; 38: 649. doi:10.1016/j.jct.2005.11.006

[59] Hwang IC, Park SJ, Lee SY, Ahn HS. Isothermal vapor liquid equilibrium at 333.15 K and excess molar volumes at 298.15 K for the ternary system di-isopropyl ether + n-propyl alcohol + toluene and its binary subsystems. Fluid Phase Equilibria. 2008; 270: 103–108. doi:10.1016/j.fluid.2008.06.019

[60] Letcher TM, Mercer-Chalmers J, Govender UP, Radloff S. Excess molar enthalpies and excess molar volumes of binary mixtures of 1-alkenes with 1-propanol and 2-propanol. Thermochimica Acta. 1993; 224: 33–38. doi:10.1016/0040-6031(93)80151-Y

[61] Treszczanowicz A, Pawlowski TS, Treszczanowicz T, Szafranski AM. Excess volume of the 1-propanol + 1-alkene systems in terms of an equation of state with association. Journal of Chemical and Engineering Data. 2010; 55: 5478–5482. doi:10.1021/je100595s

[62] Kumar A, Pradhan SD, Katti SS. Volume change on mixing of isomeric butanols with normal heptane. Proc. Indian Acad. Sci. (Chem. Sci.). 1979; 88: 249–255. DOI:10.1007/BF02880931

[63] Berro C, Peneloux A. Excess gibbs energies and excess volumes of 1-butanol-n-heptane and 2-methyl-1-propanol-n-heptane binary systems. Journal of Chemical and Engineering Data. 1984; 29: 206–210. doi:10.1021/je00036a033

[64] Zielkiewicz J. Excess volumes of (heptane + propane-2-ol or butan-1-ol or 2-methylpropan-1 -ol or 2-methylpropan-2-ol or pentan-1-ol) at the temperature 313.15 K. Journal of Chemical Thermodynamics. 1994; 26: 959–964. doi:10.1006/jcht.1994.1112

[65] Nath J. Speeds of sound in and isentropic compressibilities of (n-butanol + n-pentane, or n-hexane, or n-heptane, or n-octane, or 2,2,4-trimethylpentane, or carbon tetrachloride) at T=293.15 K. Journal of Chemical Thermodynamics. 1997; 29: 853–863. doi: 10.1006/jcht.1997.0200

[66] Nath J, Pandey JG. Binary mixtures of butanol + pentane, +hexane, +heptane, +octane, +2,2,4-trimethylpentane, and + carbontetrachloride. 1. Excess molar volumes at 288.15 K and 298.15 K and refractive indexes at 298.15 K. Journal of Chemical and Engineering Data. 1997; 42: 128–131. doi:10.1021/je960229n

[67] Westwood BM, Kabadi VN. A novel pycnometer for density measurements of liquids at elevated temperatures. Journal of Chemical Thermodynamics. 2003; 35: 1965–1974. doi:10.1016/j.jct.2003.08.005

[68] Vijande J, Pineiro MM, Garcia J, Valencia JL, Legido JL. Density and surface tension variation with temperature for heptane + 1-alkanol. Journal of Chemical and Engineering Data. 2006; 51: 1778–1782. doi:10.1021/je060179e

[69] Kijevcanin ML, Radovic IR, Serbanovic SP, Tasic AZ. Experimental determination and modelling of densities and excess molar volumes of ternary system (1-butanol + cyclohexylamine + n-heptane) and corresponding binaries from 288.15 to 323.15 K. Thermochimica Acta. 2009; 496: 71–86. doi:10.1016/j.tca.2009.07.002

[70] Moravkova L, Troncoso J, Machanova K, Sedlakova Z. Volumetric behaviour of the (2,2,4-trimethylpentane + methylbenzene + butan-1-ol) ternary system and its binary sub-systems within the temperature range (298.15-328.15) K. Journal of Chemical Thermodynamics. 2013; 64: 137–150. doi:10.1016/j.tca.2009.07.002

[71] French HT. Thermodynamic functions of the systems 1-butanol, 2-butanol, and t-butanol + cyclohexane. Journal of Solution Chemistry. 1983; 12: 869. doi:10.1007/BF00643927

[72] Artigas H, Sanz ML, Mainar AM, Royo FM, Urieta JS. Excess volumes and excess viscosities of binary mixtures of cyclohexane + an isomer of butanol at several temperatures. Physics and Chemistry of Liquids. 1995; 30: 17–28. doi: 10.1080/00319109508028430

[73] Mascato E, Mosteiro L, Pineiro MM, Garcia J, Iglesias TP, Legido JL. Thermodynamic properties of mixing for (1-alkanol + an-alkane + a cyclic alkane) at T = 298.15 K. I. (n-Hexane + cyclohexane + 1-butanol). Journal of Chemical Thermodynamics. 2001; 33: 269–285. doi:10.1006/jcht.2000.0732

[74] Haro M, Gascón I, Cea P, Lafuente C, Royo FM. Volumetric and acoustic properties of the ternary system (1-butanol+1,4-dioxane+cyclohexane). Journal of Thermal Analysis and Calorimetry. 2005; 79: 51–57. DOI: 10.1007/s10973-004-0561-x.

[75] Bebek K, Strugala-Wilczek A. Acoustic and thermophysical properties of binary liquid mixtures of primary butanols with hexane and cyclohexane at 293.15 K. International Journal of Thermophysics. 2010; 31: 8–15. doi:10.1007/s10765-009-0665-4

[76] Torín-Ollarves GA, Martín MC, Chamorro CR, Segovia JJ. Densities, viscosities, and isobaric heat capacities of the system (1-butanol + cyclohexane) at high pressures. Journal of Chemical Thermodynamics. 2014; 74: 153–160. doi:10.1016/j.jct.2014.01.020

[77] Alonso R, Corrales JA. Excess molar volumes of (methylcyclohexane + an alkanol) at 298.15 K. I. Results for n-butanol, n-pentanol, and n-hexanol. Journal of Chemical Thermodynamics. 1989; 21: 515–518. doi:10.1016/0021-9614(89)90168-7

[78] Oswal SL, Prajapati KD, Ghael NY, Ijardar SP. Speeds of sound, isentropic compressibilities and excess molar volumes of an alkanol + cycloalkane at 303.15 K. II. Results for

alkan-2-ols + cyclohexane and alkan-1-ols + methylcyclohexane and theoretical interpretation. Fluid Phase Equilibria. 2004; 218: 131–140. doi:10.1016/j.fluid.2003.11.012

[79] Rodríguez V, Lafuente C, López MC, Royo FM, Urieta JS. Excess molar volumes and vapor pressures of (benzene + each of several isomers of butanol). Journal of Chemical Thermodynamics. 1993; 25: 679–685. doi:10.1006/jcht.1993.1065

[80] Yu CH, Tsai FN. Excess volumes of binary mixtures of benzene + 1-alkanols at 298.15 and 308.15 K. Journal of Chemical and Engineering Data. 1994; 39: 441–443. doi: 10.1021/je00015a008

[81] Bhardwaj U, Maken S, Singh KC. Excess molar volumes of (an isomer of butanol + benzene or toluene) at the temperature 308.15 K. Journal of Chemical Thermodynamics. 1996; 28: 1173–1177. doi:10.1006/jcht.1996.0103

[82] Ali A, Nain AK, Lal B, Chand D. Densities, viscosities, and refractive indices of binary mixtures of benzene with isomeric butanols at 30°C. International Journal of Thermophysics. 2004; 25: 1835–1847. doi:10.1007/s10765-004-7738-1

[83] Smiljanić JD, Kijevčanin ML, Djordjević BD. Temperature dependence of densities and excess molar volumes of the ternary mixture (1-butanol + chloroform + benzene) and its binary constituents (1-butanol + chloroform and 1-butanol + benzene). International Journal of Thermophysics. 2008; 29: 586–609. doi:10.1007/s10765-008-0390-4

[84] Litkenhous EE, Van Arsdale JD, Hutchison IW. The physical properties of the ternary system butyl alcohol-ethyl acetate-toluene. Journal of Physical Chemistry. 1940; 44: 377–388. doi:10.1021/j150399a010

[85] Moravkova L, Troncoso J, Skvorova M, Havlica J, Petrus P, Sedlakova Z. Volumetric behavior of the ternary system (methyl tert-butyl ether + methylbenzene + butan-1-ol) and its binary sub-system (methyl tert-butyl ether + butan-1-ol) within the temperature range (298.15 to 328.15) K. Journal of Chemical Thermodynamics. 2015; 90: 59–70. doi: 10.1016/j.jct.2015.06.009

[86] Torín-Ollarves GA, Segovia JJ, Martín MC, Villamañán MA. Density, viscosity, and isobaric heat capacity of the mixture (1-butanol + 1-hexene). Journal of Chemical and Engineering Data. 2013; 58: 2717–2723. doi:10.1021/je301301j

[87] Radovic IR, Serbanovic SP, Djordjevic BD, Kijevcanin ML. Experimental determination of densities and refractive indices of the ternary mixture 2-methyl-2-propanol + cyclohexylamine + n-heptane at T = (303.15 to 323.15) K. Journal of Chemical and Engineering Data. 2011; 56: 344–349. doi:10.1021/je100484n

[88] Peña MP, Martínez-Soria V, Montón JB. Densities, refractive indices, and derived excess properties of the binary systems tert-butyl alcohol + toluene, +methylcyclohexane, and + isooctane and toluene + methylcyclohexane, and the ternary system tert-butyl alcohol

+ toluene + methylcyclohexane at 298.15 K. Fluid Phase Equilibria. 1999; 166: 53–65. doi: 10.1016/S0378-3812(99)00284-8

[89] Bernatova S, Wichterle I. Isothermal vapour-liquid equilibria in the ternary system tert-butyl methyl ether + tert-butanol + 2,2,4-trimethylpentane and the three binary subsystems. Fluid Phase Equilibria. 2001; 180: 235–245. doi:10.1016/S0378-3812(01)00349-1

[90] Wang CC, Chen HW, Tu C-H. Densities, viscosities, and refractive indices for binary and ternary mixtures of ethanol, 2-methylpropan-2-ol, and 2,2,4-trimethylpentane. Journal of Chemical and Engineering Data. 2005; 50: 1687–1693. doi:10.1021/je0501639

[91] Rived F, Roses M, Bosch E. Densities, refractive indices, absolute viscosities, and static dielectric constants of 2-methylpropan-2-ol + hexane, + benzene, + propan-2-ol, + methanol, + ethanol, and + water at 303.2 K. Journal of Chemical and Engineering Data. 1995; 40: 1111–1114. doi:10.1021/je00021a017

[92] Gurukul SMKA, Raju BN. Isobaric vapor liquid equilibria of the 1-propanol-n-heptane system. Journal of Chemical and Engineering Data. 1966; 11: 501–502. doi:10.1021/je60031a013

[93] Sipowska J, Wieczorek S. Vapor pressures and excess gibbs free energies of propan-1-ol + n-heptane between 278.164 and 303.147 K. Journal of Chemical Thermodynamics. 1980; 12: 459–464. doi:10-1016/0021-9614(80)90059-2

[94] Zielkiewicz J. (Vapor + liquid) equilibria in (propan-1-ol + n-hexane + n-heptane) at the temperature 313.15 K. Journal of Chemical Thermodynamics. 1991; 23: 605–612. doi: 10.1016/S0021-9614(05)80103-X

[95] Zielkiewicz J. (Vapor + liquid) equilibria in (propan-1-ol + heptane + octane) at the temperature 313.15 K. Journal of Chemical Thermodynamics. 1992; 24: 455–462. doi: 10.1016/S0021-9614(05)80117-X

[96] Pradhan AG, Bhethanabotla VR, Campbell SW. Vapor-liquid equilibrium data for etanol-n-heptane-1-propanol and etanol-n-heptane-2-propanol and their interpretation by a simple association model. Fluid Phase Equilibria. 1993; 84: 183–206. doi: 10.1016/0378-3812(93)85123-4

[97] Olson JD. Thermodynamics of hydrogen-bonding mixtures II. GE, HE, and SE of 1-propanol + n-heptane. International Journal of Thermophysics. 1995; 16: 215–226. doi: 10.1007/BF01438972

[98] Chen G, Wang Q, Zhang L-Z, Bao J, Han S-J. Study and applications of binary and ternary azeotropes. Thermochimica Acta. 1995; 253: 295–305. doi: 10.1016/0040-6031(94)02078-3

[99] Dobrjakov YG, Balashova IM, Maurer G. Experimental results for the limiting activity coefficients in some binary and ternary mixtures of organic components. Journal of Chemical and Engineering Data. 2000; 45: 185–193. doi:10.1021/je990204z

[100] Vrbka P, Dohnal V, Arlt W. Limiting activity coefficients by comparative tensimetry: 1-propanol and 1-butanol in heptane and in octane. Journal of Chemical and Engineering Data. 2004; 49: 867–871. doi:10.1021/je0302295

[101] Hiaki T, Takahashi K, Tsuji T, Hongo M, Kojima K. Vapor-liquid equilibria of 1-propanol or 2-propanol with 2,2,4-trimethylpentane at 101.3 kPa. Journal of Chemical and Engineering Data. 1994; 39: 602–604. doi:10.1021/je00015a047

[102] Hiaki T, Takahashi K, Tsuji T, Hongo M, Kojima K. Vapor-liquid equilibria of etanol + 2,2,4-trimethylpentane at 333.15 K and 1-propanol + 2,2,4-trimethylpentane at 343.15 K. Journal of Chemical and Engineering Data. 1994; 39: 605–607. doi:10.1021/je00015a048

[103] Garces SI, Faneite AM, Soto D, Álvarez CA, Urdaneta MR. Isothermal vapor liquid equilibrium data of propan-1-ol + 2,2,4-trimethylpentane and butan-1-ol + 2,2,4-trimethylpentane at 318.15 K. Journal of Chemical and Engineering Data. 2011; 56: 3346–3350. doi:10.1021/je200279q

[104] Hwang S-C, Robinson Jr RL. Vapor-liquid equilibria at 25°C for nine alcohol-hydrocarbon binary systems. Journal of Chemical and Engineering Data. 1977; 22: 319–325. doi:10.1021/je60074a025

[105] Hiaki T, Tochigi K, Kojima K. Measurement of vapor-liquid equilibria and determination of azeotropic point. Fluid Phase Equilibria. 1986; 26: 83–102. doi: 10.1016/0378-38120(86)85006-3

[106] Gupta V, Maken S, Kalra KC, Singh KC. Molar excess free energy of mixing of 1-propanol or 2-propanol with cyclohexane at 298.15 and 308.15 K in terms of association model with a Flory contribution term. Fluid Phase Equilibria. 1996; 120: 195–203. doi: 10.1016/0378-3812(96)03001-4

[107] Kurihara K, Uchiyama M, Kojima K. Isothermal vapor-liquid equilibria for benzene + cyclohexane + 1-propanol and for three constituent binary systems. Journal of Chemical and Engineering Data. 1997; 42: 149–154. doi:10.1021/je9602475

[108] Castells CB, Eikens DI, Carr PW. Headspace gas chromatographic measurements of limiting activity coefficients of eleven alkanes in organic solvents at 25°C. 1. Journal of Chemical and Engineering Data. 2000; 45: 369–375. doi:10.1021/je990146h

[109] Raju BN, Rao DP. Isobaric vapor-liquid equilibrium data for the systems 1-propanol-methylcyclohexane and methylcyclohexane-1-butanol. Journal of Chemical and Engineering Data. 1969; 14: 283–286. doi:10.1021/je60042a019

[110] Nagata I. Isothermal vapour-liquid equilibria for binary mixtures of methylcyclohexane with alcohols. Thermochimica Acta. 1989; 144: 95–107. doi: 10.1016/0040-6031(89)85088-9

[111] Loras S, Monton JB, España FJ. Vapor-liquid equilibria for the binary systems of methylcyclohexane with 1-propanol, 2-propanol, 1-butanol and 2-butanol at 101.3 kPa. Journal of Chemical and Engineering Data. 1997; 42: 914–918. doi:10.1021/je970079w

[112] Britton EC, Nutting HS, Horsley LH. Vapor-liquid equilibrium diagrams of alcohol-ketone azeotropes as a function of pressure. Analytical Chemistry. 1947; 19: 601–602. doi:10.1021/ac60008a003

[113] Prabhu PS, Van Winkle M. Effect of polar components on the relative volatility of the binary system n-hexane-benzene. Journal of Chemical and Engineering Data. 1963; 8: 210–214. doi:10.1021/je60017a022

[114] Skaates JM, Kay WB. The phase relations of binary systems that form azeotropes: N-alkyl alcohol-benzene systems: Metanol through n-butanol. Chemical Engineering Science. 1964; 19: 431–444. doi:10.1016/0009-2509(64)85070-3

[115] Oracz P, Kolasinska G. Vapour-liquid equilibria-III. Total vapour pressure measurements for binary mixtures of metanol, etanol, 1-propanol and 1-butanol with benzene, toluene and p-xylene at 313.15 K. Fluid Phase Equilibria. 1987; 35(1–3): 253–278. doi: 10.1016/0378-3812(87)80016-X

[116] Rhodes JM, Griffin TA, Lazzaroni MJ, Bhethanabotla VR, Campbell SW. Total pressure measurements for benzene with 1-propanol, 2-propanol, 1-pentanol, 3-pentanol, and 2-methyl-2-butanol at 313.15 K. Fluid Phase Equilibria. 2001; 179: 217–229. doi:10.1016/S0378-3812(00)00502-1

[117] Villamañán RM, Chamorro CR, Villamañán MA, Segovia JJ. Total pressure and excess Gibbs energy for the ternary mixture di-isopropyl ether + 1-propanol + benzene and its corresponding binary systems at 313.15 K. Fluid Phase Equilibria. 2006; 239: 183–187. doi:10.1016/j.fluid.2005.11.014

[118] Gupta V, Maken S, Kalra KC, Singh KC. Molar excess free energy of 1-propanol or 2-propanol + aromatic hydrocarbons at 298.15 K in terms of an association model with a Flory contribution term. Thermochimica Acta. 1996; 277: 187–198. doi: 10.1016/0040-6031(95)02745-9

[119] Vrbka P, Rozbroj D, Vladimirov DA. Limiting activity coefficients in binary mixtures of 1-alkanols and toluene. Fluid Phase Equilibria. 2003; 209(2): 265–280. doi:10.1016/S0378-3812(03)00114-6

[120] Hwang I-C, Park S-J, Lee J-Y. Binary and ternary vapor-liquid equilibrium at 323.15 K and excess molar volumes at 298.15 K for the mixtures of propyl vinyl ether + 1-propanol + toluene. Journal of Chemical and Engineering Data. 2009; 54: 1041–1045. doi:10.1021/je800972b

[121] Vijayaraghavan SV, Deshpande PK, Kuloor NR. Isobaric vapor-liquid equilibrium of n-heptane-n-butanol system. Journal of Chemical and Engineering Data. 1966; 11: 147–149. doi:10.1021/je60029a004

[122] Vijayaraghavan SV, Deshpande PK, Kuloor NR. Vapor-liquid equilibrium data for the system n-heptane-butyl-alcohol at medium pressures. Journal of Chemical and Engineering Data. 1967; 12: 13–15. doi:10.1021/je60032a004

[123] Pividal KA, Sandler SI. Neighbor effects on the group contribution method: Infinite dilution activity coefficients of binary systems containing primary amines and alcohols. Journal of Chemical and Engineering Data. 1990; 35: 53–60. doi:10.1021/je00059a018

[124] Zielkiewicz J. (Vapor + liquid) equilibria in (heptane + propan-2-ol or butan-1-ol or 2-methylpropan-1-ol or 2-methylpropan-2-ol or pentan-1-ol) at the temperature 313.15 K. Journal of Chemical Thermodynamics. 1994; 26: 919–923. doi:10.1006/jcht.1994.1109

[125] Gmehling J, Boelts R. Azeotropic data for binary and ternary systems at moderate pressures. Journal of Chemical and Engineering Data. 1996; 41: 202–209. doi:10.1021/je950228f

[126] Powell JR, McHale MER, Kauppila AS-M, Acree WE, Flanders PH, Varanasi VG, Campbell SW. Prediction of anthracene solubility in alcohol + alkane solvent mixtures using binary alcohol + alkane VLE data. Comparison of Kretschmer-Wiebe and mobile order models. Fluid Phase Equilibria. 1997; 134: 185–200. doi:10.1016/S0378-3812(97)00039-3

[127] Prasad TEV, Satyakishore P, Ramserish GV, Prasad DHL. Boiling temperature measurements on the binary mixtures of n-heptane with some aliphatic alcohols. Journal of Chemical and Engineering Data. 2001; 46: 1266–1268. doi:10.1021/je010006j

[128] Mohsen-Nia M, Memarzadeh MR. Isobaric vapor-liquid equilibria of heptane + 1-butanol and heptane + 1-pentanol systems at (53.3 and 91.3) kPa. Journal of Chemical and Engineering Data. 2010; 55: 2140–2144. doi:10.1021/je9006629

[129] Belabbaci A, Villamañán RM, Negadi L, Martín MC, Ait Kaci A, Villamañán MA. Vapor-liquid equilibria of binary mixtures containing 1-butanol and hydrocarbons at 313.15 K. Journal of Chemical and Engineering Data. 2012; 57: 114–119. doi:10.1021/je200840e

[130] Hull A, Kronberg B, Stam JV, Golubkov I, Kristensson J. Vapor-liquid equilibrium of binary mixtures. 2. Ethanol + 2,2,4-trimethylpentane, 1-butanol + 2,2,4-trimethylpentane, and ethanol + o-xylene. Journal of Chemical and Engineering Data. 2006; 51: 2002–2008. doi:10.1021/je060005x

[131] Ghellai S, Belabbaci A, Villamañán RM, Martín MC, Villamañán MA, Negadi L. Vapour-liquid equilibria of binary and ternary mixtures containing 1-butanol, 2,2,4-trimethylpentane and 1-hexene at T = 313.15 K. Journal of Chemical Thermodynamics. 2013; 63: 164–168. doi:10.1016/j.jct.2013.04.003

[132] Ramalho RS, Delmas J. Isothermal and isobaric vapor-liquid equilibrium data and excess free energies by the total pressure method. Systems: 2,2,4-trimethylpentane-toluene, cyclohexane-1-butanol, and ethanol-n-heptane. Journal of Chemical and Engineering Data. 1968; 13: 161–164. doi:10.1021/je60037a005

[133] Thomas ER, Newman BA, Long TC, Wood DA, Eckert CA. Limiting activity coefficients of nonpolar and polar solutes in both volatile and nonvolatile solvents by gas chroma-

tography. Journal of Chemical and Engineering Data. 1982; 27: 399–405. doi:10.1021/je00030a010

[134] Trampe DM, Eckert CA. Limiting activity coefficients from an improved differential boiling point technique. Journal of Chemical and Engineering Data. 1990; 35: 156–162. doi:10.1021/je00060a018

[135] Prasad VTE, Naidu RP, Madhukiran D, Prasad DHL. Boiling temperature measurements on the binary mixtures of cyclohexane with some alcohols and chlorohydrocarbons. Journal of Chemical and Engineering Data. 2001; 46: 414–416. doi:10.1021/je000139p

[136] Gascón I, Martín S, Artigas H, López MC, Lafuente C. Isobaric vapour liquid equilibrium of binary and ternary mixtures containing cyclohexane, n-hexane, 1,3-dioloxane and 1-butanol at 40.0 and 101.3 kPa. Chemical Engineering Journal. 2002; 88: 1–9. doi:10.1016/S1385-8947(01)00258-3

[137] Allen BB, Lingo SP, Felsing WA. Total and partial pressures of binary solutions of the butyl alcohols in benzene at 25°C. Journal of Physical Chemistry. 1939; 43: 425–430. doi:10.1021/j150391a004

[138] Mann RS, Shemilt LW, Waldichuck M. Vapor-liquid equilibria at atm. pressure II. 1-butanol-benzene system. Journal of Chemical and Engineering Data. 1963; 8: 502–504. doi:10.1021/je60019a008

[139] Chen G, Wang Q, Ma Z-M, Yan X-H, Han S-J. Phase equilibria at superatmospheric pressures for systems containing halohydrocarbon, aromatic hydrocarbon, and alcohol. Journal of Chemical and Engineering Data. 1995; 40: 361–366. doi:10.1021/je00018a003

[140] Maken S, Park JJ, Bhardwaj U, Singh KC, Park J-W, Han SD, Deshwal BR. (Vapour + liquid) equilibria of (1-butanol + benzene, or toluene, or o-, or m-, or p-xylene) at T = 308.15 K. Journal of Chemical Thermodynamics. 2004; 36: 309–315. doi:10.1016/j.jct.2004.01.001

[141] Villamañán RM, Martín MC, Chamorro CR, Villamañán MA, Segovia JJ. Phase equilibrium properties of binary and ternary systems containing di-isopropyl ether + 1-butanol + benzene at 313.15 K. Journal of Chemical Thermodynamics. 2006; 38: 547–553. doi:10.1016/j.jct.2005.07.004

[142] Mann RS, Shemilt LW. Vapor-liquid equilibria at atm. pressure. 1-butanol-toluene system. Journal of Chemical and Engineering Data. 1963; 8: 189–190. doi:10.1021/JE60017a013

[143] Darwish NA, Al-Khateib AA. Isobaric vapor-liquid equilibria of the system toluene + 1-butanol at 94.0, 70.5, and 56.4 kPa. Fluid Phase Equilibria. 1997; 132: 215–223. doi:10.1016/S0378-3812(97)00006-x

[144] Belabbaci A, Ghezouali C, Villamañán RM, Segovia JJ, Villamañán MA, Negadi L. Isothermal vapor liquid equilibrium and molar excess Gibbs energies of two ternary

systems containing either 1-butanol or 2-butanol + 1-hexene + methylbenzene at 313.15 K. Fluid Phase Equilibria. 2015; 386: 1–6. doi:10.1016/j.fluid.2014.11.010

[145] Janaszewski B, Oracz P, Goral M, Warycha S. Vapor-liquid equilibria: I. An apparatus for isothermal total vapor pressure measurements. Binary mixtures of ethanol and t-butanol with n-hexane, n-heptane and n-octane at 313.15 K. Fluid Phase Equilibria. 1982; 9: 295–310. doi:10.1016/0378-3812(82)80025-3

[146] Zong ZL, Yang XH, Zheng XY. Determination and correlation of vapor-liquid equilibria of alcohol solutions. Journal of Chemical Engineering of Japan. 1983; 16: 1–6. doi: 10.1252/jcej.16.1

[147] Cheung Y-S, Chen Y-J, Ng CY, Chiu S-W, Li W-K. Combining theory with experiment: Assessment of the thermochemistry of SFn, SFn+, and and SFn-, n = 1-6. Journal of the American Chemical Society. 1995; 117: 9725–9733. doi:10.1021/ja00143a016

[148] Martínez-Soria V, Peña P, Montón JB. Vapor-liquid equilibria for the binary systems tert-butyl alcohol + toluene, + isooctane, and methylcyclohexane at 101.3 kPa. Journal of Chemical and Engineering Data. 1999; 44: 148–151. doi:10.1021/je980165b

[149] Prasad TEV, Sriram N, Raju AN, Prasad DHL. (Vapor + liquid) equilibria of binary mixtures formed by iso-octane with a variety of compounds at 95.8 kPa. Journal of Chemical Thermodynamics. 2006; 38: 119–122. doi:10.1016/j.jct.2005.04.006

[150] Tripathi RP, Krishna S, Gulati IB. Isobaric binary vapor-liquid equilibria in cyclohexane-tert-butyl alcohol and 2,4-dimethylpentane-tert-butyl alcohol systems. Journal of Chemical and Engineering Data. 1976; 21: 44–47. doi:10.1021/je60068a013

[151] Triday JO, Veas C. Vapor-liquid equilibria for the system cyclohexane-tert-butyl alcohol. Journal of Chemical and Engineering Data. 1985; 30: 171–173. doi:10.1021/je00040a011

[152] Young S, Fortey EC. LXXV. The properties of mixtures of the lower alcohols with benzene and with benzene and water. Journal of the Chemical Society, Transactions. 1902; 81: 739–752. doi:10.1039/CT9028100739

[153] Tongberg CO, Johnston F. Vapor-liquid equilibria for n-hexane-benzene mixtures. Industrial and Engineering Chemistry. 1933; 25: 733–735. doi:10.1021/ie50283a006

[154] Brown I, Fock W, Smith F. The thermodynamic properties of solutions of normal and branched alcohols in benzene and n-hexane. Journal of Chemical Thermodynamics. 1969; 1: 273–291. doi:10.1016/0021-9614(69)90047-0

[155] Govindaswamy S, Andiappan AN, Lakshmanan SM. Vapor-liquid equilibria of the binary and ternary systems containing n-hexane (1) + benzene (2) + tert-butyl alcohol (3) at 760 mmHg pressure. Journal of Chemical and Engineering Data. 1977; 22: 264–269. doi:10.1021/je60074a008

[156] Bhardwaj U, Singh KC, Maken S. Excess molar Gibbs energies of (2-methylpropan-2-ol + benzene, or toluene, or toluene, or o-, or m-, or p-xylene) at the temperature 308.15 K. Journal of Chemical Thermodynamics. 1998; 30: 109–115. doi:10.1006/jcht.1997.0281

[157] Fortier JL, Benson GC. Excess heat capacities of binary liquid mixtures determined with a Picker flow calorimeter. Journal of Chemical Thermodynamics. 1976; 8: 411–423. doi: 10.1016/0021-9614(76)90061-6

[158] Kalinowska B, Jedlinska J, Stecki J, Woycicki W. Heat capacities of liquids in the temperature interval between 90 and 300 K and at atmospheric pressure II. Heat capacities and excess heat capacities of propan-1-ol +n-heptane and +n-hexane. Journal of Chemical Thermodynamics. 1981; 13: 357–366. doi:10.1016/0021-9614(81)90025-2

[159] Roux AH, Roux-Desgranges G, Grolier JPE. Excess molar heat capacities and enthalpies for 1-alkanol + N-alkane binary mixtures. New measurements and recommended data. Fluid Phase Equilibria. 1993; 89: 57–88. doi:10.1016/0378-3812(93)85045-N

[160] Torin-Ollarves GA, Segovia JJ, Martin MC, Villamañán MA. Thermodynamic characterization of the mixture (1-butanol + iso-octane): Densities, viscosities, and isobaric heat capacities at high pressures. Journal of Chemical Thermodynamics. 2012; 44: 75–83. doi:10.1016/j.jct.2011.08.012

[161] Cobos JC, Garcia I, Casanova C, Roux AH, Roux-Desgranges G, Grolier JPE. Excess heat capacities of 1-butanol + toluene from 298 to 368 K. Fluid Phase Equilibria. 1991; 69: 223–233. doi:10.1016/0378-3812(91)90035-6

Chapter 19

Photocatalytic Reforming of Lignocelluloses, Glycerol, and Chlorella to Hydrogen

Masahide Yasuda

Additional information is available at the end of the chapter

http://dx.doi.org/10.5772/64901

Abstract

Bioethanol, biodiesel, and biogas have gained much attention as sustainable energy alternatives to petroleum-based fuels. Bioethanol production is the most typical method to provide liquid fuel. Recently cellulosic materials have been recognized as one of the promising sources for bioethanol, since they are not directly in competition with food sources. However, ethanol concentration is usually too low to separate by distillation at a low-energy cost. Gaseous H_2 is spontaneously isolated without operation to separate. Therefore, H_2 production is an economical approach to biofuels. Photocatalytic H_2 production over a Pt-loaded TiO_2 is initiated by the charge separation. Electrons reduce water to generate H_2, while holes oxidize hydroxide to hydroxyl radicals. Generally, the use of sacrificial agents remarkably accelerates the H_2 production since the hydroxyl radical is consumed by them. This chapter deals with the photocatalytic H_2 production (PR) using sacrificial water-soluble materials derived from lignocelluloses, lipids, and Chlorella. Lignocellulosic Italian ryegrass (2.00 g) was turned into H_2 (78.7 mg) through alkali treatment, hydrolysis, and PR processes. The PR process of glycerol (10.4 g) and methanol (11.3 g), which were by-products in biodiesel synthesis, formed H_2 (3.10 g). Dried Chlorella (10 g) was turned into H_2 (578 mg) by protease hydrolysis and PR.

Keywords: TiO_2, sacrificial agents, lignocelluloses, BDF, Chlorella

1. Introduction

Plants collect sunlight energy through photosynthesis and store it as a variety of polymeric saccharides. Polymeric saccharides are converted into monomeric saccharides, which are then converted into energy in all living organisms. Thus, saccharides are energy-storage substances which are produced from CO_2 and easily converted to energy along with CO_2 emission.

Therefore, saccharides have highly potential resources to produce renewable energy. Bioethanol production from starch of maize, sugarcane, and sugar sorghum is the most typical method to provide renewable liquid fuel [1, 2]. Recently in order to avoid the direct competition with food sources, cellulosic materials have been widely recognized as one of the promising sustainable resources to produce second-generation bioethanol [3]. However, the ethanol concentrations (<5.0 %) were still too low to separate by distillation at a low-energy cost [4]. On the other hand, gaseous H_2 is spontaneously isolated from reaction mixtures without operations to separate. Therefore, H_2 production from saccharides and biomass-derived materials is one of the economical approaches to biofuels [5].

In this chapter, I will show the photocatalytic reforming over titanium dioxide (TiO_2) using saccharides, glycerol, and amino acids, which are derived by hydrolysis of lignocelluloses, lipids, and Chlorella, respectively. This will lead to construct the sustainable energy system alternatives to petroleum-based fuels.

2. Outline of photocatalytic biomass reforming

A general procedure of biomass reforming is started by the production of water-soluble materials from biomass through biological treatment as well as chemical reaction. The resulting water-soluble materials are converted to biofuels such as ethanol, methane, and hydrogen through various catalytic reactions involving methane fermentation and steam reforming. It was demonstrated that the photocatalytic H_2 production from biomass-derived materials had an advantage compared with other thermal catalytic reforming by Shimura and Yoshida in their review in 2011 [6].

Our biomass reforming was performed in aqueous solution through enzymatic and chemical hydrolysis of biomass (lignocelluloses, lipids, and Chlorella) followed by photocatalytic reaction of water-soluble materials (saccharides, glycerol, and amino acids) over TiO_2 under UV-irradiation (**Figure 1**). Saccharides were produced by enzymatic hydrolysis of lignocelluloses using cellulase and xylanase. Glycerol was obtained by transesterification of lipid with methanol. Amino acids were obtained from hydrolysis of Chlorella by protease. These water-soluble materials were served as sacrificial agents for the photocatalytic H_2 production in aqueous solution. Details of each process were described in the following sections.

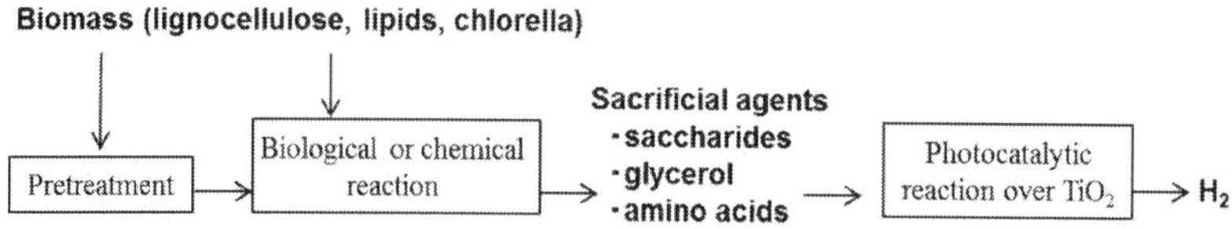

Figure 1. Outline of photocatalytic reforming of biomass.

http://dx.doi.org/10.5772/64901

3. Biological reactions

For biological reaction, a cellulase from *Acremonium cellulolyticus* (Acremozyme KM, Kyowa Kasei, Osaka, Japan) [7] was selected among commercially available cellulases. A xylanase from *Trichoderma longibrachiatum* (reesei) (Sumizyme X, Shin Nihon Chemicals, Anjyo, Japan) was selected from commercially available enzymes. Proteins were hydrolyzed by protease (protease A AmanoSD, Amano enzyme, Nagoya) at 50°C in a phosphate buffer (0.1 M, pH 7.6) which was prepared by dissolving Na_2HPO_4 (2.469 g) and NaH_2PO_4 (0.312 g) in 100 mL of water.

The cell suspension of *Saccharomyces cerevisiae* was prepared as follows. *S. cerevisiae* NBRC 2044 was grown at 30°C for 24 h in a basal medium consisting of glucose (20.0 g/L), bactotryptone (1.0 g/L, Difco), yeast extract (1.0 g/L), $MaSO_4$ (3.0 g/L), and $NaHPO_4$ (1.0 g/L) at initial pH 5.5 [7].

Cellulose and hemicellulose (holocellulose), which were composed of glucan and xylan, were hydrolyzed to glucose and xylose by the enzymatic saccharification (SA, Eq. 1). The powdered and pre-treated lignocellulose (4.0 g) was dispersed in an acetate buffer solution (80 mL, pH 5.0, 0.1 M) which was prepared by mixing 0.808 g acetic acid and 3.05 g sodium acetate in 500 mL of water. Cellulase (200 mg) and xylanase (200 mg) were added to the suspension of lignocellulose. The SA was performed by stirring the solution vigorously with a magnetic stirrer at 45°C for 120 h. After centrifugation of reaction mixture, the supernatant solution involving glucose and xylose was analyzed by HPLC and used as sacrificial agents in the following photocatalytic reaction.

$$\text{Xylan} + \text{Glucan} \xrightarrow[\text{water}]{\text{Cellulase, Xylanase}} \text{Xylose} + \text{Glucose} \qquad (1)$$

Also, lignocellulose could be turned into ethanol and xylose through simultaneous saccharification and fermentation (SSF, Eq. 2) using cellulase and xylanase as well as *S. cerevisiae* as follows [8]. An acetate buffer solution (10 mL, pH 5.0, 0.1 M) was added to pre-treated lignocelluloses (3.0 g) in the reaction vessel. The reaction vessel was autoclaved at 120°C for 20 min. After cooling, cellulase (180 mg) and xylanase (120 mg) in an acetate buffer solution (8.0 mL) and the cell suspension of *S. cerevisiae* (0.36 mL) were introduced into the reaction vessel. After air was purged with N_2 stream for 15 min, the SSF was performed at 34°C under stirring vigorously with a magnetic stirrer. The evolved CO_2 was collected by a measuring cylinder to monitor the volume of CO_2 gas. The SSF reaction was continued for about 96 h until CO_2 evolution ceased. After unreacted biomass was removed from the reaction mixture by centrifugation, the supernatant solution was analyzed by gas chromatography (GC) and high-performance liquid chromatography (HPLC) to determine the concentrations of ethanol and saccharides, respectively. Ethanol was collected from the SSF solution by evaporation under reduced pressure while the residual xylose was subjected to the photocatalytic reaction.

$$\text{Xylan} + \text{Glucan} \xrightarrow[\substack{S.cerevisiae \\ water}]{\text{Cellulase, Xylanase}} \text{Xylose} + CO_2 + C_2H_5OH \tag{2}$$

Another process to convert lignocellulose to ethanol is simultaneous saccharification and co-fermentation (SSCF). A recombinant *Escherichia coli* KO11 which can ferment xylose was used. Glucan and xylan in lignocellulose are turned to ethanol by SSCF using cellulase, xylanase, yeast, and *E. coli* KO11. An example is an SSCF process of the low-moisture anhydrous ammonia (LMAA)-treated Italian ryegrass (Section 6.1), which produced ethanol in 84.6% yield [9]. In this case, it was not necessary to undergo the photocatalytic process.

$$\text{Xylan} + \text{Glucan} \xrightarrow[\substack{S.cerevisiae, E.coli\,\text{KO11} \\ \text{water}}]{\text{Cellulase, Xylanase}} CO_2 + C_2H_2OH \tag{3}$$

4. Photocatalytic H_2 production

4.1. Titanium dioxide (TiO_2) as the photocatalyst

TiO_2 is a white powder material which is thermally stable, non-flammable, and no health hazards. Therefore, TiO_2 has been used for many years in industrial and consumer goods, including paints, coated fabrics and textiles, cosmetics, and so on. The photocatalytic H_2 production was performed by use of an anatase-type TiO_2. It has a semi-conductor structure whose band gap is known to be 3.20 eV, which corresponds to 385 nm. Therefore, TiO_2 can be excited by 366 nm-emission from a high-pressure mercury lamp. Irradiation induces charge separation into electrons and holes on the TiO_2 [10]. Electrons (e^-) reduce water to generate H_2, while holes (h^+) oxidize hydroxide anions to hydroxyl radicals (**Figure 2**) [11]. In most cases, noble metals (Pt, Pd, and Au) were loaded on TiO_2 to accelerate the reduction of water by electrons. We used a Pt-loaded TiO_2 (Pt/TiO_2) throughout the present investigation.

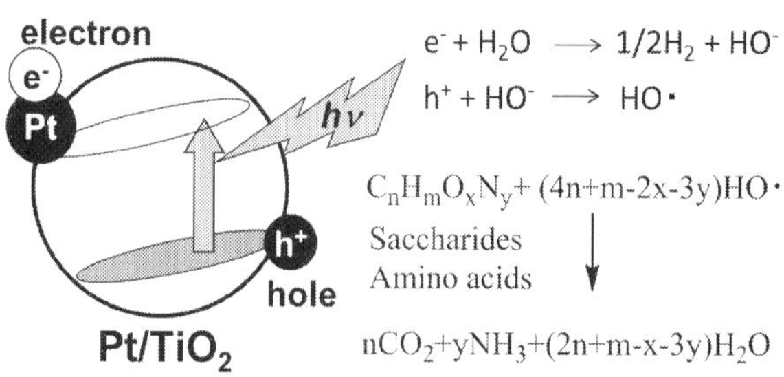

Figure 2. Hydrogen evolution on Pt/TiO_2 under irradiation.

Moreover, it was well known that the use of sacrificial agents remarkably accelerates H_2 production because the hydroxyl radical is consumed by them. Especially, we have elucidated that sacrificial agents with all of the carbon attached heteroatoms (O and N) are superior sacrificial agents because they continued to serve as electron sources until their sacrificial ability was exhausted [12, 13]. Glucose, xylose, glycerol, and glycine meet this requirement. The photocatalytic H_2 production using sacrificial agents is called "sacrificial H_2 production."

4.2. Preparation of Pt-loaded TiO_2 photocatalyst

For photocatalytic reaction, almost researches have continued to use a P25 (Degussa Co. Ltd, Germany) and a ST01 (Ishihara Sangyo Co. Ltd., Japan). The P25 is prepared through hydrolysis of $TiCl_4$ and composed of 75% of anatase and 25% of rutile, while the ST01 was prepared through hydrolysis of $TiOSO_4$ and composed of 100% of anatase.

The Pt-loaded TiO_2 (Pt/TiO_2) was prepared by the method reported by Kennedy and Datye [14] as follows. An aqueous solution (400 mL) containing TiO_2 (4.0 g, ST01, particle size 7 nm and surface area 300 m^2g^{-1}), K_2PtCl_6 (200 mg), and 2-propanol (3.06 mL) was introduced into a reaction vessel which is illustrated in Section 4.3. After O_2 was purged by N_2 gas, the solution was irradiated by a high-pressure mercury lamp with stirring for 24 h when the gas evolution reached over 100 mL. After the irradiation, water was entirely evaporated. The resulting gray precipitate was moved on a filter and washed with water and then dried and ground to produce Pt/TiO_2 powder. The Pt-content on TiO_2 was optimized to be 2.0 wt% from the photocatalytic H_2 evolution by various Pt-content TiO_2 using glucose as a sacrificial reagent. Identification of Pt/TiO_2 was usually performed by an XRD pattern and TEM image [15]. **Figure 3** shows a TEM image and an X-ray diffraction pattern of a P/TiO_2 (2.0 wt% of Pt content).

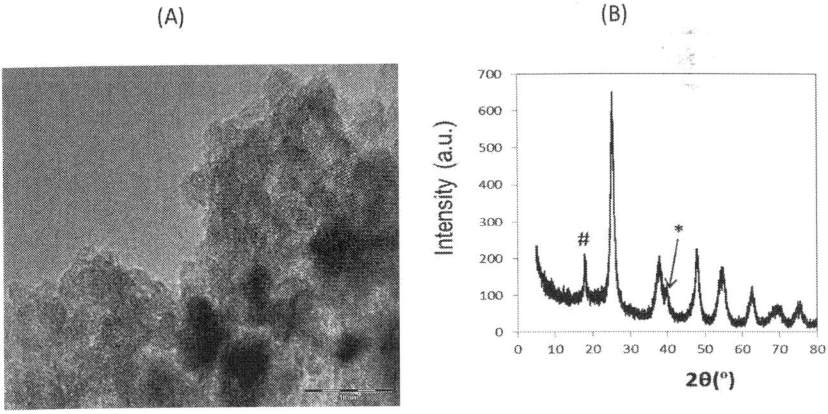

Figure 3. (A) TEM images of Pt/TiO_2 (2.0 wt% of Pt content). (B) X-ray diffraction of a P/TiO_2 (2.0 wt% of Pt content). Mark * was the peak for Pt. Mark # was the peak for impurity of Teflon removed from the stirrer chip.

4.3. Experimental method

The photocatalytic H_2 production was performed using a photo-irradiation apparatus (**Figure 4**). The catalyst (100 mg) and the given amounts of aqueous solution of sacrificial agent were introduced into a reaction vessel. The volume of the reaction solution was adjusted to 150 mL with water. The reaction vessel was connected with a measuring cylinder through a gas-impermeable fluororubber tube to collect the evolved gas. A high-pressure mercury lamp (100 W, UVL-100HA, Riko, Japan) was inserted into the reaction vessel, which was set in a water bath to keep it at a constant temperature (usually 20°C). After O_2 was purged from the reaction vessel by N_2 gas for 15 min, the reaction mixture was irradiated with vigorous stirring using a magnetic stirrer until the gas evolution ceased. The evolved gas was collected by a measuring cylinder to measure the total volume of the evolved gas. The evolved gas (0.5 mL) was obtained using a syringe and was subjected to the quantitative analysis of H_2, N_2, and CO_2, which were performed on a Shimadzu GC-8A equipped with a TCD detector at a temperature raised from 40 to 180°C using a stainless column (3 mmΦ, 6 m) packed with a SHINCARBON ST (Shimadzu). In the absence of sacrificial agents, the H_2 evolution from water was small (<2 mL).

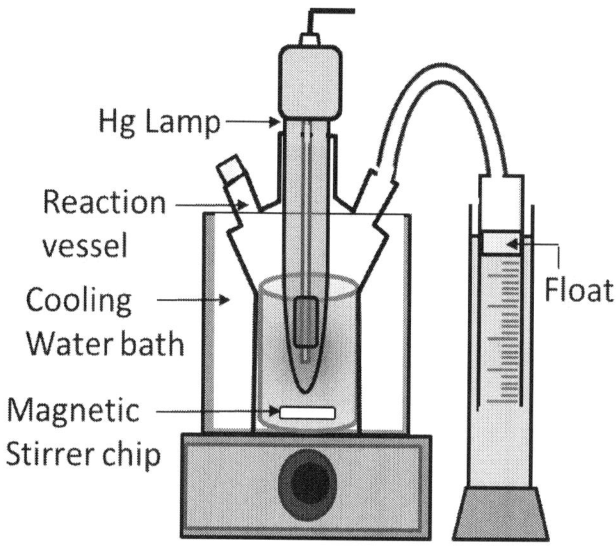

Figure 4. Apparatus for photocatalytic reaction.

4.4. Analysis of photocatalytic reaction

Theoretically, the photocatalytic reaction can convert glucose and xylose to 12 and 10 equivalents of H_2 (Eq. 4). Indeed, the photocatalytic reaction using glucose and xylose produced 11.8 and 10.0 mol of H_2 from 1 mol of glucose [15] and xylose [16], respectively.

$$C_nH_{2n}O_n + nH_2O \xrightarrow[Pt/TiO_2]{UV} nCO_2 + 2nH_2 \quad (4)$$

I show a method to determine the amounts of H_2 evolved from 1 mol of sacrificial agent. A typical example is the photocatalytic H_2 production using saccharides obtained from enzymatic saccharification of Napier grass. Although the saccharides contained not only xylose but also glucose, the evolved H_2 and CO_2 were plotted against the moles of xylose in a mixture of xylose and glucose, as shown in **Figure 5A**. Gas volumes of H_2 and CO_2 increased with the increase of xylose. However, the molar ratios of H_2 to xylose (H_2/xylose) were not constant to the amount of xylose used. It was speculated that the colored material in the solution and the carboxylic acids formed during the photocatalytic reaction may lower the activity of photocatalyst. Therefore, the H_2/xylose ratio was plotted against the molar ratio of xylose to catalyst (xylose/catalyst), as shown in **Figure 5B**. As the xylose/catalyst ratios decreased, the H_2/xylose ratios increased. The intercept of the plots was equaled to H_2^{max}, which is the limiting mole amount of H_2 produced from one mole of xylose (sacrificial agent) at an infinite amount of catalyst [17]. Thus, the total molar amount of H_2 was calculated by the equation: H_2^{max} × (moles of sacrificial agent).

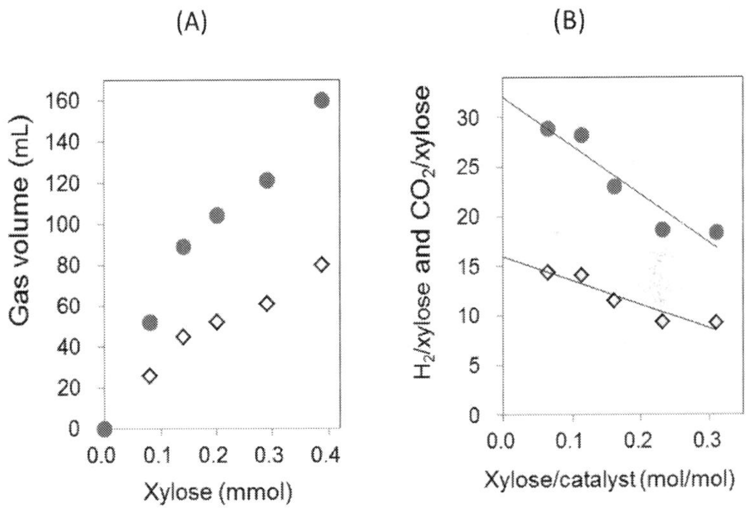

Figure 5. The TiO_2-photocatalytic H_2 production using a mixture of xylose and glucose obtained from the enzymatic saccharification of Napier grass. (A) Dependence of volumes of H_2 (●) and CO_2 (◇) against the mole of xylose. (B) Plots of H_2/xylose (●) and CO_2/xylose (◇)) against xylose/catalyst.

Similar plots of CO_2/xylose against the xylose/catalyst were performed, giving the CO_2^{max} values from the intercept of the plots. Other gasses such as methane and CO were not detected in evolved gas.

5. Energy recovery efficiency (E_{ff})

Total energy recovery efficiency (E_{ff}) from biomass to biofuels was calculated using combustion energy: $E_{ff} = 100H_F/H_0$ where H_0 and H_F were the combustion energies of biomass and biofuels, respectively. The combustion energies of sacrificial agents such as glucose, xylose, and glycerol are 2803 [18], 2342 [19], and 1654 kJ/mol [18], respectively. The combustion energies of biofuels such as ethanol and H_2 are 285 and 1367 kJ/mol [18], respectively. In the case of lignocellulose, the H_0 value was combustion energy of glucose and xylose at the complete hydrolysis of glucan and xylan which were determined by the National Renewable Energy Laboratory (NREL) [20].

6. Practical photocatalytic biomass reforming

6.1. Lignocelluloses

Lignocellulosic biomass was composed of cellulose, hemicellulose, lignin, and other components. The components of glucan, xylane, lignin, ash, and others in non-treated lignocelluloses are summarized in **Table 1**. Since the contents of cellulosic components in lignocelluloses were in the range of 41.0–66.5 wt%, only a half of lignocelluloses were utilized for production of H_2. The method to determine the content of each component was shown as follows.

Lignocelluloses	Contents (wt%)			
	Holocellulose[a]	Lignin	Ash	Others
Italian ryegrass	50.1 (35.1, 15.0)	23.5	12.9	13.5
Napier grass	48.2 (31.3, 16.9)	12.6	13.9	28.7
Bamboo	66.5 (39.5, 26.4)	26.2	1.4	5.9
Rice straw	47.8 (27.9, 19.5)	20.3	17.7	14.2
Silver grass	41.0 (30.8, 10.0)	21.7	4.0	33.3

[a]The values in parenthesis are the contents of glucan and xylan in holocellulose.

Table 1. Components of non-treated lignocelluloses.

Lignocelluloses were cut by a cutter and dried at 70°C for 72 h. The dried matter was powdered by a blender until the powder passed through a sieve with 150 μm of mesh. The powdered lignocellulose (30 g) was treated with a 1% aqueous solution of NaOH (400 mL) at 95°C for 1 h. The reaction mixture was centrifuged and filtered to isolate the holocellulose (a mixture of cellulose and hemicellulose) as a pale yellow precipitate. The supernatant solution was made acidic (pH 5.0) with a dilute HCl solution to isolate dark brown precipitate which was identified as lignin. The precipitate was collected via centrifugation at 10,000 rpm for 10 min.

The contents of saccharides in holocellulose were analyzed according to the methods published by NREL [20]. Sulfuric acid (72 wt%, 3.0 mL) was added slowly to holocellulose (300 mg) in a reaction vessel and kept at 30°C for 1 h. Water (84 mL) was added to the reaction vessel so that the concentration of sulfuric acid became 4.0 wt%. Acid hydrolysis was performed by autoclaving at 121°C for 1 h in an autoclave. The treated solution was neutralized with $CaCO_3$ and was centrifuged. The supernatant solution (ca. 87 mL) was concentrated to 30 mL by evaporation. The solution was analyzed by HPLC to determine the amounts of glucose and xylose. The amounts of glucan and xylan were determined from the amounts of glucose and xylose. The ash component in lignocellulose was obtained by the burning of the lignocellulose (2.0 g) in an electric furnace (KBF784N1, Koyo, Nara, Japan) for 2 h at 850°C.

The pre-treatments to promote an enzymatic digestibility of the cellulosic components and to remove the lignin component were usually performed. Alkali (AL) treatment is a popular method to remove lignin from lignocelluloses [21]. A powdered lignocellulose (30 g) was added to a 1% aqueous solution of NaOH (400 mL). The mixture was heated under stirring at 95°C for 1 h. The reaction mixture was subjected to centrifugation at 10,000 rpm for 10 min. The lignin remained in the supernatant solution. The holocellulose, which is a mixture of cellulose and hemicellulose, is isolated as a pale yellow precipitate, which was washed by dispersion in water to remove the contaminated lignin. After the pH adjustment to 7.0, the washed precipitate was collected by centrifugation and dried. Thus, lignin-removed holocellulose was obtained. The AL treatment is effective for saccharification of the lignocellulose with higher lignin contents. However, in the case of lignocelluloses with low lignin content such as Napier grass, the AL treatment retarded the yeast-fermentation rate because AL treatment removed not only lignin but also nutrients to help yeast fermentation [22].

Another useful pretreatment of lignocelluloses is LMAA (low-moisture anhydrous ammonia pretreatment), described as follows [23]. Dry powdered lignocelluloses (100 g, volume 320 mL) were mixed homogeneously with water (100 g) in a flask (1 L). The flask containing wet lignocellulose was evacuated with a pump and then gaseous NH_3 was introduced into the flask repeatedly until the atmosphere inside the flask was entirely replaced with NH_3 gas. The moist powdered lignocellulose was kept under an NH_3 gas atmosphere at room temperature for 28 days. After NH_3 was removed with an evaporator, the treated lignocellulose was washed with water to liberate the brownish aqueous alkali solution of the lignin. This washing operation was continued until the pH became below 7.7. The treated lignocellulose was dried at 60°C. Here, NH_3 served for transformation of the cellulose crystal phase to a highly reactive structure toward enzymatic degradation rather than the removal of lignin [24]. As a special pretreatment method, TiO_2-photocatalytic pretreatment was developed by our group [25].

The photocatalytic reforming was applied to lignocelluloses such as Italian ryegrass [26], Napier grass [26], bamboo [27], rice straw [27], and silver grass [27]. The results are summarized in **Table 2**. The SA→PR method is a process through the enzymatic saccharification (SA) of the pretreated lignocelluloses into glucose and xylose which were then used as sacrificial agents for the photocatalytic H_2 production over Pt/TiO_2 (PR). For example, the dried Italian ryegrass (2.00 g) was subjected to the AL treatment to give the AL-treated Italian ryegrass (1.00 g) which was turned into 554 mg of glucose and 193 mg of xylose by SA. The

SA of xylan was more inefficient than that of glucan. Glucose and xylose were turned into H_2 (78.7 mg) by PR. As a result, the total energy recovery efficiency (E_{ff}) from AL-treated Italian ryegrass to H_2 was calculated to be 71.9% (**Figure 6**). In the case of Napier grass, dried Napier grass (2.075 g) was subjected to the AL treatment to give the AL-treated Napier grass (1.00g) which was turned into 487 mg of glucose and 197 mg of xylose by SA. The PR of glucose and xylose gave 84.0 mg of H_2, which corresponded to 77.0% of E_{ff}.

Biomass					**Process**[a]	**Biofuels**			
Lignocellulose	**PT**[b]	W_G[c] **(mg)**	W_X[c] **(mg)**	H_0[d] **(kJ)**		**EtOH (mg)**	H_2 **(mg)**	H_F[e] **(kJ)**	E_{ff}[f] **(%)**
Italian ryegrass	LMAA	480	206	11.96	SSCF	333	0	9.90	82.7
Italian ryegrass	AL	700	300	15.58	SA→PR	0	78.7	11.20	71.9
Italian ryegrass	LMAA	480	206	11.96	SSF→PR	250	17.3	9.89	82.7
Napier grass	AL	651	350	15.60	SA→PR	0	84.0	12.0	77.0
Napier grass	LMAA	398	214	10.68	SSF→PR	177	21.0	8.25	77.2
Bamboo	AL	594	396	17.30	SSF→PR	213	44.5	12.68	73.4
Rice straw	AL	756	238	17.30	SSF→PR	364	34.8	15.76	91.1
Silver grass	AL	749	253	17.28	SSF→PR	323	35.7	14.68	85.0

[a]SSF = simultaneous saccharification and fermentation using cellulase and yeast. SA = enzymatic saccharification. PR = photocatalytic H_2 production over Pt/TiO_2. SSCF= Simultaneous saccharification and co-fermentation using cellulase, yeast, and recombinant *E. coli* KO11. Referred from reference [9].
[b]PT, pretreatment; LMAA, low moisture anhydrous ammonia pretreatment; AL, alkali pretreatment.
[c]W_G and W_X were the amounts of glucan (G) and xylan (X) per 1 g of the pretreated lignocellulose.
[d]The total combustion energies (H_0) of xylose and glucose theoretically derived from 1.0 g of the pretreated lignocelluloses were calculated according to the following equation: $H_0 = 2803 \times W_G/162 + 2342 \times W_X/132$.
[e]Total combustion energy (H_F) of biofuels (ethanol and hydrogen).
[f]Energy recovery efficiency (E_{ff}) = $100 \times H_F/H_0$.

Table 2. Biofuel production from lignocelluloses.

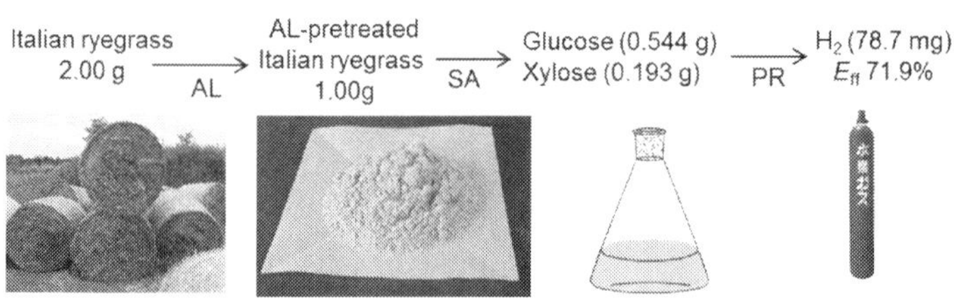

Figure 6. AL→SA→PR process of Italian ryegrass.

In the case of the SSF→PR method, the LMAA treatment of the dried Italian ryegrass (1.458 g) gave the LMAA-treated Italian ryegrass (1.0 g) which was turned into ethanol (250 mg), xylose

(121 mg), and glucose (19 mg) by SSF process. Ethanol was removed from SSF solution, whereas the residual xylose and glucose were converted to H_2 (17.3 mg) by PR. The E_{ff} value of H_2 combined with ethanol was 82.7% from the LMAA-treated Italian ryegrass. We have reported the ethanol production through an SSCF process of Italian ryegrass [9]. The E_{ff} value was 82.7%. These E_{ff} values showed similar values. In the cases of Napier grass, the LMAA treatment of the dried Napier grass (1.637 g) gave the LMAA-treated Napier grass (1.0 g) which was turned into ethanol (177 mg), xylose (167 mg), and glucose (13 mg) by SSF process. After ethanol was removed from SSF solution, the residual xylose and glucose were converted to H_2 (21.0 mg) by PR. The E_{ff} value of H_2 combined with ethanol was 77.2% from the LMAA-treated Napier grass. In the cases of bamboo, rice straw, and silver grass, the AL treatment of bamboo (1.656 g), rice straw (2.092), and silver grass (2.439 g) produced the AL-treated lignocelluloses (1.00 g). They were turned into ethanol and H_2 by the SSF→PR process with E_{ff} of over 73.4%.

6.2. Glycerol

Biodiesel (BDF) is one of new sustainable energy alternatives to petroleum-based fuels. BDF market has significantly increased in Europe to adhere energy and climate policies [28]. BDF (methyl alkanoate) is produced by transesterification of vegetable oil or animal fats with methanol under basic conditions [29]. However, glycerol as co-production and unreacted methanol was not utilized and went to waste. Glycerol has a potential to produce H_2 in maximum theoretical yield of seven equivalents (Eq. 5). Also methanol can produce three equivalents of H_2. Hydrogen transformation of glycerol and unreacted methanol isolated from the BDF synthesis was performed by sacrificial H_2 production over a Pt/TiO_2 [30].

$$C_3H_8O_{3n}\,(\text{glycerol}) + 3H_2O \xrightarrow[\text{Pt / TiO}_2]{\text{UV}} 3CO_2 + 7H_2 \qquad (5)$$

As starting material, we used vegetable oil which was mainly composed of oleic acid ($C_{17}H_{33}CO_2H$) triglyceride. The average molecular weight of vegetable oil was thought to be 884 g/mol. Vegetable oil (150 mL, 136.5 g, 0.154 mol) was set in a reaction vessel. Methanol (30 mL, 23.8 g, 0.743 mol) was mixed with NaOH (0.485g, 0.012 mol). About half of the mixture of methanol and NaOH was poured into a reaction vessel and then kept at 61°C for 1 h. Moreover, the remaining mixture of methanol and NaOH was added into the reaction vessel and the reaction mixture was kept at 61°C for another 1 h. After cooling, the reaction mixtures were separated into a lower layer and an upper layer. The procedure of the follow-up process is shown in **Figure 7**. The lower layer (GL layer) contained glycerol (GL, 0.113 mol) and methanol (0.214 mol). The upper layer (BDF layer) was washed with water (300 mL) to give BDF (114.5 g, 0.387 mol) and the aqueous washing solution which contained 0.137 mol of methanol. The total recovery yield of unreacted methanol was 47.5%. The yields of GL and BDF were 73.3 and 83.7%, respectively.

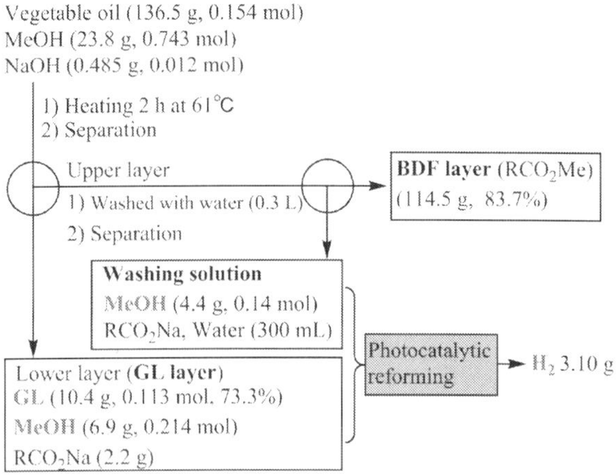

Figure 7. Outline for preparation of BDF and the follow-up process.

The photocatalytic reaction was performed by irradiation of aqueous solution containing Pt/TiO_2 powder (100 mg, 1.25 mmol) and GL layer, which was added to the reaction vessel so that the amounts of GL became 0.25, 0.50, 0.75, 1.00, and 1.25 mmol. The limiting mole amount of H_2 (H_2^{max}) per 1 mol of GL was obtained from the plots of the H_2/GL against the GL/catalyst. Similarly the photocatalytic reaction was performed for the washing solution, which contained methanol. Using H_2^{max} values, it was calculated that 2.82 and 0.28 g of H_2 was obtained from the GL layer and washing solution, respectively. The E_{ff} value of H_2 was determined to be 100.8% using H_F of H_2 (444 kJ) and the sum of combustion energy of glycerol (H_0 = 187 kJ) and unreacted methanol (H_0 = 255 kJ).

6.3. Chlorella

Chlorella is single-cell green algae with 2–10 μm diameter and multiplies rapidly, requiring only carbon dioxide, water, sunlight, and a small amount of minerals [31]. Chlorella is mostly composed of proteins (45%), lipids (20%), saccharides (20%), and minerals (10%). Thus, the content of saccharides is low, suggesting that ethanol production is inefficient.

We examined the photocatalytic H_2 production from Chlorella [32]. The frozen Chlorella was thawed and dried in a drying machine and then ground. Gas evolution did not occur from the non–enzymatic-treated solution, which was prepared by magnetic stirring of the Chlorella powder (10 g) in a phosphate buffer (60 mL) for 48 h at 50°C. Therefore, the enzymatic hydrolysis of Chlorella powder (10 g) was performed using protease (1.0 g) in a phosphate buffer (0.1 M, pH 7.6, 60 mL) under stirring at 50°C for 48 h to give the enzymatic hydrolyzed solution. The solution was subjected to centrifugation to remove the precipitate. The supernatant solution (EH solution) was collected. The EH solution was subjected to freezing-drying in order to weigh the water-soluble components in the EH solution. It was determined to be 117 g/L. Since the weight of the solid was 167 g/L before hydrolysis, more than 70% of the solid

http://dx.doi.org/10.5772/64901

was hydrolyzed into water-soluble components. The EH solution was composed of 98.0 g/L of amino acids and 18.3 g/L of glucose which were determined by colorimetric analysis using ninhydrin and by HPLC analysis, respectively.

The photocatalytic H_2 production was performed using the EH solution (0.10 – 0.50 mL) over a Pt/TiO_2 (100 mg) in 150 mL of water. The limiting volume of H_2 per 1 mL of the EH solution (H_2^{max}) was determined to be 119 mL/mL from the plots of the H_2/(EH solution) against the (EH solution)/catalyst. We successfully produced 579 mg of H_2 from 10.0 g of dry Chlorella (**Figure 8**). This yield is higher than 394 mg for the H_2 production through AL treatment, saccharification, and photocatalytic H_2 production from non-treated Italian ryegrass (10.0 g) [25, 27]. Thus, the photocatalytic reforming is applicable to not only saccharides but also amino acids.

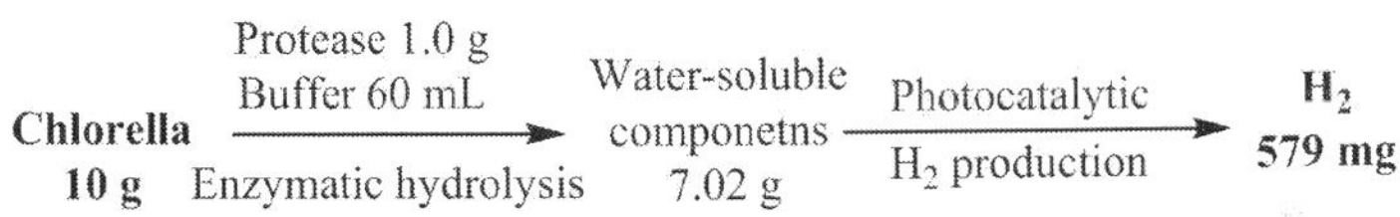

Figure 8. Mass balance for the H_2 production from Chlorella.

Chlorella includes colored materials such as chlorophyll which may disturb the light absorption by the catalyst. Therefore, dried Chlorella (20 g) was subjected to refluxing in ethanol (100 mL) for 6 h to remove the colored materials. Almost all amount of colored materials remained in the ethanol solution. However, the decolorization did not affect the amount of H_2 but could shorten the irradiation time.

7. Conclusions

We examined photocatalytic H_2 production using sacrificial saccharides, glycerol, and amino acid derived from lignocelluloses, lipids, and Chlorella. As a conclusion, the photocatalytic reforming of biomass has the following features:

1. The photocatalytic reforming can be performed in aqueous solution as well as in biological treatment.

2. Gaseous H_2 can spontaneously isolate from aqueous reaction mixtures without operations to be separated.

3. Although it is not easy to produce ethanol from saccharides other than glucose since they are not fermented by yeast (*S. cerevisiae*), sacrificial hydrogen production is applicable to a variety of water-soluble materials.

4. The photocatalytic reforming of biomass is one of the promising approaches because biomass is abundant, clean, and renewable. If sacrificial H_2 evolution is practically accomplished by the use of solar radiation, this method will provide new ways to produce sustainable energy.

In this chapter, biohydrogen production was discussed from the viewpoints of feedstock and methodology to transform biomass to fuels. This will help life recycle assessment (LCA) to evaluate CO_2 emission during cultivation, transportation, and manufacturing, as performed for bioethanol from cellulose [33].

Acknowledgements

This study was supported by a Grant-in-Aid for Scientific Research (C) No 24610055 from the Ministry of Education, Culture, Sports, Science, and Technology of Japan.

Author details

Masahide Yasuda

Address all correspondence to: yasuda@cc.miyazaki-u.ac.jp

Department of Applied Chemistry, Faculty of Engineering, University of Miyazaki, Gakuen-Kibanadai Nishi, Miyazaki, Japan

References

[1] Ward OP, Singh A. Bioethanol technology: Development and perspectives. Adv. Appl Microbiol . 2002;51:53–80.

[2] Öhgren K, Rudolf A, Galbe M, Zacchi G. Fuel ethanol production from steam-pretreated corn stover using SSF at higher dry matter content. Biomass Bioenergy. 2006;30:863–869.

[3] Balat M, Balat H. Recent trends in global production and utilization of bio-ethanol fuel. Appl Energy. 2009;86:2273–2282.

[4] Yasuda M, Ishii Y, Ohta K. Napiergrass (*Pennisetum purpureum* Schumach) as raw material for bioethanol production: Pretreatment, saccharification, and fermentation. Biotechnol Bioprocess Engineer. 2014;19:943–950.

[5] Ni M, Leung DY, Leung MK, Sumathy K. An overview of hydrogen production from biomass. Fuel Process Technol. 2006;87:461–472.

[6] Shimura K, Yoshida H. Heterogeneous photocatalytic hydrogen production from water and biomass derivatives. Energy Environ Sci. 2011;4:467–481.

[7] Yasuda M, Miura A, Shiragami T, Matsumoto J, Kamei I, Ishii Y, Ohta K. Ethanol production from non-pretreated napiergrass through a simultaneous saccharification

and fermentation process followed by a pentose fermentation with *Escherichia coli* KO11. J Biosci Bioeng. 2012;114:188–192.

[8] Yasuda M, Nagai H, Takeo K, Ishii Y, Ohta K. Bio-ethanol production through simultaneous saccharification and co-fermentation (SSCF) of a low-moisture anhydrous ammonia (LMAA)-pretreated napiergrass (*Pennisetum purpureum* Schumach). Springer Plus. 2014;3:333.

[9] Yasuda M, Takenouchi Y, Nitta Y, Ishii Y, Ohta K. Italian ryegrass (*Lolium multiflorum* Lam) as a high potential bio-ethanol resource. BioEnerg Res. 2015;8:1303–1309.

[10] Fujishima A, Rao TN, Tryk DA . Titanium dioxide photocatalysis. J Photochem Photobiol C Rev. 2000;1:1–21.

[11] Galinska A, Walendziewski J. Photocatalytic water splitting over Pt-TiO_2 in the presence of sacrificial agents. Energy Fuels. 2005;19:1143–1147 .

[12] Shiragami T, Tomo T, Matsumoto T, Yasuda M. Structural dependence of alcoholic sacrificial agents on TiO_2-photocatalytic hydrogen evolution. Bull Chem Soc Jpn. 2013;86:382–389.

[13] Yasuda M, Tomo T, Hirata S, Shiragami T, Matsumoto T. Neighboring hetero-atom assistance of sacrificial amines to hydrogen evolution using Pt-loaded TiO_2-photocatalyst. Catalysts. 2014;4:162–173.

[14] Kennedy III JC, Datye AK. Photochemical heterogeneous oxidation of ethanol over Pt/TiO_2. J Catal. 1998;179:375–389.

[15] Yasuda M, Kurogi R, Matsumoto T. Quantum yields for sacrificial hydrogen generation from saccharides over a Pt-loaded TiO_2 photocatalyst. Res Chem Interm. 2015;8:1303–1309.

[16] Shiragami T, Tomo T, Tsumagari H, Yuki R, Yamashita T, Yasuda M. Pentose acting as a sacrificial multi-electron source in photocatalytic hydrogen evolution from water by Pt-doped TiO_2. Chem Lett. 2012;41:29–31.

[17] Shiragami T, Tomo T, Tsumagari H, Ishii Y, Yasuda M. Hydrogen evolution from napiergrass by the combination of biological treatment and a Pt-loaded TiO_2-photocatalytic reaction. Catalysts. 2012; 2:56–67.

[18] Atkins PW. Physical Chemistry, 5th ed. Oxford University Press: Oxford, United Kingdom, 1994:922–926

[19] Ribeiro Da Silva MAV, Ribeiro Da Silva MDMC, Lobo Ferreira AIMC, Shi Q, Woodfield BF, Goldberg RN. Thermochemistry of α-D-xylose(cr). J Chem Thermodyn. 2013;58:20–28.

[20] Sluiter A, Hames B, Ruiz R, Scarlata C, Sluiter J, Templaton D, Crocker D (2010). Determination of structural carbohydrates and lignin in biomass, Technical Report NREL/TP-510-42618, National Renewable Energy Laboratory, Golden, CO.

[21] Silverstein RA, Chen T, Sharma-Shivappa RR, Boyette MD, Osborne J. A comparison of chemical pretreatment methods for improving saccharification of cotton stalks. Bioresour Technol. 2007;98:3000–3011.

[22] Yasuda M, Takeo K, Matsumoto T, Shiragami T, Matsushita Y, Sugamoto K, Ishii Y (2013). Effectiveness of lignin-removal in simultaneous saccharification and fermentation of napiergrass, rice straw, silvergrass, and bamboo with different lignin-contents. Sustainable Degradation of Lignocellulosic Biomass—Techniques, Applications and Commercialization, Edited by A. K. Chandel and S. Silverio da Silva, InTech, Croatia.

[23] Yasuda M, Takeo K, Nagai H, Uto T, Yui T, Matsumoto T, Ishii Y, Ohta K. Enhancement of ethanol production from napiergrass (*Pennisetum purpureum* Schumach) by a low-moisture anhydrous ammonia pretreatment. J Sustain Bioenergy Syst. 2013;3:179–185.

[24] Igarashi K, Wada M, Samejima M. Activation of crystalline cellulose to cellulose IIII results in efficient hydrolysis by cellobiohydrolase. FEBS J. 2007;274:1785–1792.

[25] Yasuda M, Miura A, Yuki R, Nakamura Y, Shiragami T, Ishii Y, Yokoi H. The effect of TiO_2-photocatalytic pretreatment on the biological production of ethanol from lignocelluloses. J Photochem Photobiol. A: Chem. 2011;220 :195–199.

[26] Yasuda M, Misriyani, Takenouchi Y, Kurogi R, Uehara S, Shiragami T. Fuelization of Italian ryegrass and Napier grass through a biological treatment and photocatalytic reforming. J Sustain Bioenergy Syst. 2015;5:1–9.

[27] Yasuda M, Kurogi R, Tsumagari H, Shiragami T, Matsumoto T. New approach to fuelization of herbaceous lignocelluloses through simultaneous saccharification and fermentation followed by photocatalytic reforming. Energies. 2014;7:4087–4097.

[28] Onwudili JA, Williams PT. Catalytic pyrolysis of low-density polyethylene over alumina-supported noble metal catalysts. Fuel. 2010;89:501–509.

[29] Ma F, Hanna MA. Biodiesel production: A review. Bioresour Technol. 1999;70:1–15.

[30] Yasuda M, Kurogi R, Tomo T, Shiragami T. Hydrogen production from residual glycerol from biodiesel synthesis by photocatalytic reforming. J Jpn Inst Energy. 2014;93:710–715.

[31] Handbook of algae. Their diversity and utilization. Ed. by M. Watanabe, NTS press, Tokyo (2012).

[32] Yasuda M, Hirata S, Matsumoto T. Sacrificial hydrogen production from enzymatic hydrolyzed Chlorella over a Pt-loaded TiO_2 photocatalyst. J Jpn Inst Energy. 2016;95:599–604

[33] Fu GZ, Chan AW, Minns DE. Life cycle assessment of bio-ethanol derived from cellulose. Int J LCA. 2003;8:137–141.

Chapter 20

Renewable Hydrocarbons from Triglyceride's Thermal Cracking

Vinicyus R. Wiggers, Ramon F. Beims, Laércio Ender,
Edésio L. Simionatto and Henry F. Meier

Additional information is available at the end of the chapter

http://dx.doi.org/10.5772/65498

Abstract

This chapter gives an overview of renewable hydrocarbon production through triglyceride's thermal-cracking process. The influence of feedstock characteristics and availability is discussed. It also presents issues about the reaction, the effect of operational conditions, and catalysts. A scheme of the reaction is presented and discussed. The composition and properties of bio-oil is presented for both thermal and catalytic cracking. The high content of olefins and the high acid index are drawbacks that require downstream processes. The reactor design, kinetics, and scale-up are opportunities for future studies. However, the similarity of bio-oil with oil turns this process attractive.

Keywords: waste fatty acids, triglyceride, pyrolysis, biofuels, green chemicals

1. Introduction

Nowadays, the search for processes that aims to reduce the use and the dependence of fossil fuels is imperative. Decrease in the emission of greenhouse gases might be a global effort. In this way, the biomass appears to be the logical choice to produce solid, liquid, and gaseous fuels, once it is abundant and available all over the world [1]. There are many technological processes applied to different kinds of biomass being studied and proposed by scientific community [2]. One thing is for sure, there will not be only one technology that will solve all the issues, but different technological routes taking into account the specific characteristics of the source region and the feedstock.

Besides the fact that these new technologies to produce biofuels must be environmentally friendly, they are facing some obstacles to overcome economic and technical viability, high scale, and stable production. Specifically on liquid biofuels, another technical barrier is the fact that almost every machine and vehicle was designed for fossil fuels usage. These fossil fuels have several regulations and quality parameters that must be attended for commercialization. In this way, it is a "sine-qua-non" condition that the new generation on liquid biofuels shall be compatible with actual standard of the engines. The modern electronic fuel injection systems make possible the use of different fuels maintaining a good combustion in the engine. However, how higher the similarity of the biofuel with fossil fuels, higher is the possibility for its commercial application. In this way, the organic liquid product produced by thermal cracking of vegetable oils and waste fats appears with high potential of oil substitute in the refineries [3].

The objective of this chapter is to provide a brief overview about the thermal conversion of triglycerides into a liquid fraction, called bio-oil, rich in hydrocarbons, presenting its properties.

2. Thermal cracking of triglycerides

The production of bio-oil through thermal cracking of biomass is easily found in literature [4–32]. The bio-oil is defined as a dark brown viscous corrosive fuel obtained from biomass pyrolysis [33], but it is very important to highlight that the bio-oil has different properties according to the feedstock. If it is produced from lignocellulosic materials, the bio-oil has significant amount of water and oxygen content, decreasing its gross calorific value and its stability [34, 35]. On the other hand, if the feedstock is triglycerides, the oxygen and water content is low and the high heating value is comparable to the fossil fuels [6, 36]. Another important characteristic is the similar chemical composition, based on hydrocarbons [37]. So, based on these issues, the bio-oil produced from triglycerides appears like one of the most promising technologies for biofuels production [38].

2.1. Feedstock

The triacylglycerol, also known as triglyceride (TAG) is an ester derived from glycerol and three fatty acids [38]. It can be found in edible and nonedible vegetable oils, animal fats, and used oils. The most abundant vegetable oils are soybean, palm, canola, sunflower, rapeseed, among others. From animals, the main sources are pork lard, poultry fat, fish oil, and beef tallow [39]. Waste greases or tap greases are found in cooking oils and sewage scum [40].

In general, they have similar physical properties and chemical structure. They differ in the composition of the fatty acids, in the acidity and content of saturated fatty acids [39]. The acidity of the oil is evaluated through the acid index determination (ASTM D974) which gives the free fatty acids (FFAs) content in the oil. Waste oils are classified in yellow and brown greases according to the content of FFA. Oils with lower than 15% (w/w) are classified as yellow greases, while if it has more than 15% it is brown greases.

The iodine index (pr EN 14111) provides the number of double bonds in the fatty acids. Oils with high content of unsaturated acids are liquid in ambient conditions; however, oils with high content of saturated acids are solid or semisolid in the same conditions.

The fatty acid composition is provided by the fatty acid methyl ester (FAME) determination [41]. It is a chromatographic analysis, which is a well-accepted method for its determination. The fatty acids composition of various TAGs can be found in the literature [34, 41, 42].

One fact that must be pondered over, when one talks about biofuels production using TAGs, is the feedstock availability [42]. In this way, we have two subjects to consider: the use of edible oils and the logistic to join the wasted ones.

In the first case, we need to consider the food versus fuels issues. The main concern is based on the assumption that biofuel feedstocks tend to be more profitable than food feedstocks, which may lead to food shortages. Thus, it must be carefully pondered in order to efficiently attend both markets [43].

In the second case, it is possible to consider the waste-cooking oils, the animal fats, and the sewage scum. From cooking oils, its generation varies to each country, as it depends on the vegetable oil consumption. The estimated generation in the European Union (EU) is about 700,000–1,000,000 tons/year [44]. Only the UK generates an amount of approximately 250,000 tons per year [45]. Canada produces around 135,000 tons of yellow grease every year [46]. Mexico's generation is about 840,000 tons every year, similar to Malaysia. Japan produces around 450,000–570,000 tons/year [47]. Hong Kong generates approximately 20,000 tons/year [48]. The USA's generation is about 1,000,000 tons/year [47]. Even so, it is estimated that the general worldwide generation is around 4.1 kg per habitant per year [49].

Animal fats availability is also related to the region. It is well known that China, the USA, and Brazil are large producers of meat. Only in 2013, the US industry processed 180,000 tons of meat and poultry [50]. The fish industry also plays an important role. In 2014, the world fish production was about 146 million tons of fish [51]. As the amount of oil ranges from 40 to 65% [52], it represents around 70.8 million tons of waste fish oil.

Thus, these numbers show that it is possible to use biofuels production as a final destination to these wastes. It is important to highlight the complex logistic to use it and that these amounts will not replace the oil, but they can be a viable alternative.

2.2. Process and reaction

The thermal-cracking reaction is defined as thermal decomposition of the organic chains by heat in an atmosphere free of oxygen, with or without the aid of a catalyst. **Figure 1** presents a basic scheme of the triglycerides thermal-cracking process. As one can see in the scheme, the reaction will generate always a solid fraction, generally called coke, a liquid product named as bio-oil, and a gaseous stream known as biogas.

This reaction is affected by the feedstock characteristics and the pair temperature-residence time [34]. The higher the temperature and the residence time, the higher the yield of the gas product. Lower temperatures and higher residence times improve the coke formation.

Moderate temperatures with short residence times yield the liquid product. This last operational condition is called fast pyrolysis [5]. The fast pyrolysis process is gaining attention due to the possibility to obtain high amounts of bio-oil, which can be used as fuel. **Figure 1** shows that independent of the operational conditions, the solid fraction called coke will appear, and this product will not be easily removed from the reactor. In general, this product formation is associated with clogging [53]. One possibility to remove it is to proceed a controlled burning in the heated reactor through feeding air instead of biomass, for a certain period of time, promoting the combustion of the coke.

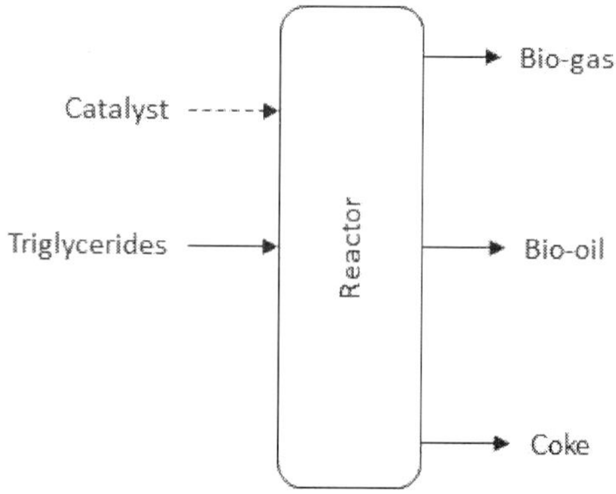

Figure 1. A general scheme of thermal-cracking process.

The reactor design is the heart of the process [54]. Different configurations have been proposed in the literature for several researches. It is possible to find batch [9, 10, 12, 16, 21, 22, 24, 31, 32] and continuous configurations [4–8, 11, 13, 17–20, 23, 25–27, 55]. In general, the batch reactors are used to evaluate the reaction mechanism, kinetics, yields, and chemical characterization. As it works with lower capacities, they are not appropriated for industrial applications. The continuous ones are in a higher sizes, bench or pilot, testing different reactor designs and operational conditions, evaluating the kinetics, yields, characterization, energy consumption, and economic evaluation, aiming the scale-up studies [26].

The irreversible reaction is highly endothermic and requires high heat transfer rates. The possibility to run the process in an autothermal operation promotes an advantage over other processes. This condition can be reached burning a fraction of the products to produce the thermal energy required for the reaction. An energy balance of the TAGs thermal cracking was presented by [5].

Due to the complexity of the organic reactions, there is no complete knowledge about all the reactions involved, just proposals for the principal ones. A simplified reaction step for the thermal cracking of triglycerides is presented in **Figure 2**. The reaction starts with the decomposition of the triglyceride molecule forming heavy oxygenated hydrocarbons. Some of the saturated fatty acids formed may not suffer any subsequent breaking. The decarboxylation and decarbonylation reactions (2) are favored by unsaturations and compete with the C-C bond cleavage reaction (3). The CO and CO_2 are formed by the deoxygenation reactions in (2) and (4). The isomerization, polymerization, dehydrogenation, and cyclization are responsible for dienes, acetylenes, cycloparaffins, and polyolefins (5). The Diels-Alder addition of dienes to olefins also produce cyclo-olefins (8) resulting in hydrogen formation. The hydrogenation of cyclo-olefins to cycloparaffins and the reverse reaction occurs in steps (6) and (7). Hydrogen also comes from steps (9) and (10). The solid product coke is produced directly from triglyceride (12), by the polycondensation of heavy hydrocarbons and saturated fatty acids (11) and aromatics (10). The polymerization of olefins can also lead to coke (13). Considering the reaction scheme in **Figure 2**, it is very important to advance the cracking at least to the point which deoxygenation reaction occurs, eliminating the oxygen by CO and CO2. It is also important avoid coke formation (steps 10 and 13 in the **Figure 2**). As a first conclusion, for thermal cracking, the temperature-residence time is the key factor for this process.

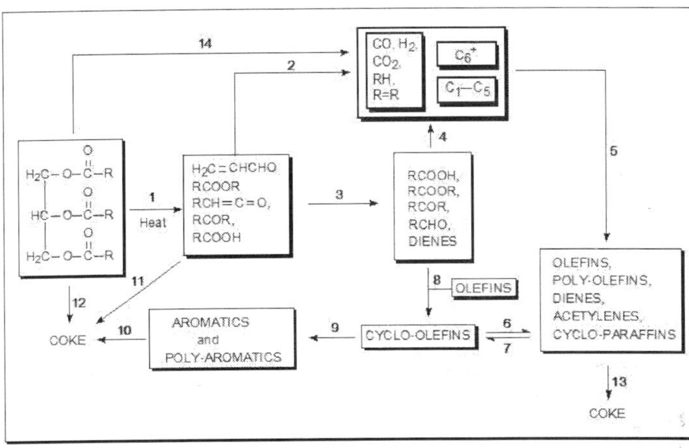

Figure 2. Proposed reaction scheme for the thermal cracking of vegetable oil and animal fats (triglyceride). Adapted from [13, 26, 38, 56]. (1) Initial cracking, thermolysis of triglyceride molecule ester bond; (2) decarboxylation/decarbonylation of long-chain oxygenated hydrocarbons; (3) C-C bond cleavage of unsaturated oxygenated hydrocarbons; (4) decarboxylation/decarbonylation of short-chain oxygenated hydrocarbons; (5) isomerization, polymerization/dehydrogenation, cyclization to form dienes, acetylenes, cycloparaffins, and polyolefins; (6) dehydrogenations of cycloparaffins to form cyclo-olefins; (7) hydrogenations of cyclo-olefins to form cycloparaffins; (8) Diels-Alder addition of dienes to olefins to form cyclo-olefins; (9) aromatization of cyclo-olefins to form aromatics and polyaromatics hydrocarbons; (10) Coking from polyaromatics; (11) coking by polycondensation of oxygenated hydrocarbons; (12) coking by polycondensation of triglyceride molecule; (13) polymerization of olefins to form coke; (14) direct route for C1-C5 hydrocarbon formation from triglyceride molecule.

The use of catalysts aims to aid the reaction and increases the products' quality [57]. As the composition of the products may vary due to catalyst material, size, and shape [58], several works evaluate the use of many types of catalysts. **Table 1** shows the different catalysts used for the cracking of triglycerides. One of the concerns involving catalysts use relies on their stability and reutilization, which directly affect the cost of the process [31]. In general, the coke formation limits the use of heterogeneous catalysts, due to the deactivation, and this phenomenon requires a regeneration process for its reuse, making the entire process for the conversion complex. A scheme reaction for catalytic cracking was proposed by [59].

Catalyst	Reference
Fe-ZSM-5, H-Beta	[9]
H-ZSM-5	[8, 9, 17, 32, 55]
K_2CO_3	[11, 17, 18, 31, 32]
Na_2CO_3	[11, 17, 18, 21, 23, 31, 32]
NaY	[17, 31]
USY	[17]
Si-MCM-41	[17, 31, 32]
Alumina	[20]
ZSM-5, Ni/ZSM-5, Ni/h-ZSM-5 (12)	[8]
Al_2O_3, MCM-41	[11, 15]
SAPO-5, SAPO-11, MgAPO-36	[7]
Silicalite, silica, y-alumina, silica-alumina	[55]
Calcium oxide, magnesium oxide	[55]
CaO	[15, 18]
NaOH, Fe_2O_3	[18]
KOH	[18, 22]
ZnO	[18, 22]
CO_3O_4, MoO_3, NiO, V_2O_5	[22]
Metallic oxides	[25]
Zeolite REY	[27]
K_2O/Si-MCM-41, Mg-MCM-41	[31]
K_2O/Mg-MCM-41	[31, 32]
Ba-MCM-41	[32]

Table 1. Main catalysts used.

2.3. Yields, properties and characterization

The yields of the products are strongly affected by the operational conditions. **Table 2** shows the range of temperature and residence time applied in published papers, presenting the average product yields obtained in thermal [4–6, 9, 10, 13–15, 17, 18, 22, 24, 26, 31, 55] and catalytic cracking [7, 9, 11, 15, 17, 18, 20, 21–23, 25, 27, 31, 55]. In thermal-cracking processes, the temperature range is higher than catalytic. One can also note that the yield of liquid and gas products tends to be a little higher in thermal cracking. On the other hand, the coke formation is more favorable in the catalytic cracking.

Triglyceride's cracking		
Temperature range (°C)		
	Thermal	**Catalytic**
Max	600	550
Min	300	320
Residence time (s)		
	Thermal	**Catalytic**
Max	1800	1800
Min	1	10
Yields average (%)		
Liquid	63.20 ± 16.45	56.67 ± 20.55
Gas	28.77 ± 21.06	26.19 ± 15.68
Coke	8.22 ± 7.27	15.39 ± 13.01

Table 2. Average products yielding obtained with thermal and catalytic cracking of triglycerides.

The liquid fuels have fundamental importance in final energy consumption, especially due to its energy density. So, in this way, most of the researches are being conducted in the way to maximize the organic liquid product. No less important are the properties and the characterization of this product. **Table 3** presents average properties of the bio-oil presented in the literature for thermal [4, 5, 9, 10, 12–15, 18, 19, 22, 24, 31, 55] and catalytic cracking [7–9, 11, 15, 17, 18, 20–22, 55]. The elementary chemical composition for bio-oil does not vary so much and the sulfur content is low. The high heating value (HHV) is also comparable to the fossil fuels. The acidity of the bio-oil is higher for the thermal cracking compared to catalytic, but in both cases, the bio-oil requires a reduction in this property for processing and usage. The esterification reaction and reactive distillation were performed by [11] and [60] to reduce the acid index.

The content of olefins in the liquid can also be problematic, once its content is associated with poor stability, which may lead to gum or insoluble materials formation. To saturate the double bonds, the hydrorefining process can be applied [61]. The direct hydrocracking also can be an option [62–64].

Figure 3 presents typical chromatograms from two samples of bio-oil produced through fast pyrolysis of soybean oil and waste-cooking oil. For comparison, the chromatograms of an n-alkane sample and an oil sample are shown together. The samples were injected at the same conditions. The oil and bio-oil samples are complex mixtures containing hundreds of compounds and this turns difficult to determine the complete composition and physicochemical properties.

Properties of bio-oil		
	Thermal	Catalytic
Variable		
Carbon (%)	75.15 ± 4.43	79.96 ± 5.58
Hydrogen (%)	11.46 ± 0.68	12.20 ± 0.85
Nitrogen (%)	0.29 ± 0.59	1.83 ± 0.50
Sulfur (%)	0.02 ± 0.03	0.24 ± 0.20
Oxygen (%)	13.07 ± 5.34	9.78 ± 4.96
Ash (%)	0.51 ± 0.70	0.02 ± 0.02
HHV (MJ/kg)	33.38 ± 15.34	40.75 ± 2.38
Density (kg/m^3)	865.98 ± 22.87	858.25 ± 22.69
Water content (%)	1.39 ± 1.00	2.29 ± 1.48
Acid index (mg KOH/g)	132.08 ± 35.56	59.46 ± 26.74
Iodine index (chI$_2$/g)	64.00	-
Hydrocarbon groups		
Aliphatic (%)	3.70 ± 0.88	10.87 ± 6.46
Aromatic (%)	38.99 ± 15.41	36.73 ± 18.16
Oxygenated (%)	4.83 ± 0.74	14.86 ± 8.49
Unknown (%)	50.72 ± 14.96	29.54 ± 17.81

Table 3. Average properties of bio-oil.

One way of characterizing these liquid fuels is the distillation curve, used to plot the true boiling point (TBP) versus distilled volume fraction. In general, a simple distillation is performed according to ASTM D86 and ASTM D1160 methods and data obtained are converted to TBP according to correlations outlined in [65]. Process simulators also can be used for this conversion and to predict the thermophysical properties of the oil and its fractions [66]. The bio-oil characterization using distillation curves applying the oil correlations was presented by [34]. The authors showed that it is possible to use this method, but it requires more studies to confirm the results.

A chemical characterization was performed by [37] in the distilled fractions of the bio-oil produced by [4]. The purified products, light bio-oil and heavy bio-oil, were obtained in the range of the gasoline and diesel oil, respectively. The detailed hydrocarbon analysis (DHA) performed in light fraction showed that it was composed by aromatics (16.86%), i-paraffins (8.31%), naphthenes (6.07%), olefins (26.56%), paraffins (4.48%), C14+ (5.3%), oxygenates (0.06%), and unclassified (32.38%). The main composition of heavy bio-oil was formed by olefins, aromatics, and carboxylic acid residues. In a continuation of the study [60], samples of the bio-oil were submitted to a reactive distillation process to produce light and heavy bio-oil cuts, with lower acid index.

http://dx.doi.org/10.5772/65498

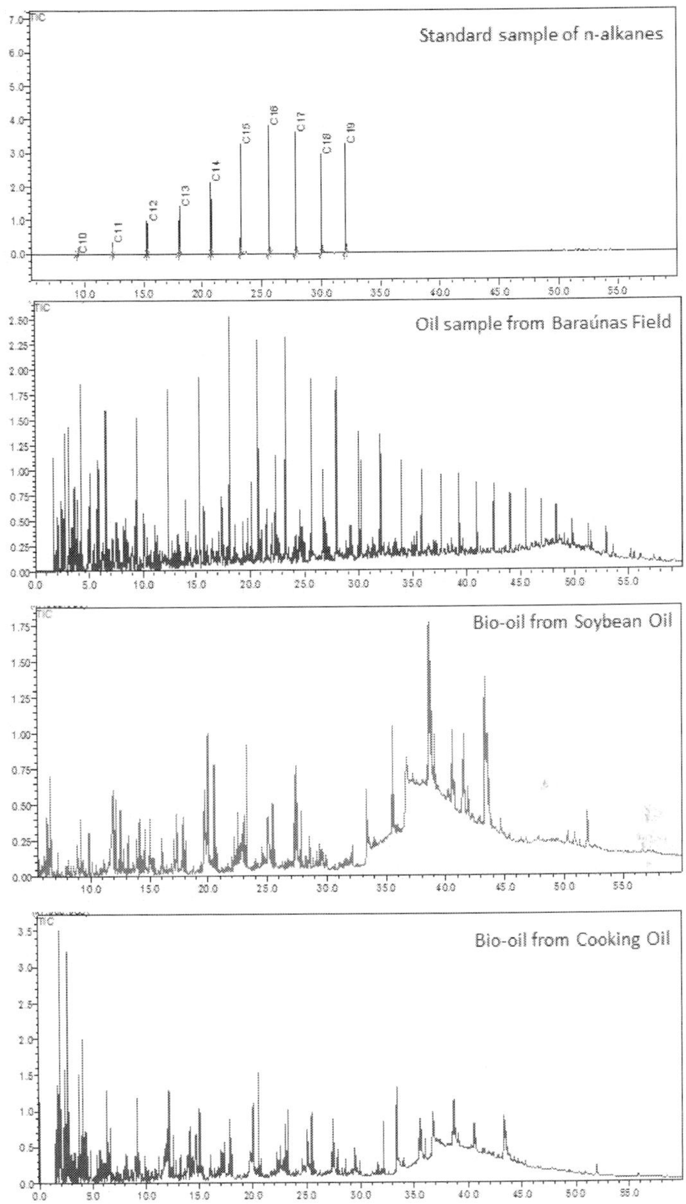

Figure 3. GC-FID chromatogram of n-alkanes sample, an oil sample, bio-oil from soybean oil, and a bio-oil from waste-cooking oil.

The gaseous products have great importance as liquids, once it has short hydrocarbons and a high HHV and it can be fuel source for the thermal energy required by the endothermic reaction. **Table 4** presents the average composition of biogas from thermal [5, 10, 13, 55] and catalytic cracking [7, 8, 17, 23, 55]. Using this average composition, the HHV is estimated in 46.6 MJ/kg (thermal cracking) and 46.3 MJ/kg (catalytic cracking). The high content of ethene also makes this product interesting for petrochemical industries.

Biogas (v/v %)		
Component	**Thermal**	**Catalytic**
CO	4.47 ± 3.58	6.02 ± 8.15
CO_2	4.15 ± 2.74	4.42 ± 7.61
H_2	1.88 ± 1.24	3.88 ± 5.68
CH_4	13.40 ± 5.34	6.31 ± 4.61
C_2H_4	29.32 ± 3.12	12.69 ± 10.19
C_2H_6	9.64 ± 1.02	4.89 ± 2.91
C_3H_8	2.82 ± 2.53	6.89 ± 7.52
C_4H_{10}	10.16 ± 1.73	5.99 ± 11.55

Table 4. Average composition of the biogas produced from thermal and catalytic cracking.

2.4. Kinetics

One of the technical difficulties to scale up the process is the determination of the reaction kinetics. Once the process has hundreds, maybe thousands of reactions, it is very difficult to determine an accurate kinetic mechanism. In these cases, the first step is to use the lumping method to propose simplified mechanisms. The lumping strategy consists in join groups of products according to some similar property, the boiling range, for example. The works of [67–69] presented the first kinetic lumped models for TAG's thermal cracking. **Table 5** shows the kinetic models proposed in the literature. The model proposed by [67] is simpler than the other models once it has fewer lumps, but it can predict the solid fraction. The study of the kinetic of cracking of TAGs is increasing and soon more models shall appear.

2.5. Challenges

The continuous availability of the feedstock is an issue that requires a complex logistic to solve the high-scale collection. In certain regions, staying close to animal-rendering facilities can be an option [70].

The industrial application of the thermal/catalytic-cracking technology has some obstacles to overcome [71]. The first is related to reactor design and scale-up. With the improvement of the kinetics, the simulation using computational fluid dynamics shall help to deal with this issue. A short work presented by [72] deals with the simulation of TAG's thermal-cracking reactor aiming scale-up studies.

Kinetic-lumped models								
Mechanism	**Kinetic parameters**				**Arrhenius parameters**			**Reference**
	Constant rate (min^{-1})	**500°C**	**525°C**	**550°C**	**Activation energy (J/mol)**	**Frequency factor (min^{-1})**		
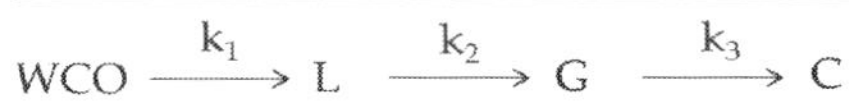	k_1	1.900E-02	2.810E-02	6.880E-02	5.84E+04	k_{10}	2.20E+04	[67]
	k_2	1.900E-02	2.810E-02	5.880E-02	5.14E +04	k_{20}	4.14E+03	
	k_3	9.070E-03	7.150E-03	3.380E-03	4.47E+04	k_{30}	4.73E+06	
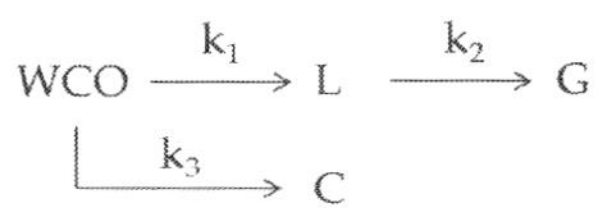	k_1	1.780E-02	3.680E-02	4.740E-02	4.51E+04	k_{10}	9.79E+02	[67]
	k_2	5.670E-03	5.970E-03	7.810E-03	1.45E+04	k_{20}	5.62E+00	
	k_3	1.730E-03	2.740E-03	3.050E-03	2.62E+04	k_{30}	1.01E+00	
	Constant rate (s^{-1})	475°C	500°C	525°C	Activation energy (J/mol)	Frequency factor (s^{-1})	Reference	
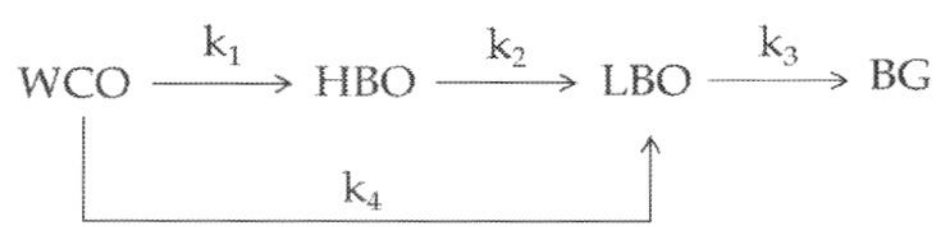	k_1	2.681E-02	2.783E-02	4.441E-02	4.96E+04	k_{10}	7.27E+01	[68]
	k_2	1.410E-02	1.426E-02	1.513E-02	6.97E+03	k_{20}	4.29E-02	
	k_3	9.580E-03	4.400E-03	1.037E-02	6.17E+03	k_{30}	1.98E-02	
	k_4	2.697E-02	3.906E-02	8.035E-02	1.08E+05	k_{40}	8.76E+05	
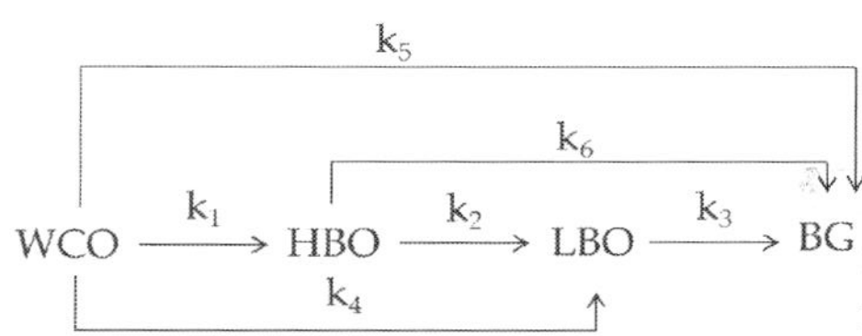	k_1	1.831E-02	3.342E-02	5.874E-02	1.16E+05	k_{10}	2.20E+06	[69]
	k_2	4.425E-03	8.504E-03	1.569E-03	1.26E+05	k_{20}	2.63E+06	
	k_3	1.058E-03	2.175E-03	4.274E-03	1.38E+05	k_{30}	5.05E+06	
	k_4	2.418E-02	4.374E-02	7.624E-02	1.14E+05	k_{40}	2.20E+06	
	k_5	2.473E-03	4.816E-03	8.994E-03	1.28E+05	k_{50}	2.20E+06	
	k_6	1.952E-03	3.852E-03	7.285E-03	1.31E+05	k_{60}	2.63E+06	

Table 5. Kinetic-lumped models.

The products upgrading is required also, especially to deal with the acid index and olefins content. The acidity reductions, mainly caused by carboxylic acids, using the esterification reaction and neutralization, are opportunities for this issue. The reduction of alkenes content can be done through hydrotreatment reactions, widely used in oil refineries. The use of actual sites for oil refining can be suitable for this biofuel production, once most of polishing processes are present.

3. Conclusions

The thermal and/or catalytic-cracking processes are a promising technique to produce renewable source for hydrocarbon production. The product similarity with fossil fuels turns its usage and development attractive. However, some obstacles such as feedstock availability, reactor design, scale-up, and products upgrading require more studies. The thermal/catalytic cracking of triglycerides will not completely substitute the oil, but it can reduce our dependence and be a suitable environmental option.

Acknowledgements

The authors are grateful to the University of Blumenau (FURB), FAPESC, and the Brazilian governmental agencies ANP, CAPES, CNPq, and FINEP for the financial support.

Author details

Vinicyus R. Wiggers[1*], Ramon F. Beims[1], Laércio Ender[1], Edésio L. Simionatto[2] and Henry F. Meier[1]

*Address all correspondence to: vwiggers@furb.br

1 Chemical Engineering Department, University of Blumenau, Blumenau, Santa Catarina, Brazil

2 Chemistry Department, University of Blumenau, Blumenau, Santa Catarina, Brazil

References

[1] Herbert G.M.J., Krishnan A.U. Quantifying environmental performance of biomass energy. Renewable and Sustainable Energy Reviews. 2016;59:292–308. DOI: 10.1016/j.rser.2015.12.254.

[2] Sharma A., Pareek V., Zhang D. Biomass pyrolysis – A review of modelling, process parameters and catalytic studies. Renewable and Sustainable Energy Reviews. 2015;50:1081–1096. DOI: 10.1016/j.rser.2015.04.193.

[3] Hua T., Chunyi L., Chaohe Y., Honghong S. Alternative processing technology for converting vegetable oils and animal fats to clean fuels and light olefins. Chinese Journal of Chemical Engineering. 2008;16(3):394–400. DOI: 10.1016/S1004-9541(08)60095-6.

[4] Wiggers V.R., Wisniewski Jr A., Madureira L.A.S., Barros A.A.C., Meier, H.F. Biofuels from waste fish oil pyrolysis: Continuous production in a pilot plant. Fuel. 2009;88:2135–2141. DOI: 10.1016/j.fuel.2009.02.006.

[5] Wiggers V.R., Meier H.F., Wisniewski Jr A., Barros A.A.C., Wolf Maciel M.R. Biofuels from continuous fast pyrolysis of soybean oil: A pilot plant study. Bioresource Technology. 2009;100:6570–6577. DOI: 10.1016/j.biortech.2009.07.059.

[6] Wiggers V.R., Zonta G.R., França A.P., Scharf D.R., Simionatto E.L., Ender L., Meier H.F. Challenges associated with choosing operational conditions for triglyceride thermal cracking aiming to improve biofuel quality. Fuel. 2013;107:601–608. DOI: 10.1016/j.fuel.2012.11.011.

[7] Katikaneni S.P.R., Adjaye J.D., Bakhshi N.N. Performance of aluminophosphate molecular sieve catalysts for the production of hydrocarbons from wood-derived and vegetable oils. Energy and Fuels. 1995;9:1065–1078.

[8] Botas J.A., Serrano D.P., García A., Ramos R. Catalytic conversion of rapeseed oil for the production of raw chemicals, fuels and carbon nanotubes over Ni-modified nanocrystalline and hierarchical ZSM-5. Applied Catalysis B: Environmental. 2014;145:205–215. DOI: 10.1016/j.apcatb.2012.12.023.

[9] Giannakopoulou K., Lukas M., Vasiliev A., Brunner C., Schnitzer H. Low pressure catalytic co-conversion of biogenic waste (rapeseed cake) and vegetable oil. Bioresource Technology. 2010;101:3209–3219. DOI: 10.1016/j.biortech.2009.12.032.

[10] Biswas S., Mohanty P., Sharma D.K. Studies on co-cracking of Jatropha oil with bagasse to obtain liquid, gaseous product and char. Renewable Energy. 2014;63:308–316. DOI: 10.1016/j.renene.2013.09.045.

[11] Junming X., Jianchun J., Yanju L., Jie C. Liquid hydrocarbon fuels obtained by the pyrolysis of soybean oils. Bioresource Technology. 2009;100:4867–4870. DOI: 10.1016/j.biortech.2009.04.055.

[12] Lima D.G., Soares V.C.D., Ribeiro E.B., Carvalho D.A., Cardoso E.C.V., Rassi F.C., et al. Diesel-like fuel obtained by pyrolysis of vegetable oils. Journal of Analytical and Applied Pyrolysis. 2004;71:987–996. DOI: 10.1016/j.jaap.2003.12.008.

[13] Idem R.O., Katikaneni S.P.R., Bakhshi N.N.. Thermal cracking of canola oil: reaction products in the presence and absence of steam. Energy & Fuels. 1996;10:1150–1162.

[14] Iha O.K., Alves F.C.S.C., Suarez P.A.Z., Silva C.R.P. Potential application of *Terminalia catappa* L. and *Carapa guianensis* oils for biofuel production: Physical-chemical properties of neat vegetable oils, their methyl-esters and bio-oils (hydrocarbons). Industrial Crops and Products. 2014;52:95–98. DOI: 10.1016/j.indcrop.2013.10.001.

[15] Junming X., Jianchun J., Jie C., Yunjuan S. Biofuel production from catalytic cracking of woody oils. Bioresource Technology. 2010a;101:5586–5591. DOI: 10.1016/j.biortech.2010.01.148.

[16] Junming X., Jianchun J., Yunjuan S., Jie C. Production of hydrocarbon fuels from pyrolysis of soybean oils using a basic catalyst. Bioresource Technology. 2010b;101:9803–9806. DOI: 10.1016/j.biortech.2010.06.147.

[17] Li L., Quan K, Xu J., Liu F., Liu S., Yu S., et al. Liquid hydrocarbon fuels from catalytic cracking of rubber seed oil using USY as catalyst. Fuel. 2014;123:189–193. DOI: 10.1016/j.fuel.2014.01.049.

[18] Junming X., Guomin X., Yonghong Z., Jianchun J. Production of biofuels from high-acid-value waste oils. Energy Fuels. 2011;25:4638–4642. DOI: 10.1021/ef2006723.

[19] Ben Hassen-Trabelsi A., Kraiem T., Naoui S., Belayouni H. Pyrolysis of waste animal fats in a fixed-bed reactor: Production. Waste Management. 2014;34:210–218. DOI: 10.1016/j.wasman.2013.09.019.

[20] Boocock D.G.B., Konar S.K., Mackay A., Cheung P.T.C., Liu J. Fuels and chemicals from sewage sludge 2. The production of alkanes and alkenes by the pyrolysis of triglycerides over activated alumina. Fuel. 1992;71:1291–1297. DOI: 0016-2361/92/111291-07.

[21] da Mota S.A.P., Mancio A.A., Lhamas D.E.L., de Abreu D.H., da Silva M.S., dos Santos W.G. Production of green diesel by thermal catalytic cracking of crude palm oil (Elaeis guineensis Jacq) in a pilot plant. Journal of Analytical and Applied Pyrolysis. 2014;110:1–11. DOI: 10.1016/j.jaap.2014.06.011.

[22] Yigezu Z.D., Muthukumar K. Catalytic cracking of vegetable oil with metal oxides for biofuel. Energy Conversion and Management. 2014;84:326–333. DOI: 10.1016/j.enconman.2014.03.084.

[23] Junming X., Jianchun J., Zhang T., Weidi D. Biofuels production from catalytic cracking of triglyceride materials followed by esterification reaction in a scale-up reactor. Energy & Fuels. Forthcoming. DOI: 10.1021/ef3018173

[24] Biswas S., Sharma D.K. Studies on cracking of Jatropha oil. Journal of Analytical and Applied Pyrolysis. 2013;99:122–129. DOI: 10.1016/j.jaap.2012.10.013.

[25] Botton V., Riva D., Simionatto E.L., Wiggers V.R., Ender L., Meier H.F., et al. THERMO-CATALYTIC CRACKING OF THE MIXTURE OF USED FRYING OIL - TEXTILE STAMPING SLUDGE FOR THE PRODUCTION OF OIL WITH LOW ACIDITY INDEX. Quimica Nova. 2012;35(4):677–682.

[26] Shirazi Y., Viamajala S., Varanasi S. High-yield production of fuel- and oleochemical-precursors from triacylglycerols in a novel continuous-flow pyrolysis reactor. Applied Energy. 2016;179:755–764. DOI: 10.1016/j.apenergy.2016.07.025.

[27] Tamunaidu P., Bhatia S. Catalytic cracking of palm oil for the production of biofuels: Optimization studies. Bioresource Technology. 2007;98:3593–3601. DOI: 10.1016/j.biortech.2006.11.02.

[28] Junming X., Jianchun J., Jiaping Z. Thermochemical conversion of triglycerides for production of drop-in liquid fuels. Renewable and Sustainable Energy Reviews. 2016;58:331–340. DOI: 10.1016/j.rser.2015.12.315.

[29] Hilten R., Speir R., Kastner J., Das K.C. Production of aromatic green gasoline additives via catalytic pyrolysis of acidulated peanut oil soap stock. Bioresource Technology. 2011;102:8288–8294. DOI: 10.1016/j.biortech.2011.06.049.

[30] Chen G., Zhang X., Ma W., Yan B., Li Y. Co-pyrolysis of corn-cob and waste cooking oil in a fixed bed reactor with HY upgrading process. In: Energy Procedia, editor. The 6th International Conference on Applied Energy – ICAE2014, Taipei City, Taiwan; Elsevier Ltd; 2014. p. 2363–2366. DOI: 10.1016/j.egypro.2014.12.005.

[31] Lu L., Kejing Q., Junming X., Fusheng L., Shiwei L., Shitao Y., Congxia X., Xiaoping G. Preparation of basic mesoporous molecular sieves K2O/Mg-MCM-41 and its catalytic performance on the cracking of soybean oils. Journal of Analytical and Applied Pyrolysis. 2014;110:313–317. DOI: 10.1016/j.jaap.2014.09.019.

[32] Lu L., Kejing Q., Junming X., Fusheng L., Shiwei L., Shitao Y., et al. Mesoporous molecular sieves K2O/Ba(Ca or Mg)-MCM-41 with base sites as heterogeneous catalysts for the production of liquid hydrocarbon fuel from catalytic cracking of rubber seed oil. Green Chemistry. 2013;15(9):2573–2578. DOI: 10.1039/C3GC40796A.

[33] Bridgwater A.V. Review of fast pyrolysis of biomass and product upgrading. Biomass and Bioenergy. 2012;38:68–94. DOI: 10.1016/j.biombioe.2011.01.048.

[34] Stedile T., Ender L., Meier H.F., Simionatto E.L., Wiggers V.R. Comparison between physical properties and chemical composition of bio-oils derived from lignocellulose and triglyceride sources. Renewable and Sustainable Energy Reviews. 2015;50:92–108. DOI: 10.1016/j.rser.2015.04.080.

[35] Xiu S., Shahbazi A. Bio-oil production and upgrading research: A review. Renewable and Sustainable Energy Reviews. 2012;16:4406–4414. DOI: 10.1016/j.rser.2012.04.028.

[36] DeOliveira E., Quirino R.L., Suarez P.A.Z., Prado A.G.S. Heats of combustion of biofuels obtained by pyrolysis and by transesterification and of biofuel/diesel blends. Thermochimica Acta. 2006;450:87–90. DOI: 10.1016/j.tca.2006.08.005.

[37] Wisniewski Jr A., Wiggers V.R., Simionatto E.L., Meier H.F., Barros A.A.C., Madureira L.A.S. Biofuels from waste fish oil pyrolysis: Chemical composition. Fuel. 2010;89:563–568. DOI: 10.1016/j.fuel.2009.07.017.

[38] Maher K.D., Bressler D.C. Pyrolysis of triglyceride materials for the production of renewable fuels and chemicals. Bioresource Technology. 2007;98:2351–2368. DOI: 10.1016/j.biortech.2006.10.025.

[39] Bankovic-Ilic I.B., Stojkovic I.J., Stamenkovic O.S., Veljkovic V.B., Hung Y. Waste animal fats as feedstocks for biodiesel production. Renewable and Sustainable Energy Reviews. 2014;32:238–254. DOI: 10.1016/j.rser.2014.01.038.

[40] Manara P., Zabaniotou A. Towards sewage sludge based biofuels via thermochemical conversion – A review. Renewable and Sustainable Energy Reviews. 2012;16:2566–2582. DOI: 10.1016/j.rser.2012.01.074.

[41] Ramos M.J., Fernández C.M., Casas A., Rodríguez L., Pérez A. Influence of fatty acid composition of raw materials on biodiesel properties. Bioresource Technology. 2009;100:261–268. DOI: 10.1016/j.biortech.2008.06.039.

[42] Adewale P., Dumont M.J., Ngadi M. Recent trends of biodiesel production from animal fat wastes and associated production techniques. Renewable and Sustainable Energy Reviews. 2015;45:574–588. DOI: 10.1016/j.rser.2015.02.039.

[43] Hassan S.N., Sani Y.M., Abdul Aziz A.R., Sulaiman N.M.N., Daud W.M.A.W. Biogasoline: An out-of-the-box solution to the food-for-fuel and land-use competitions. Energy Conversion and Management. 2015;89:349–367. DOI: 10.1016/j.enconman.2014.09.050.

[44] Supple B., Holward-Hildige R., Gonzalez-Gomez E., Leahy J.J. The effect of steam treating waste cooking oil on the yield of methyl ester. Journal of the American Oil Chemists' Society. 2002;79(2):175–178.

[45] UKSBA. Written evidence submitted to the Environmental Audit Committee [Internet]. August 23, 2011. Available from: http://www.publications.parliament.uk/pa/cm201012/cmselect/cmenvaud/1025/1025vw08.htm [Accessed: August 1, 2011].

[46] Kulkarni M.G., Dalai A.K. Waste cooking oils an economical source for biodiesel: A review. Industrial & Engineering Chemistry Research. 2006;45:2901–2913.

[47] Sheinbaum-Pardo C., Calderon-Irazoque A., Ramírez-Suarez M. Potential of biodiesel from waste cooking oil in Mexico. Biomass and Bioenergy. 2013;56:230–238. DOI: 10.1016/j.biombioe.2013.05.008.

[48] Karmee S.K., Patria R.D., Lin C.S.K. Techno-economic evaluation of biodiesel production from waste cooking oil – a case study of Hong Kong. International Journal of Molecular Sciences. 2015;16:4362–4371. DOI: 10.3390/ijms16034362.

[49] Maddikeri G.L., Gogate P.R., Pandit A.B. Improved synthesis of sophorolipids from waste cooking oil using fed batch approach in the presence of ultrasound. Chemical Engineering Journal. 2015;263:479–487.

[50] North American Meat Institute (NAMI). The United States Meat Industry at a Glance [Internet]. Available from: https://www.meatinstitute.org/index.php?ht=d/sp/i/47465/pid/47465 [Accessed: August 10, 2016].

[51] FAO. The State of World Fisheries and Aquaculture 2016. [Internet]. Available from: http://www.fao.org/3/a-i5555e.pdf [Accessed: August 8, 2016].

[52] Arruda L.F., Borghesi R., Oetterer M. Use fish waste as silage – A review. Brazilian Archives of Biology and Technology. 2007;50(5):879–876.

[53] Yang Y., Brammer J.G., Ouadi M., Samanya J., Hornung A., Xu H.M., et al. Characterisation of waste derived intermediate pyrolysis oils for use as diesel. Fuel. 2013;103:247–257. DOI: 10.1016/j.fuel.2012.07.014.

[54] Bridgwater A.V. Renewable fuels and chemicals by thermal processing of biomass. Chemical Engineering Journal. 2003;91(2–3):87–102. DOI: 10.1016/S1385-8947(02)00142-0.

[55] Idem R.O., Katikaneni S.P.R., Bakhshi N.N. Catalytic conversion of canola oil to fuels and chemicals: Roles of catalyst acidity, basicity and shape selectivity on product distribution. Fuel Processing Technology. 1997;51:101–125.

[56] Kubátová A., Luo Y., Štavová Y., Sadrameli S.M., Aulich T. New path in the thermal cracking of triacylglycerols (canola and soybean oil). Fuel. 2011;90:2598–2608. DOI: 10.1016/j.fuel.2011.04.022.

[57] Fréty R., da Rocha M.G.C., Brandão S.T., Padilha J.F., Borges L.E.P., Gonzalez W.A. Cracking and hydrocracking of triglycerides for renewable liquid fuels: Alternative processes to transesterification. Journal of the Brazilian Chemical Society. 2011;22(7): 1206–1220. DOI: 10.1590/S0103-50532011000700003.

[58] Doronin V.P., Potapenko O.V., Lipin P.V., Sorokina T.P. Catalytic cracking of vegetable oils and vacuum gas oil. Fuel. 2013;106:757–765. DOI: 10.1016/j.fuel.2012.11.027.

[59] Katikaneni S.P.R., Adjaye J.D., Bakhsh N.N. Catalytic conversion of canola oil to fuels and chemicals over various cracking catalysts. The Canadian Journal Of Chemical Engineering. 1995;73:484–497.

[60] Wisniewski Jr A. Wosniak L., Scharf D.R., Wiggers V.R., Meier H.F., Simionatto E.L. Upgrade of biofuels obtained from waste fish oil pyrolysis by reactive distillation. Journal of the Brazilian Chemical Society. 2015;26(2):224–232. DOI: 10.5935/0103-5053.20140251.

[61] Pstrowska K., Walendziewski J., Stolarski M. Hydrorefining of oil from rapeseed cake pyrolysis over NiMo/Al2O3 catalyst. Fuel Processing Technology. 2014;128:191–198. DOI: 10.1016/j.fuproc.2014.07.035.

[62] Zandonai C.H., Yassue-Cordeiro P.H., Castellã-Pergher S.B. Production of petroleum-like synthetic fuel by hydrocracking of crude soybean oil over ZSM5 zeolite – Improve-

ment of catalyst lifetime by ion exchange. Fuel. 2016;172:228–237. DOI: 10.1016/j.fuel.2015.12.059.

[63] Ishihara A., Fukui N., Nasu H., Hashimoto T.. Hydrocracking of soybean oil using zeolite – alumina composite supported NiMo catalysts. Fuel. 2014;134:611–617. DOI: 10.1016/j.fuel.2014.06.004.

[64] Hanafi S.A., Elmelawy M.S., Shalaby N.H., El-Syed H.A., Eshaq G., Mostafa M.S. Hydrocracking of waste chicken fat as a cost effective feedstock for renewable fuel production: A kinetic study. Egyptian Journal of Petroleum. 2015. Article in press. DOI: 10.1016/j.ejpe.2015.11.006.

[65] API – American Petroleum Institute. Technical Data Book – Petroleum Refining. Washington, DC: American Petroleum Institute; 1993.

[66] Fahim M.A., Al-Sahhaf T.A., Elkilani A.S. Fundamentals of Petroleum Refining. 1st ed. Elsevier B.V. Amsterdam, The Netherlands; 2010. 516 p.

[67] Frainer B.L.M., Wiggers V.R., Sharf D.R., Meier H.F., Ender L., Simionatto E.L. Thermal cracking of frying oil: A proposal for a kinetic mechanism based on groups of compounds. In: American Institute of Chemical Engineers, editor. 2014 Annual Meeting, November 20, 2014; Atlanta, GA: American Institute of Chemical Engineers; 2014.

[68] Meier H.F., Wiggers V.R., Zonta G.R., Scharf D.R., Simionatto E.L., Ender L. A kinetic model form thermal cracking of waste cooking oil based on chemical lumps. Fuel. 2015;144:50–59. DOI: 10.1016/j.fuel.2014.12.020.

[69] Periyasamy B. Reaction pathway analysis in thermal cracking of waste cooking oil to hydrocarbons based on monomolecular lumped kinetics. Fuel. 2015;158:479–487. DOI: 10.1016/j.fuel.2015.05.066.

[70] Baladincz P., Hancsók J. Fuel from waste animal fats. Chemical Engineering Journal. 2015;282:152–160. DOI: 10.1016/j.cej.2015.04.003.

[71] Butlera E., Devlina G., Meierb D., McDonnella K. A review of recent laboratory research and commercial developments in fast pyrolysis and upgrading. Renewable and Sustainable Energy Reviews. 2011;15(8):4171–4186. DOI: 10.1016/j.rser.2011.07.035.

[72] Jacomel C.R.R., Bastos J.C.S.C., Ender L., Wiggers V.R., Meier H.F., Noriler D. Numerical simulation of a thermal cracking continuous reactor. In: American Institute of Chemical Engineers, editor. 2015 Annual Meeting, November 9, 2015; Salt Lake City, UT: American Institute of Chemical Engineers; 2015.

Chapter 21

Biogasification of Horse Dung Using a Cylindrical Surface Batch Biodigester

Patrick Mukumba, Golden Makaka and Sampson Mamphweli

Additional information is available at the end of the chapter

http://dx.doi.org/10.5772/65373

Abstract

Anaerobic digestion of animal dung offers several benefits such as reduction of odors, pathogens, and production of renewable energy biogas. In this study, a 1 m^3-surface batch biogas digester was designed, constructed, and insulated with sawdust to minimize temperature fluctuations within the digester. The horse dung was collected from the University of Fort Hare Honey dale farm and fed into the batch biogas digester. The horse dung was analyzed for total solids (TS), volatile solids (VS), total alkalinity (TA), calorific value (CV), pH, chemical oxygen demand (COD), and ammonium-nitrogen (NH_4-N). The optimum total alkalinity, ammonium-nitrogen, and chemical oxygen demand were 6235, 901, and 24230 mg/L, respectively. The study found that horse dung produced biogas yield with an average methane yield of 51% without codigesting it with other wastes. Therefore, horse dung is a good substrate for biogas production, and its use in biogas digesters can reduce greenhouse gas emissions into the atmosphere leading to climate change.

Keywords: biogasification, digester, horse dung, biogas, methane

1. Introduction

South Africa like any developing country is overdependent on conventional energy sources such as coal and firewood. Coal is a fossil fuel and is the main source of electricity in the country. Fossil fuels have many negative impacts on the environment, which include environmental degradation, climate change, and human health problems [1]. Biogas production would benefit mainly the rural population by providing a clean fuel from renewable substrates and help to end energy poverty.

After the ratification of the United Nations Framework Convention on Climate Change (UNFCCC) and Kyoto Protocol in August 1997 and July 2002, respectively, the South African government embarked on numerous projects related to climate change, including projects that have been intended as measures to reduce greenhouse gases. Furthermore, the high volatility in oil prices in the recent past resulting in turbulence in energy markets has also compelled the country to look for alternative sources of energy such as from biogasification. As a result, the South Africa Integrated Resource Plan (SAIRP) approved by the parliament in 2010 sets a target of 40% renewable energy contribution by the year 2030 [2].

Biogasification is a method by which biogas is produced from organic material through the action of microbes. It is therefore a biological process in which organic materials break down in an environment that is sufficiently warm and oxygen-free. The end product of biogasification is biogas, mainly methane, carbon dioxide, and digestion sludge. The sludge can be used as a fertilizer. Biogas is a combustible gas, which is generally used as a source for cogeneration, for producing electricity and heat by means of gas or dual fuel engines [3].

Biogas from anaerobic digestion (AD) can be viewed as one of the vehicles to reduce rural poverty and could lead to rural development. The process produces less greenhouse gases than waste treatment processes such as composting [4] and land filling [5]. It can substitute fossil fuels and can decrease environmental pollution including acid rains and global warming [6]. Therefore, AD has wide flexibility and can be modified to fulfill the precise requirements in management of agricultural farms [7]. When compared to other renewable energy sources, such as solar and wind power, the methane component of biogas can be easily stored in bio-bags. Furthermore, the biogas digesters are not prone to theft unlike solar panels and wind turbines. In addition, biogas production would benefit mainly the rural population by providing a clean fuel and reduce energy poverty.

According to Ref. [8], biogas is an overlooked source of fuel in spite of the excitement surrounding the use of biofuels as an alternative source of energy. The use of biogas digesters can improve the lives of the people in rural areas in many ways; it reduces deforestation, reduces greenhouse emissions, and controls unpleasant odors from human or animal wastes and reduction of workload and marginalization of women who collect firewood.

The most common types of biogas digesters are fixed dome digester, balloon-type digester, and floating drum digester. The two most familiar types in developing countries are fixed dome and floating drum digesters. The three main digesters are discussed and their advantages and disadvantages are also given.

The fixed dome digester is the most popular digester; its archetype was developed in China as early as 1936. The fixed dome digester is shown in **Figure 1**.

It is a closed dome shape digester with an immovable, rigid gas-holder and a displacement pit (compensating tank). The biogas produced by methanogenic bacteria in the biogas digester is captured in the gas holder and the slurry is displaced in the compensating tank. When gas is consumed, slurry enters back into the digester from the overflow tank. As a result of these movements, a certain degree of mixing is obtained. The more the gas is produced, the higher the level at the slurry outlet [10].

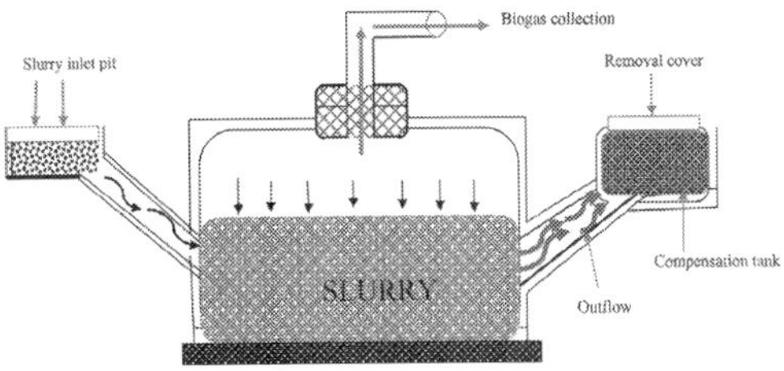

Figure 1. Chinese fixed dome digester, adapted with permission from [9].

The fixed dome digester has some advantages that include: relatively cheap and durable, no moving parts, and well insulated [11]. However, the fixed dome digester has disadvantages that include: high technical skills are required for a gas tight construction, special sealant is required for the gasholder, difficult to construct in high water table areas, requires more excavation work, and enormous structural strength is required for construction [11, 12].

A balloon digester (bag digester) is a plastic or rubber bag combining the gas holder and digester. This is a plug-flow-type reactor. This design was developed in the 1960s in Taiwan. Gas is collected in the upper part and manure in the lower part. The inlet and outlet are attached to the skin of the bag. The pressure of the biogas is adjustable by laying stones on the bag [13]. **Figure 2** shows a balloon digester. The biogas is collected in the balloon.

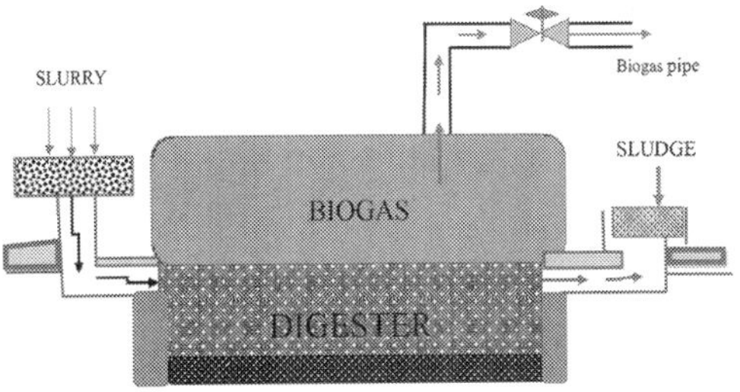

Figure 2. Balloon digester, adapted with permission from [10].

The advantages of the bag digesters include: low cost, simple technology, and easy to clean. However, the disadvantages include: short lifespan, susceptible to physical damage, hard to repair, need high quality plastic, and difficult to insulate [14].

Floating drum digesters are common in India. The digesters have a moving floating gas-holder, or drum. The gas holder floats either directly in the fermenting slurry or in a separate water jacket. The drum in which the biogas collects has an internal or external guide frame that provides stability and keeps the drum upright. When the biogas is produced the drum moves up, and when the gas is consumed, the gas holder sinks back.

The floating drum digesters have advantages which include: the operation of the plant is easy to understand, the gas drum is air tight, and there is constant gas pressure as a result of the weight of the drum [15]. However, it does also have disadvantages which are: steel drum is relatively expensive and needs regular maintenance (priming, painting, and coating) and the effect of low temperature during winter is high [11]. A floating drum digester is shown in **Figure 3**.

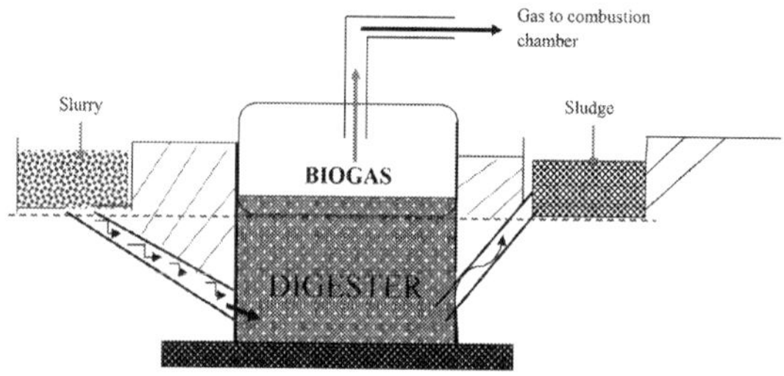

Figure 3. Indian-type digester, adapted with permission from [9].

The main aim of the project was to measure methane content in horse dung using a designed and constructed 1 m^3 cylindrical batch biogas digester.

2. Objectives

The research was carried out with the following objectives in mind:

(i) To design and construct a 1 m^3 batch biogas digester.

(ii) To determine the impact of substrate properties on methane production.

(iii) To determine the effect of insulation of the digester on biogas yield.

(iv) To measure the methane and carbon dioxide in horse dung.

3. Methodology

3.1. Design of the digester

Biogas digester design plays a crucial role in digester performance and a number of considerations are taken into account. The following aspects were considered during the design process: durability, air tightness, availability of local materials and easy operation. The design parameters included:

http://dx.doi.org/10.5772/65373

3.1.1. *Total solid (TS) contents of organic materials*

The total solid (TS) contained in a substrate is usually used as the material unit to indicate the biogas production rate of the materials. The most favorable TS value is 8% for better biogas production [16].

3.1.2. *Favorable temperature, pH value, and carbon/nitrogen ratio for good fermentation*

The mesophilic temperature between 25 and 35°C was chosen in the design. The digester was insulated to keep the temperature within the latter mesophilic range to optimize mesophilic bacterial activity. The pH value selected ranged from 6.8 to 7.8 because the methanogens prefer a neutral atmosphere with pH between 6.8 and 7.5.

The carbon/nitrogen ratio considered ranged from 20:1 to 30:1. Carbon and nitrogen are the main nutrients required by microorganisms. Therefore, a C/N ratio of 20–30:1 was considered for optimum anaerobic digestion, based on biodegradable organic carbon [17, 18]. Hence, codigestion experiments were done to maintain the carbon/nitrogen ratio within the desired range.

3.1.3. *Hydraulic retention time (HRT)*

For mesophilic digestion where temperature varies from 25 to 35°C, the HRT was greater than 20 days. In the thermophilic environment, HRT is usually less than 10 days [19]. Shortening retention time can lead to increase in the volatile fatty acids (VFA) [20], and this is why mesophilic digestion was considered. A surface cylindrical biogas digester was chosen because it was easy to feed, insulate, clean, and easy to construct and remove slurry after every hydraulic retention period. In addition, the batch digester was easy to agitate. **Figure 4** shows the cylindrical batch digester body with all the calculated values indicated and **Table 1** shows calculated volume and geometrical dimensions of the batch biogas digester.

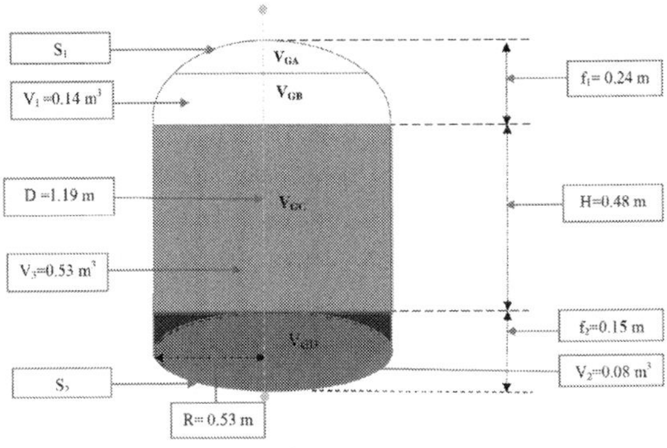

Figure 4. Geometrical dimensions of the cylindrical shaped biogas digester body. KEY: Volume of gas collecting chamber = V_{GA}; Volume of gas storage chamber = V_{GB}; Volume of fermentation chamber = V_{GC}; Volume of the sludge layer = V_{GD}; Total volume of the digester, $V = V_{GA} + V_{GB} + V_{GC} + V_{GD}$.

Assumptions: For volume	For geometrical dimensions
$V_{GD} = 5\%V$	$D = 1.3078 \times V^{1/3}$
$V_{GB} + V_{GC} = 80\%V$	$V_1 = 0.0827\ D^3$
$V_{GD} = 15\%V$	$V_2 = 0.05011\ D^3$
$V_{GB} = 0.5\ [V_{GB} + V_{GC} + V_{GD}]\ k$	$V_3 = 0.3142\ D^3$
$k = 0.4\ m^3/day$	$R_1 = 0.725\ D;\ R_2 = 1.0625\ D$
	$f_1 = D/5;\ f_2 = D/8$
	$S_1 = 0.911\ D_2;\ S_2 = 0.8345\ D^2$

Table 1. Assumptions for volume and geometrical dimensions [21].

Figure 5 shows a more detailed diagram of the batch biogas digester with the monitoring sensors' positions.

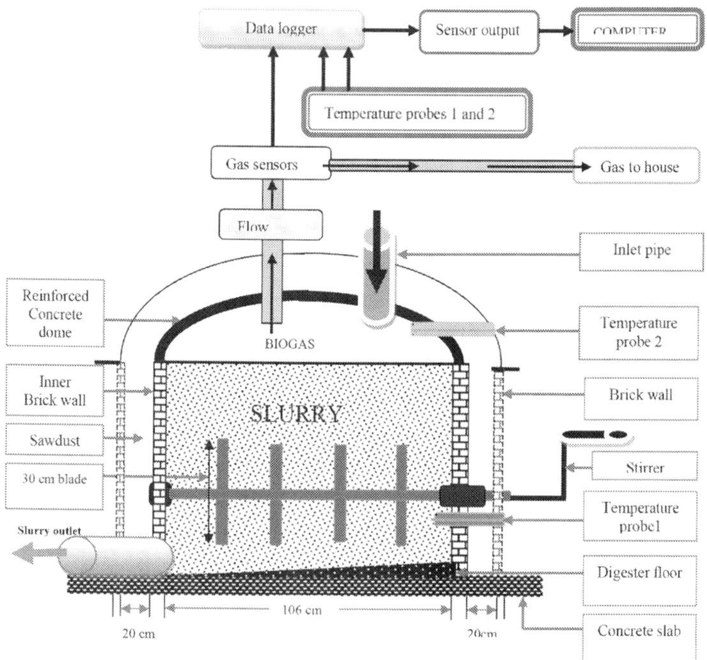

Figure 5. Detailed diagram of the designed batch biogas digester with various sensors positions.

3.2. Construction of the biogas digester

A number of issues were considered during the construction of the biogas digester to ensure nonleakages and minimization of influence of ambient temperatures on the substrate temperature. The construction of the digester was done in the following stages:

- Selection of construction material
- Site selection and layout
- Excavation
- Inserting the mechanical stirrer
- Inserting the curved top of the digester
- Second wall construction of the biogas digester
- Insulation of the biogas digester

The main building materials for the biogas digester included clinker bricks, sand, concrete stones, and Portland cement. The concrete stones were free of soil and organic material. Furthermore, the sand used was clean in order to increase the strength of the digester.

The klinker bricks were first soaked in clean water for 5 minutes in order to remove dust and to prevent the bricks from sucking moisture from the mortar thus allowing a strong bonding. Clinker bricks were used in the construction because they offered the following advantages;

- Acid proof
- Wear proof
- Anticorrosive
- High compressive strength
- Low thermal conductivity of 0.67 W/(m.K)
- Relatively cheaper than other types of bricks

Portland cement was used because it has a low thermal conductivity of 0.29 W/(m.K) compared to masonry cement with a thermal conductivity of 0.5 W/(m.K) and epoxy was used for painting the inside of the batch biogas digester because it has a high water proofing and low thermal conductivity of 0.30 W/(m.K). Sawdust was selected for insulation because of its availability in the area and low thermal conductivity of 0.08 W/(m.K) compared to dry sand, which has a thermal conductivity of 0.15–0.25 W/(m.K). Therefore, the heat transfer in materials with low thermal conductivity, for example, sawdust, is very low.

The biogas digester was constructed and reinforced concrete dome was placed on top of the plastered biogas digester with mortar smeared on its top surface to ascertain tight air seal as shown in **Figure 6**.

Figure 6. Biogas digester with a reinforced concrete dome.

3.3. Second wall construction of the biogas digester and insulation of the biogas digester

An outer wall of 1 inch (115 mm) was constructed with bricks to make the biogas digester two walled as shown in **Figure 7**. The separation gap for the two walls was 200 mm. Sawdust, an insulating material, was then put after the plastering of the outer wall.

Sawdust was selected for insulation because of its availability in the area and low thermal conductivity of 0.08 W/(m.K) compared to dry sand which has a thermal conductivity of 0.15–0.25 W/(m.K).

Figure 7. Constructed second wall of the biogas digester.

3.4. Source of substrate and mixing proportions

Fresh horse dung was collected from University of Fort Hare Honey dale farm. The horse dung, before fed into the 1 m^3 biogas digester, was chopped with a compost chopper to accelerate biogasification.

3.5. Substrate parameters

The following parameters in horse dung were determined: pH, total solids (TS), volatile solids (VS), ammonium-nitrogen (NH_4-N), total alkalinity (TA), temperature (T), and caloric value (CV). All the analytical determinations were performed according to the standard methods for examination of water and wastewater [22].

3.6. Biogas analysis

The biogas composition was analyzed using the biogas analyzer consisting of nondispersive infrared (NDIR) sensor for sensing methane and carbon dioxide and palladium/nickel (Pd/Ni) sensor for sensing hydrogen and hydrogen sulfide. The data for biogas composition was recorded by a CR1000 data logger at a time interval of 2 minutes. The biogas analyzer and the CR1000 data logger were powdered by a 12V DC battery that was connected to a 20 W photovoltaic module. The slurry and ambient temperatures were measured using type K thermocouples connected to the same CR1000 data logger as the biogas sensors. The data logger was interfaced to a computer.

The data acquisition system which consisted of a palladium-nickel and nondispersive infrared sensors is shown in **Figure 8**.

Figure 8. The data acquisition system.

4. Results and discussion

The substrate characteristics of the horse dung is shown in **Table 2**, namely, total solids (TS), volatile solids (VS), total alkalinity (TA), pH, caloric value (CV) and ammonium-nitrogen (NH_4-N).

Parameter	Horse dung
Total solids (mg/L)	269,515.67
Volatile solids (mg/L)	172,934.47
Total alkalinity (mg/L)	6234–6256
pH [average]	7.34–7.36
Calorific value (MJ/g)	26.40
Ammonium-nitrogen (mg/L)	1101–1289

Table 2. Substrate characteristics of horse dung.

The average slurry temperature of the batch biogas digester after the insulation was 30°C. Therefore, it can be concluded that the insulation of the batch biogas digester was advantageous because the desired temperature for optimum biogas production was achieved. **Figure 9** shows the biogas yield from horse dung. The biogas production increased until it reached the peak and then began to decline.

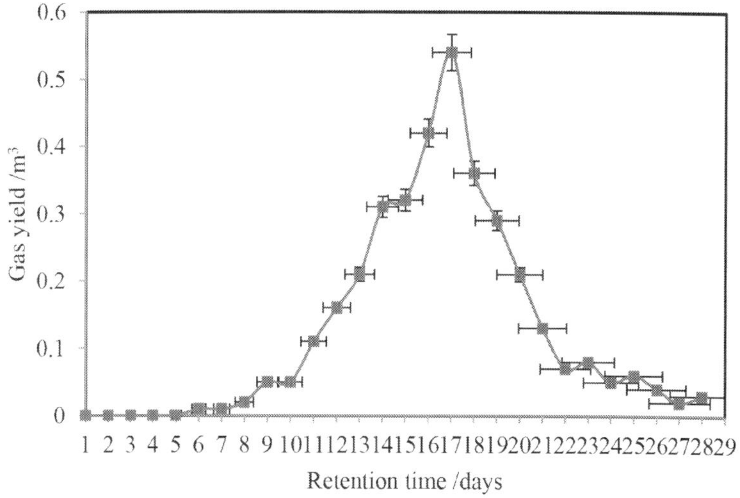

Figure 9. Biogas yield for horse dung.

The horse dung had a peak biogas yield of 0.51 m^3 on day 18. The linear biogas production of horse dung from day 12 (0.11 m^3) to day 17 (0.54 m^3) can be approximated by the equation:

$$Y = 0.0686t - 0.6643 \qquad \text{for } 12 \le t \le 17 \tag{1}$$

The decay of biogas production for horse dung from the peak (day 17) to day 22 is represented by:

$$Y = -0.977t - 2.188 \qquad \text{for } 17 \le t \le 22 \tag{2}$$

The relationship between biogas yield and pH in horse dung is shown in **Figure 10**. The maximum biogas yield of 0.54 m^3 was produced at pH 6.9. The initial pH of was 7.9 and the pH was observed to decline with time, attaining a minimum value of 6.9 where an optimum biogas production of 0.54 m^3 was achieved. The decline is due to the conversion of the substrate to acids during acidogenesis and acetogenesis stages of methane production. As from day 17, the pH was seen to increase as the acids produced were converted to methane by the methanogens.

Relationship between biogas yield and chemical oxygen demand (COD) in horse dung is shown in **Figure 11**. The highest biogas yield of 0.54 m^3 was produced on day 17 where the COD value was 24,230 mg/L as shown in **Figure 11**.

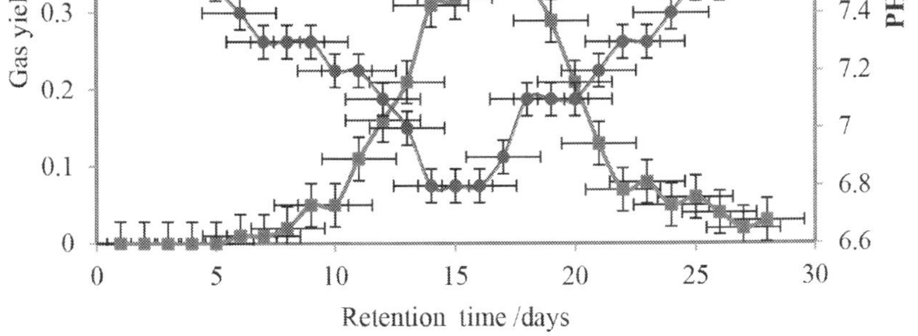

Figure 10. Relationship between biogas yield and pH in horse dung.

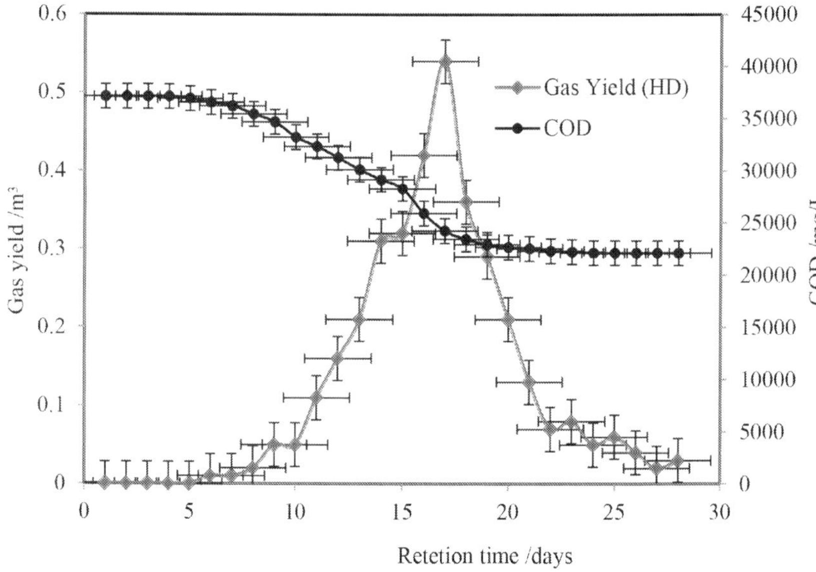

Figure 11. Relationship between biogas yield and COD in horse dung.

The initial COD for horse dung was 37,110 mg/L and the final COD was 22,110 mg/L. The highest COD destruction was between days 15 and 19. The concentration of COD destroyed from day 22 to day 28 was very low indicating a low biogas yield and a higher COD destruction means a high biogas yield.

The relationship between biogas yield and NH_4-N in horse dung is shown **Figure 12**. The highest biogas yield of 0.54 m^3 was produced at NH_4-N concentrations of 901 mg/L. The initial NH_4-N concentration was 1196 mg/L and the final NH_4-N concentration was 962 mg/L. It was observed that the concentrations NH_4-N for horse dung were between 850 and 1196 mg/L. However, there was no inhibitory effect of the ammonium ion because the NH_4-N concentrations were below 1500 mg/L. In the experiment, it was observed that higher NH_4-N values corresponded with lower biogas production

The inhibiting concentrations of NH_4-N are reported to be above 1500 mg/L [23–25]. The stability of NH_4-N levels in the substrate improved biogasification. Relationship between biogas yield and total alkalinity is shown in **Figure 13**.

From the graph it was observed that the total alkalinity for horse dung was between 6190 and 6256 mg/L. The higher the alkalinity, the greater the buffering capacity in the anaerobic digestion process which in turn promoted a stable pH value (**Figure 10**) and this resulted in an increase in the biogas yield. It was observed that total alkalinity changes were directly proportional to changes in the pH.

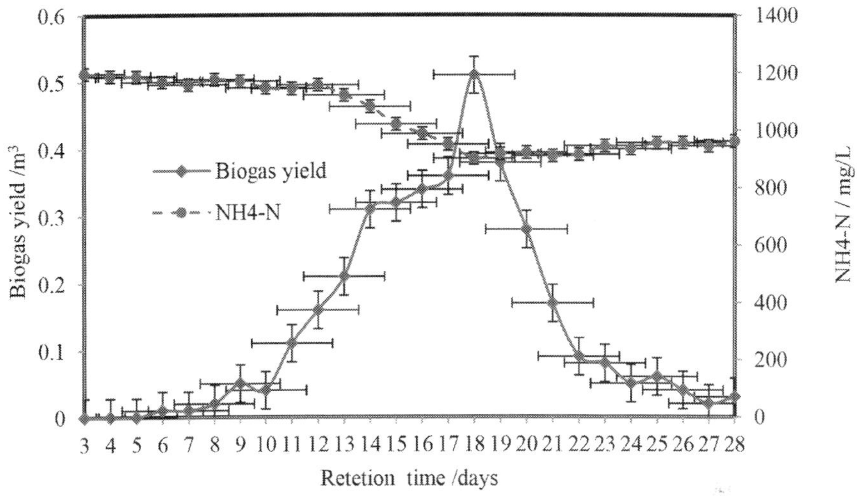

Figure 12. Relationship between biogas yield and NH_4-N in horse dung.

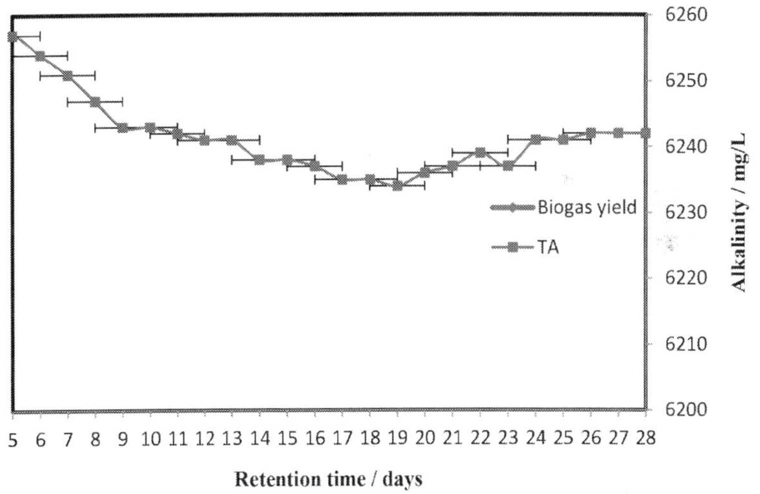

Figure 13. Relationship between biogas yield and total alkalinity.

The composition of biogas in horse dung is shown in **Table 3**. The methane content horse dung was 51% and the carbon dioxide content was 43%. Theoretically, horse dung produces more biogas than cow dung because of its carbon/nitrogen ratio of 25:1 [10].

Gases	% yield
Methane (CH_4)%	51
Carbon dioxide (CO_2)%	43
Carbon monoxide (CO)	0
Hydrogen (H_2)% and other gases	6

Table 3. Composition of biogas in horse dung.

The biogas from horse dung with methane content of above 50% can be used by rural communities for electricity production and as fuel for stoves, refrigerators, and generators, thereby replacing liquid petroleum gas (LPG) as fuel. In addition, the total lifecycle environmental impacts of the produced biogas are decreased via anaerobic digestion, since methane from biogas can be used as fuel for diesel or petrol engines. However, for biogas from horse dung to be used as fuel for vehicles, it should follow processes such as purification, upgrading, compression, and storage. The digestate of horse dung from the biogas digester is subsequently collected and used mainly to replace mineral fertilizers. Any overflow of the effluent from the storage tanks is discharged directly into the aquatic environment, as nutrient for vegetable crops. From **Table 3**, biogas from horse dung has a hydrogen content of 6%. The hydrogen from anaerobic digestion of horse dung can be used as fuel in hydrogen fuel cells. Hydrogen fuel cells have the following benefits: have a higher efficiency than diesel or gas engines, operate silently compared to internal combustion engines, fuel cells have no "memory effect" when they are getting refueled, and finally, the maintenance of fuel cells is simple since there are few moving parts in the system.

5. Conclusion

The 1 m^3 batch biogas digester was successfully designed, constructed, and insulated with sawdust and fed with horse dung. The batch biogas digester designed, constructed, and insulated with sawdust was easy to feed and clean as compared to underground fixed dome digesters which are not easy to clean and stir to agitate biogas production. Therefore, the designed biogas digester could ease energy problems if installed in rural communities of South Africa that have energy crisis. The results of the batch anaerobic digestion experiment show that horse dung is a good substrate for biogas production. The total methane potential of the horse dung was 51%, while the carbon dioxide content was 43%. In the experiment, it was observed that the methane yield from the horse dung increased exponentially with time and ceased after certain days. Furthermore, the study found that the optimum total alkalinity, ammonium-nitrogen, and chemical oxygen demand for horse dung are 6235, 901, and 24,230 mg/L, respectively.

It can also be concluded that anaerobic digestion of horse dung and other biogas digester substrates in the country would improve the country's service delivery and serve as a local solution to the world energy crisis caused by deletion of fossil fuels. From the research findings, no previous experiments to measure methane content in horse dung were done using field

http://dx.doi.org/10.5772/65373

scale digesters. The current study would be the first study to operate with a large digester insulated with sawdust in an outdoor setting operating at an average temperature of 30°C.

6. Acknowledgements

The authors would like to thank ESKOM for funding this project and University of Fort Hare for providing facilities for this project.

7. Author contributions

Patrick Mukumba designed, constructed, and fed the biogas digester with horse dung. Golden Makaka and Sampson Mamphweli supervised the research project. All the authors contributed to preparing and approving the final manuscript.

Author details

Patrick Mukumba[1*], Golden Makaka[1] and Sampson Mamphweli[2]

*Address all correspondence to: pmukumba@ufh.ac.za

1 Physics Department, University of Fort Hare, Alice, South Africa

2 Institute of Technology, University of Fort Hare, Alice, South Africa

References

[1] Elaiyaraju, P. and Partha, N. Biogas production from co-digestion of orange peel waste and jatropha de-oiled cake in an anaerobic batch reactor. African Journal of Biotechnology, 2012; 11: 3339–3345.

[2] South Africa Resource Plan; SAIRP. Department of Minerals and Energy [Online]. Available at: http:www.doe.Irp.co.za/content/IRP2010-2030-Final-Report-20110325pdf (Accessed: 5 April 2012).

[3] Ward, A. J., Hobbs, P. J., Holliman, P. J. and Jones, D. L. Optimisation of the anaerobic digestion of agricultural resources. Bioresource Technology, 2008; 99: 7928–7940.

[4] Walker, L., Charles, W. and Cord-Ruwisch, R. Comparison of static, in vessel composting of MSW with thermophilic anaerobic digestion and combinations of the two processes. Bioresource Technology, 2009;100: 3799–3807.

[5] Lou, X. F. and Nair, J. The impact of land filling and composting on greenhouse gas emissions–A review. Bioresource Technology, 2009; 100: 3792–3798.

[6] Chynoweth, D. P., Owens, J. M. and Legrand, R. Renewable methane from anaerobic digestion of biomass. Renewable Energy, 2001; 22: 1–8.

[7] Klimiuk, E., Pokój, T., Ski, W. B., and Dubis, B. Theoretical and observed biogas production from plant biomass of different fiber contents. Bioresource Technology, 2010; 101: 9527–9535.

[8] AGAMA Biogas. Integrated Biogas Solutions. Green Energy solutions for Africa Energy; 2007. Available from: http://www.agama.co.za/. [Accessed: 2015-06-22]

[9] Florentino, H. Mathematical tool to size rural digesters. Scientia Agricola, 2003; 60: 185–190.

[10] Fulford, D. Running a Biogas Programme. ITDG Publishing, United Kingdom; 2001.

[11] Nijaguna, B.T. Biogas Technology. New Age International (P) Limited, Publishers, New Delhi; 2002.

[12] Sharma, N. and Giuseppe, P. Anaerobic biotechnology and developing countries – Technical status. Energy Conversions, 1991; 32: 447–469.

[13] FAO. Biogas Technology. A Training Manual for Extensionz. Consolidated Management Services Nepal (P) Ltd, UN; 1996.

[14] APCAEM. Recent Developments in Biogas Technology for Poverty Reduction and Sustainable Development, APCAEM, Beijing; 2001.

[15] ISAT and GTZ. Biogas – Application and Product Development. Information and Advisory Services on Appropriate Technology (ISAT) and German Technical Cooperation (GTZ) GmbH; 1999. Available from: www.sswn.Info/library/1002.pdf [Accessed: 2015-09-20]

[16] Ituen, E. E., John, N. M. and Bassey, B. E. Biogas production from organic waste in Akwa Ibom State of Nigeria. Appropriate Technologies for Environmental Protection in the Developing World, 2007; 207: 17–19.

[17] Mshandete, A., Kivaisi, A., Rubindamayugi, M. and Mattiasson, B. Anaerobic batch co-digestion of sisal pulp and fish wastes. Bioresource Technology. 2004; 95: 19–24.

[18] Yen, H. W. and Brune, D. E. Anaerobic co-digestion of algal sludge and waste paper to produce methane. Bioresource Technology, 2007: 98: 130–134.

[19] Demetriades, D. Thermal Pre-treatment of Cellulose Rich Biomass for Biogas Production. Master thesis. Swedish University of Agricultural Sciences, Uppsala, Sweden; 2008.

[20] Kobayashi, T. and Li, Y. Y. Performance and characterisation of a newly developed self–Agitated anaerobic reactor with biological desulfurization. Bioresource Technology, 2011; 102: 5580–5588.

[21] Design of Biogas Plant, Bio-gas Project, LGED: Preparing this training material all the important information have been collected from the booklets & research materials of Biogas Training Center (BRC) Chendu, Sichuan, Chaina. Available from: www.sswm.info/sites/default/files/pdf. [Accessed: 2015-07-12]

[22] American Public Health Association (APHA). Standard Methods for Examination of Water and Waste Water. WEF, Washington, DC; 2005.

[23] Calli, B., Mertoglu, B., Inac, B. and Yenigun, O. Effects of free ammonia concentrations on the performance of anaerobic bioreactors. Process Biochemistry, 2005; 4: 1285–1292.

[24] Chen, Y., Cheng, J. J. and Creamer, K. S. Inhibition of anaerobic digestion process: A review. Bioresource Technology, 2008; 99: 4044–4064.

[25] Fricke, K., Santen, H., Wallman, R., Hütter, A. and Dichtl, N. Operating problems in anaerobic digestion plants resulting from nitrogen in MSW. Waste Management, 2007; 27: 30–43.

Chapter 22

Refractory Materials for Biofuel Boilers

Valentin Antonovič, Jacek Szczerba,

Jadvyga Keriene, Rimvydas Stonys and Renata Boris

Additional information is available at the end of the chapter

http://dx.doi.org/10.5772/65550

Abstract

The energy equipment usable for solid biofuel incineration usually operates upon aggressive conditions. The internal structures (lining) of the equipment are made of refractory materials that are affected by combined loads: thermal, mechanical and chemical (i.e. high temperature–up to 1200°C, chemical impact of alkaline compounds and slag, repeating thermal shocks, abrasive effect caused by solid particles and so on). A majority of traditional refractories usable for lining in such equipment are not durable. Upon certain conditions of use (such as high local temperatures, influence of alkaline biofuel combustion products and so on), durability of the traditional materials is 1–2 years only. The opportunities of new refractory materials application should be set upon taking into account the conditions of operation for biofuel boilers of specific types. In this section - the data on the peculiarities of using refractory materials in biofuel boilers are reviewed, and the impact of aggressive operating conditions of such thermal equipment on the properties of refractory materials is discussed. In addition, the investigations results of refractory castables alkali resistance and its explosive spalling are discussed. The recommendations for use of refractory materials in biofuel boilers are also presented.

Keywords: biofuel boiler, refractory materials, refractory castables, alkali resistance

1. Introduction

Biofuel boilers are used in modern fuel combustion systems, with the function to ensure high energy conversion efficiency and comply with environmental standards applied on such devices. Despite the fact that energy equipment may have different designs, their operation is based on a standard three-level process diagram (**Figure 1**) [1].

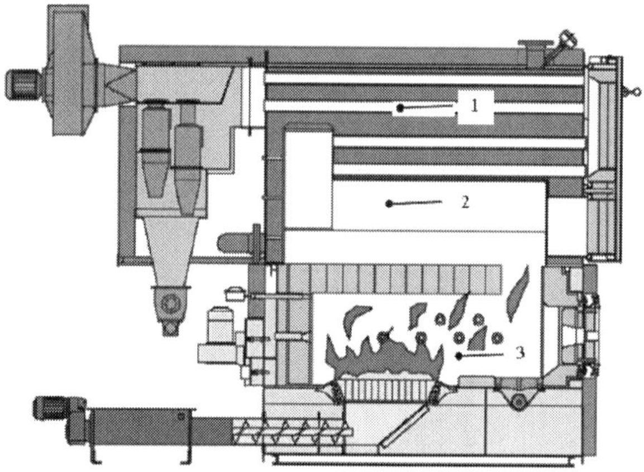

Figure 1. Solid biofuel incinerator scheme: 1—primary combustion chamber with fuel feeding system; 2—secondary combustion chamber; 3—exhaust gas and the heating system [1].

Refractory materials are used for biofuel boilers' internal structures (lining). These materials are non-metallic inorganic materials, which do not melt and do not decompose at high temperatures (600–2000°C). Main elements of lining (**Figure 2**) are made of shaped (bricks, blocks, etc.) and unshaped (concrete, mortar, coatings, etc.) refractories. They can be classified by general (chemical or mineral composition, refractoriness, porosity, etc.) and specific (type of binder and main raw material, forming method, etc.) features.

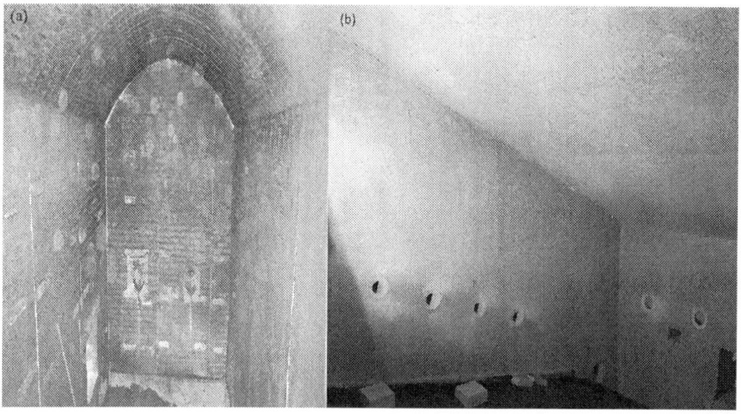

Figure 2. Biofuel boiler lining made of (a) fired bricks and (b) castables.

One of the most important groups in the unshaped materials classification is refractory castables. These are mixtures of refractory aggregates and bond(s), mainly supplied dry and

http://dx.doi.org/10.5772/65550

used after addition and mixing with water or another liquid. They are placed by casting with vibration, by casting without vibration (self-flowing), by rodding, by shotcreting or when necessary by tamping. Based on the standard EN ISO 1927-1:2012, they can be dense or insulating and divided into chemically and hydraulically bounded; the latter are then divided into regular and deflocculated. The above-mentioned standard contains reference castable classification scheme (**Figure 3**), which can be used when reviewing the variety of refractory castables.

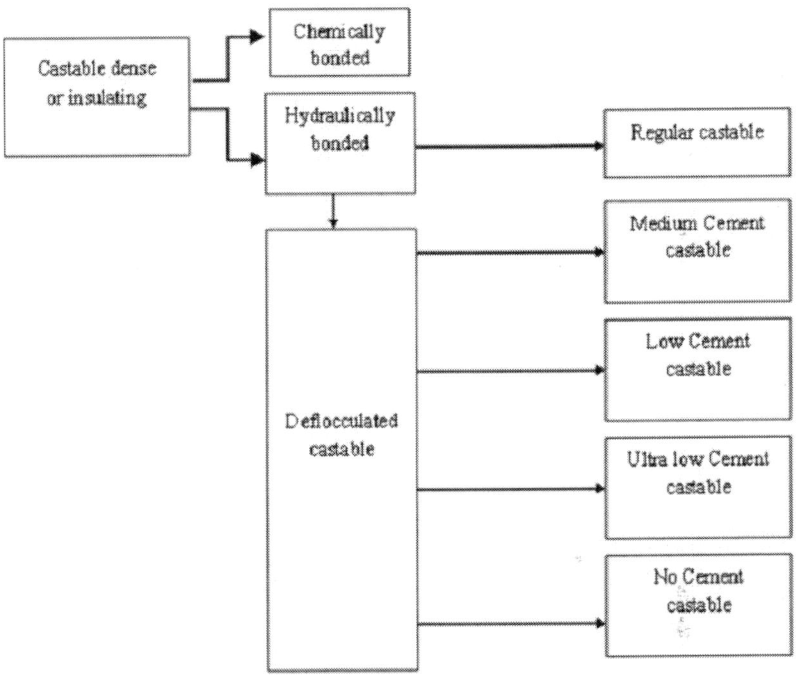

Figure 3. General classification scheme of dense or insulating castables according to standard EN ISO 1927-1:2012.

In **Figure 3**, refractory castables with hydraulic binders (calcium aluminate cement) are divided into groups according to the scheme depending on the amount of cement. Regular castable (RC) group includes castables with calcium aluminate cement content of up to 15–30%, medium cement content (MCC) group—8–15%, castables with low amount of cement (LCC) group—4–8%, castables group with ultra-low content of cement (ULCC)—1–3% and no cement group includes cast castables without cement (NCC). MCC, LCC and ULCC types of castables, compared with RC, contain special ultra-fine particles (less than 1 μm), and various deflocculants (soluble compound (usually an electrolyte) which, when added even in very small quantities, will reduce the water content in castable). Hydraulically bonded castables set and harden at ambient temperatures.

The manufacture of modern LCC or ULCC refractory castables with very low cement content often involves a number of process difficulties. Among them the problem of the loss of workability of castables, since these castables are sensitive to environmental temperature changes during manufacture, water quantity and quality, mixing parameters and other factors. For example, increase in water content by 2% in LCC and ULCC type of castable with chamotte filler reduces the cold crushing strength (CCS) of those castables after drying at 110°C and firing at the temperature of 1100°C from 80 and 90 MPa to 60 and 20 MPa, respectively [2]. MCC type refractory castables with performance characteristics much better than the traditional concrete are not as "sensitive" to the conditions of production; they are attractive to use in a variety of thermal equipment linings.

Not only castables with hydraulic binders, but also with other types of binders as well as gunning and ramming materials can be used in biofuel boilers. Another bond may be:

- a ceramic bond with hardening by sintering during firing;
- a chemical bond (inorganic or organic-inorganic) with hardening by chemical, but not hydraulic, reaction at ambient temperature or at a temperature lower than that of a ceramic bond; and
- an organic bond with binding or hardening at ambient temperature or at higher temperatures.

When choosing refractory linings for biofuel boiler, it is necessary to know the effect of loads attributable to the material during operation and to adapt it to the service properties of the materials [3] (**Table 1**). Before selecting the material, it is also important to check the material bond, in order to know about the need of special heat treatment. This will help to avoid discussions about setting times, progress of work, and the date of taking the boiler into operation [4].

Type of load	Properties
Thermal	Pyrometric cone equivalent (refractoriness), refractoriness under load, thermal expansion under load (creep), hot modulus of rupture, thermal expansion, reheat change (after shrinkage and after expansion) and thermal shock resistance
Thermo-technical	Thermal conductivity, specific heat, bulk density, thermal capacity and temperature conductivity
Mechanical	Crushing strength, abrasion resistance, cold modulus of rupture and deformation modulus, porosity and density
Chemical	Chemical composition, mineralogical composition and crystal formation, pore size distribution and types of pores, gas permeability, resistance to slag, glass melts, gases and vapours

Table 1. Important service properties of refractory materials [3].

Information about properties of refractory materials and their determination methods can be found in various works that are subject to EN or ASTM standards [4, 5]. Some of these information will be presented in this chapter.

2. Influence of operation conditions of biofuel boilers on durability of refractory materials

It is noted that thermal equipment that use solid biofuel experience significant increase in thermal and mechanical loads and chemical effects on lining and refractory materials. In some cases, sudden spalling of lining is observed in biofuel boilers as soon as after 1–2 years of use. The observation of the lining and the investigations of refractory materials used in various biofuel boilers show that the cause of poor durability of refractory materials is a combined impact of negative factors, such as high temperatures, an aggressive chemical effect of alkali compounds, an abrasive effect caused by solid particles, repeating thermal shocks and mechanical loads. It is noted that the risk of failure of materials highly increases with the increase in alkali when changing the type of fuel.

In biofuel boilers, depending on the type and sort of firewood used, different levels of ash and alkali metals (Na, K) are produced during combustion that adversely affects refractory materials. **Table 2** shows the content of alkaline oxides in different types of wood ashes [6]. Wood ash contains much more potassium than sodium; moreover, potassium diffusion to refractory material is faster than that of sodium. Therefore, while investigating the lining materials that spall in solid biofuel combustion devices, potassium compounds are found in corrosion products.

Types of wood	Ash content, %	Chemical composition of ash, %					
		SiO_2	CaO	Na_2O	K_2O	MgO	P_2O_5
Oak	0.51	0.01	0.37	0.02	0.05	0.02	0.03
Birch	0.26	0.01	0.15	0.02	0.03	0.02	0.02
Beech	0.55	0.03	0.31	0.02	0.09	0.06	0.03
Pine	0.29	0.04	0.14	0.01	0.04	0.03	0.02
Larch	0.25	0.01	0.07	0.02	0.04	0.07	0.03

Table 2. Various types of wood ash chemical composition (of total dry mass of wood) [6].

Although ash content in wood fuel or other solid fuels is low (up to several per cent), ash fusion characteristics have an impact on the properties of refractory materials because ash melt easily penetrates to the structure of the material. Ash melting behaviour depends on the type of fuel. With the ASTM D1857 standard, the changes in the shape of a standard ash cone by burning it in acidifying (oxidizing) environment are defined: initial deformation—IT; softening temperature—ST; the point of hemisphere formation—HT; and flow temperature—YP.

Table 3 presents fusibility characteristics of ash of some wood types [7]. As we can see, pine sawdust ash may adhere to the material of lining when the boiler's operating temperature reaches the ash ST of around 1180°C. Such ash begins to melt and flow at temperature of 1225°C. This means that pine sawdust fuel greatly increases the chemical effects on the refractory materials given the boiler operation if the local temperature (e.g. in the secondary combustion chamber) is around 1200°C.

Fuel	Values of melting characteristics, °C			
	IT	ST	HT	YP
Woodchips total, pine	1210	1225	1250	1275
Slashings	1175	1205	1230	1250
Sawdust, pine	1150	1180	1200	1225
Bark, spruce	1405	1550	1650	1650
Bark, pine	1340	1525	1650	1650

Table 3. Fusibility characteristics of wood ash [7].

2.1. Alkali effects at high temperatures

In the combustion chamber, under reducing environment alkali metals react with the refractory lining material. There are two different types of alkaline reactions with refractory materials: in dry conditions under the influence of alkali vapour or in humid environment, when melt ash is formed on the surface of the refractory material.

Potassium or sodium released during combustion reacts with CO gas [8]:

$$2K + CO \rightarrow K_2O + C \quad (1)$$

$$2K + 3CO \rightarrow K_2CO_3 + 2C \quad (2)$$

$$K_2CO_3 + CO \rightarrow 2K + 2CO_2 \left(> 930°C;\ \text{potassium} - \text{vapour}\right) \quad (3)$$

Potassium vapour over time can penetrate into the refractory material to the depth of more than 100 mm [8].

When refractory material is exposed to alkali vapour or melt, it may form the following compounds: kalsilite ($K_2O \cdot Al_2O_3 \cdot 2SiO_2$), leucite ($K_2O \cdot Al_2O_3 \cdot 4SiO_2$), feldspar ($KAlSi_3O_8$, $NaAlSi_3O_8$, $CaAl_2Si_2O_8$) and others [9]. Formation of this type of minerals in the refractory material increases its volume by 15–30% and sometimes even 55%, compared to the initial volume of the material. This promotes the formation of porous structure (**Figure 4**) [10], micro-cracks in refractory material and the spalling degradation due to alkali effects.

Under wet conditions when the melt forms on the surface of the refractory material, it can lead to reactions that reduce the temperature of melt formation [11]:

$$K_2CO_3 + SiO_2 \rightarrow K_2O \cdot SiO_2(\text{liquid}) + CO_2(\text{first melt} < 1000°C) \quad (4)$$

$$K_2CO_3 + 3Al_2O_3 \cdot 2SiO_2 \rightarrow K_2O \cdot Al_2O_3 \cdot 2SiO_2 + 2Al_2O_3 + CO_2(\text{first melt} > 850°C) \quad (5)$$

Further, $4K_2O \cdot CaO \cdot 10SiO_2$ first melt <950°C.

Thus, spalling of refractory material due to the effect of alkali may be intensive even with reduced operating temperature of the boiler.

Figure 4. Mullite brick structure: (a) undamaged layer and (b) porous layer affected by alkali [10].

It should be noted that the possibilities of the melt penetration into the structure of the material depend on the porosity of refractory materials, effective potential of pores and capillaries, etc. Thermal expansion coefficient of melt is significantly different from the thermal expansion coefficient of the refractory material. Therefore, with cooling material (e.g. when the boiler is stopped), expansion differences of unaffected refractory materials and its areas saturated with melt, cause stresses leading to the layering, crumbling and destruction of the product. **Figure 5** shows characteristic nature of disintegration of shaped refractories when affected by ash melt [12].

Resistance of refractory materials to alkaline compounds is often measured with the crucible method [9, 13]. When analysing, the alkali resistance of castables or fired bricks with test samples with a cylindrical cavity are made. The cylindrical cavity is filled with certain alkali salt (K_2CO_3, K_2SO_4, etc.), and the samples are heated for some time at the temperature of ≥1000°C. After multiple tests (each time anew by adding a fixed amount of salt), the samples are visually inspected, capturing the occurrence of micro-cracks. Some of the specimens are cut along the cylindrical axis into two parts, and the depth of the material affected by alkaline substances is evaluated.

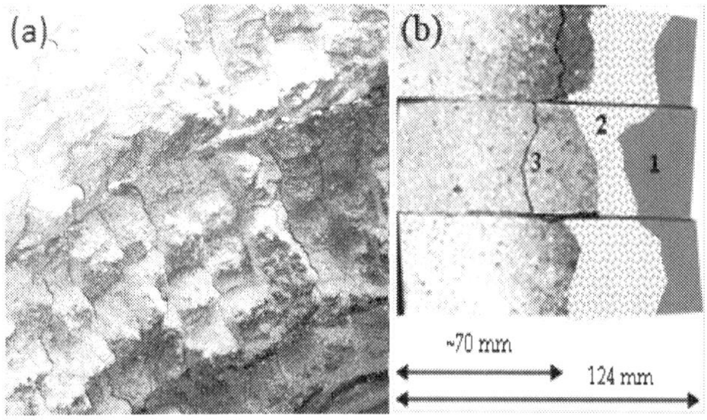

Figure 5. Typical fragmentation of bricks affected by ash melt (a) and scheme of a bricks' step-by-step degradation (b): 1,2,3—steps of degradation.

2.2. Resistance to the impact of carbon monoxide (CO)

Incompletely burned carbon compound products, the main of which is carbon monoxide (CO) can penetrate (diffuse) in the material and react with refractory materials containing iron oxide. In such a case, four-step reaction occurs in which one of the end-products is Fe_3C [3]:

$$3Fe_2O_3 + CO \rightarrow 2Fe_3O_4 + CO_2 \tag{6}$$

$$Fe_3O_4 + CO \rightarrow 3FeO + CO_2 \tag{7}$$

$$FeO + CO \rightarrow Fe + CO_2 \tag{8}$$

$$3Fe + 2CO \rightarrow Fe_3C + CO_2 \tag{9}$$

Fe_3C can react with CO:

$$20Fe_3C + 14CO \rightarrow 3Fe_{20}C_9 + 7CO_2 \tag{10}$$

$$3Fe_{20}C_9 \rightarrow 20Fe_3C + 7C \tag{11}$$

If the refractory material contains metallic iron and/or iron oxides, CO in the temperature interval of 400–800°C produces carbon: $2CO \rightarrow CO_2 + C$. Mechanical stresses caused by crystallization of carbon deposited in local areas may cause complete disintegration of the material. It has been found [14] that degradation of certain types of aluminosilicates due to CO exposure is a result of two interrelated processes: reduction in iron oxides and volume changes and carbon formation and its accumulation in the material structure.

It has been observed that when excessive CO has been formed in the boiler for extended periods (disrupted boiler operational mode), disintegration of refractory castable with high iron oxide content (>4.4%) due to the general effect of CO and alkali occurred already after 8 months of operation (**Figure 6**) [15].

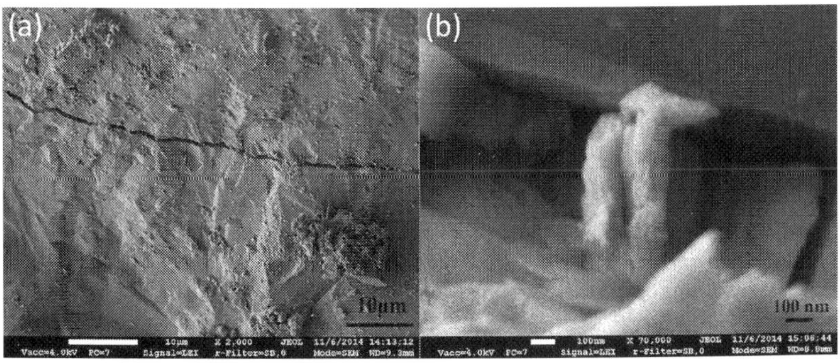

Figure 6. Cracking (a) and degradation of aluminosilicate materials due to the formation of new compounds (carbon and leucite) in its structure (b) [15].

Risks of refractories degradation due to CO can be reduced by using materials with as low amount of Fe_2O_3 as possible (<1%).

2.3. Thermal shock resistance

This indicator shows the ability of refractory materials to resist thermal stresses in its structure from temperature gradients. Such temperature gradients cause degradation of refractory materials when boiler is often stopped (material is cooled) and start to operate (material is heating up). Burning of solid biofuels generates a lot of fly ash that cause fouling of heat transfer surfaces. As a result, boilers must be frequently stopped for cleaning and therefore linings experience repeated thermal shocks. Different countries apply different methods [16] to determine thermal shock resistance (number of cycles) of refractory materials, which vary by sample size, heating temperature and sample cooling method (water, air, water-cooled panels). It is noted that the thermal shock resistance of the refractory material may differ depending on the selected method [17]. Where it is difficult to evaluate test results obtained in one or another method, thermal shock resistance criteria R_4 and R_{st} are calculated [18, 19].

The thermal shock resistance of refractories can be evaluated not only by calculating the thermal shock resistance criteria, but also by the refractory material surface appearance after thermal shocks—test sample heating and cooling cycles. In the case of a low thermal shock resistance refractory castable, a network of long cracks appears on the surface (**Figure 7a**).

Meanwhile, in the case of high thermal shock resistance refractory castable, a network of short cracks is formed (**Figure 7b**). Such fragmental structure of the castable compensates its thermal extensions and relaxes its stress. Therefore, when the number of cycles was increased the cracks slightly widened but the castable did not collapse.

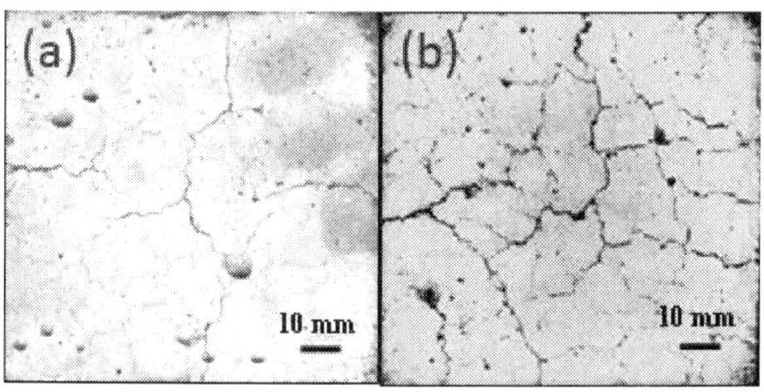

Figure 7. Surface of castable before break up of sample: (a) which withstood 9 cycles (water—800°C) [20] and (b) 45 cycles [21].

Refractory materials that have structures with built-in micro-cracks show better thermal shock resistance than rigid systems. In some refractory materials, the bond possesses micro-structural defects or cracks that provide better thermal shock resistance [4].

2.4. Abrasion resistance

Abrasion resistance is a feature of material to it surface that resists external mechanical effect when solid particles fly at a high speed and mechanically rubs the material surface.

Refractory materials used in chemical and cement plants, when process products intensively circulate and rubs the surface of refractory material, must have a high abrasion resistance. Abrasion resistance is determined according to standard ASTM C-704:1999. The abrasion resistance rate of materials used under the above-mentioned conditions must not esceed 5–6 cm^3.

In biofuel combustion plants, abrasion resistance of refractory materials is relevant when the fluidized bed system (movement of a mixture of sand and fuel) is used in the technology and also when during boiler pipe blowing off (clean procedure) ash particles fly at a high speed. **Figure 8** shows a fragment of cross section of a fireclay brick where the surface in the bottom part of the picture has been exposed to high speed particles flying at the direction marked with the arrow.

Abrasion resistance and compressive strength are correlated with each other: the higher the compressive strength, the greater its abrasion resistance. In this regard, strength characteristics of refractory materials used in biofuel combustion equipment must be maximally high.

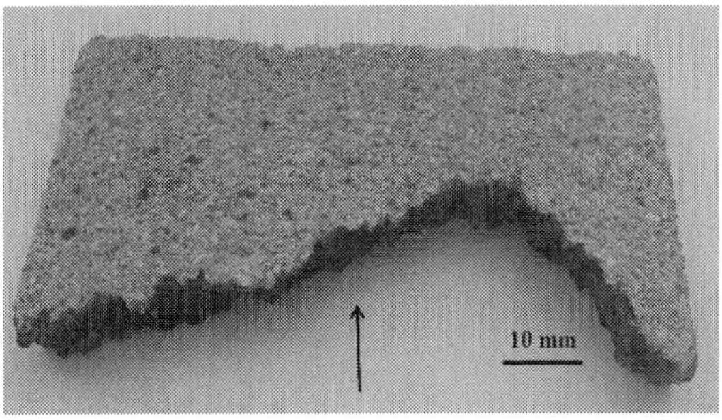

Figure 8. Sectional fragment of fireclay brick after 3 months of exploitation in solid biofuel combustion lining. Particles flying at the direction marked with the arrow mechanically affected the brick surface in the bottom part of the picture.

2.5. Carbonation of calcium aluminate cement-bonded regular refractory castable

The observations showed that the lining of domestic boilers made by using regular castables do not have long durability. Having been exploited for some time, it destructs. One of the reasons that cause this destruction might be the so-called "carbonation" of calcium aluminate cement hydration products. It is known that the main hydration products formed during the reaction between calcium aluminate cement and water are as follow: CAH_{10} (forms at the temperature <21°C), C_2AH_8 and AH_3 (21–35°C) and C_3AH_6 and AH_3 (>35°C) [22]. The carbonation of calcium aluminate cement hydration products is thought to occur by the following reactions [23]:

$$CAH_{10} + CO_2 + xH_2O \rightarrow CaCO_3 + Al_2O_3 \times yH_2O + (10 + x - y)H_2O \quad (12)$$

$$C_2AH_8 + 2CO_2 + xH_2O \rightarrow 2CaCO_3 + Al_2O_3 \times yH_2O + (8 + x - y)H_2O \quad (13)$$

$$C_3AH_6 + 3CO_2 + xH_2O \rightarrow 3CaCO_3 + Al_2O_3 \times yH_2O + (6 + x - y)H_2O \quad (14)$$

Carbonation causes a large-scale destruction of calcium aluminate cement materials [24] when Na^+, K^+ ions participate in the so-called "alkaline hydrolysis" [25]. CO_2, alkalis and H_2O environment is typical for domestic boilers during often stopping and starting of operations. After the calcium aluminate cement hydration products dehydration at the temperature of 500–800°C, $C_{12}A_7$ is formed, which, after heating at 1000°C is converted to CA, CA_2. If the operation temperature is less than 1000°C (usually in domestic boiler), $C_{12}A_7$ in humid environment (in moment of stopping and starting of boiler operation) is repeatedly hydrated. Then carbonation of hydrates occurs (**Figure 9**), and the destruction of castable will start. It was established that the additive of micro-silica (SiO_2) in regular castable increases its resistance to the carbonation [25].

Figure 9. C_3AH_6 hydrates (a) and its carbonation products (CP) (b).

2.6. Destruction of SiC-based refractory materials

Studies have shown that castables with SiC filler resist much better the effects of alkali compounds than those with aluminosilicate filler (fireclay and mullite) [13]. It should be noted, however, that in the oxidizing atmosphere at >900°C SiC castable filler can oxidise resulting in the formation of SiO_2 and higher volume of minerals. The reaction takes place according to the following scheme [26]:

$$2SiC + 3O_2 \rightarrow 2SiO_2 + 2CO \tag{15}$$

Reverse reaction in castable with SiC may occur under reducing environment [11]:

$$SiO_2 + 3H_2 + CO \leftrightarrow SiC + 3H_2O \tag{16}$$

Because of mineralogical changes of structural elements of refractory material with SiC, the strength is critically reduced.

3. Materials for working layer of linings of biofuel boilers and its investigations

Over the last decade it has been noted that the use of shaped products is reducing, while the use of unshaped materials such as refractory castables is constantly growing. This is due to the shortcomings of shaped products: long duration of installation of the lining in thermal equipment, complex repairs, complex design and manufacturing technologies of thermal equipment from shaped products and higher cost of production of shaped products.

Research shows [9, 13] that in alkali-resistant castables, under the influence of alkali on the surface of the material, a layer of glass of high viscosity is formed, which prevents further penetration of alkali into the material.

The aim of investigations [26] was to evaluate the resistance to potassium compounds' attack on refractory castables, modified and unmodified, by additive of milled quartz sand (SiO_2). The findings are presented below.

Unmodified commercial fireclay castable (B0) and unmodified clinker castable (B1) and modified clinker castables (B2, B3), in which ground quartz sand was used to increase alkali resistance, were tested. Chemical composition (mass %) of castables B0, B1, B2, B3 was as follows: B0—Al_2O_3 45.7; SiO_2 43.6; CaO 7.6; Fe_2O_3 1.50; B1—Al_2O_3 42.9; SiO_2 25.5; CaO 27.3; Fe_2O_3 1.77; B2—Al_2O_3 41.8; SiO_2 27.3; (2.5% of this quantity has ground quartz sand additive); CaO 26.6; Fe_2O_3 1.79; B3—Al_2O_3 40.9; SiO_2 29.0 (5.0% of this quantity has quartz sand additive); CaO 26.0; Fe_2O_3 1.80 [26]. **Table 4** presents technical characteristics of castables used in alkali tests with potassium carbonate salt by crucible method.

Characteristics	The mark of the castables			
	B0	B1	B2	B3
Cold crushing strength, MPa	53	115	115	114
Open porosity, %	27.4	14	14	14
Bulk density, kg/m^3	2070	2460	2450	2420
Shrinkage, %	0.2	0.28	0.28	0.28
Thermal shock resistance (950°C—water), cycles	19	13	10	8

Table 4. Technical characteristics of fireclay and clinker refractory castables after firing at the temperature of 1100°C [26].

Macroscopic assessment of samples is presented in **Table 5**. It was found that the samples of commercial fireclay castable B0, affected by K_2CO_3, cracked after 1 cycle (**Figure 10a**) and after 2 cycles split into multiple fragments. The analysis of the surface view of the sample cut along the cylinder bore axis (**Figure 10b**) shows changed zones because of alkaline impact (penetration depth ~11 mm) [26].

Clinker-based castable B1 without additives during the alkali test split into separate fragments after 3 cycles (**Table 5**), while clinker castable with ground quartz sand additive (B2, B3), depending to its quantity, split after 6–8 cycles.

The analysis of the surfaces of sawn samples of clinker castable B1 without additives and B3 with ground quartz sand additive (**Figure 11**) shows that decomposition products of potassium carbonate salt already in the first cycle are easily penetrated into the structure of castable without additive (similar as with commercial fireclay castable, **Figure 10b**). Potassium carbonate salt decomposition products penetrated the structure of the castable B3, modified with ground quartz sand additive, with more difficulty. After 3 cycles, a protective layer of 2–

3 mm was observed (in some places up to 8 mm), capturing the penetration of potassium carbonate salt decomposition products into the material (deeper) (**Figure 11b**). This increased the resistance of castable samples to alkaline compounds—samples cracked just after 8 cycles [26].

Macroscopic assessment	The mark of the castables			
	B0	B1	B2	B3
The number of cycles that caused appearance of cracks of a width over 0.4 mm	1	1	3	5
The number of cycles that caused disintegration of the specimen to two or more fragments	2	3	6	8

Table 5. Macroscopic assessment of fire clay and clinker refractory castables, affected by K_2CO_3 [26].

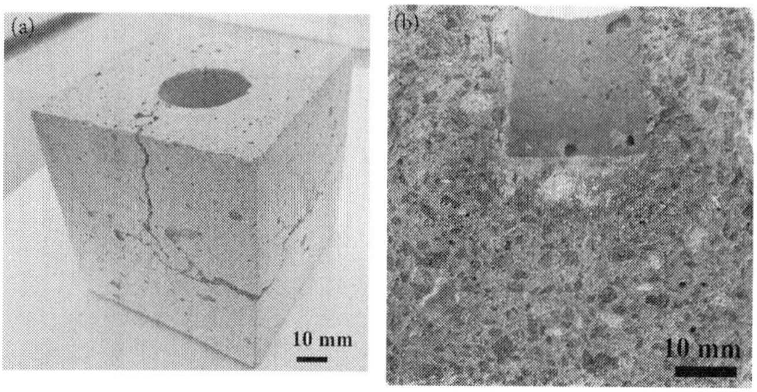

Figure 10. The view of specimens of commercial fireclay castable after the tests with alkali compounds: (a) appearance of over 0.4 mm wide cracks and (b) the section view of the specimens after one cycle [26].

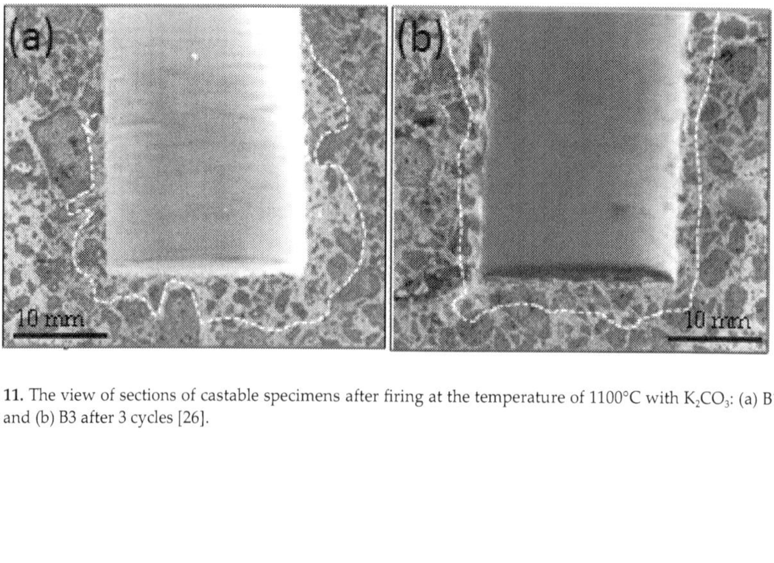

Figure 11. The view of sections of castable specimens after firing at the temperature of 1100°C with K_2CO_3: (a) B1 after 1 cycle and (b) B3 after 3 cycles [26].

The phase composition of substances formed during the reaction with K_2CO_3 was found with the tablet method [13]. The results are provided in **Table 6**. For comparison, the table also contains the phase composition of products formed in fired castables at the temperature of 1100°C in the absence of the effect of K_2CO_3. These data show that commercial fireclay castable B0 contains the following minerals after firing at the temperature of 1100°C: gehlenite (C_2AS), mullite($3Al_2O_3 \cdot 2SiO_2$) and quartz (SiO_2). The resistance test to alkaline compounds allowed to identify new products in this castable—feldspars and leucite. In clinker castables, without additive (B1) and with ground quartz sand additive (B2, B3), a new product leucite was also identified. Test results of the tablet method suggest that in all cases, both in absence and presence of quartz sand additive in clinker castable, during the reaction of its compounds with K_2CO_3 decomposition products, leucite is formed. However, the tablet method, which allows to identify the chemical composition of compounds occurring from the reaction, does not allow to assess a very important factor of castable corrosion—diffusion rate of corrosion-causing substances deeper into the castable. So, a comparison of the penetration depth of fireclay castable B0 without ground quartz sand additives (**Figure 10b**) and clinker B1 (**Figure 11a**), with the penetration depth of castable with quartz additives B3 (**Figure 11b**) shows that in the case of castable B3, diffusion was stopped. Apparently, the reaction of grounded quartz with decomposition products of K_2CO_3 resulted in a viscous layer of this reaction product inhibiting the penetration of alkaline compounds deeper into the sample. Therefore, a destruction and disintegration of the specimens caused by formation of corrosion products and different thermal expansion coefficients of the initial material and zone saturated with the melt in the castable with ground quartz sand additive appeared considerably later.

The mark of composition	The treatment method	The phase identified
B0	1100°C	Gehlenite, mullite, quartz, hematite
	After test	Gehlenite, mullite, quartz, hematite, feldspars, leucite
B1	1100°C	Gehlenite, CA_2, CA, anorthite, corundum
	After test	Gehlenite, CA_2, CA, anorthite, corundum, leucite
B3	1100°C	Gehlenite, CA_2, anorthite, corundum
	After test	Gehlenite, CA_2, anorthite, corundum, leucite

Table 6. The phase composition of fireclay and clinker refractory concretes before and after test with K_2CO_3 upon applying the tablet method and firing at the temperature of 1100°C [26].

The above test results show that often traditional fireclay materials used in biofuel boilers are not resistant to the effects of alkaline compounds. Refractory materials recommended for biofuel boiler lining should be examined in laboratories to evaluate the alkaline salt penetration into the material.

3.1. Explosive spalling of refractory castable

Calcium aluminate cement-based refractory castable should be dried and heated up after curing for moisture removal. In the process of heating, the temperature is gradually raised until the operational temperature (1000–1200°C) of the boiler is achieved. During heating up of the castable, chemical and physical processes causing the removal of chemically bound water and formation of new crystalline phases take place. All these processes also cause great changes in the micro-structure of a castable and pose a threat of its explosive spalling [27, 28]. In **Figure 12**, part of the structure of the heating unit used in oil refinery, damaged by explosive spalling, is shown.

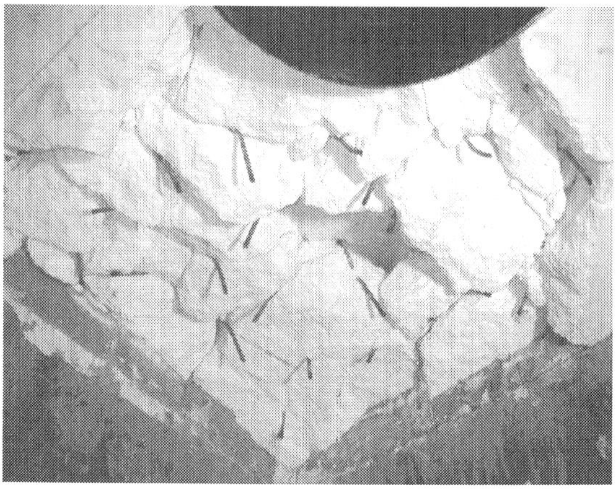

Figure 12. The part of the combustion zone structure in the heating unit used in oil refineries damaged by explosive spalling (metal anchors can be seen on the photograph) [29].

Explosive spalling is usually caused by water vapour pressure, which builds up when chemically bound water is turned into free water. The risk of explosion of the structure is greatly increased, if the following types of castable are used: MCC, LCC and ULCC. To avoid explosive spalling of refractory castable due to the pressure of water vapours developed at the initial stage of heating, new produced linings of thermal equipment are dried and heated up for the first time in a very careful way [29]. But in the case of biofuel boilers, in practice, it is hardly technically possible to perform the procedure of castable drying accurately. Therefore, in order to reduce a risk of explosive spalling, when castable drying and the initial heating modes are not rigorously controlled, various additives (e.g. aluminium powder, polymer fiber, etc.), which increase castable permeability by forming a capillary system for removing water vapour without damaging the castable, are used. Aluminium powder reacts with water in the alkaline medium, releasing hydrogen, which causes the formation of open porosity in castable and makes it more easily permeable to water vapours. However, though the addition of aluminium powder increases castable permeability to water vapours, a loose structure is formed; therefore, the mechanical properties of the castable is decreased.

It has been found that the additive of polypropylene fibres (PPF) (**Figure 13a**) is well suited for decreasing the risk of explosive spalling of refractory castables [29]. A positive effect of this additive, with regard to its ability to decrease the risk of explosive spalling, is explained by the fact that PPF disintegrates at the temperature of 150–180°C, leading to the formation of micro-channels (**Figure 13b**), allowing water vapours to pass through, and help to avoid a dangerous rise of pressure.

The testing of cylindrical MCC-type castable specimens, for their resistance to explosive spalling [29], has shown that the MCC-type specimen without of PPF additive explode at the temperature of 600°C under the conditions when temperature is raised to 1000°C at a rate of 40°C/min (**Figure 14a**). The specimen with PPF additive does not explode when the temperature is raised at the same range in heating up to 1000°C (**Figure 14b**).

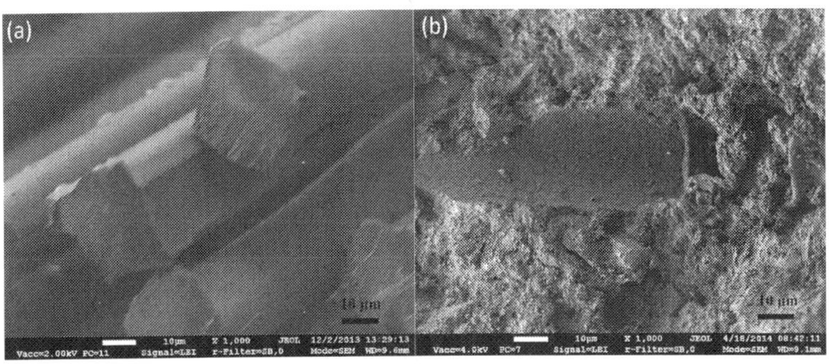

Figure 13. SEM micrographs of the PPF (a) and the micro-structure of refractory material with burned PPF after heating at 170°C (b) [29].

Figure 14. Castable specimens tested for explosive spalling, when the temperature was raised at the rate of 40°C/min: (a) castable sample without PPF additive exploded, when the temperature was raised to 600°C and (b) castable sample with PPF additive that did not explode, when the temperature was raised to 1000°C.

In order to simplify the drying and the first heating procedure, and to reduce the risk of explosive spalling, expensive NCC-type castables [30] are used. Such castables considerably reduce the time of drying and the first heating procedure.

3.2. Recommendations for use of refractory materials in biofuel boilers

Due to the aggressive operating conditions in biofuel incineration plants, manufacturers of refractory materials use the following specific shaped and unshaped materials: fireclay with a small amount of iron oxide, silicon carbide (SiC), mullite, zirconia, andalusite and chrome (**Table 7**) [11]. Fireclay, mullite and andalusite materials belong to the Al_2O_3–SiO_2 (aluminosilicate) system.

Material base	Castables						Fired bricks			
	Fireclay	Alumina, mullite, SiC	Alumina-Zirconia-Silica		Mullite	Alumina Chrome	Andalusite	Alumina-Zirconia-Silica	Alumina Chrome	
Recommended application, °C	≤1000	≤1300	≤1200	≤1200	≤1200	≤1600	≤1200	≤1300	≤1400	≤1600
SiO_2, %	<40	6	16	16	13	<4	36	18	6	1.3
Al_2O_3, %	>50	60	27	56	58	>90	61	>50	87	>90
Fe_2O_3, %	<1	<1	<1	<1	<1	<0.5	<1	<0.5	<0.5	<0.5
ZrO_2, %	–	–	–	26	–	–	–	28	–	3.5
Cr_2O_3, %	–	–	–	–	–	5	–	–	5	5
SiC, %	–	30	>55	–	>25	–	–	–	–	–
Bulk density, kg/m^3	2200	2800	2500	3000	2800	3100	2600	3100	3000	3200
CCS, MPa	60	80	55	110	90	60	100	140	100	100
Thermal shock resistance DIN 51068/1	30	120	120	100	120	30	120	100	120	120

CO resistance ASTM C 288 always A, not relevant for alumina-chrome materials.

Table 7. Lining recommendations for biomass combustion furnaces [11].

It is stated [11] that refractory materials, suitable for use in biofuel boilers installations, should be dense (>2200 kg/m^3), with CCS of at least 50 MPa, thermal shock resistance >30 cycles (under DIN 51068-1:1976 standard) and iron oxide content less than 1%.

However, it should be noted that the recommended high-grade materials with zirconium and chromium fillers are considerably more expensive than with fireclay and andalusite fillers. These materials are very dense (≥3000 kg/m^3), their heat transfer coefficient is high and reaches up to 1.6–2.5 W/(m·K); therefore, the lining increases the need for insulating materials. After the lining operation time, refractory materials containing chromium oxide must be disposed of in hazardous waste landfills because their processing is complicated.

In practice, up till now the most widely used materials in solid biofuel combustion plant linings are bricks and castables with aluminosilicate fillers such as fireclay and andalusite. Selection of this type of material for biofuel boiler linings must belong to the alkali resistance class of materials and their general characteristics should be no less than that specified in **Table 7**. In **Figure 15(a)**, a view of lining made of non-alkali resistance fire clay bricks, damaged by alkali attack after 6 months of boiler operation, is shown, and in **Figure 15(b)**, a view of non-damaged lining made of alkali resistance fire clay bricks after 8 months of boiler operation is shown. The main reason of the difference between resistance of these linings was the different alkali resistance class of the materials used for their production.

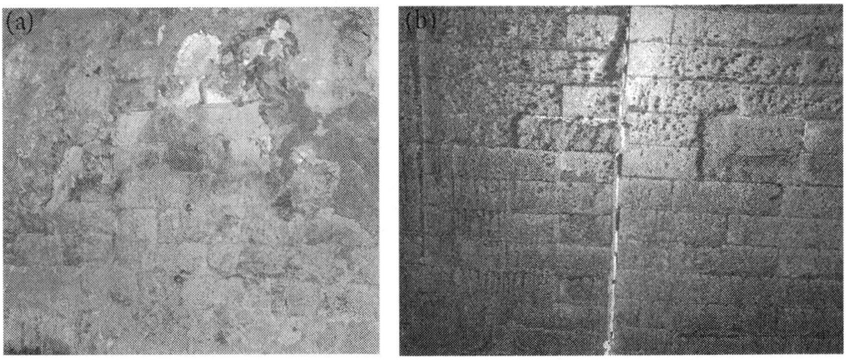

Figure 15. View of lining made of not alkali resistance fire clay bricks after 6 months of boiler operation (a) and made of alkali resistance fire clay bricks after 8 months of boiler operation (b).

It is also necessary to note that not only quality, but also suitability for high durability lining of used materials is of great importance. The correct installation of lining is very important as well. Especially in the case of installation of monolithic lining such key quality control elements such as installation monitoring, as-installed testing, pre-dryout inspection, dryout monitoring and post-dryout inspection are necessary. Some standards [31, 32] can be useful for organization of quality control for the installation of biofuel boilers lining.

Author details

Valentin Antonovič[1*], Jacek Szczerba[2], Jadvyga Keriene[1], Rimvydas Stonys[1] and Renata Boris[1]

*Address all correspondence to: valentin.antonovic@vgtu.lt

1 Vilnius Gediminas Technical University, Vilnius, Lithuania

2 AGH University of Science and Technology, Kraków, Poland

References

[1] Firewood and wood chips for heating. Available from: http://www.carmen-ev.de/files/informationen/hackschn.pdf. [Accessed 2016-07-19].

[2] Parr C, Bier T A, Bunt N E, Spreafico E. Calcium aluminate cement (CAC) based castables for demanding applications. Proceedings of the 1st International Conference on monolithics, Tehran, Iran 1997: 11 p.

[3] Didier-Werke AG. Refractory materials and their properties. Germany: Vulkan Verlag GmbH; 1989: 103 p.

[4] Schacht CA. Refractories handbook. Florida, USA: CRC Press, Taylor and Francis Group; 2004: 520 p.

[5] Routschka G, Wuthnow H. Refractory materials: structure, properties and testing methods: directory. Essen, Germany: Vulkan Verlag GmbH; 2010: 392 p.

[6] Polubojarinov OI. The density of wood. Moskva: Lesnaja promyshlennost; 1976: 160 p.

[7] Vares V, Kask U, Muiste P, Pihu T, Soosaar S. Manual for biofuel users. Manual. Tallinn, Estonia: TUT Press; 2005: 178 p.

[8] Riedl R, Dahl J, Obernberger J, Narodoslawski M. Corrosion in fire tube boilers of biomass combustion plants. Proceedings of the China International Corrosion Control Conference. Paper Nr. 90129. Beijing, China 1999: pp. 1–5.

[9] Alibasic E, Oldin J, Kannabiran S, Yiakoumi A. Design of castables and their relevance to alkali resistance applications. 57th International Colloquium on Refractories: September 24th and 25th, EUROGRESS, Aachen, Germany. ECREF; 2014 pp. 67–69.

[10] Antonovič V, Keriene J, Stonys R, Boris R, Škamat J. Study of the Structure and Phase Composition of the Refractory Comprising the Lining of a Boiler Operating on Solid Biofuel, Refractories and Industrial Ceramics. New York: Springer International Publishing; 2015, Vol. 56(2): pp. 188–193.

[11] Klinger W, Weber W, Zimmermann H. Application concepts of shaped and unshaped refractories for combustion plants. Proceedings of the Unified International Technical Conference on Refractories (UNITECR). Dresden, Germany; 2007: pp. 440–443.

[12] Goberis S, Antonovič V. The monolithic refractory lining of boiler fired by wood fuel. Novye Ogneupory 9; 2003: pp. 13–15

[13] Szczerba J, Boczon A, Studencka B. Studies of impact of alkali salts on alumina based castables without additives and with them. Ceramika 88; 2005: pp. 158–172.

[14] Pitac NV, Shapalov VS, Azarenkova LE, Privalova NG. Influence of carbon monoxide and its decomposition catalysts on the destruction of blast-furnace aluminosilicate refractories. Refractories 1991; 32: pp. 578–584.

[15] Keriene J, Boris R, Antonovič V, Stonys R, Škamat J. Action of the products of biofuel combustion on the phase composition and structure of refractory material, glass and ceramics. New York: Springer; 2016, Vol. 72(9–10): pp. 345–350.

[16] Goberis S. Thermal stability of unshaped refractory materials. Refractories and Industrial Ceramics 2003; 44(62): pp. 427–430.

[17] Antonovič V, Šukšta M, Pundiene I, Stonys R. Procedural elements in estimation of the thermal shock resistance of different types of refractory concrete based on chamotte filler. Refractories and Industrial Ceramics 2011; 52(1): pp. 70–74.

[18] Hasselman DPH. Unified theory of thermal shock fracture initiation and crack propagation in brittle materials. Journal of the American Ceramic Society 1969; 52(11): pp. 600–604.

[19] Kingery WD, Bowen HK, Uhlmann DR. Introduction to ceramics. New York: John Wiley; 1979: 1032 p.

[20] Goberis S, Antonovič V. Properties of refractory castable based on the Portland cement and with the addition of microsilica, Novye Ogneupory 5; 2002: pp. 33–36.

[21] Antonovič V, Goberis S. The properties of refractory concrete with liquid glass binder and chamote aggregate. Materials Science 2002; 8(2): pp. 189–192.

[22] Majumdar SD, Sarkar R, Vajifdar PP, Narayanan S, Cursetji RM, Chatterjee AK. User friendly high refractory calcium aliuminate cement. Proceedings of the International Conference of Calcium Aluminate Cements; 2001: pp. 467–476.

[23] Sakamoto S, Kudo E. Carbonation of alumina cement-bonded castable refractories. Journal of Technical Association of Refractories 2000; 20(1): pp. 18–23.

[24] Pundiene I, Goberis S, Antonovič V, Špokauskas A. Carbonation of alumina cement-bonded conventional refractory castable in fireplace. Materials Science 2006; 12(4): pp. 333-337.

[25] Garcia Alcocel E, Garces P, Chinchon S. General study of alkaline hydrolysis in calcium aluminate cement mortars under a broad range of experimental conditions. Cement and Concrete Research 2000; 30: pp. 1689–1699.

[26] Boris R. The investigation on the structure and properties of medium cement refractory castable with clinker aggregate (doctoral dissertation). Vilnius: Technika; 2016: 105 p.

[27] Vadasz P, Kamoda O, Čipkova I. Corrosion of refractory concrete monoliths on the basic of Al_2O_3-SiO_2 and Al_2O_3-SiO_2-SiC in the process of municipal waste incineration. Interceram Refractories Manual 2009; 3: pp. 41–46.

[28] Gitzen WH, Hart LD. Explosive spalling of refractory castables with calcium aluminate cement. American Ceramic Social Bulletin 1961; 40(8): pp. 503–507.

[29] Boris R, Keriene J, Antonovič V, Madej D. Characterization of microstructural evolution and mechanical properties of refractory composite. In: Composite Structures. Oxford: Elsevier; 2015. Vol. 134: pp. 811–819.

[30] Luz AP, Braulio MAL, Pandolfelli VC. Refractory castable engineering. Leipzig: Deutsche Bibliothek; 2015: 734 p.

[31] Standard specification 3-24-4: Vibracast refractory lining V?anchors. UOP LLC. Illinois, USA; 2011: 38 p.

[32] API Standard 936: Refractory installation quality control. Inspection and testing. Monolithic refractory linings and materials, Third Edition. American Petroleum Institute. Washington, USA; 2008: 49 p.

Chapter 23

Power Form Agripellets

Claudia Santibáñez Varnero and

Marcela Vargas Urrutia

Additional information is available at the end of the chapter

http://dx.doi.org/10.5772/110404

Abstract

Currently, the production of thermal energy by biomass has shown a clear trend toward densified biofuels (pellets). This is due to their consistent size and shape that can be more easily delivered to homes, businesses, and power plants and can be automatically fed into advanced pellet boilers in a controlled and calibrated way. The use of densified biofuels also reduces the costs associated with handling and transportation, due to the increase in density involved by densification process. Demand for wood pellets is currently growing at a faster rate than supply in Europe. It is estimated that pellet market is growing to 50 Mt year^{-1} by 2025; however, most wood waste is already committed for pressed wood products and pellets, therefore more supply of raw materials are needed. With the possible shortage of woody raw materials for pellet production and considering the low forestry residues potential in several countries, agricultural residues could be largely used in the future for fuel pellets manufacturing. Agricultural pellets, as well known as "agripellets", are emerging and promising. However, they have certain differences compared to conventional wood pellets.

Keywords: agripellets, biomass, solid biofuels, agricultural residues

1. Introduction

The growing domestic and industrial demand of biomass for heat and power production in Canada, United States, Europe, and China has resulted in a strong growing global pellet marked during the last decades, and continuous growth of the market is predicted for the next years [1]. It is estimated that the demand for pellets will be triples from 2012 to 2020, rising from 16 to 46 million metric tons per year [2]. In the pellet production, there is a shortage of woody raw materials, and the price of the wood raw material increases. Considering that only woody pellets from forestry residues have already successfully established technologies and markets for production and consumption in these countries, it is necessary to focus on studying the pelletization of new sources of raw materials. Agricultural residues can be one of the potential alternative feedstock since it is abundantly available and at low cost. In the near future, agricultural residues have a tremendous potential in biomass pellets industry. It is therefore of great interest to study the characteristics of this new category of raw material, paying special attention to the problems that they may trigger both at production and utilization level. At a technical level, the main difference between wood pellets and agripellets is the somehow higher friability, the slightly lower energy content and the higher ash content of the latter.

The future of agripellets as a carrier of energy from biomass appears to be promising. This is influenced by many factors. Popularity of biomass is motivated by aspects of various types, such as:

- Economic: energy production from biomass is the least capital intensive process compared with other forms of renewable energy. Besides, the production of pellets can help to stimulate economic activity and reduce unemployment.
- Policy: European Union defines the share of renewables in the energy balance of the Member States, and also in the United States, pellet is expected to significantly increase its importance.
- Social: the growing environmental awareness of citizens contributes to the popularization of biofuels in industrial and individual scale.
- Environmental: biomass is characterized by carbon neutrality and lower emissions of harmful elements in comparison with the use of fossil fuels.
- Energy: a large and stable technical potential of biomass as energy source.

2. Agricultural residues for energy purposes: agripellets

Nowadays, fuel pellets are mainly made from sawdust, wood chips, and wood shavings. The supply of wood materials can be limited because producers of fiberboard, particleboard, and oriented strand board compete for the same forestry and mill residues as pellet producers [3]. This competition and the current increased demand for wood pellets, both for residential and industrial use, have led to a shortage of sawdust and wood shavings. If demand and prices continue to rise, other biomass wastes than sawdust, wood chips, and wood shavings should be considered for pellet production. Agricultural residues are among those future new raw materials. Agricultural residues refer all the organic materials which are produced as by-products from harvesting and processing of agricultural crops. These residues can be further categorized into primary residues and secondary residues. Primary or field-based residues are those generated in the field at the time of harvest (e.g., straw, stalks, and leaves that are left over after harvest), whereas those coproduced during processing are called secondary or processing-based residues (e.g., sugar beet pulps, cotton mill wastes, peanut shells).

Availability of primary residues for energy application is usually low since collection is difficult, and they have other uses as fertilizer, soil conservation, animal feeding, and litter. The amount of secondary residues varies widely depending on the crop and processing methods used [4].

Currently, large amounts of agricultural residues are left in the field to rot or are burned in the open air, ultimately releasing carbon dioxide to the atmosphere. This biomass could be used to produce pellets which are a form of solid fuel. The most important reason for using agricultural residues for energy purposes is that it is carbon neutral, that is, the carbon emitted during their combustion is taken up in the regrowth of the biomass used to produce them and therefore does not add to greenhouse gas emissions. Further, any consumption of fossil fuels replaced by biomass will lower CO_2 emissions.

Agricultural residues are available in large quantities and can be utilized for sustainable heat and power production, when used as fuel. However, they have low energy density (MJ m^{-3}) and low yield per unit area (dry tons ha^{-1}) [5]. Often, long distances have to be bridged between the biomass place of origin and the place of its utilization, resulting in expensive handling and transportation. Transportation costs of low-density and high-moisture agricultural residues which increase the total biomass-processing cost are a major constraint to their use as an energy source [6]. To increase the density of the biomass, it can be compressed into pellets using a mechanical process in which pressure is applied to the biomass to crush its cellular structure, and thereby increasing its density. Densified biomass, especially pellets, has drawn attention due to its superiority over raw biomass in terms of its physical and combustion characteristics [7]. Many materials originated by agriculture could be used for the production of densified biomass fuels: straw, grain hull waste, tree pruning, fruit stones, dry fruit waste, grain, cork, cotton, and other wastes.

The main characteristics of some selected agripellets are summarized in **Table 1**. The higher heating value (HHV) of agripellets is high, being even higher than pine sawdust pellets (in the case of pellets made of olive pomace and tomato peels and seed). The ash content confirms the necessity of blending agricultural residues with sawdust or other woody material. In fact, the ash contents for agripellets (3.3–12%) are significantly higher comparing to pine sawdust (2.5%).

According to Colley [14], pellets durability is regarded as high if exceeding 80%, medium if measured as 70–80%, and low if values do not reach 70%. Therefore, agripellets show high quality in terms of durability.

2.1. Olive mill residues

The olive oil industry produces significant quantities of solid olive residues (pieces of skin, pulp, and stones) which due to their characteristics can be utilized for the production of cleaner energy [15]. According to Barbanera et al. [16], it can be assumed that 1 ha of olive tree produces about 2500 kg of olives and about 35 kg of olive pomace. Several studies have been conducted during the last two decades that examined the thermochemical characteristics and performance of solid olive residues [17–21]. The results obtained from these studies suggest that these solid residues constitute a promising biomass resource because their thermochemical characteristics provide the opportunity for their potential utilization for energy purposes, offering at the same time a solution to the management problems [20].

Residue	HHV (MJ kg^{-1})	Ash content (%)	Pellets durability	References
Vineyard pruning	17.8	4.4	98.8	[8, 9]
Olive pomace	22.0	5.6	91.4	[10]
Peels and seeds of tomato	27.1	4.9	91.2	[11]
Grape marc	19.5	12.0	85.8	[9, 12]
Olive pruning	19.6	3.3	91.7	[13]
Wheat straw	18.3	9.1	94.4	[9]
Pine sawdust	19.7	2.5	97.2	[9]

Table 1. Characteristics of some agripellets.

The solid fraction combustion of olive residues indicates good combustion behavior of olive kernels and the residual olive pomace, with suitable efficiency and a reduced presence of unburned fraction. However, lower combustion efficiencies are observed during pulp processing [10]. The olive pomace, once it has been subjected to a drying process, can be used as a fuel. However, their oleaginous characteristics limit the densification during the pelletizing process. Moreover, their high concentrations of certain components such as nitrogen and ashes exceed the specifications given by pellets quality standards [20]. Therefore, it is necessary to blend the olive oil by-products with other biomass residues that must present suitable characteristics for an ideal pelletization [22]. Barbarena et al. [16] reported that adding olive tree pruning to olive pomace, the chemical composition of pellet blends respects the standard requirements in terms of mechanical durability and N and Cu content. In addition, the bulk density was enhanced allowing a reduction of transport and storage economic cost. On the other hand, Brlek et al. [23] suggest that limitations regarding combustion of olive pomace pellets can be established due to elevated nitrogen content and higher percent of abrasion. These constrains can possibly be diminished by adding wood biomass to pelletization blend.

2.2. Vine residues

The wine industry produces huge amounts of residues every year. Marculescu and Ciuta [24] estimate that for every kilogram of grape processed for wine, more than 20% is residue. The current use of the wood residues produced by the annual pruning activity is generally eliminated through crushing in the vineyard and then spread along the soil, in order to reduce erosion and recycle nutrients that can be incorporated into the soil [25].

Another important residue generated in the production of wine is grape marc. This residue is the skins and pips that remain after the grapes have been crushed. These residues are highly wet (more than 60% of moisture content on wet basis) and have low pH. Also, they present high contents of phosphorous (P), potassium (K), organic matter, phytotoxic, and antibacterial phenolic substances, which make them resistant to biological degradation. These wastes have high contents of lignin and tannin. Hence, they are not appropriate as a nutritional supplement for animals [26]. In addition, due to their high C/N ratio, their recovery as soil fertilizer presents difficulties. Another important constraint related to the management and disposal of wineries and distilleries processing industries is the generation of large quantities in which discharges are usually centralized and seasonal in a short period of the year (3–4 months).

Fernandéz-Puratich et al. [8] studied the use of vine wastes for the production of pellets and concluded that the use of these biomass residues is a viable alternative option in terms of economy as well as energy.

Marculescu and Ciuta [24] studied the thermal degradation of grape marc in a laboratory furnace. They found that grape marc has high energy content (19.7 kJ kg^{-1}) and they have recommended for energy production.

Kraiem et al. [12] recommend blending these residues with pine sawdust. They found that a blend with sawdust leads to the decrease of ash contents while densification leads to the increase of the energy densities. Combustion tests of pellets prepared from these residues indicate that boiler and combustion efficiencies are comparable to wood pellets. However, gaseous and particulate emissions are higher and are strongly affected by the operating parameters of the domestic boilers.

2.3. Industrial tomato residues

The industrial processing of tomato leads to a great variety of output products. Large volumes of residual biomass (mainly peels and seeds) are generated by tomato industrial processing plants. Currently, industrial tomato residues do not generate so many benefits for industries, in particular for storage and preservation issues. The accumulation of these residues, predominantly in the warm periods, promotes uncontrolled anaerobic fermentations leading to environmental problems [27]. In this way, fast consumption is advised in order to prevent fermentation processes, which are favored by high temperatures during the industrial processing period. Those residues have a high moisture content, which leads to some storage difficulties, and are generated in large quantities in which discharges are usually centralized and seasonal in a short period of the year. Also, important costs are derived from transport of wastes with significant moisture content, thus impeding their reasonable use. To avoid added costs related to disposal process, tomato manufacturing companies often give their production residues for free to other companies that generally use them for feeding livestock [11] or as soil amendment [28]. However, Rossini et al. [29] found that tomato waste could be suitable for combustion, but the relatively higher nitrogen content can generate environmental problems in terms of NO_X emissions. In addition, the high chlorine and sulfur contents may lead to the corrosion of the combustion systems. Therefore, as a solution, the authors proposed to separate tomato waste into peels for combustion and seeds for vegetable oil production.

González et al. [30] studied the tomato waste combustion in a mural boiler. They found that tomato residues give higher boiler efficiency than other biomasses (forest residues, sorghum, almond pruning, and reed). Ruiz-Celma et al. [11] studied tomato seeds and peels pellets. They reported a high heating value of these pellets and an energy density (approaching 8 GJ/m^3) similar to that of other biomass pellets, regardless their low bulk density values.

3. Techniques for biomass densification

Biomass is densified via two main processes: pelletizing (mechanical densification) and torrefaction. Pelletizing involves applying pressure to mechanically densify the material, while torrefaction involves heating the biomass in the absence of oxygen.

3.1. Pelletizing

The low density of agricultural residues poses a challenge for the handling, transportation, storage, and combustion processes. Those problems are mainly related to the high bulk volume, which results in high transportation costs and demands for large storage capacities, and to the high moisture content which results in freezing and blocking the in-plant transportation systems, as well as in biological degradation. In addition, variations in moisture content make difficult optimal plant operation and process control. All these problems may be addressed through densification, a process that produces solid fuel with denser and more uniform properties than the raw biomass [31].

The main advantages of densified compared to non-densified fuels are the following:

- Reduced cost of transportation due to increased bulk density (from 80–200 to 600–700 kg m^{-3}).
- An increased energy density, resulting higher energy efficiency.
- Simplified mechanical handling and feeding.
- Uniform combustion in boilers resulting in lower emissions during combustion.
- Reduced dust production.
- A lower moisture content (lower than 10%), favoring a long conservation and less loss of product during storage.
- Reduced possibility of spontaneous combustion in storage.
- Simplified storage and handling infrastructure, lowering capital requirements at the combustion plant.

All these factors make pellets one of the more attractive forms of biomass-based energy. The major disadvantage to biomass densification technologies is the relative high energy cost for the pelleting process, increasing the price of the end product. In addition, it is important to have in mind that agricultural residues are highly dispersed and may be over long distances

http://dx.doi.org/10.5772/110404

from the pelleting facilities. An appropriate solution could be to carry the pelletizing mill to the raw material. A few mobile pelletizing mills already exist. Such a pelletizing mill can meet a specific demand, for instance several farmers wanting to share the investment cost of a pelleting equipment which after can be moved from a place to another [32].

The process of pellet manufacturing was first developed for the livestock feed industry. The process consists of a few basic sub-processes: comminuting of the raw material, drying, pelletizing, and cooling. The raw material is first cleaned of contaminant such as rocks, metals and other foreign material, and then grinded in a hammer mill or a chipping machine. The particle size is adjusted to a uniform maximum dimension and should have proper size and be consistent. The moisture content in the raw material can be considerably high and are usually up to 50–60% which should be reduced. Rotary drum dryer is the most common equipment used for this purpose, where the moisture content of the uniformly dimensioned particles is reduced to about 10–15% (w.b.). Drying increases the efficiency of biomass, and it produces almost no smoke on combustion. The feedstock should not be over dried, as a small amount of moisture helps in binding the biomass particles. The drying process is the most energy intensive process and accounts for about 70% of the total energy used in the pelletization process. Thereafter, raw material can be conditioned according to legal specifications (i.e., steam or organic binding agents can be added). The particles are then moved by conveyor to a pellet mill, where the pellets are compressed against a heated metal plate (known as die) using a roller. Due to the high pressure, frictional forces increase, leading to a considerable rise in temperature (90–100°C), and are immediately air quenched down to 25°C. High temperature causes the lignin, and resins present in biomass to soften which acts as a binding agent between the biomass fibers. This sets up the lignin and hardens the product, and contributes to maintain its quality during storage and handling. On the outer side of the latter, a knife cut off the pellets at the desired length. Residual moisture in the feedstock turns to steam during compression and helps to lubricate the compression die [33].

Finally, the pellets are packed into bags using an overhead hopper and a conveyor belt. Pellets are then ready for storage (in a silo) or for automatic packing (in 25 kg bags or big bags—1 to 1.5 m^3). Commercial pellet mills and other pelletizing equipment are widely available worldwide.

3.2. Torrefaction

Torrefaction is a very promising technology for improving the fuel properties of solid biomass (e.g. pellets). This technology is a version of slow pyrolysis processes that comprise the heating of biomass in the absence of oxygen and air [30, 34] in which the goal is to dry, embrittle, and waterproof the biomass. This is accomplished by heating the biomass in an inert environment at temperatures of 200–320°C. During the treatment, biomass starts to decompose and releases combustible volatile matter, mainly composed by organic compounds, together with moisture. Biomass loses most of the low-energy content material in the form of gaseous and condensable liquids. Common events that occur during torrefaction include drying, depolymerization and recondensation, limited devolatilization and carbonization, and extensive devolatilization and carbonization [35]. Several studies have been conducted to evaluate a combined torrefaction-pelletization process possible in a commercial scale. Reed and Bryant [36] first considered the combination of torrefaction and pelletization to produce a new type of high energy density and water-resistant pellets. Bergman [37] proposed and demonstrated a combined torrefaction and pelletization process for the production of high energy density wood pellets. The addition of pelletization to torrefaction would potentially create a bio-based fuel with similar energy density to coal, prompting the adoption of this product for replacing coal in heat and power facilities. Currently, a number of torrefaction pilot plants have been designed, under construction, or publicly announced [38]. Carapeda [35] reports that if the biomass is torrefied before being densified, the energy consumption during the pelleting process is reduced by a factor of 2 and the throughput is increased, also by a factor of 2.

The main advantages of torrefaction of raw biomass feedstock include:

- Increase in energy content and heating value of the final torrefied product by reducing O/C and H/C ratios.
- Decrease in moisture content, which provides two main benefits: (i) reduced transportation costs associated with moving unwanted water and (ii) the prevention of biomass decomposition and moisture absorption (biomass becomes hydrophobic) during storage and transportation, which in turns helps in preserving the quality of the product.
- Improved grindability and friability (80–90% less energy consumption for grinding).
- Torrefied pellets have more strength (1.5–2 times impact load and does not disintegrate easily during handling and storage).
- Overall improvement in the chemical composition of the biomass (smoke-producing compounds removed).

The main disadvantage of torrefaction of raw biomass feedstock includes additional cost, energy, and equipment required for processing. In addition, ash content is not removed so ash content would likely increase per unit of weight [35]. Therefore, torrefaction can be considered as one of the major pretreatment technologies for improving the properties of agricultural residues, in order to deal with such problems as high bulk volume, high moisture content, and poor grindability.

4. Combustion of agripellets

Physical and chemical properties vary significantly within and between the different agricultural raw materials. Depending on the type of application, these variations may be critical and may affect the performance of the system. Physical properties, such as bulk density, moisture content, particle size and distribution, and durability, are important for the choice of processes and equipment. On the other hand, chemical properties are of great importance for the energy efficiency, environmental pollution, and ash-related operating problems.

Agripellets combustion triggers several major obstacles regarding emissions (gas, dust, and aerosols), deposit formation (slagging, fouling), and corrosion. Another problem is that the ash content of agripellets is higher than wood pellets (about 2–10 times higher than that of wood pellets). All those problems not only depend on the fuel characteristics but also on the design of the combustion equipment and the way it is operated. Recently, Kraiem et al. [12] reported that silicon (Si), potassium (K), calcium (Ca), magnesium (Mg), phosphorous (P), and aluminum (Al) are the major elements of agripellets. Compared to wood pellets, a typical feature of agripellets is their higher content in nitrogen (N), sulfur (S), chlorine (Cl), and K, increased by the use of fertilizers and pesticides/herbicides in agriculture. The presence of those elements leads to relatively important emissions of NOx, SOx, and HCl compared to wood pellets. In addition, K influences both particulates emission and slagging (by lowering the softening temperature of the fuel) of an increased ash volume. Besides, a high Cl content results both in corrosion problem on the surfaces of the boiler and in formation of dioxins and furans. Finally, for a large-scale use, in relation with the high ash content and the low melting point, it has been stated that straw pellets could present better results with grate combustion or fluidized bed systems. Those problems can be overcome by the use of multi-fuel boilers in the range of 10–60 kW which is more suitable for burning agripellets; co-firing of agricultural residues with fossil fuel; cleaning the agricultural residues before pelletizing them into agripellets to make them with less ash content; and to add in specific anti-slagging agents (e.g., kaolin) or mix in some sawdust to change the fuel characteristic.

Environmental and technical features of combustion technologies indicate that pellets made from agricultural residues should be used primarily in large-scale combustion plants equipped with sophisticated combustion control systems and flue gas cleaning systems, whereas wood pellets should be preferred for small-scale heating systems. In the future, the main technical challenges regarding agripellets are the production of a high-quality fuel, and technological improvement for small-scale combustion devices.

4.1. Emissions

During combustion, N, S, and Cl in the fuel (present in higher proportion in agripellets than in wood pellets) may lead to atmospheric pollutants such as nitrogen oxides (NOx), sulfur dioxide (SO_2), hydrogen chlorine (HCl), and chlorinated hydrocarbons. Moreover, Cl favors the formation of dioxins and furans. The incomplete combustion of agripellets is mainly the result of low combustion temperatures, short residence times, oxygen shortage, or combinations of these effects. Incomplete combustion results in emissions of carbon monoxide (CO) and volatile organic compounds (VOC), particles, tar, and polycyclic aromatic hydrocarbons (PAH). Zeng et al. [39] demonstrated that the emission of NOx, SO_2, HCl, and total particulate matter can be reduced by blending agricultural raw materials with woody biomass though substantial reduction potential was only observed for blends with at least 50 wt% wood.

In addition, ashes formed during agripellets combustion can generally be divided into bottom ashes, coarse fly ashes, and aerosols (fine fly ash). These fractions differ significantly concerning their particle size and chemical composition as well as their formation mechanisms.

The bottom ash is the ash fraction remaining in the furnace after combustion of the fuel and is then removed by the de-ashing system. Coarse fly ashes are particles entrained from the fuel bed with the flue gas. They mainly consist of refractory species (such as Ca, Mg, Si as well as small amounts of K, Na, and Al), and their particle sizes can vary between some μm and 100 μm. Particles that are small enough to follow the flue gas on its way through the furnace and the boiler finally form the coarse fly ash emission at the boiler outlet. Aerosols are formed by gas-to-particle conversion processes in the furnace and in the boiler. Some of the aerosol particles coagulate with coarse fly ashes due to collisions [33]. During combustion of agripellets, part of the volatile compounds is released from the fuel to the gas phase: aerosols are then formed by condensation or nucleation of these volatiles compounds. Aerosols are much smaller than coarse fly ash (typical particle size significantly <1 μm). Aerosol emissions present high concentrations of heavy metals and sometimes of organic compounds. By their dimension, they can remain suspended in the air for long period of time and enter into the inner parts of lungs. In small-scale pellet furnaces and boilers, the main ash fraction is bottom ash. Furnace and boiler ash form the major share of coarse fly ash, which is usually precipitated and mixed with the bottom ash. A small amount of course fly ash is emitted with the flue gas [33].

4.2. Deposit formation

Biomass boiler issues regarding slagging, fouling, and corrosion are related to alkali species present in agricultural residues. These alkali species are released as gaseous alkali chlorides, hydroxides, and/or sulfates during combustion. Alkali chlorides/sulfates later condense on cold boiler surfaces enhancing fouling and corrosion. This is referred to as slagging when the deposits are in a molten or highly viscous state, or fouling when the deposits are built up largely by species that have vaporized and then condensed. Slagging is often found in the radiant section of the furnace, while fouling occurs in the cooler furnace regions where the heat exchanger equipment is located [40]. The negative effects of slagging and fouling are high furnace material wear, heat transfer efficiency reduction with pressure drop, and increased corrosion of the boiler.

Potassium and sodium compounds are present in all agricultural residues. During combustion, these alkali compounds combine with silica and causes slagging and fouling problems in conventional combustion equipment designed for burning wood at higher temperatures. Volatile alkali also lowers the fusion temperature of ash; combustion of agricultural residue causes slagging and deposits on heat transfer surfaces. In order to overcome this problem, special boilers have been designed with lower furnace exit temperatures or low operation temperature. These designs can reduce slagging and fouling from combustion of agripellets.

Hence, this underlines the necessity of a careful treatment of raw materials so as to avoid mineral contamination. Deposit formation related problems affecting agripellets deserve a special attention because they lead to reduced accessibility of the appliances, and also to bad publicity for the agripellet market.

4.3. Corrosion

The presence of even a small concentration of Cl in fuel will result in the formation of alkaline chloride compounds on boiler surfaces. Chlorine can influence the corrosion of superheater tubes in many ways. Gases containing Cl_2, HCl, NaCl, and KCl may cause a direct corrosion by accelerating the oxidation of the metal alloys. Such gases may also influence the corrosion caused by other mechanisms, such as molten alkali sulfate corrosion of superheater alloys and sulfidation of water walls. In addition, Cl may also deposit on superheater tubes and thereby influence its corrosion [41]. Chlorine corrosion could be prevented by co-firing aluminum silicates containing fuel, such as coal or peat. When those fuels are co-fired with agricultural residues, chloride formation can be avoided. In addition, a parameter that has been often referred to is the sulfur-to-chlorine atomic ratio (S/Cl) in fuel or fuel blend. It has been suggested that if this ratio in fuel is less than two, there is a high risk for superheater corrosion. When the ratio is at least four, the blend could be regarded as noncorrosive. However, the best way to prevent the molten phase corrosion is to keep superheater metal temperature below the first melting temperature of deposits, in practice below 500°C when firing agripellets [32].

Several field studies have shown that the main contributor to superheater corrosion in boilers is Cl, in particular alkali chlorides (NaCl, KCl). The relatively low sulfur content in most agricultural residues may introduce corrosion problems in the superheaters.

4.4. Ash recycling for agricultural applications

The rapidly growing number of pellet heating installations illustrates an increased interest in environmentally friendly heating systems. The problems associated with the use of agripellets are essentially linked to the ash management. Thus, the recycling or storage of agripellets ash deserves a special attention.

Ash is the inorganic uncombustible part of fuel left after complete combustion and contains the bulk of the mineral fraction of the original biomass [42]. In wood pellets, ash represents less than 2%, while in agripellets, it can be 5–10% and up to 30–40% in rice husks and milfoil [43].

The first and direct consequence for small scale stoves and boilers of the increased ash residue with agripellets is that there the ash storage under the furnace will have to be emptied more frequently, which is quite negative as far as the convenience of users is concerned. Considering that the ash storage should normally be emptied once every 5–15 days with wood pellets depending on the consumption, agripellets would oblige to remove ashes more frequently (almost daily). On the other hand, James et al. [43] suggest that inefficiencies in boilers and furnaces result in high percentages of unburned organic matter in ash. This carbon content may be recycled to the boiler or furnace to improve energy output and increase the process efficiency.

Part of the ash is taken out in the bottom of the boiler and is called bottom ash while the remainder is composed of the fine particles that are driven out of the boiler with the flue gases. This part of the ash is called fly ash. Each ash fraction has different composition. Filter fly ash tends to accumulate the largest part of heavy metals. Ashes from agripellets are produced in higher quantities, but the content in heavy metals for each fraction seems to be lower.

The collected bottom ash and fly ash from the combustion of agripellets should be disposed of in a safe way. These ashes contain nutrients, primarily potassium, and other soil-fertilizing elements like magnesium, phosphorus, and calcium and can therefore be applied in agriculture as fertilizer. It seems that ash is strongly alkaline (pH of 11–12) and could cause sharp increase of pH and ion concentration in the soil after spreading [32]. Thus, ash should not be used unless a soil pH test has been done. Such a phenomenon would be harmful with respect to plant growth. Consequently, ash could be treated (e.g., granulated) in some way to reduce impact on soil. In regard to acidic soil correction, agripellets ashes as a garden amendment are a much more convenient means than the traditionally used ground limestone, bearing in mind that it is an absolutely costless resource.

In order to utilize the nutrients from the fly ash, a utility owned plant has developed a method for washing them leaving heavy metals behind in a fraction to be stored. The product is a valuable fertilizer, and the process could be carried out centrally using fly ash from both utility owned plants and district heating plants.

According to Gomez-Barea et al. [44], the utilization of ash has also seen its application in the construction industry. Fly ash can be used as a cement replacement in concrete, for soil stabilization, as a road base, structural filler in asphalt and asphalt base products, lightweight bricks and synthetic aggregate.

5. Pellets quality standards

The combustion of densified biomass fuels in fully automatic heating systems for residential sector and small-scale furnaces requires high fuel quality. However, high quality is not necessary if these fuels are used in larger industrial furnaces because they are equipped with more sophisticated flue gas cleaning, combustion, and process control systems. Pellets are a standardized fuel, which simplifies construction and operation of burners. For pellets, producers are very important to have quality standardization because it increases the customer confidence. However, the quality of the pellets should be defined in terms of the heating technology, since different heating systems require different fuel qualities. For example, large heating plants are not demanding in terms of pellets durability or amount of fines. In contrast, pellets stoves require an extremely durable pellet, which does not produce too much dust in the storage bunker, and do not cause technical problems in the feeding and combustion unit. From this perspective, the different quality of agripellets would suggest a different use, preferentially in large-scale systems [45].

In the present, the development of quality standards for wood pellets is set on five different levels in Europe: (1) European Commission and European Committee for Standardization; (2) EU member state governments; (3) European Biomass Association and European Pellet Council (which represent the European biomass sector); (4) Wood Pellet Buyers Initiative (which represents end users of biomass); and (5) standards developed by individual private companies [46]. Standards exist on national (e.g., DIN), European (EN), and on international level (ISO).

Usually, standards developed by standards organizations are voluntary but can become mandatory if adopted by a government or business contract.

http://dx.doi.org/10.5772/110404

According to European Standard ENplus which is related to wood pellets for nonindustrial use and which will gradually supersede all national standards (e.g., ÖNORM M1735, or DINplus), wood pellets can be classified into three basic categories: ENplus A1, ENplus A2, and ENplus B. The quality requirements of ENplus are based on an international standard: ISO 17225-2. This international standard has replaced the European Standard EN 14961-2. The ENplus certificate requires stricter quality criteria. This quality seal stands for low emissions and trouble-free heating with high energy value.

Class A1 includes wood pellets originating from stem wood, without chemical additives and with low ash and Cl content. Class A2 considers wood pellets with slightly higher ash and/or Cl content. Wood pellets derived from reused wood, residues, or bark are included into class B. **Table 2** shows the specifications of the three wood pellet classes according to ENplus in comparison with German (DIN-plus) and Austrian (ÖNORM M7135) standards.

Property	DIN plus	ÖNORM M 7135	ENplus A1	ENplus A2	ENplus B
Diameter (mm)	$4 \le d \le 10$	$4 \le d < 10$	6 (±1)	6 (±1)	6 (±1)
Length (mm)	$\le 5 \times d$	$5 \times d$	$3.15 \le L \le 40$	$3.15 \le L \le 40$	$3.15 \le L \le 40$
Ash content (wt%)	<0.5*	<0.5*	≤0.7	≤1.5	≤3.0
Ash melting behavior (°C)	Not specified	Not specified	≥1200	≥1100	≥1100
Moisture (%)	<10	<10	≤10	≤10	≤10
Net calorific value (MJ kg^{-1})	≥18	>18*	≥16.5	≥16.5	≥16.0
Fines (wt%)	≤1	Not specified	≤1	≤1	≤1
Mechanical durability (wt%)	≥97.7	Not specified	≥97.5	≥97.5	≥95.5
Chlorine (wt%)	≤0.02	Not specified	≤0.02	≤0.03	≤0.03
Arsenic (mg kg^{-1})	Not specified	Not specified	≤1	≤1	≤1
Cadmium (mg kg^{-1})	Not specified	Not specified	≤0.5	≤0.5	≤0.5

* dry mass

Table 2. Comparison of regulations related to pellet quality.

Generally, limit values for ash content, moisture, net calorific value, and chlorine are fairly similar. All standards prohibit the use of binding agents. Austrian and German standards do not mention the amount of fines, while in Enplus, fines must not be more than 0.5–1.0%. German and Austrian standards do not define durability or mechanical stability despite the importance of these attributes. This is because during transport in tankers and the pneumatic filling of storage bunkers, mechanical strain on pellets is high. In addition, pellets with poor mechanical stability produce large amounts of dust. Hence, small heating systems require very high pellet quality. In these systems, the amount of fines in fuel pellets is of special importance. In contrast, combustion units in large heating systems are not affected by the amount of fines. The different requirements of small and large combustion systems make necessary definition of different groups of standards [32].

6. Conclusions

The increasing competition for solid biomass, such as wood pellets, will create space for relatively novel biomass sources to enter the market, among which agricultural residues have the greatest potential. In comparison with wood, agricultural residues present high ash, N, K, and Cl content. The underlying problems are higher related emissions, deposit formation and corrosion. Many techniques are currently used, while others are under improvement stage to overcome the inherent drawbacks of agripellets composition. These techniques include agricultural practices, fuel preparation, combustion technologies, flue gas cleaning systems, and the possibility of co-combustion of agripellets with solid fossil fuels. ENplus quality certification is a major step toward establishing biomass pellets as a widely used energy source. However, the high nitrogen and ash contents strongly limit the certification of pure agripellets. In this regard, several studies have shown that blending agricultural residues with sawdust before pelleting could help to meet ENplus certifications.

Consequently, the use of agripellets in the residential heating sector cannot be recommended at present, because small-scale pellet furnaces are not specially designed for this kind of fuel. Therefore, for small-scale heating systems, which require high-quality fuels, the use of wood pellets is recommended. Agripellets should be used primarily in large-scale combustion plants equipped with sophisticated combustion control systems and flue gas cleaning systems.

Author details

Claudia Santibáñez Varnero* and Marcela Vargas Urrutia

*Address all correspondence to: claudia.santibanez@umayor.cl

Department of Agronomy, Faculty of Sciences, Univesidad Mayor, Santiago, Chile

References

[1] Verma V.K., Brams S., Delattin F., Laha P., Vandendael O., Hubin A., De Ruyck J. Agro-pellets for domestic heating boilers: standard laboratory and real life performance. Applied Energy. 2012;90(1):17–23. doi:10.1016/j.apenergy.2010.12.079

[2] Taylor R.E., Butzelaar P., Leeuwen G.V., Palmer A., Keyes J., Gimenez C., MacDonald B. Wood pellet market outlook. In: Wood Market International Monthly report. 2013;18(1).

[3] McKeown M., Trabelsi S., Tollner E. Effects of temperature and material on sensing moisture content of pelleted biomass through dielectric properties. Biosystems Engineering. 2016;149:1–10.

[4] Cushion E., Whiteman A., Dieterle G. Solid biomass. In: Bioenergy development: issues and impacts for poverty and natural resource. Washington DC: The World Bank; 2010. p. 41–87.

[5] Kumar A., Cameron J.B., Flynn P.C. Biomass power cost and optimum plant size in Western Canada. Biomass and Bioenergy. 2003;24(6):445–464.

[6] Searcy E., Flynn P., Ghafoori E., Kumar A. The relative coast of biomass energy transport. Applied Biochemistry and Biotechnology. 2007;137–140(1–12):639–652. doi: 10.1007/s12010-007-9085-8

[7] Obernberger I., Thek G. Physical characteristic and chemical composition of densified biomass fuels with regard to their combustion behavior. Biomass and Bioenergy. 2004;27:653–669.

[8] Fernandéz-Puratich H., Hernández D., Tenreiro C. Analysis of energetic performance of vine biomass residues as an alternative fuel for Chilean wine industry. Renewable Energy. 2015;83:1260–1267.

[9] Miranda T., Montero I., Sepúlveda F., Arranz J., Rojas C., Nogales S. A Review of pellets from different sources. Materials. 2015;8:1413–1427. doi:10.3390/ma8041413

[10] Miranda M.T., Cabanillas A., Rojas S., Montero I., Ruiz A. Combined combustion of various phases of olive wastes in a conventional combustor. Fuel. 2007;86:367–372.

[11] Ruiz Celma A., Cuadros F., López-Rodríguez F. Characterization of pellets from industrial tomato residues. Food and Bioproducts Processing. 2012;90:700–706.

[12] Kraiem N., Lajili M., Limousy L., Said R., Jeguirim M. Energy recovery from Tunisian agri-food wastes: evaluation of combustion performance and emissions characteristics of green pellets prepared from tomato residues and grape marc. Energy. 2016;107:409–418.

[13] Garcia-Maraver A., Rodriguez ML., Serrano-Bernardo F., Diaz LF., Zamorano M. Factors affecting the quality of pellets made from residual biomass of olive trees. Fuel Processing Technology. 2015;129:1–7.

[14] Colley Z. Compaction of switchgrass for value-added utilization [thesis]. Aurburn: Auburn University.

[15] Christoforou E., Fokaides P. A review of olive mill solid wastes to energy utilization techniques. Waste Management. 2016;49:346–363.

[16] Barbanera M., Lascaro E., Stanzione V., Esposito A., Altieri R., Bufacchi M. Characterization of pellets from mixing olive pomace and olive tree pruning. Renewable Energy. 2016;88:185–191.

[17] Ramachandran S., Singh S.K., Larroche C., Soccol C.R., Pandey, A. Oil cakes and their biotechnological applications—a review. Bioresource technology. 2007;98:2000–2009.

[18] Arvanitoyiannis, I.S., Kassaveti, A. Olive oil waste management: treatment methods and potential uses of treated waste. In: Arvanitoyiannis, I.S, editors. Waste management for the food industries. New York: Elsevier; 2008. p. 453–568.

[19] Abu Tayeh, H., Najami, N., Dosoretz, C., Tafesh, A., Azaizeh, H. Potential of bioethanol production from olive mill solid wastes. Bioresources Technology. 2014;152:24–30.

[20] Santibáñez C., Vargas M. Solid fuels produced from olive pomace. In: Santibáñez C, editor. Proceedings of 3er Iberoamerican Congress on Biorefinery; Concepción, Chile. University of Concepcion:2015. p. 278.

[21] Kinab, E., Khouri, G. Management of olive solid waste in Lebanon: from mill to stove. Renewable and Sustainable Energy Reviews. 2015;52:209–216.

[22] Vourdoubas J. Possibilities of using olive kernel wood for pellets production in Crete —Greece. In: Proceedings of 16th European Biomass Conference & Exhibition; Valencia, España. 2008.

[23] Brlek T., Pezo L., Voća N., Krička T., Vukmirović D., Čolović R., Bodroža-Solarov M. Chemometric approach for assessing the quality of olive cake pellets. Fuel Processing Thechnology. 2013;116:250–256.

[24] Marculescu C., Ciuta S. Wine industry waste thermal processing for derived fuel properties improvement. Renewable Energy. 2013;57:645–652.

[25] Badalíkov B., Cervinka J. Effect of ploughing-down of grapevine chips on soil structure when using special agricultural machinery. Acta Universitatis Agriculturae et Silviculturae Mendelianae Brunensis. 2012;60:9–14.

[26] Bustamante M.A., Moral R., Paredes C., Perez-Espinosa A., Moreno-Caselles J., Perez-Murcia M.D. Agrochemical characterisation of the solid by-products and residues from de winery and distillery industry. Waste Management. 2008;28:372–380.

[27] Toscano G., Pizzi A., Pedretti E., Rossini G., Ciceri G., Martignon G., Duca D. Torrefaction of tomato industry residues. Fuel. 2015;143:89–97.

[28] Zanón M.J., Font M.I., Jordá C. Use of tomato crop residues into soil for control of bacterial wilt caused by *Ralstonia solanacearum*. Crop Protection. 2011;30:1138–1143.

[29] Rossini G., Toscano G., Duca D., Corinaldesi F., Foppa Pedretti E., Riva G. Analysis of the characteristics of the tomato manufacturing residues finalized to the energy recovery. Biomass Bioenergy. 2013;51:177–182.

[30] González JF., González Garcia CM., Ramiro A., Gañan J., Ayuso A., Turegano J. Use of energy crops for domestic heating with a mural boiler. Fuel Processing Technology. 2006;87:717–726.

[31] Uslu A., Faaij A.P.C., Bergman P.C.A. Pre-treatment technologies and their effect on international bioenergy supply chain logistics. Techno-economic evaluation of torrefaction, fast pyrolysis and pelletisation. Energy. 2008;33:1206–1223.

[32] Pastre O. The European Biomass Industry Association [Internet]. 2002. Available from: http://www.pelletcentre.info/resources/1093.pdf [Accessed: 2016]

[33] Obernberger, I., Thek, G. Pellet production and logistics. In: The Pellet handbook. London, Washington DC: Earthscan; 2012. p. 85–129.

[34] Shang L., Nielsen N.P.K., Dahl, J. Stelte W. Quality effects caused by torrefaction of pellets made from Scots pine. Fuel Processing Technology. 2012;101:23–28.

[35] Capareda S. Torrefaction. In: Introduction to biomass energy conversions. Boca Raton, London, New York: CRC Press; p. 289–316.

[36] Reed T, Bryant B. Densified biomass: a new form of solid fuel. In: Densified biomass: a new form of solid fuel. Colorado: Solar Energy Research Institute; 1978.

[37] Bergman P.C.A. Combined torrefaction and pelletization the TOP process. In: Energy Research Centre of the Netherlands Report. Netherlands: Energy Research Centre of the Netherlands; 2005.

[38] Peng J.H., Bi X.T., Sokhansanj S., Lim C.J. Torrefaction and densification of different species of softwood residues. Fuel. 2013;111:411–421.

[39] Zeng T., Weller N., Pollex A., Lenz V. Blended biomass pellets as fuel for small scale combustion appliances: influence on gaseous and total particulate matter emissions and applicability of fuel indices. Fuel. 2016;184:689–700.

[40] Werkelin J., Skrifvars B.J. Zevenhoven M., Holmbom B., Hupa M. Chemical forms of ash-forming elements in woody biomass fuels. Fuel. 2010;89(2):481–493.

[41] Nielsen H.P., Frandsen F.J., Dam-Johansen K., Baxter L.L. The implications of chlorine-associated corrosion on the operation of biomass-fired boilers. Progress in Energy and Combustion Science. 2000;26:283–298.

[42] Khan A., Jong W., Jansens P., Spliethoff H. Biomass combustion in fluidized bed boilers: Potential problems and remedies. Fuel Processing Technology. 2009;90:21–50.

[43] James A., Thring R., Helle S., Ghuman H. Ash management review—applications of biomass bottom ash. Energies. 2009;5:3856–3873. doi:10.3390/en5103856

[44] Gomez-Barea A., Vilches L., Campoy M., Fernandez-Pereira C. Plant optimization and ash recycling in fluidised waste gasification. Chemical Engineering Journal. 2009;146:227–236.

[45] Goliński T., Foltynowicz Z. Pellet—a key to biomass energy. International Journal of Economic Practices and Theories. 2012;2:197–204.

[46] Hoeft. Report on market research, national incidents, problems and lacks within national guidelines. In: Project Report: Report on market research, national incidents, problems and lacks within national guidelines. Swedish: 2013.

Chapter 24

SWOT Analysis Applied to Wheat Straw Utilization as a Biofuel in Mexico

Gisela Montero, Conrado García,
Marcos A. Coronado, Lydia Toscano,
Margarita Stoytcheva, Ricardo Torres,
Ana M. Vázquez and Daniela G. Montes

Additional information is available at the end of the chapter

http://dx.doi.org/10.5772/65496

Abstract

Wheat is one of the main crops worldwide with a production of 733 million of tons by 2015. By 2013, the wheat grain production in Mexico was 3,357,307 t. Wheat straw is generated as a biomass waste once the wheat is harvested. However, the agricultural biomass waste has acquired international relevance as a source of bioenergy. The utilization of bioenergy has significant environmental benefits, and also economic benefits because the biomass waste is valorized as biofuel. The use of wheat straw as raw material for any productive process presents diverse factors that must be considered. Among those factors are the low density of biomass, handling and high transportation cost, an attractive heating value, and the physicochemical characterization. Therefore, the aim of this work was to apply the SWOT analysis to wheat straw utilization as a biofuel in Mexico. The main findings highlighted an estimation of 4,612,950.23 t of wheat straw generated. The experimental results of proximate analysis were 64.42% volatile matter, 19.49% fixed carbon and 16.09% ash. The higher heating was 14.86 MJ/kg. An energy potential of 69 PJ per agricultural cycle was calculated, equivalent to 19% of the biomass energy share reported in Mexico's National Energy Balance, by 2014.

Keywords: biofuel, SWOT analysis, wheat straw, biomass, agricultural waste

INTECH
open science | open minds

1. Introduction

Wheat is one of the main crops of the world with an annual production of 733 million tons reported in 2015 [1]. In Mexico, it was ranked 7th among all the crops in 2013, with a harvested surface of 683,044.42 ha [2]. The most common harvested wheat varieties are *Triticum aestivum* and *Triticum durum*. Note that 90% of wheat production is obtained in the fall-winter season, and the remaining 10% corresponds to the spring-summer season. The harvest season is done mainly in May and June [3]. The Northwest region is the largest area for wheat crop production. Hence, the major quantity of wheat straw is generated in that region. It was estimated that 4,612,950.23 t of wheat straw was generated in Mexico. Approximately, 85% of that biomass waste is burned as a traditional practice performed by farmers at the end of the agricultural cycle [4]. The burning of agricultural waste biomass is regulated by NOM-015-SEMARNAT/SAGARPA-2007 that establishes the technical specifications of methods on the use of fire in forest land and agricultural land use, in order to prevent and reduce forest fires. However, it is not applied because of the lack of human resource to inspect and supervise those events [5]. In **Figure 1**, the burning practices and their impact on the environment are depicted. Generally, the burning practice is done in uncontrolled and unsafe conditions that cause air pollution, soil organic nutrients loss, elimination of microorganisms, and pH soil modification. However, wheat straw is a biomass resource that can be valorized and used as a biofuel because it has an attractive heating value.

Figure 1. Wheat straw burning practices in Mexico [source: by the authors].

http://dx.doi.org/10.5772/65496

Using wheat straw as a raw material for any productive process presents diverse factors that must be considered. Among those factors are the constant supply of wheat straw, the low density, handling and high transportation cost, the higher heating value, and the physico-chemical characterization. Due to the type of the different factors involved, the SWOT (strengths, weaknesses, opportunities, and threats) methodology is a useful tool for analyzing such factors. Therefore, the aim of this work was to apply the SWOT analysis to wheat straw utilization as a biofuel in Mexico.

1.1. Power generation through wheat straw

Traditionally, biomass has been used for heating in open fireplaces or stoves. Currently, the biomass utilization as fuel for electricity generation has gained more importance internationally. It is a productive alternative to the exploitation of waste biomass generated in the agriculture and forestry, annually.

With the purpose of reducing greenhouse gases (GHG) emissions and depletion of nonrenewable energy sources, there is an increase in the share of renewables worldwide. Bioenergy plays a vital role in the reduction of GHG and climate change.

In 2012, the installed biomass power generation capacity reached 83 gigawatt-electric (GWe), equivalent to 1.5% of global power generation capacity [6]. Denmark is a pioneer in developing power plants using agricultural wastes; the first commercial straw power plant, Haslev, has been developed since 1989. Four power plants were developed and operated with wheat straw as the sole fuel. Moreover, large-scale straw power plants also have been commissioned in the United Kingdom (38 MW, Ely in 2002) and Spain (25 MW, Sangüesa in 2002) [7]. The biggest advantage of using straw in the energy sector is that it is a CO_2 neutral fuel, which does not contribute to an increase of the atmosphere's content of greenhouse gases.

1.2. Biomass share in the Mexican energy matrix

By 2014, the biomass share in the energy matrix of Mexico was 4.07%, and it represented the highest among all the renewable energy sources [8]. The biomass considered by the National Energy Balance was only firewood and sugarcane bagasse. The energetic use of biomass in Mexico is limited to food cooking processes in rural places and as a fuel in power generation plants in sugar refineries. The biomass electricity generation has a total capacity of 634 MW [9].

The current economic situation of the energy sector of Mexico is leading to many opportunities to increase the renewable energy share in the energy matrix. Renewable energy is an alternative to a petroleum-based economy that in the recent years has shown high prices fluctuations of crude oil. The regulatory framework was already established, and it is comprised in the laws of energy transition, promotion and development of bioenergy, and renewable energy from Baja California. However, there is the need to create mandates accompanied with the right energy policy to encourage and increase the renewable energy market in Mexico.

1.3. Renewable energy regulation in Mexico

In 2014, the energy situation in Mexico had experienced a radical change with the approval of the energy reform. Its aim to maintain the energy security of the country and economic connectivity and make energy as a motto of the Mexican economy to create jobs and attract investments and technology. The main structural changes are established in the reform, such as the opening of the electricity market. These changes are reflected in the modifications performed to the articles 27 and 28 of the Mexico's Constitution [10]. The article 27 establishes that the planning and control of the national electricity system, energy transmission, and distribution are exclusive functions of the nation. It is forbidden to provide concessions to private companies related to the mentioned functions. However, it allows the State to have contracts with the private sector on behalf of the nation, to carry out financing, maintenance, management, operation, and expansion of the necessary infrastructure to provide the public service of transmission and distribution of electricity. The elimination of the exclusivity to generate electricity by the State was the main modification to the article 28. Nevertheless, the planning and control of the national electricity system and the public service of electricity transmission and distribution are exclusive areas of the State.

The energy reform allows the private electricity producers to sell energy, not only in the self-supply modality as previously, but openly. Also, it removed entry barriers of the energy sector, allowing greater flexibility for private sector investment and promoting equitable and competitive conditions for all private generations including the Federal Electricity Commission. This reform represents an opportunity for the increment of the biomass share in the national energy matrix.

In 2015, the Law of Energy Transition was enacted. The purpose of this law is to regulate the sustainable use of energy as well as the obligations of clean energy and reduction of pollutant emissions from the electricity industry while maintaining the competitiveness of the productive sectors [11]. In this law, it is established that power consumption is met by a portfolio of alternatives that include energy efficiency and an increasing proportion of clean energy generation in conditions of economic viability. Through clean energy and energy efficiency goals, the Secretariat of Energy will encourage electricity generation from clean energy sources to reach the levels established in the Mexican General Law on Climate Change. The Secretariat should consider the biggest boost to energy efficiency and clean energy generation that can be supported in a sustainable way under the economic conditions and the electricity market in the country. Policies and measures to boost energy efficiency and renewable resources to replace fossil fuels in final consumption will be considered.

2. Wheat straw generation in Mexico

2.1. Wheat producers in Mexico

Table 1 shows the wheat producers in Mexico, the wheat harvested area, and the wheat straw generated. Sonora and Baja California were responsible for the production of 61.69% of wheat straw.

http://dx.doi.org/10.5772/65496

No.	State	Wheat harvested area (ha)	Wheat straw generated (t)
1.	Sonora	304,547.50	2,223,196.75
2.	Baja California	86,731.00	633,136.30
3.	Tlaxcala	33,912.00	247,557.60
4.	Jalisco	30,676.00	223,934.80
5.	Guanajuato	30,626.50	223,573.45
6.	Chihuahua	28,522.05	208,210.97
7.	Michoacán	25,213.07	184,055.41
8.	Nuevo León	24,876.20	181,596.26
9.	Sinaloa	17,670.57	128,995.16
10.	Oaxaca	10,324.00	75,365.20
11.	Estado de México	9,239.00	67,444.70
12.	Zacatecas	7,748.00	56,560.40
13.	Coahuila	7359.82	53,726.69
14.	Baja California Sur	4,786.00	34,937.80
15.	Puebla	4,183.30	30,538.09
16.	Durango	3,365.17	24,565.74
17.	Hidalgo	2,130.81	15,554.91
	Total	**631,910.99**	**4,612,950.23**

Table 1. Wheat producer's states in Mexico in 2013 [12].

The states of Mexico that produce wheat were ranked and localized geographically through the analysis of the statistical information system of crops. **Figure 2** illustrates the location of the states from Mexico that produced wheat in 2013.

Based on data reported by the Secretariat of Energy, a generation index of 7.3 t/ha [13] of wheat straw was used for the estimation of the wheat straw availability. The experimental determinations were performed to *Triticum aestivum* that is one of the most common wheat varieties that is harvested in Mexico. The proximate analysis and higher heating value determinations were applied to the wheat straw. The proximate analysis was conducted according to ASTM E870-82, and the heating value was determined following the ASTM E711. Based on the wheat straw estimation and the experimental results, the SWOT methodology was applied to evaluate the internal and external factors affecting the utilization of wheat straw as biofuel in Mexico.

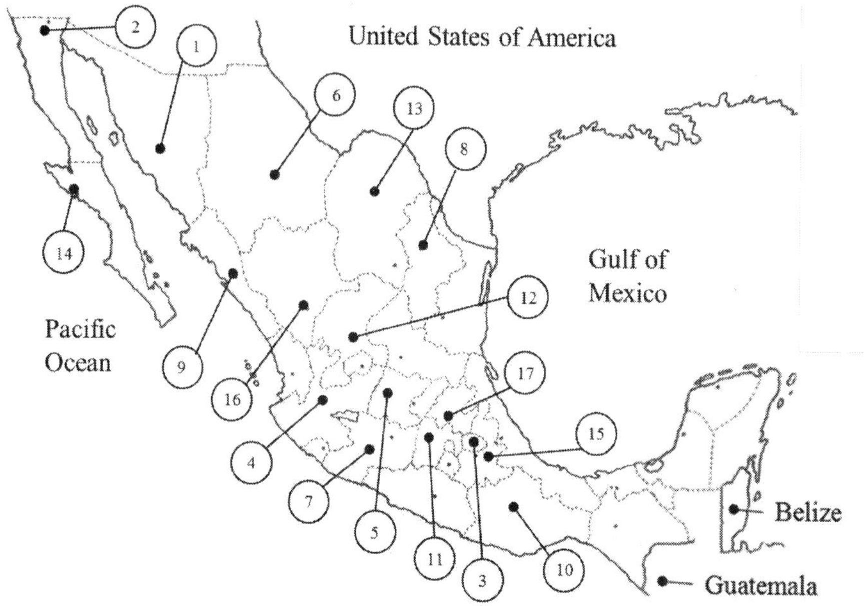

Figure 2. Location of wheat producer's states in Mexico.

2.2. Wheat straw experimental determinations

The proximate analysis and higher heating value determinations were applied to the wheat straw. The analysis procedures were conducted according to ASTM E870-82 (2006), and the heating value was determined according to ASTM E711 [14, 15].

2.3. Proximate analysis

Among the analysis for physicochemical characterization of the biomass, the proximate analysis is the one with less complexity. It does not require sophisticated laboratory equipment. The proximate analysis allows determining the weight percentages of moisture (M), volatile matter (VM), fixed carbon (FC), and ash of the biomass. With the results obtained from this analysis, it is possible to define the most suitable biomass conversion process, e.g., biological or thermochemical processes. It also permits establishing fuel quality criteria, among others [16].

2.4. Higher heating value

The heating value is an important parameter that must be determined in the evaluation of any fuel and to analyze and design bioenergy systems [17]. It is a measure of the amount of energy that can be released per unit mass, through an oxidation reaction. It is one of the most important characteristics to define the suitability of a solid biomass as a fuel. The heating value was determined experimentally by employing an adiabatic calorimetric bomb IKA WERKE; model C2000 basic.

2.5. SWOT analysis

The SWOT analysis evaluates the strengths, weaknesses, opportunities, and threats related to the development of a project. The strengths and weaknesses of the project are internal characteristics and are controllable while opportunities and threats are external factors but can react at a determining moment in their favor [18]

The implementation of the SWOT analysis allows understanding the strengths of a project and to exploit its opportunities and plan based on them. Also, it contributes to recognize treat or avoid the weaknesses and protect against any threat known [19]. The SWOT methodology was applied to evaluate the internal and external factors affecting the utilization of wheat straw as biofuel in Mexico.

3. Results

The main findings highlighted an estimation of 4,612,950.23 t of wheat straw generated in Mexico. The states of Sonora and Baja California were responsible for 61.69% of the wheat straw generation.

The results of proximate analysis experimentally obtained were 64.42% volatile matter, 19.49% fixed carbon, and 16.09% ash.

The experimental higher heating of wheat straw determined was 14.86 MJ/kg. Based on these results, an energy potential of 69 PJ per agricultural cycle was calculated, equivalent to 19% of the biomass energy share reported in Mexico's National Energy Balance, in 2014.

Table 2 depicts the results of the SWOT analysis applied to evaluate the internal and external factors affecting the utilization of wheat straw as biofuel in Mexico.

The main strength identified for the use of wheat straw as biofuel in Mexico was its higher heating value and high intensive activity in the agricultural sector, specifically, wheat harvesting. The higher heating value of the wheat straw is an attractive and the most important characteristic from the energy point of view. The amount of wheat straw generated annually in the Mexican agriculture is considerable and highlights high resource availability. It is an important aspect because it can contribute to ensuring the biomass supply. The valorization of wheat straw for energy applications can foster the economic development of the agricultural sector of Mexico, provide to energy security, and reduce the fossil fuel use. It is a sustainable alternative that helps to control and reduce the pollutant emissions by avoiding the open burning practices of waste agricultural biomass. In the global market, there are proven technologies for the utilization of agricultural waste as biofuel. In Mexico, there is experience in power generation by waste biomass from the agriculture. It is an advance regarding the learning curve.

About the weaknesses found, the wheat straw has a low density. It is an issue that requires physical conditioning and densification of the biomass to facilitate its collection, handling, transportation, and storage. The addition of these preprocessings increases the costs due to the implementation of specialized equipment, labor, and fuel consumption. The wheat straw is not concentrated in one place. Therefore, the long distances involved between the wheat straw generation fields represent a challenge to collect it.

Strengths	Weaknesses
A higher heating value of wheat straw.	Low bulk density of biomass.
Intensive agriculture activity in Mexico.	Large distance between biomass generation places.
Sustainable exploitation of residual biomass.	High handling cost.
Proven technologies for the utilization of agricultural waste as biofuel.	High transportation cost.
Fostering the economic development of agriculture sector.	Biomass is not concentrated.
Greenhouse gas emissions reduction by fossil fuels replacement.	The requirement of specialized equipment for the densification and handling of biomass.
Pollutant emissions reduction by avoiding open burning of crop residues.	Biomass market inexistent in Mexico.
Substitution and reduction of fossil fuels use.	
Ensure energy security.	
Opportunities	**Threats**
National Energy Transition Law.	Traditional open burning crop residues practices.
Recent Energy Reform.	Lack of public policy to foster the utilization of agricultural residues as biofuels.
Ambitious goals of the energy sector to increase fuels from renewable sources.	Ensuring the constant supply of waste biomass.
Favorable policies for the development of renewable energy in Mexico.	Price of residual biomass.
Research and development infrastructure available.	Crop harvested surface variations.

Table 2. SWOT analysis results.

The current situation in the energy sector of Mexico provides opportunities for the use of wheat straw as biofuel. The recent energy reform, the Law of Energy Transition, and the goals to increase the participation of renewable energy in Mexico are setting the platform to favor and to encourage the exploitation of waste biomass for energy applications.

Among the main threats analyzed are the biomass supply ensuring the annual crop harvested surface variations, the price of residual biomass, and the lack of public policy that promotes the valorization of waste biomass. There is another one, related to a sociocultural aspect, and it is the traditional burning practices of wheat straw performed by farmers across the country. Therefore, it is necessary to gain the social acceptability of farmers and the rural communities strategically to avoid burning of wheat straw and to assure the constant supply of wheat straw.

4. Conclusion

Due to the current situation that Mexico is facing concerning energy security, decreasing of dependence to conventional energetics, as well as the reduction of greenhouse gases emissions, it is necessary to find alternatives to diversify the energy sources. For that reason the Mexican government committed to sustainability has empowered the Secretariat of Energy based on international trends that postulate changing patterns of production and use of energy, to develop a national strategy for the energetic transition due to environmental, social, and economic issues. The energetic transition involves major changes, including the promotion of renewable energy sources, e.g., solar, wind, biomass, hydraulic, and the rational use of energy as key strategic actions. The main goal of the Law of Energy Transition is to increase the share of clean energy production to 25% by 2018, 30% by 2021, 35% by 2024, and 40% by 2035. The biomass and waste biomass can play a key role in the energy transition because the high intensive activity in the agriculture sector. The wheat straw standout as an abundant biomass residue generated in Mexico and it has an important energy potential estimated at 69 PJ per agricultural cycle. The valorization and utilization of wheat straw for bioenergy purposes is equivalent to 19% of the biomass energy share reported in Mexico's National Energy Balance, in 2014.

The results of the SWOT analysis applied to evaluate the internal and external factors affecting the utilization of wheat straw as biofuel in Mexico depicted nine strengths, seven weaknesses, five opportunities, and five threats. The development of the SWOT analysis provided the consideration of the main factors for the utilization of wheat straw as a biofuel in Mexico. The actual conditions in Mexico are favorable for the exploitation of wheat straw as a biofuel.

Author details

Gisela Montero[1*], Conrado García[1], Marcos A. Coronado[1], Lydia Toscano[2], Margarita Stoytcheva[1], Ricardo Torres[1], Ana M. Vázquez[3] and Daniela G. Montes[1]

*Address all correspondence to: gmontero@uabc.edu.mx

1 Universidad Autónoma de Baja California, Instituto de Ingeniería, Mexicali, Baja California, México

2 Academia de Química y Bioquímica, Instituto Tecnológico de Mexicali, Mexicali, Baja California, México

3 Universidad Autónoma de Baja California, Escuela de Ingeniería y Negocios Guadalupe Victoria, Mexicali, Baja California, México

References

[1] FAO. Cereal Supply and Demand Brief (2016). Available at: http://www.fao.org/worldfoodsituation/csdb/en/

[2] Agrifood and Fisheries Information Service. Statistical Yearbook of Agricultural Production (2013). Available at: www.siap.gob.mx/cierre-de-la-producción-agrícola-por-estado/

[3] Current situation and overview of wheat production in Mexico (1990–2000). Available at: http://www.campomexicano.gob.mx/portal_siap/Integracion/Estadistica-Derivada/ComercioExterior/Estudios/Perspectivas/Trigo90-00.pdf9

[4] Moncada A, Quintero N. Pollution and control of agricultural burning in Imperial, California and Mexicali, Baja California. Región y Sociedad. 2008; 20:03–24.

[5] SEMARNAT/SAGARPA. NOM-015-SEMARNAT/SAGARPA-2007, It establishes the technical methods of use of fire in forest land and agricultural land use specifications. Official Journal of the Federation (2007). http://www.dof.gob.mx/nota_detalle_popup.php?codigo=5077654

[6] IRENA. Global bioenergy supply and demand projections. International Renewable Energy Agency (2014). Available at: http://irena.org/remap/IRENA_REmap_2030_Biomass_paper_2014.pdf

[7] Suramaythangkoor T, Gheewala S H. Potential alternatives of heat and power technology application using rice straw in Thailand. Applied Energy 2010; 87(1):128–133.

[8] SENER. National energy balance 2014. (2015). http://www.gob.mx/cms/uploads/attachment/file/89382/Balance_Nacional_de_Energ_a_2014.pdf

[9] National Inventory of Renewable Energy (2016). Available at: http://inere.energia.gob.mx/version4.5

[10] Official Journal of the Federation. Political Constitution of the United States of Mexico (1917). http://www.diputados.gob.mx/LeyesBiblio/pdf/1_150816.pdf

[11] Official Journal of the Federation. Law of Energy Transition (2015). http://www.diputados.gob.mx/LeyesBiblio/pdf/LTE.pdf

[12] García C, Montero G, Coronado M, Torres R, Vázquez A, Ayala J, León J, Sagaste C. Power generation estimation from wheat straw in Mexico. WIT Transactions on Ecology and The Environment 2015;195:101–110. WIT Press. Available at: http://www.witpress.com/elibrary/wit-transactions-on-ecology-and-the-environment/195/34392

[13] SENER. Secretaría de Energía. (2006). Potenciales y viabilidad del uso del bioetanol y biodiesel para el transporte en México. Retrieved from: http://www.bioenergeticos.gob.mx/bio/descargas/SENER-BID-GTZ-Biocombustibles-en-Mexico-Estudio-completo.pdf

[14] ASTM E870 – 82. Standard Test Methods for Analysis of Wood Fuels. Vol.11.06; 2013.

[15] ASTM E711-87. Standard Test Method for Gross Calorific Value of Refuse-Derived Fuel by the Bomb Calorimeter (2004).

[16] Montero G, Coronado M, Torres R, Jaramillo B, García C, Stoytcheva M, Vázquez A, León J, Lambert A, Valenzuela E. Higher heating value determination of wheat straw from Baja California, Mexico. Energy 2016;109:612–619. Available at: http://dx.doi.org/10.1016/j.energy.2016.05.011

[17] Nhuchhen D, Abdul P. Estimation of higher heating value of biomass from proximate analysis: A new approach. Fuel. 2012; 99:55–63.

[18] Coronado M, Montero G, García C, Torres R, Vázquez A, Ayala J, León J, Perez L, Romero E. Cotton stalks for power generation in Baja California, Mexico by SWOT analysis methodology. WIT Transactions on Ecology and The Environment 2015;195:75–86. WIT Press. Available at: http://www.witpress.com/elibrary/wit-transactions-on-ecology-and-the-environment/195/34390

[19] Zhao Z, Yan H. Assessment of the biomass power generation industry in China. Renewable Energy 2012;37(1):53–60.

Chapter 25

Use of Corn Dried Distillers Grains (DDGS) in Feeding of Ruminants

Ewa Pecka-Kiełb, Andrzej Zachwieja, Dorota Miśta,
Wojciech Zawadzki and Anna Zielak-Steciwko

Additional information is available at the end of the chapter

http://dx.doi.org/10.5772/66357

Abstract

Bioethanol is the product of fermentation of starch contained in renewable resources, such as corn, wheat, rye and rice. Depending on the technology used for its production, dried distillers decoction may exist in different forms: dried distillers grain (DDG); dried distillers grain with solubles (DDGS) and high-protein dried distillers grains (HPDDG), as well as wet distillers grain (WDG), wet distillers grain with solubles (WDGS), and high-protein wet distillers grains HPWDG). Research conducted in recent years has demonstrated the possibilities of corn DDG as feed for livestock due to its high content of valuable protein, high calorific value and bioelements. Distillers grain has been used as feed for beef and dairy cattle, sheep, swine and poultry. In case of ruminants, it is important that distillers grain is foodstuff high in ruminal undegradable protein, with beneficial fibre content that does not cause rumen acidosis. DDGS has positive influence on milk yield and its fat and protein content. Research on rumen fermentation has proven that DDGS positively affecs processes in forestomachs: methanogenesis, ammonia emission and volatile fatty acids profile. Reprocessing of agri-food industry by-products may well be an alternative for traditional methods of feeding animals and utilizing valuable nutrients that they contain.

Keywords: dried distillers grains with solubles (DDGS), corn, ruminants, animal production

1. Introduction

The prospective exhaustion of non-renewable energy sources and the negative influence their burning has on the environment enforces the search for alternative fuels coming from the plant biomass. Nowadays, in an attempt of replacing conventional fuels, bioethanol coming from the plant-based biofuels is used [1]. Bioethanol is obtained from the fermentation of starch contained in renewable material, such as, corn, wheat, rye, rice, and similar things. Its production involves fermentation of raw material and its distillation followed by dehydration. The byproduct of ethanol production is a decoction.

2. Bioethanol production process and types of distillation decoctions

The chemical formula for ethyl ethanol is C_2H_5OH and its calorific value equals 19.6 MJ/L. There exist two technological processes for obtaining ethanol from corn: the wet and the dry ones [2, 3] (**Figure 1**). The processes are alike, but they result in different byproducts. Dry grinding allows for greater volume of bioethanol, but the only other product is the animal feed (WDGS, DDGS). When the wet technology is used, apart from ethanol and animal feed corn oil, corn syrup and gluten are obtained. The production of ethanol requires a two-stage fermentation of starch: the first stage is the decomposition of starch into glucose and maltose, the second stage is yeast fermentation during which disaccharides and monosaccharides are converted into ethanol. The polymeric structure of starch is destroyed by enzymes and temperature. Two enzymes are used in the industrial process. The first one is α-amylase, whose function is to hydrolyze polymers to produce shorter chains (dextrins), which remain in the solution: this is the condensation stage. Then, due to the activity of glucoamylase in the saccharification stage, dextrins convert to simple sugars, glucose and maltose (dimer of α-1–4 glucose) [2]. The obtained solution undergoes yeast fermentation (*Saccharomyces cerevisiae*): one molecule of glucose is converted to two molecules of carbon dioxide. Post fermentation liquid is then distilled and the decoction is separated into the solid and liquid phases with the use of decanters. The liquid phase is then evaporated and condensed into syrup, which is mixed with the solid phase from the decanters. The resulting decoction is centrifuged, dried and finally granulated.

Depending on the technology used for ethanol production, different types of decoctions may be obtained [4]: dried distillers grains (DDG) obtained from distilling ethanol from yeast production; dried distillers grains with soluble (DDGS)—the most widely used—obtained from wet corn residues (DG) mixed with condensed liquid phase in the form of syrup (CDS) and dried; and high-protein dried distillers grains (HPDDG)—bran and germ (rich in fiber and fat) are removed before distillation allowing for the production of dried decoction with high-protein content [5]. Foodstuffs in hydrated form containing dry mass between 5% and 8% (WDG, WDGS, and HPWDG) are cheaper but difficult to transport and to store. That is the reason why dried distillers grain is the most widely used product of this kind [1].

The output of bioethanol depends on the plant used (**Table 1**). Yield per hectare, soil and weather conditions make corn the main resource for ethanol production. According to Food and Agricultural Policy Research Institute [6], in 2021, the production of ethanol will reach 141,28 metric tons in the USA and 7174 in the UE (**Figure 2**).

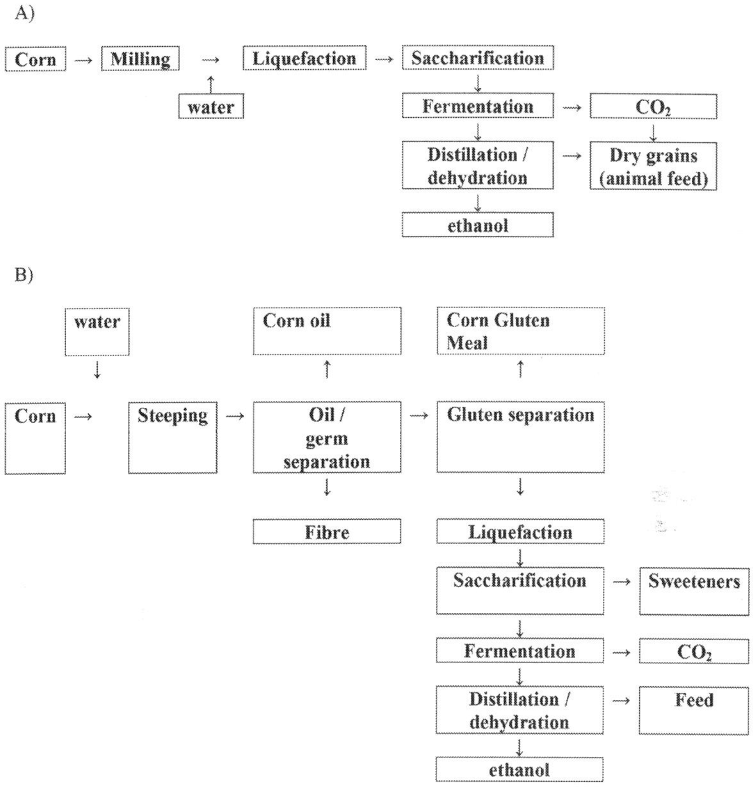

Figure 1. Ethanol production from corn: (A) Dry mill process and (B) wet mill process [3].

Raw material	Production (t/ha)	Yield (l/t)	Yield (l/ha)
Corn	3.0	380	3420
Sugercane	80.0	75	6000
Cassava	20.0	180	1379
Potato	20.0	130	2600

Table 1. Alcohol production rates from different raw materials [7].

According to Lee and Bressan [7], it is possible to produce 308 l of ethanol and 329 kg of DDGS from 1 ton of corn [8]. The numbers suggest that in 2021, the production of DDGS will be 146,784 metric tons in the United States and 7453 metric tons in the EU.

More and more restrictive laws concerning the disposal of biofuel byproducts make it necessary to utilize the decoctions in an alternative way: using them as feed for livestock is a good solution. Additionally, corn dried distillers grain may reduce prices of nutritive fodder used for feeding animals [9]. In July 2006, according to IndexMundi [10], the price of a ton of corn in the USA was $ 119.69, soybean—$ 349.33, oats—$ 161.6, barley—$ 140.19, and DDGS—$ 120.00. Price analysis suggests that replacing 20% of soy would reduce the cost of feed by 13%.

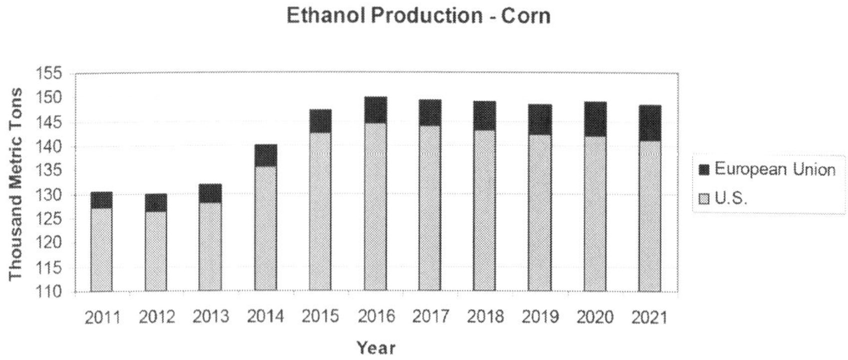

Figure 2. Production of ethanol from corn in the United States and the European Union. Food and Agricultural Policy Research Institute (FAPRI) [6].

3. Nutritional characteristics of DDGS

Corn is one of the most frequently grown agricultural crops in the world and increasingly important not only in food but also in chemical and energy industries. The earliest research into dried distillers grain as possible sources of protein in feeding livestock took place in 1945 [11]. In recent years, the possibility of using dried distillers grains as a feed for farm animals, especially cattle (both beef cattle and dairy cattle), pigs, and poultry, has been demonstrated [12, 13].

Corn dried distillers grain is high-protein feed: on average it contains 28–36% total protein (BO) in dry matter [14] which is characterized by the low rate of decomposition in the rumen, resulting in high content of ungradable fraction (RUP)—from 47% to 63% BO (55% on average) [4]. The presence of dead yeast cells gives the protein better amino acids composition and very good nutritive value.

Because of the high content of insoluble fibre, DDGS has positive influence on digestion and lowers the pH in the digestive system. This results in the reduction of pathogen population and diminishes the occurrence of diarrhoea in young animals. DDGS is also a good source of protein and energy for lactating cows [15].

DDGS has lower level of energy than the soybean meal (by 4%), barley (by 17%), and wheat (by 25%), but higher than the rapeseed meal (20–40%). Energy values for different feeds in swine, poultry, and ruminants are shown in **Table 2** [16]. The tabular content of energy for corn DDGS, except for gross energy (heat of combustion), is lower than in grains. Technological improvements in the ethanol production have made it possible for the net energy of lactation in the decoction to equal the concentration of energy in the grains.

		Wheat	Barley	Maize gluten feed	Wheat DDGS	Extract of soya bean meal
Net energy—pig	MJ/kg	10.61	9.66	6.89	8.00	8.44
Metabolisable energy—poultry	MJ/kg	13.00	11.80	8.2	9.94	10.30
Metabolisable energy—ruminant	MJ/kg	11.90	11.35	11.5	11.2	12.30

Table 2. Wheat DDGS in comparison to other major types of feed [16].

The main source of energy in corn grain is starch, which is almost completely fermented in the process of biofuel production. In DDGS, the main carrier of energy is fat and neutral detergent fibre (NDF). Ether extract constitutes 8.2–11.7% of dry matter, with nonsaturated acids accounting for 80% of fatty acids [14]. Therefore, appropriate introduction of DDGS to feed rations assumes not only the concentration of energy coming from fat but also the improvement of fat composition in milk (enrichment in nonsaturated acids). On the other hand, 5% is the upper limit of fat concentration in the dry mass of feed ratio and it should not be exceeded because of DDGS. Neutral detergent fibre (NDF), which constitutes 40–45% of dry matter, has low content of lignin and easily ferments in the rumen to produce volatile fatty acids. So, DDGS is a foodstuff that does not threaten the rumen acidosis. However, because of large fragmentation of the decoction, it cannot be treated as a source of physically efficient NDF.

The content of amino acids in DDGS is higher than in corn (**Table 3**). It is worth stressing that corn DDGS has the lowest content of lysine, since it is produced from the grain which is poor in this amino acid. Yeast present in the decoction not only improves the composition of amino acids in its protein but also the taste. The ingestion of TMR with DDGS content is usually higher in cows [17]. Nevertheless, it is necessary to balance lysine and methionine, in poultry mainly. The content of protein and fat in DDGS is relatively high, but its composition is slightly different depending on the source (**Table 4**).

Amino acid	Corn	DDGS
Arginine	0.54	1.05
Histidine	0.25	0.70
Isoleucine	0.39	1.52
Leucine	1.12	2.43
Lysine	0.24	0.77
Methionine	0.21	0.54
Phenylalanine	0.49	1.64
Threonine	0.39	1.01
Thryptophan	0.09	0.19
Tyrosine	0.43	0.76
Valine	0.51	1.63

Table 3. Comparison of essential amino acid content (g/100 g of dry matter) in DDGS and corn [17].

Dry matter content (%)	Ash (%)	Crude protein (%)	Crude fiber (%)	Crude fat (%)	NDF (%)	ADF (%)	Author
91.82	5.01	24.87	8.72	11.2	36.71	11.86	Pecka-Kiełb et al. [18]
88.9	5.8	30.2	8.8	10.9	42.1	16.2	Spiehs et al. [19]
93.3	5.3	24.3	7.5	10.5	Na	Na	Szulc et al. [20]
86.0	3.90	29.1	Na	9.80	Na	Na	Lumpkins et al. [13]
88.5	Na	29.3	Na	9.55	37.3	18.5	Janicek et al. [21]

Na—no data available in the sources.

Table 4. Composition of DDGS.

Corn dried distillers grain is rich in phosphorus (0.43–0.83% of dry matter) and its level depends mostly on the content of condensed syrup (CDS)—the carrier of phosphorus compounds. In our research [22], it has been determined that high producing dairy cows may show symptoms of subclinical hypophosphatemia, which is often accompanied by postpartum paralysis [23]. Very low level of phosphorus in cows' blood serum is probably underrated in diagnosing postpartum paresis (milk fever). For cows, DDGS may be a valuable source of phosphorus in postpartum period preventing hypophosphatemia.

Another element whose concentration in DDGS is visibly higher than in other foodstuffs is sulphur. According to Shurson [24], corn decoction may contain from 0.31% to 1.93% S in dry matter. Its high concentration is partly due to sulphates from sulphuric acid used for cleaning brewery installations. High content of sulphur in DDGS is yet another argument for utilizing it in the postpartum period. The decoction may provide sulphur anion reducing cation-anion balance of feed ration (DCAB), which is recommended for preventing postpartum paralysis. It is essential that feed ratio does not contain more than 0.4% of dry matter (NCR, 2001). Feeding diets with higher concentration of this element may result in disorders of the nervous system, and disturb absorption and metabolism of copper and selenium (the so-called antagonism of elements syndrome)

4. Influence of corn dried distillers grains on health and productivity of animals

Literature abides in research results concerning the addition of wet (WDGS) [25] and dried (DDGS) [26] distillers grains to TMR. Distillers grains is used as a substitute for the postextraction of soy meal, or as an additive to TMR mixture in the ratio of 10% to 20% [25]. According to Janicek et al. [21], this ratio of DDGS in compound feeds for cattle influences the growth of milk yield and the content of fat and protein in it. Powers et al. [26] showed that the use of DDGS and WDGS in feeding high producing dairy cows gives positive result irrespective of the type of decoction, i.e., dried and wet.

The percentage of fat in milk increases slightly in livestock fed TMR with the addition of DDGS and WDGS. However, feeding the wet decoction causes substantial growth of FAT percentage in milk, probably due to access to fibre in WDGS.

Other authors [25, 26] demonstrated that the use of mixtures with dried distillers grains decreases the ratio of n6/n3 fatty acids in milk, which improves its dietary properties.

One of the possible reasons may be the reactions of lipolysis, hydrogenation, and synthesis of fatty acids in the rumen, so their volume depends on the ratio and changes in the profile during fermentation. Analyzing conversions of fatty acids in cow and sheep rumen and their flow to duodenum, Beam et al. [27] and Jenkins [28], assert the amount of fat obtained from the feed. The compositions of DDGS show high levels of nonsaturated fatty acids (**Table 5**), which has a beneficial influence on their profile in the rumen digesta. The level of C18:1n9c and C18:2n6c acids in the rumen digesta in the *in vitro* examination increases with the addition of DDGS. However, the levels of C15:0, C16:0, and C20:0 saturated fatty acids and nonsaturated C14:1 in the rumen digesta during *in vitro* fermentation does not change [29].

Fatty acids	DDGS
C14.0[1]	0.03
C16:0[1]	10.58
C18:0[1]	1.92
C20:0[1]	0.40
C14:1[1]	0.04
C18:1n9c[1]	28.54
C18:2n6c[1]	54.31
C18:3n3[1]	1.16

[1]g/100 g fat

Table 5. Percentage of fatty acids in DDGS [29].

According to Al-Suwaiegh et al. [25] and Anderson et al. [30], the percentage of protein and lactose in milk in cows fed DDGS is similar to those fed WDGS. The growth in milk production and the percentage of casein fraction after the inclusion of DDGS in the feed ration of milking cows in the early stages of lactation have been shown (**Figure 3**) [20].

The use of 10–15% of DDGS dry feed in cows in the postpartum period increases the general protein and immunoglobulin level in colostrum. DDGS has no impact on the content of amino acids in colostrum per 1 g of protein. However, with increased DDGS content in cow diet, some physico-chemical properties of colostrum deterirate (decrease in thermal stability and shortening of coagulation time under the influence of rennet). Yet, despite the deterioration of the values of technological properties, DDGS demonstrates the beneficial use of colostrum components, and in consequence the level of total protein and immunoglobulin in the serum of calves [29].

In sheep, 10–20% of DDGS in the feed does not have negative effect on the production of milk. It slightly lowers the protein, dry matter, and the fat content [31]. In sheep and goats, the use of DDGS may increase the level of PUFA acids in kefir produced from their milk [32]. Available literature does not present research on the influence of DDGS on the composition and quality of goat's milk. It might be the consequence of a small number of these animals as compared to the dairy cows. Studies by other authors show that the DDGS and WDGS have limited usage in optimizing feeding of livestock. Heavy doses are harmful because of the high level of fat, which reduces the digestibility of feed ratios. About 20–25% of DDGS in compound feeds is considered to be safe and optimal for dairy cows, whereas for beef cows it is safe up to 50%.

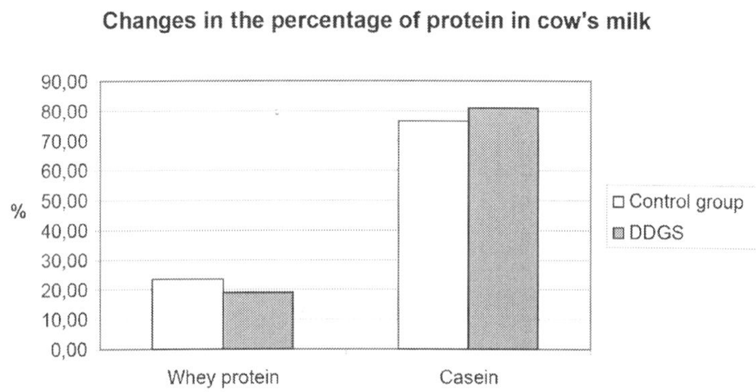

Figure 3. Changes in the percentage of whey protein and casein in cow's milk [20].

In feeding sheep, DDGS has no impact on the condition of animals or milk productivity [33]. About 21.20% of DDGS in dry matter of feed for ewes decreases glucose level in blood and increases the level of insulin. The authors have also confirmed that DDGS is a good source of nutrient for sheep, which has a positive influence on the mass growth of newborn lambs, and has no impact on their mortality [34].

Şahin et al. [35] demonstrated that the inclusion of DDGS in the diet of 3-month-old lambs did not have negative impact on their growth, forage consumption or rumen parameters. According to the authors, only when DDGS constitute 20% of forage, it can be considered as a good source of protein in lambs' diet, as digestibility is considerably lower when DDGS constitutes 10% of the feed. Schauer et al. [36] asserted that 60% inclusion of DDGS in the diet does not show considerable impact on the productivity in growing slaughter lambs (**Table 6**). In goats, however, DDGS may completely replace the soybean meal and up to 31% of corn in addition to the dry matter of the diet. Also, other authors did not observe negative influence of DDGS on productivity in slaughter animals, carcass components, or fatty acids profile in sheep meat [37]. Literature of the subject does not contain research on the influence of DDGS on the quality of goat meat. As in the case of goat's milk, it may be linked with small number of animals and insignificant influence of goat meat on the world economy.

In beef cows, 35% content of DDGS in feed ration results in the growth of PUFA and CLA acids in meat [38]. Other authors asserted that 25% content of dried distillers grain does not harm the quality of meat (**Table 7**) [39].

Recent research has demonstrated the impact of DDGS on fermentation in the rumen. It increases the pH amplitude of ruminal fluid and extends the time in which the pH falls below 5.8 [15], and the concentration of acetate and its proportions to propionate decrease [40]. Corn dried distillers grain results in linear decrease of methane production in the rumen of cows in proportion to the growth in DDGS contents in the diet *in vivo* [41] as well as *in vitro* [42].

	Treatment[1]		
Item	**Control**	**20%**	**60%**
Initial weight (lbs)	68.00	70.00	70.00
Final weight (lbs)	132.00	137.00	137.00
ADG (lbs/day)	0.58	0.62	0.62
Intake (lbs/hd/d)	3.69	3.91	4.20
Mortality	0.75	0.25	0.00
Leg score	10.30	10.50	10.50
Conformation score	10.30	10.30	10.50
Fat depth (in)	0.29	0.32	0.32
Quality grade	10.30	10.80	11.00
Yield grade	3.26	3.57	3.55

[1]Control = 0% replacement of barley with dried distillers grains; 20% = 20% dried distillers grains in ration replacing barley; 60% = 60% dried distillers grains in ration replacing barley.

Table 6. The influence of dried distillers grains on feedlot lamb [36].

Item	**Control**	**25% DDGS**
Initial BW- Body Weight (kg)	379	377
Final BW- Body Weight (kg)	494	485
ADG (kg/day)	9.01	8.52
Yield grade	2.62	2.74

Table 7. The influence of corn dried distillers grains on feedlot beef cattle [39].

In sheep, a 3% decrease in the production of methane in the group fed 30% of DDGS is observed. This decrease in the rumen fermentation is considered good for animals. The emission of methane is a waste of energy, which may result in drop in milk productivity of ruminants [18, 43]. In cows, the growing proportion of DDGS in the substrate in the *in vitro* study causes a significantly reduced ammonia production in the rumen digesta; after 24 hours of fermentation, the amount of ammonia is more than five times lower with DDGS (22.4 mmol/l) in comparison to control, where TMR was the substrate (124.6 mmol/l). The *in vitro* studies showed that the use of DDGS reduced the acetate and propionate levels in lambs [44]. In sheep and cows, the contents of DDGS in forage reduce the production of acetate in the rumen and increases the ratio of propionate [18, 42].

In the *in vivo* conditions, DDGS does not change SCFA concentration in the ruminal fluid of cows, but it lowers the content of acetate in SCFA in groups of animals fed DDGS (57.4 mol% in the group of 10% DDGS in dry matter ration, 53.1 in 15% DDGS in dry matter ration, and 63.5 in 30% DDGS in dry matter ration) as compared to control group where traditional TMR (65.7 mol%) was used. DDGS increases the levels of propionate. The SCFA utilization factor expressed as the ratio of nonglycogenic to glycogenic SCFA acids (that is NGR) decreases in animals fed DDGS [45].

The obtained results show beneficial impact of DDGS on the content of the most important volatile acids in the rumen digesta. The use of DDGS as a substrate in the *in vitro* fermentation of the rumen digesta in cows as well as sheep changed the levels of butyric and isovaleric acids: their levels were decreasing with the augmented ration of DDGS

The consequence of the drop in production of isoacids reduced the decomposition of protein in the rumen, which is desirable in this group of animals [18, 42].

Research results suggest the possibility of using corn dried distillers grain as an addition or a substitute for other compound feeds in feeding lactating dairy cows. In recent years, studies of corn DDGS in feed rations for cows in the dry period showed that it may be included in TMR in this phase of the production cycle. The dry period is connected with the significant physiological, metabolic, and nutritional changes. Feeding cows determine possible problems in the postpartum period, define their metabolic status, and in consequence their health condition, fertility, value of functional traits, which affect the efficiency of milk production. Proper inclusion of DDGS in the feed ration allows for the assumption that not only the concentration of energy from fat, but also the improvement of milk composition (it will be richer in non-saturated fats). DDGS does not threaten the incidence of rumen acidosis. It may be an important source of phosphorus for cows in postpartum period and its use may prevent hypophosphatemia. High content of sulphur in DDGS is yet another argument for its use around the calving period. The decoction is an important source of sulphur anion which diminishes the cation-anion balance of the feed ration (DCAD). When DDGS was used in the last three weeks before calving as 10%, 15%, and 20% of the dry matter of feed ration, respectively, the experiment results showed a drop of DCAD of the feed ration from 189 (TMR without the addition of DDGS) to 10 mEq/kg when 20% of DDGS was used [29].

When 10% DDGS is used, in cows immediately after calving, the level of liver enzyme aspartate aminotransferase type (AST) grows and the level of triglycerides drops, which suggests the development of subclinical ketosis. However, the 15–20% DDGS content does not increase ketogenesis or alkalosis. Large doses of 20% DDGS cause excessive increase of AST activity after calving. Feeding 20% DDGS to cows in the postpartum period favourably influences the content of Ca and P in the serum after calving, and physiological hypocalcaemia is not observed in this period. The decrease of the total protein and G type immunoglobulins in the blood of cows receiving larger amounts of DDGS in their feed rations simultaneously causes a slight decrease in the level of albumins. It may indicate the possibility of more intense transmission of immunoglobulins to mammary gland before calving [29].

The results of the study in their overwhelming majority confirm the possibility of safe and efficient inclusion of DDGS in nutritional programme of ruminants, but—as in the case of all types of feed—standard precautions are necessary. Changeable/varying content of particular nutritional components, physical and chemical properties connected with the method of fermentation in the production of bioethanol or storing of the decoction may pose problems.

The process of drying DDGS is of significant importance for the production of DDGS—too fast and in too high temperature causes negative changes in protein: denaturation of protein takes place and products of Maillard's reaction are created, which results in the growth of nitrogen insoluble in acidic detergent (ADIN), indigestible fraction of the total protein [24]. In the studies by Cromwell et al. (1993) [46], the percentage of ADIN in total nitrogen of DDGS is between 8.8% and 36.9%. It shows that with inappropriate drying of the decoction, almost 40% of the total protein may have no nutritional value. Colour is a good marker of a correct/accurate drying process of DDGS—good quality decoction should be light orange in colour.

Another problem encountered when using DDGS is a big variability in the molecular size. American studies determined that the average size of decoction molecules is 1282 μm and the range was from 612 to 2125 μm. Such large differences in the molecular size cause spontaneous stratification of DDGS components during the transport and storage of the feedstuff. The smallest precipitated fraction has strong caking properties and may result in dangerous suspension of hard mass in storing silos. Additional factors enlarging the problem is high temperature (summer period), increased water content (secondary moisture), and fat concentration [24].

Distillers grains may also be a source of mycotoxin. If bioethanol is produced from low-quality mould-infected grain, it may pose great threat to animal health and determine the quality of animal products. The concentration of mycotoxins in DDGS is on average three times higher than in the mould-infected grain from which it comes. In Austrian research [19], where 89 samples of DDGS (70% from the USA, 30% from Asia) were tested, it was shown that the biggest/most serious problem was the presence of zearalenone (ZON), B1 and B2fumonisins (FUM), and deoxynivalenol (DON). Mycotoxins were discovered in 91, 85, and 57% of samples. Aflatoxin B1 (AFB1) and T-2 toxins were smaller threats. Their highest mean and maximum levels of concentration were observed for DON (1405 and 12000 g/kg) and FUM (935 and over g/kg). It was also demonstrated that only 1% of samples were free from any mycotoxins. Because of the confirmed threat of toxic compounds, it is recommended that every batch of DDGS reaching farmsteads is examined in reference laboratories (with the use of chromatographic techniques) for the presence of mycotoxins.

One of the great advantages of DDGS is the possibility of storing it even for a year; WDGS may be stored for 3–7 days. However, there may be difficulties in balancing the diet combining different components of the feed rations with distillers grains [37]. DDGS contains relatively large amounts of elements such as sulphur, phosphorus, and nitrogen, which enter the environment as a consequence of excretion process. High content of phosphorus in the organisms of ruminants may lead to disturbances in phosphate-calcium balance between phosphate and calcium [33, 45, 47]. Because of the need to balance DDGS as a feed additive containing different proportions of nutrients, every batch of DDGS requires standard chemical analyses performed on all compound feeds by the manufacturers [37, 45]. **Figure 4** shows the recommended chemical analyses of distillers grains [2].

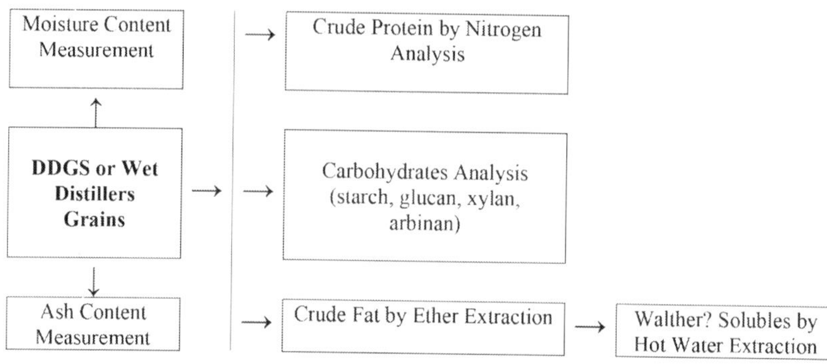

Figure 4. Diagram of biomass analysis [2].

5. Conclusions

Existing research results suggest the effective use of dry and wet distillers grains in livestock nutrition and especially the inclusion of corn dried distillers grains (DDGS) in feed rations for cows, sheep, swine, poultry, and even rabbits. Reprocessing the byproducts of agriculture and food industry is likely an alternative for traditional nutrition of animals. It is also a good way of utilizing the valuable nutrients that these byproducts contain. Compared to other feeds, DDGS is cheaper but its use poses problems, as it is a changeable composition, which requires technological procedures to standardize it.

The growing demand for renewable sources of energy will parallel the supply of biofuels, whose byproducts are the alternative feed materials, rich in energy and protein. The production of corn dried distillers grains (DDGS) acquired in the process of bioethanol making is relatively large and it may lead to problems utilizing it without negative impact on the environment. One of the eco-friendly alternatives for using dried distillers grains is feeding the livestock. Corn DDGS is particularly a valuable feed for dairy cows in the postpartum period, when its use prevents the postpartum paralysis (it is a good source of phosphorus and sulphur), diminishes the negative balance of energy (large fat content) and the threat of rumen acidosis (favourable composition fibre fraction NDF), as well as improves the feed intake (yeast content). The decoction may partly substitute the soybean meal in feed rations for high producing lactating cows. Optimum addition of DDGS for dairy cows is 10–15% of dry matter in the feed ration. One of the beneficial effects of DDGS as a component in feed rations is the decrease in methane production. Another one may be lowering the costs of feed for animals as it is relatively cheap. Propagation of DDGS reprocessing as animal feed will significantly reduce potential threats for the environment.

Author details

Ewa Pecka-Kiełb[1*], Andrzej Zachwieja[2], Dorota Miśta[1], Wojciech Zawadzki[1] and Anna Zielak-Steciwko[1]

*Address all correspondence to: ewa.pecka@up.wroc.pl

1 Department of Biostructure and Animal Physiology, Wroclaw University of Environmental Life Sciences, Wroclaw, Poland

2 Department of Cattle Breeding and Milk Production, Wroclaw University of Environmental Life Sciences, Wrocław, Poland

References

[1] Zachwieja A, Kupczyński R, Bodarski R, Paczyńska K, Miśta D, Pecka E, Zawadzki W, Tumanowicz J, Adamski M. An influence of dried distillers grains with soluble (DDGS) application on dietary cation-anion balance, changes in rumen fermentation profile, metabolic profile of cows and physico-chemical features of colostrums. Przemysł Chemiczny. 2013;**92**:1050–1055.

[2] Kim Y, Mosier NS, Hendrickson R, Ezeji T, Blaschek H, Dien B, Cotta M, Dale B, Ladisch MR. Composition of corn dry-grind ethanol by-products: DDGS, wet cake, and thin stillage. Bioresource Technology. 2008;**99**:5165–5176. Doi: 10.1016/j.biortech.2007.09.028.

[3] Piscina PR, Homs N. Use of biofuels to produce hydrogen (reformation processes). Chemical Society Reviews. 2008;**37**:2459–2467 DOI: 10.1039/B712181B.

[4] Schingoethe DJ. Utilization of DDGS by cattle. In: Proceedings of 27th Western Nutrition Conference; 19–20 September 2006; Winnipeg, Manitoba, Canada. 2006; pp. 61–74.

[5] Hubbard KJ, Kononoff PJ, Geham AM, Kelzer JM, Karges K, Gibson ML. The effect of feeding high-protein distillers dried grains on milk production of Holstein cows. Journal of Diary Sciences. 2009;**92**:2911–2914. DOI: 10.3168/jds.2008-1955

[6] Food and Agricultural Policy Research Institute (FAPRI) [Internet]. 20v12. Available from: http://www.fapri.iastate.edu/outlook/2012/[Accessed: 2016-08-10].

[7] Lee TSG, Bressan EA. The potential of ethanol production from sugarcane in Brazil. Sugar Tech. 2006;**8**:195–198. DOI: 10.1007/bf02943556

[8] Świątkiewicz M, Hanczakowska E, Olszewska A. Distiller's dried grains with solubles (DDGS) in swine nutrition. Zootechnic news. 2014;4:141–153.

[9] Gurung NK, Solaiman SG, Rankins DL, McElhenney WH. Effects of distillers dried grains with solubles on feed intake, growth performance, gain efficiency and carcass quality of growing Kiko x Spanish male goats. Journal of Animal and Veterinary Advances. 2009;**8**:2087–2093. DOI: javaa.2009.2087.2093

[10] IndexMundi, Commodity Prices [Internet]. 2006. Available from: http://www.indexmundi.com/[Accessed: 2016-08-14].

[11] Hughes CW, Hauge SM. Nutritive value of distillers' dried solubles as a source of protein. Journal of Nutrition. 1945;**30**:245–258.

[12] Liu K. Particle size distribution of distillers dried grains with solubles (DDGS) and relationships to compositional and color properties. Bioresource Technology. 2008;**99**:8421–8428. DOI: 10.1016/j.biortech.2008.02.060.

[13] Lumpkins BS, Batal AB, Dale NM. Evaluation of distillers dried grains with solubles as a feed ingredient for broilers. Poultry Science. 2004;**83**:1891–1896.

[14] Schingoethe DJ, Kalscheur KF, Garcia AD. Distillers grains for dairy cattle. South Dakota State University Cooperative Extension Service [Internet]. 2002. Available from: http://pubstorage.sdstate.edu/AgBio_Publications/articles/ExEx4022.pdf[Accessed: 2016-08-03]

[15] Zhang SZ, Penner GB, Yang WZ, Oba M. Effects of partially replacing barley silage or barley grain with dried distillers grains with solubles on rumen fermentation and milk production of lactating dairy cows. Journal of Diary Sciences. 2010;**93**:3231–3242. DOI: 10.3168/jds.2009-3005.

[16] Hazzledine M, Pine A, Mackinson I, Ratcliffe J, Salmon L. Estimating displacement ratios of wheat DDGS in animal feed rations in Great Britain. A report commissioned by the International Council on Clean Transportation. Working Paper 2011–8. [Internet]. 2011. Available from: http://www.fapri.iastate.edu/outlook/2012/[Accessed: 2016-08-10]

[17] Belyea RL, Rausch KD, Tumbleson ME. Composition of corn and distillers dried grains with solubles from dry grind ethanol processing. Bioresource Technology. 2004;**94**:293–298. DOI: doi.org/10.1016/j.biortech.2004.01.001.

[18] Pecka-Kiełb E, Zawadzki W, Zachwieja A, Michel O, Mazur M, Miśta D. In vitro study of the effect of corn dried distillers grains with solubles on rumen fermentation in sheep. Polish Journal of Veterinary Sciences. 2015;**18**:751–758. DOI: 10.1515/pjvs-2015-0097.

[19] Spiehs MJ, Whitney MH, Shurson GC. Nutrient database for distillers' dried grains with solubles produced from new ethanol plants in Minnesota and South Dakota. Journal of Animal Science. 2002;**80**:2639–2645.

[20] Szulc T, Zachwieja A, Demkowicz M, Newlacil I, Pecka E. The efficiency of the replacement of protein meal by DDGS—dried distiller grains with solubles in the ration for dairy cows during early lactation. Electronic Journal of Polish Agricultural Universities. Series Animal Husbandry. 2011;**14**(2):03. http://www.ejpau.media.pl/volume14/issue2/art-03.html

[21] Janicek BN, Kononoff PJ, German AM, Doane PH. The effect of feeding dried distillers grains plus solubles on milk production and excretion of urinary purine derivatives. Journal of Diary Sciences. 2008;**91**:3544–3553. DOI: 10.3168/jds.2007-0777.

[22] Bodarski R., Kinal S., Preś J., Krzywiecki S., Słupczyńska M., Twardoń J., Mordak R. Potassium, calcium and magnesium salts indispensable In kation-anion balance regulation of Leeds and mixed rations.[Sole potasu, wapnia i magnezu niezbędne w regulacji bilansów kationowo-anionowych pasz i dawek TMR]. Przemysł Chemiczny. 2010; 89: 939–944..

[23] Preś J, Mordak R, editors. Wybrane elementy żywienia a problemy zdrowotne krów mlecznych, 1st ed., MedPharm, Wrocław; 2010:222p.

[24] Shurson J. Quality characteristics and nutritional profiles of DDGS [Internet]. 2006. Available from: http://www.biofuelscoproducts.umn.edu/sites/biodieselfeeds.cfans.umn.edu/files/cfans_asset_414236.pdf[Accessed: 2016-08-03]

[25] Anderson JL, Schingoethe DJ, Kalscheur KF, Hippen AR. Evaluation of dried and wet destillers grains included at two concentrations in the diets of lactating dairy cows. Journal of Diary Sciences. 2006;**89**:3133–3142. DOI: 10.3168/jds.S0022-0302(06)72587-5

[26] Powers WJ, Vanv Horn HH, Harris B Jr, Wilcox CJ. Effects of variable sources of distiller dried grains plus solubles on milk yield and composition. Journal of Diary Sciences. 1995;**8**:388–396. DOI: 10.3168/jds.S0022-0302(95)76647-4.

[27] Beam TM, Jenkins TC, Moate PJ, Kohn RA, Palmquist RA. Effects of amount and source of fats on the rates of lipolysis and biohydrogenation of fatty acids in ruminal contents. Journal of Diary Sciences. 2000;**83**:2564–2573. DOI: 10.3168/jds.S0022-0302(00)75149-6.

[28] Jenkins TC. Lipid metabolism in the rumen. Journal of Diary Sciences. 1993;**76**:3851–3863. DOI: 10.3168/jds.S0022-0302(93)77727-9.

[29] Zachwieja A, Bodarski R, Kupczyński R, Zawadzki W, Miśta D, Adamski M, Pecka E, Paczyńska K, Tumanowicz J. The impact of dried corn decoction on the balance of cationic-anionic feed ration in the dry period of dairy cows, thier health status, colostrum quality and calves passive immunity. Research report 2014. NN311537540, 1–105.

[30] Al-Suwaiegh S, Fanning KC, Grant RJ, Milton CT, Klopfenstein TJ. Utilization of distillers grains from the fermentation of sorghum or corn in diets for finishing beef and lactating dairy cattle. Journal of Diary Sciences. 2002;**80**:1105–1111.

[31] Wertz-Lutz A, Zelinsky R, Held J. Effects of increasing the energy density of a lactating ewe diet by replacing grass hay with soybean hulls and dried distillers grains with solubles. Animal Science Field Day Proceedings and Research Reports. 2007. Available from: http://openprairie.sdstate.edu/cgi/viewcontent.cgi?article=1004&context=sd_sheepreport_2006[Accessed: 2016-08-10]

[32] Cais-Sokolińska D, Wójtowski J, Pikul J, Danków R, Majcher M, Teichert J, Bagnicka E. 2015. Formation of volatile compounds in kefir made of goat and sheep milk with high polyunsaturated fatty acid content. Journal of Diary Sciences. 2015;**98**:6692–6705. DOI: 10.3168/jds.2015-9441.

[33] Dimova N, Ivanova I, Mihailova M, Todorov N, Naydenova N. Wheat distiller's grains as a source of protein in dairy sheep. Bulgarian Journal of Agricultural Science. 2009;**15**:574–582.

[34] Radunz AE, Fluharty FL, Zerby HN, Loerch SC. Winter-feeding systems for gestating sheep I. Effects on pre- and postpartum ewe performance and lamb progeny preweaning performance. Journal of Animal Science. 2011;**89**:467–477. DOI: 10.2527/jas.2010-3035.

[35] Şahin T, Kaya Ö, Aksu Elmali D, Kaya İ. Effects of dietary supplementation with distiller dried grain with solubles in growing lambs on growth, nutrient digestibility and rumen parameters. Revue de Médecine Vétérinaire. 2013;**164**:173–178.

[36] Schauer CS, Stamm MM, Maddock TD, Berg PB. Feeding of DDGS in lamb rations. Feeding dried distillers grains with solubles as 60 percent of lamb finishing rations results in acceptable performance and carcass quality. Sheep and Goat Research Journal. 2008;**23**:15–19.

[37] Pezzanite L, Gunn P, Neary M, Lemenager R, Lake S. Value of distiller's grains as a feed for sheep. Purdue University Cooperative Extension Service [Internet]. 2010. Available from: https://www.extension.purdue.edu/extmedia/AS/AS-603-W.pdf/[Accessed: 2016-08-14]

[38] Łozicki A, Dymnicka M, Arkuszewska E. Evaluation of selected parameters of quality and nutritive value of meat of fattened bulls fed diet with distillers dried grains with solubles (DDGS) from wheat or maize as a source of protein. Polish Journal of Food and Nutrition Sciences. 2013;**63**:179–186. DOI: 10.2478/v10222-012-0075-3.

[39] May ML, Quinn MJ, Depenbusch BE, Reinhardt CD, Gibson ML, Karges KK, Cole NA, Drouillard JS. Dried distillers grains with solubles with reduced corn silage levels in beef finishing diets. Journal of Animal Science. 2010;**88**:2456–2463. DOI: 10.2527/jas.2009-2637

[40] Kelzer JM, Kononoff PJ, Gehman AM, Tedeschi LO, Karges K, Gibson ML. Effects of feeding three types of corn-milling coproducts on milk production and ruminal fermentation of lactating Holstein cattle. Journal of Diary Sciences. 2009;**92**:5120–5132. DOI: 10.3168/jds.2009-2208

[41] Benchaar C, Hassanat F, Gervais R, Chouinard PY, Julien C, Petit HV, Massé DI. Effects of increasing amounts of corn dried distillers grains with solubles in dairy cow diets on methane production, ruminal fermentation, digestion, N balance, and milk production. Journal of Diary Sciences. 2013;**96**:2413–2427. DOI: 10.3168/jds.2012-6037.

[42] Miśta D, Pecka E, Zachwieja A, Zawadzki W, Bodarski R, Paczyńska K, Tumanowicz J, Kupczyński R, Adamski M. Effect of corn dried distillers' grains with solubles used as replacement of concentrate components on *in vitro* ruminal fluid fermentation. Folia Biologica (Krakow). 2014;**62**:345–351. Doi: 10.3409/fb62_4.345.

[43] Patra AK. The effect of dietary fats on methane emissions, and its other effects on digestibility, rumen fermentation and lactation performance in cattle: a meta-analysis. Livestock Science. 2013;**155**:244–254. DOI: 10.1016/j.livsci.2013.05.023.

[44] Behlke EJ. Attenuation of ruminal methanogenesis [thesis]. University of Nebraska-Lincoln, USA; 2007.

[45] Wysocka O, Pecka E, Zawadzki W. The potential of using corn dried distillers grains with solubles (DDGS) in order to improve the fermentation profile in sheep. Folia Pomeranae Universitatis Technoligiae Stetinensis. 2015;**35**:109–120.

[46] Cromwell GL, Herkelman KL, Stahly TS. Physical, chemical and nutritional characteristics of distillers dried grains with solubles for chicken and pigs. Journal of Animal Science. 1993;**71**:679–686.

[47] Strzetelski J. New energy-protein by-products of agri-food industry in cattle feeding. Bulletin of Polish Association of Feed Manufacturers, Warsaw, 2007